Made Simple ADDITIONAL

This new instructive series
has been created
primarily for self-education
but can equally well
be used as
an aid to group study.
However complex the subject,
the reader is taken
step by step,
clearly and methodically,
through the course. Each volume
has been prepared by experts,
using throughout the
Made Simple technique of teaching.
Consequently the gaining
of knowledge now becomes
an experience to be enjoyed.

In the same series

Accounting
Acting and Stagecraft
Additional Mathematics
Advertising
Anthropology
Applied Economics
Applied Mathematics
Applied Mechanics
Art Appreciation
Art of Speaking
Art of Writing
Biology
Book-keeping
British Constitution
Chemistry
Childcare
Commerce
Commercial Law
Computer Programming
Cookery
Cost and Management Accounting
Economics
Electricity
Electronic Computers
Electronics
English
French
Geology
German
Human Anatomy
Italian
Journalism
Latin
Law
Management
Marketing
Mathematics
New Mathematics
Office Practice
Organic Chemistry
Philosophy
Physics
Pottery
Psychology
Rapid Reading
Russian
Salesmanship
Spanish
Statistics
Typing

ADDITIONAL MATHEMATICS Made Simple

Patrick Murphy, M.Sc., F.I.M.A.

Made Simple Books
W. H. ALLEN London
A division of Howard & Wyndham Ltd

Made and printed in Great Britain
by Richard Clay (The Chaucer Press), Ltd., Bungay, Suffolk
for the publishers W. H. Allen & Company Ltd.,
44 Hill Street, London W1X 8LB

ISBN 0 491 01262 4 Casebound
ISBN 0 491 01272 1 Paperbound

Preface

It is unneccessary to have either a total commitment or a considerable natural ability to enjoy the activity of mathematics. Most of us who now enjoy the subject have needed, at some stage, to move through an encouraging sequence of comparatively minor successes to sustain those extra efforts which must be made to master the more involved development of topics we feel we ought to understand. Nothing succeeds like success no matter how small, and so the author has attempted to present the worked examples and exercises in a style which enables the reader to appreciate the mathematics of the enterprise at an intuitive level—long before the elements of the formal methods are formally identified.

The order of the chapters in the book seeks to emphasise a strong algebraic development, via solving equations, before dealing with, for example, the elements of Coordinate Geometry in Chapters 5, 6 and 7. The reader may, if he wishes, forgo this algebraic build up and will find Chapters 5, 6 and 7 reasonably self-contained. Similarly, Chapter 9 on Determinants may be considered in conjunction with Chapter 8 as a self-contained section since the author does not force the methods into other sections where it is not necessary.

There are many differing points of view on what should properly precede the introduction of the Calculus and how this subject should fit into the content of a book of which it forms only a part. There are those who like to build up to the Calculus and then there are others who like to absorb it gradually into earlier sections and stages. Bearing in mind the very wide readership of this book the author suggests that, if a reader wishes to read or revise the Calculus section separately in Chapters 16 to 20 inclusive, he would do well to give at least some preparation time to the notion of limits in Chapter 11 and the section on radian measure in Chapter 13.

The subject matter of the book bridges the gap between O- and A-level G.C.E. of most of the examination boards in this country and is frequently referred to under the heading of *Additional Mathematics*. It also provides all the subsidiary mathematics necessary for supporting courses in Building, Engineering, Agriculture as well as some subsections of A-level G.C.E. general syllabuses.

A particular feature of the book is the provision of all the necessary tables of logarithms, trigonometric functions, squares, square roots and reciprocals, which avoids the inconvenience of forgetting to have tables at hand—and moments of exasperation are neither good for the blood pressure nor mathematics. At all times the reader must work with patience for in many ways this alone brings success when inspiration dies.

PATRICK MURPHY

[illegible]

it is unnecessary to have [illegible] adding to [illegible] subject have been [illegible] [illegible]

The [illegible] of the chapters [illegible] development, [illegible] elements of [illegible] in Chapters [illegible] reasonably [illegible]

[illegible]

The subject [illegible]

A particular feature of the book is the [illegible]

[illegible]

Table of Contents

CHAPTER ONE

EQUATIONS AND INEQUATIONS

Classifications of Equations

By what methods may a given mathematical problem be solved? How many and what kind of solutions may the problem have, assuming that it has any solutions at all? Answers to questions like these depend upon the types of equations involved and the constraints which are to be placed on the acceptance of the solutions.

Thus a problem might yield an equation such as $x + 11 = 7$, for which $x = -4$ is the solution. But this solution may not be acceptable if the wording of the problem only allows us to accept positive values of x as in the following case: *Find the present age of a person who will be seven years old in eleven years from now*. Likewise the equation $4x = 13$ gives the solution $x = 3\frac{1}{4}$, and this may not be acceptable if we only wish to consider whole number solutions to the problem as in the following case: *Share thirteen new-laid eggs equally between four people.*

There are other occasions when we obtain *more* than one solution to a problem. Thus consider being asked to find a number which when squared gives a result of 9. We can represent this as equivalent to finding the solution to the equation $x^2 = 9$, from which we see that x may be equal to $+3$ or -3; that is, we have two acceptable solutions to the problem. If, however, the problem had been to find the length of the side of a square with an area of 9 cm^2, then although the equation $x^2 = 9$ would still apply, only the solution of 3 cm would be acceptable since it is not possible to have a real square with a side which is -3 cm long. So we must always be aware of the possibility that the solution of an equation may not be a solution of the problem from which the equation arose.

Equations are classified according to the **number and degree** of the unknowns, i.e. x is of the first degree, x^2 is of the second degree and so is xy or yz, x^3 is of the third degree and so is x^2y, xy^2 or xyz. Thus equations such as the following are said to be of the first degree:

(i) $$7x = 14$$
(ii) $$3y - 8 = y$$
(iii) $$7x = 14y$$
(iv) $$3y - 8z = y$$
(v) $$5x - y + z = 2$$

Both the equations (i) and (ii) are called equations of the **first degree** in one unknown. In (i) the unknown is x and in (ii) the unknown is y. Both the equations (iii) and (iv) are called equations of the first degree in two unknowns. In (iii) we have the two unknowns x and y while in (iv) we have the two unknowns y and z. Equation (v) is called an equation of the first degree in the three unknowns x, y and z.

The following equations are all of the **second degree:**

(i) $3x^2 - 16 = 0$
(ii) $2x^2 + 5x - 3 = 0$
(iii) $xy + 4x + 3y + 1 = 0$
(iv) $3x^2 + y^2 + 7xy + x + y + 4 = 0$

Both equations (i) and (ii) are called equations of the second degree in one unknown. Both equations (iii) and (iv) are called equations of the second degree in two unknowns.

Remembering the names of the equations is not important, but it is useful to be able to recognise the differences between them and to anticipate the number of possible solutions which may arise. Thus, knowing that $2x^2 + 3x + 8 = 0$ is called a quadratic equation, or that

$$x^3 + 7x^2 + x - 5 = 0$$

is called a cubic equation, will not help us to solve them but recognising their type is the first step towards doing so.

Replacement Sets

Consider finding the solution or solutions to the equation $x + y = 7$ by suggesting numerical values to replace x. We could try the simplest ones which come to mind—namely, 0, 1, 2, 3, 4 and so on. Each value for x will produce a corresponding value for y and this will enable us to list some of the solutions as follows:

(i) $x = 0, y = 7$
(ii) $x = 1, y = 6$
(iii) $x = 2, y = 5$
(iv) $x = 3, y = 4$
(v) $x = 4, y = 3$

Clearly there is no end to the number of solutions we can find by continuing the pattern of the list. We could also obtain many more solutions 'in between' those in the list. For example:

(i) $x = \frac{1}{4}, y = 6\frac{3}{4}$
(ii) $x = \frac{1}{2}, y = 6\frac{1}{2}$
(iii) $x = \frac{3}{4}, y = 6\frac{1}{4}$
(iv) $x = 1\frac{1}{4}, y = 5\frac{3}{4}$
(v) $x = 3\frac{1}{2}, y = 3\frac{1}{2}$

Suppose that we now make x and y satisfy another equation such as $x + 2y = 12$, at the same time as they satisfy the equation $x + y = 7$. In this case we shall be trying to solve the two equations simultaneously. Under these conditions we call them **simultaneous equations.** We could start by making a list of solutions for $x + 2y = 12$ by suggesting a set of replacement values for x as we did before. Some of the solutions would be as follows:

(i) $x = 0, y = 6$
(ii) $x = 1, y = 5\frac{1}{2}$
(iii) $x = 2, y = 5$
(iv) $x = 3, y = 4\frac{1}{2}$
(v) $x = 4, y = 4$

Comparing the two lists of solutions we notice that the solution $x = 2, y = 5$ is a member of both lists and so we are able to say that $x = 2, y = 5$ satisfies both equations simultaneously.

Exercise 1.1

Using the replacement set {0, 1, 2, 3, 4, 5, 6} for x, obtain the solutions to the following pairs of simultaneous equations by comparing two lists of possible solutions as in the previous section:

1. $x - y = 1$ $x + y = 5$	2. $x + y = 5$ $x - y = 3$	3. $2x + y = 7$ $x + y = 1$
4. $3x - y = 6$ $x + 7y = 2$	5. $3x + 2y = 7$ $x + y = 3$	6. $2x + 3y = 10$ $3x - 2y = 2$

Solution of Simultaneous Equations (Two Linear)

Clearly it would be an advantage to try to develop a method of solving simultaneous equations which avoids the considerable work in making lists of possible solutions, some of which are never used.

To start with, let us examine the information which is given in one of the two equations. For example, in Exercise 1.1, No. 1, the first equation is $x - y = 1$. This equation can be rearranged to read $x = y + 1$, and in this form we see straightaway that the equation enables us to replace x by $y + 1$. If we make this replacement in the second equation $x + y = 5$, we shall obtain $y + 1 + y = 5$, which leads to $2y + 1 = 5$, then $2y = 4$ and finally $y = 2$. But as we have already seen, x is always equal to $y + 1$ so, when $y = 2$ it follows that $x = 3$.

A more formal presentation of this method would be as follows:

Example 1. Solve the simultaneous equations

$$x - y = 1 \quad \text{(i)}$$
$$x + y = 5 \quad \text{(ii)}$$

Solution	*Explanation*
Name the given equations (i) and (ii)	
From (i) $x = y + 1$ (iii)	Call this new equation, (iii)
Replace x by $y + 1$ in (ii)	
$\therefore\ y + 1 + y = 5$	
$2y + 1 = 5$	Collect the y's together
$2y = 4$	Subtract 1 from both sides of the equation
Hence $y = 2$	Divide both sides of the equation by 2
Replace y by 2 in (iii)	
$\therefore x = 2 + 1$	
Hence $x = 3$	

Check: $3 - 2 = 1$ (i); $3 + 2 = 5$ (ii) Answer: $x = 3, y = 2$

Example 2. Solve the simultaneous equations

$$4x + 3y = 8 \quad \text{(i)}$$
$$4x + y = 16 \quad \text{(ii)}$$

Solution	*Explanation*
Name the given equations (i) and (ii)	
From (ii) $y = 16 - 4x$ (iii)	Call this new equation (iii)
Replace y by $16 - 4x$ in (i)	
$\therefore\ 4x + 3(16 - 4x) = 8$	
$4x + 48 - 12x = 8$	Multiply the bracket out
$-8x + 48 = 8$	Collect the x's together
$-8x = -40$	Subtract 48 from both sides of the equation
Hence $x = 5$	Divide both sides of the equation by -8

Replace x by 5 in (iii)

$\therefore \quad y = 16 - 20$ Replace $4x$ by 20

Hence $y = -4$

Check: $20 - 12 = 8$ (i); $20 - 4 = 16$ (ii) Answer: $x = 5, y = -4$.

As always there is a danger in slavishly applying a method at the expense of the ideas which lie behind it. For instance, if we look at Example 2 which has been worked out above we notice that $4x$ occurs in both of the equations and a rearrangement of them will give the following results

$$4x = 8 - 3y \qquad \text{from (i)}$$
$$4x = 16 - y \qquad \text{from (ii)}$$

So it follows that

$$8 - 3y = 16 - y$$
$$8 - 3y - 16 = -y \qquad \text{(Subtract 16 from both sides of the equation.)}$$
$$-8 - 3y + 3y = 3y - y \quad \text{(Add } 3y \text{ to both sides of the equation.)}$$
$$-8 = 2y$$

Hence $-4 = y$ (Divide both sides of the equation by 2.)

Having found that $y = -4$ we can now replace y by -4 in (ii), so that

$$4x = 16 - (-4)$$
$$4x = 20$$
$$x = 5$$

Answer: $x = 5, y = -4$ as before.

In the next exercise the reader should try to find similar alternatives to the method outlined in the two worked examples above.

Exercise 1.2

1. $x + 2y = 10$, $x + 3y = 12$
2. $x + 3y = 16$, $2x + 3y = 20$
3. $4x - y = 1$, $4x + 3y = 13$
4. $2x + 9y = 13$, $2x - 9y = -5$
5. $x + y = 4$, $3x + y = 7$
6. $\dfrac{1}{x} + y = 2$, $\dfrac{1}{x} + 7y = 11$

Each of the examples in Exercise 1.2 is solved by using a replacement or substitution which eliminates one of the unknowns, either x or y. This substitution links the two given equations.

There are other ways of linking the equations. For example, we could add them together or subtract one from the other. Possibly we could multiply them by numbers before the addition or subtraction, always provided that the final result is a simplification. Before we study a worked example, try to prepare for the method by following the instructions in Exercise 1.3.

Exercise 1.3

Answer the questions in this Exercise using the following four equations:

(i) $3x + y = 9$ (ii) $x - y = 7$ (iii) $x - 5y = 11$ (iv) $x + y = 3$

1. Find (i) + (ii)
2. Find (ii) + (iv)
3. Find (i) − (iv)
4. Find (iii) − (ii)

5. Multiply (ii) by 5 and add the result to (iii)
6. Multiply (i) by 5 and add the result to (iii)
7. Multiply (iii) by 3 and then subtract (i) from the result

Now let us consider how to apply these ways of combining the simultaneous equations in order to solve them.

Example 1. Solve the simultaneous equations

$$4x + 7y = 2 \qquad \text{(i)}$$
$$3x + 8y = 7 \qquad \text{(ii)}$$

Solution		*Explanation*
Multiply (i) by 3 and (ii) by 4 to get		
$12x + 21y = 6$	(iii)	
$12x + 32y = 28$	(iv)	Rename the equations (iii) and (iv)
$-11y = -22$		Subtract (iv) from (iii)
$y = 2$		Divide both sides of the equation by -11
Replace y by 2 in (i)		
$4x + 14 = 2$		
$4x = -12$		Subtract 14 from both sides of the equation
$x = -3$		Divide both sides of the equation by 4

Check: $-12 + 14 = 2$ (i); $-9 + 16 = 7$ (ii) Answer: $x = -3, y = 2$

Alternatively: Multiply (i) by 8 and (ii) by 7 to get		
$32x + 56y = 16$	(iii)	
$21x + 56y = 49$	(iv)	Rename the equations (iii) and (iv)
$11x = -33$		Subtract (iv) from (iii)
$x = -3$		Divide both sides of the equation by 11
Replace x by -3 in (ii)		
$-9 + 8y = 7$		
$8y = 16$		Add 9 to both sides of the equation
$\therefore\ y = 2$		Divide both sides of the equation by 8

Answer: $x = -3, y = 2$ as before

Example 2. Solve the simultaneous equations

$$5x - 7y = 9 \qquad \text{(i)}$$
$$2x + 3y = -8 \qquad \text{(ii)}$$

Solution

Multiply (i) by 2 and (ii) by 5 to get — The reader is left to insert the explanations

$$10x - 14y = 18 \qquad \text{(iii)}$$
$$10x + 15y = -40 \qquad \text{(iv)}$$
$$-29y = 58$$
$$y = -2$$

Replace y by -2 in (ii)

$$2x - 6 = -8$$
$$2x = -2$$
$$\therefore\ x = -1$$

Check: $-5 - 7(-2) = 9$ (i); $-2 - 6 = -8$ (ii) Answer: $x = -1, y = -2$

Exercise 1.4

Solve the simultaneous equations in questions 1, 2 and 3:

1. $6x + y = 12$, $3x + 2y = 15$
2. $3x + 7y = 20$, $2x + 3y = 10$
3. $3x - 2y = 12$, $5x - y = 13$
4. Two numbers x and y have a sum of 19. When 8 is added to x the result is twice as big as y. Write down the two equations which represent these statements and then solve them for x and y.
5. A woman is five years younger than her husband. If their combined ages are eighty-nine years find their respective ages now. (Start by letting the woman be x years of age.)
6. Solve the simultaneous equations in (a), (b) and (c):

   (a) $3x + 2y = 32$, $2x + 3y = 23$ (b) $4x + 7y = 47$, $7x + 4y = 74$ (c) $5x + 9y = 59$, $9x + 5y = 95$
7. Compare the three pairs of simultaneous equations in question 6 and write down another pair of the same type using the numbers 3 and 8.
8. By applying the ideas of questions 6 and 7, try to solve the following three simultaneous equations:

$$3x + 4y + 5z = 345$$
$$4x + 3y + 5z = 435$$
$$5x + 4y + 3z = 543$$

9. Write down three simultaneous equations in the three unknowns x, y and z just like those in question 8 but using only the numbers 2, 5 and 7.

Dependent and Independent Equations

One particular feature of the simultaneous equations so far is that they all have a solution. Indeed, they each have only one solution, in which case we say that the solution is unique. The first three questions of Exercise 1·4 are examples of pairs of **independent** equations.

Consider the simultaneous equations

$$2x + 3y = 14 \qquad \text{(i)}$$
$$4x + 6y = 28 \qquad \text{(ii)}$$

Here we observe that if equation (i) is multiplied by 2 it becomes the same as (ii). Thus equation (ii) gives no more information than equation (i), i.e. the equations are interdependent. We may select any value for x and work out the corresponding value for y as follows:

$$\text{If } x = 0 \text{ then } y = 4\tfrac{2}{3}$$
$$\text{If } x = 1 \text{ then } y = 4$$
$$\text{If } x = 2\tfrac{1}{2} \text{ then } y = 3$$

We see, therefore, that the equations have as many different solutions as we please, i.e. there is no unique solution.

A very different situation is suggested in the following two simultaneous equations

$$2x + y = 7 \qquad \text{(i)}$$
$$2x + y = 4 \qquad \text{(ii)}$$

Since equation (i) says that $2x + y$ is always equal to 7 then the second equation cannot possibly be true. In this case we say that the two equations are **inconsistent.** If we had not seen this straightaway but proceeded along the lines of the usual method of solution then, by subtracting equation (ii) from equation (i), we would get $0 = 3$; and clearly there is something wrong somewhere. A problem which leads to a pair of inconsistent equations is said to be **indeterminate.**

Example 3. A mother is 20 years older than her son. In 5 years' time the son will be 20 years younger than his mother. Find their present ages.

Solution. Let the mother be m years of age and the son s years of age. Since the mother is 20 years older than her son we may write $m - s = 20$ (i). In 5 years' time their ages will be $m + 5$ and $s + 5$ and since the son will be 20 years younger than his mother we may write $(m + 5) - (s + 5) = 20$
Multiplying the brackets out we have

$$m + 5 - s - 5 = 20$$
$$\therefore \quad m - s = 20 \qquad \text{(ii)}$$

Equation (ii) is the same as (i), so we may find more than one solution to the problem. Some possible solutions of the equation are

$$m = 30, s = 10$$
$$m = 31, s = 11$$
$$m = 42, s = 22$$

One point to note is that although we can obtain as many solutions to the equation as we please, they will not necessarily be solutions to the problem. For example, $m = 1220$ and $s = 1200$ is a solution to the equation, but it is unacceptable as a solution to the problem. So we must always examine the solutions to the equations we obtain to see if they are really solutions to the problem.

Example 4. Liquid begins to flow into an empty tank at a rate which raises its level by 4 cm per min. A second tank is already filled to a level of 32 cm but 5 minutes later liquid begins to flow out of it at a rate which lowers its level by 6 cm per min. After how many minutes from the beginning of the flow in the first tank will the level in both tanks be the same, and what will both levels be at this time?

Solution. Let the number of minutes of flow into the first tank be x. Then the level in this tank after x min will be at a height of $4x$ cm. The time of flow out of the second tank will be $x - 5$ min because the flow starts 5 min after the flow into the first tank, and since the level is already at 32 cm to begin with then after $x - 5$ min it will be at a height of $32 - 6(x - 5)$ cm. When the level in both tanks is at a height of y then we have

$$y = 4x \qquad \text{(i)}$$

and $$y = 32 - 6(x - 5) \qquad \text{(ii)}$$

Substituting for y in (ii) we get $4x = 32 - 6(x - 5)$

Multiplying the bracket out, $4x = 32 - 6x + 30$

$$\therefore \quad 10x = 62$$
$$x = 6{\cdot}2$$

Substituting for x in (i) $y = 4 \times 6{\cdot}2 = 24{\cdot}8$

This means that after 6·2 min the level of liquid in both tanks is 24·8 cm: Answer

Example 5. The relation between the load L moved by an effort E applied to the load by a particular machine is given by the equation $E = mL + c$, where m and c are constants which depend on the machine only. It is found that when $L = 100$, $E = 100$ and when $L = 400$, $E = 250$. Find the constants m and c and the effort needed to move a load of 300 units.

Solution. Substituting the values of L and E in the equation $E = mL + c$ gives

$$100 = 100m + c \quad \text{(i)}$$
$$250 = 400m + c \quad \text{(ii)}$$

Subtracting equation (ii) from equation (i) we have

$$-150 = -300m$$

Dividing both sides of the equation by -300 we get

$$\tfrac{1}{2} = m$$

Substitute for m in (i) $\quad 100 = 50 + c$

$$\therefore \quad 50 = c$$

The equation for the machine is therefore

$$E = \tfrac{1}{2}L + 50,$$

so that when $L = 300$ we get $E = 150 + 50$

Therefore $m = 0{\cdot}5$, $c = 50$, and an effort of 200 units is needed to lift a load of 300 units: Answer

Exercise 1.5

1. The manufacturing cost (C) of an article is related to its mass (M) by the equation $C = aM + b$, where a and b are constants. C is 10 when M is 1, and C is 40 when M is 16. Find a and b and the cost of an article when $M = 10$ units.
2. A girl is 5 years older than her brother and the sum of their ages is 18 years. Find their ages.
3. A farmer has 350 m of wire fencing and wishes to fence in a maximum rectangular area with a length which is twice the width. The rectangle is to be separated into two squares by a single wire fence across the centre of the rectangle. Find the length, breadth and area of the rectangle.
4. Solve (if possible) the following pairs of equations:

(a) $x + y = 9$, $x + y = 19$ (b) $x + y = 9$, $3x + 3y = 27$ (c) $6x - 4y = 7$, $9x - 6y = 15$

(d) Both x and y are whole numbers

$$7x + 4y = 16$$
$$x + 4y = 4$$

(e) Both x and y are positive:

$$5x - 8y = 26$$
$$5x - 16y = 2$$

5. The relation between two quantities y and x is given by the equation $y = mx + c$, where m and c are constants. Find m and c if $x = 4$ gives $y = 2$ and $x = 2$ gives $y = -4$. Find y when $x = 100$.
6. The sum of the present ages of two people is 100 years. Eight years ago one of their ages was twice the other's age. Find their present ages.

Inequations or Inequalities

More often than not we are involved with inequalities rather than equalities. Frequently a solution to a problem rests upon choosing the best or nearest approximation that we can get, i.e. there is a shift in emphasis away from the question which reads, *What is the cost of a certain number of stamps?* to a question like *What is the greatest number of stamps you can buy with a certain sum of money?*

Very often we are content to know that we can afford something rather than know how much we shall have left over if we buy it, i.e. we say to ourselves, *I know I have enough to go to the cinema, buy an ice cream and travel home on the bus*, believing that the sum of money in our pocket is greater than the cost of these things. Similarly we look at the petrol gauge in the car and say we have enough petrol to do the journey, i.e. we think in terms of the petrol being more than or greater than that which we need to do the journey without actually calculating the amount needed.

At first sight the reader might think that a relation such as $x > 4$ was no different from $x = 4$ as far as methods of obtaining solutions are concerned. Thus $x = 4$ implies that $2x = 8$ or $10x = 40$ and so on, obtained by multiplying both sides of the equation by 2 and 10 respectively. We may go further and say that $x = 4$ implies that $-x = -4$ and $-3x = -12$, again obtained by multiplying both sides of the equation by a non-zero negative number, in this case -1 and -3 respectively. If we do the same to the inequality $x > 4$ we can say that $x > 4$ implies that $2x > 8$ or $10x > 40$ and so on, provided that we only multiply by a positive number because $x > 4$ does **not** imply that $-x > -4$ or $-3x > -12$. If not convinced by this, try considering the numerical case of $9 > 4$. Multiplying this by positive numbers 2 and 10 gives $18 > 8$ (true) and $90 > 40$ (true) but multiplying by -1 and -3 gives $-9 > -4$ (false) and $-27 > -12$ (false).

If we multiply both sides of an inequality by a positive number, therefore, the new result is still true, but if we multiply both sides of an inequality by a negative number then the new result is false. However, if we multiply by a negative number we can make the false result a true one by reversing the inequality sign, i.e. multiplying $x > 4$ by -1 produces $-x > -4$ (false), $-x < -4$ (true) and multiplying $10 > 2$ by -3 produces $-30 > -6$ (false), $-30 < -6$ (true).

Similar results hold when we divide inequalities by positive or negative numbers. For instance, division of $21 > 4$ by 3, yields $\frac{21}{3} > \frac{4}{3}$ which is true, but division of $21 > 4$ by -3, yields $\frac{21}{-3} > \frac{4}{-3}$ or $-\frac{21}{3} > -\frac{4}{3}$ which is false. More generally if $x > 4$ then $\frac{x}{3} > \frac{4}{3}$ is true but $-\frac{x}{3} > -\frac{4}{3}$ is false. Again we may change the false statement to a true one by reversing the inequality sign, i.e. if $x > 4$ then $-\frac{x}{3} < -\frac{4}{3}$. We have not introduced any new ideas by considering division separately because we may always consider division to be equivalent to multiplication by a fraction. In the examples just given we really multiplied by $\frac{1}{3}$ and $-\frac{1}{3}$.

We may summarise the above remarks in the following rules for any two numbers a and b:

(i) If $a = b$ and m is any number then $ma = mb$
(ii) If $a > b$ and m is any positive number then $ma > mb$
(iii) If $a > b$ and m is any negative number then $ma < mb$

(Note: m, a, b may be fractions as well as integers.)
There are one or two difficulties to be considered when dealing with inequalities or inequations. For example, if $x = y$ and $p = q$ then we may conclude

that $x - p = y - q$ and $x + p = y + q$, which is the basis for the method of solving simultaneous equations which we have already discussed: that is, given that $2x + 3y = 8$, $2x + y = 4$ then $(2x + 3y) - (2x + y) = 8 - 4$, which leads to $2y = 4$ and $y = 2$, etc. However, if we have $x > y$ and $p > q$ then we may always conclude that $x + p > y + q$ but not that $x - p > y - q$. This is easier to see numerically as follows.

Suppose we consider $14 > 9$ and $12 > 6$; then $14 + 12 > 9 + 6$, but $14 - 12$ is not greater than $9 - 6$. In symbols we would write this last result as $14 - 12 \ngtr 9 - 6$. On the other hand, if we consider $14 > 9$ and $12 > 8$ then again $14 + 12 > 9 + 8$, but this time $14 - 12$ is greater than $9 - 8$. It is clear that on these points alone, we need to be extra careful where inequalities are concerned. Let us see what may happen if we try to solve a pair of simultaneous inequalities in two unknowns such as

$$x + y > 4 \qquad \text{(i)}$$
$$x - y > 3 \qquad \text{(ii)}$$

We know that we may add the inequalities to get $2x > 7$; and that we may divide both sides of the inequalities by a positive number without changing the direction of the inequality, so dividing by 2 gives the result of $x > 3\frac{1}{2}$.

To obtain y we cannot substitute $3\frac{1}{2}$ for x in (i) because x must be greater than $3\frac{1}{2}$, so we must give a little more thought to inserting the result for x in the inequality in order to get some information about y.

Since we now need y we shall rewrite (i) to read $y > 4 - x$ by subtracting x from both sides, and (ii) to read $x - 3 > y$ by adding $y - 3$ to both sides. Putting these two results together, we have $x - 3 > y > 4 - x$, meaning that whenever a value for x is chosen greater than $3\frac{1}{2}$ then y must be simultaneously less than $x - 3$ and greater than $4 - x$. For example, if $x = 9$ then y is less than 6 but greater than -5, i.e. $6 > y > -5$.

Note that in rewriting (i) to get $y > 4 - x$ we used the fact that we may add or subtract the same number from either side of an inequality **without changing its truth.** In symbols this is represented by saying that if $a > b$ then $a + x > b + x$ or $a - x > b - x$. A numerical example is given by $9 > -3$, hence $9 + 10 > -3 + 10$, i.e. $19 > 7$ (true) or $9 - 10 > -3 - 10$, i.e. $-1 > -13$ (true).

The difficulty arises from being able to find as many solutions as we please, e.g.

$$x = 4, y = \tfrac{1}{2};\ x = 4, y = \tfrac{3}{4};\ x = 4, y = \tfrac{7}{8}$$

or $$x = 5{\cdot}1, y = 0{\cdot}1;\ x = 5{\cdot}1, y = 1{\cdot}1;\ x = 5{\cdot}1, y = 2, \text{etc.}$$

are all possible solutions. We were, however, fortunate to get this far only because the example was so arranged that we could eliminate one of the unknowns—in this case y—by adding the two inequalities. Consider the more difficult combinations of

$$2x + 3y > 9 \qquad \text{(i)}$$
$$x + y > 4 \qquad \text{(ii)}$$

We could start by multiplying (ii) by 2 to get

$$2x + 2y > 8 \qquad \text{(iii)}$$

Comparing (i) and (iii), does this mean that $y > 1$? i.e. may we subtract (iii) from (i) and get a new inequality which is always true? The answer is no because $x = 5$ and $y = 0$ satisfies (i), (ii) and (iii) yet y is not greater than 1. But since we went against the advice on page 9 we should not expect to be correct with $y > 1$.

Here again we are able to obtain as many solutions to the inequalities as we please, e.g. $x = 2$, $y = 3$; $x = 5$, $y = 2$; $x = 3$, $y = 1{\cdot}1$ and so on, but as before we seem unable to obtain the 'borderline' solutions, i.e. where is the dividing line between the solutions such as $x = 1$, $y = 3{\cdot}1$ and $x = 3$, $y = 1{\cdot}1$ and the non-solutions such as $x = 1$, $y = 3$ and $x = 3$, $y = 1$? We shall return to this problem in Chapter 5.

Exercise 1.6

1. If $x + y > 6$ what can you say about y when:
   (i) $x > 2$ (ii) $x < 2$ (iii) $x > 6$ (iv) $x < 6$?
2. If $x - y > 1$ what can you say about x when:
   (i) $y > 1$ (ii) $y < 1$ (iii) $y = 1$ (iv) $y > 2$?
3. The two numbers x and y are related by the following inequalities:
$$2x - y > 1$$
$$x + y > 5$$
   Prove that $x > 2$
4. The two numbers x and y are restricted by the inequalities $0 < x < 3$, $1 < y < 7$. Find the least and the greatest numerical values of the following:
   (i) $x + y$ (ii) $3x - 2y$ (iii) $3y - 4x$ (iv) $\dfrac{1}{x+1} + \dfrac{1}{y}$ (v) $\dfrac{1}{x+1} - \dfrac{1}{y}$
5. The two numbers x and y are restricted by the inequalities
   $-5 < x < -2$, $-3 < y < -1$.
   Find the least and the greatest numerical values of the following:
   (i) $\dfrac{1}{x} + \dfrac{1}{y}$ (ii) $\dfrac{1}{y} - \dfrac{1}{x}$ (iii) xy (iv) $\dfrac{x}{y}$ (v) $\dfrac{1}{x+y}$

CHAPTER TWO

EQUATIONS OF THE SECOND DEGREE

Variables, Constants and Coefficients

So far we have referred to the quantities x, y and z in equations as either known or unknown. For instance, in an equation such as $2x + 3y = 14$ the x and y were called the unknown quantities, as distinct from the known quantities 2, 3 and 14, which remain constants regardless of the values given to x and y. At this level of mathematics, if we write an equation such as $ax + by = c$ we call x and y the unknowns and a, b and c the constants. That is, we would continue to consider that this was an equation in the two unknowns x and y, even though we have not been told the values of a, b or c. This is because we usually assume (unless told otherwise) that the letters at the end of the alphabet, such as x and y, represent the unknowns, whereas the letters at the beginning of the alphabet such as a, b and c represent constants. Of course this is not very convincing when the reader meets the subject for the first time, but experience usually makes the situation clearer. In an attempt to clarify one or two ideas on notation we shall consider some definitions.

The unknown quantities are unknown or unnamed members of the set of numbers. For instance, all the equations so far have had solutions which have been **whole numbers** $\{0, 1, 2, 3, 4, \ldots\}$ or **integers** $\{\ldots -4, -3, -2, -1, 0, +1, +2, +3, +4, \ldots\}$ or **rational numbers** (fractions) $\{-\frac{3}{4}, -\frac{1}{2}, -\frac{1}{4}, -\frac{1}{5}, -\frac{1}{6}, \ldots 0, \frac{2}{3}, \frac{4}{7}, \frac{8}{9}, \ldots\}$. We shall call these numbers together with numbers like $\sqrt{2}$, $\sqrt{3} \ldots$ the **set of real numbers.** Then the x and y unknowns are unnamed members of the set of real numbers, and since they are allowed to vary inasmuch as we are free to suggest values for x and work out corresponding values for y then we shall call them **variables.** More specifically, if we first allocate a definite value to x in order to find the corresponding value of y, then x is called the **independent variable** and y is called the **dependent variable.**

By calling x the independent variable we mean that a value of x is allocated independent of any consideration of y, but having done this the value of y will be dependent upon that of x. In this respect y is often referred to as a function of x, an idea which is sufficiently common and important to merit its own symbolic form of $y = f(x)$. This is read 'y is a function f, of x'. Note that the symbol '$f(x)$', does not mean that a quantity f is to be multiplied by another quantity x in the manner of $7(4) = 28$.

For example, we may write $y = f(x) = 3x + 1$, to be read as 'y is the function f of x, equal to $3x + 1$'.

A selection of values of the independent variable x given by the replacement set $\{0, 2, 5\}$ would be indicated as follows

$$f(0) = 3 \times 0 + 1 = 1$$
$$f(2) = 3 \times 2 + 1 = 7$$
$$f(5) = 3 \times 5 + 1 = 16$$

These would be read as 'for $x = 0$, the function f of x, is 1; for $x = 2$, the function f of x, is 7; and for $x = 5$, the function f of x is 16'.

When two different functions of the same independent variable appear in the same problem they may be distinguished from each other by separate letters or subscripts. Thus we may write

$$y = f(x) = 3x + 1 \text{ and } y = g(x) = 5x + 7,$$

keeping them separate by using a different letter g, or we may write $y = f_1(x) = 3x + 1$ and $y = f_2(x) = 5x + 7$, and keep them separate by using f with different subscripts 1 and 2. These are read as 'y is the function f one of x, equal to $3x + 1$ and y is the function f two of x, equal to $5x + 7$'.

A second-degree equation in one variable is an equation which can be written in the form $ax^2 + bx + c = 0$, where a, b, c are any constants, with the exception that $a \neq 0$ (i.e. a is never zero). This is also called a quadratic equation: $ax^2 + bx + c$ alone is referred to as a '**quadratic expression**' or just a 'quadratic'.

Similarly, we speak of a second-degree function of one variable as a function which can be written in the form

$$y = f(x) = ax^2 + bx + c \ (a, b, c \text{ constants and } a \neq 0)$$

The numbers like a and b which multiply x^2 and x are called **coefficients.** For example, in the equation $4x^2 - 3x + 2 = 0$ we refer to 4 as being the coefficient of x^2 and -3 as being the coefficient of x. In the equation $-3x^2 + 7 = 0$ the coefficient of x^2 is -3 and since there is no term in x it follows that the coefficient of x must have been 0 (zero).

Since the word 'coefficient' occurs so frequently a further statement is desirable. In the case of any product such as 9×4 each of the factors 9 and 4 is said to be the coefficient of the other. In the product $9 \times 4 \times 5$ we would say that the coefficient of 5 is 36 and the coefficient of 4 is 45. For algebraic products such as pqr we state similarly that the coefficient of r is pq, the coefficient of q is pr and so on.

Solving Quadratic Equations

The simplest form of the quadratic equation may be considered to be something like

$$x^2 = 4, x^2 = 9, x^2 = 13, \text{etc.}$$

which have the solutions

$$x = +2 \text{ or } -2, x = +3 \text{ or } -3, x = +\sqrt{13} \text{ or } -\sqrt{13}, \text{ respectively.}$$

These solutions are also written as $x = \pm 2$, $x = \pm 3$, $x = \pm\sqrt{13}$. With the sign $\pm$ being read as 'plus or minus'. In general, finding the solutions to these equations often involves the extraction of square roots and, because of this, the solutions are often called **roots of the equation** or just **roots.**

Example 1. Solve the equation $18x^2 = 50$

Solution.	*Explanation*
$x^2 = \frac{50}{18}$	Divide both sides of the equation by 18
$\therefore\ x^2 = \frac{25}{9}$	Simplify the fraction

$$x = \pm\sqrt{\frac{25}{9}} = \pm\frac{5}{3} \qquad \text{Extract the square root}$$

Check: $x^2 = \frac{25}{9}$; $18x^2 = 50$

$\therefore \quad x = \pm\frac{5}{3}$ **Answer**

This means that the original equation is equivalent to $\left(x - \frac{5}{3}\right)\left(x + \frac{5}{3}\right) = 0$; or $(3x - 5)(3x + 5) = 0$.

Example 2. Solve the equation $(x + 1)^2 - 9 = 0$

Solution. *Explanation*

$(x + 1)^2 = 9$	Add 9 to both sides
$x + 1 = \pm 3$	Extract the square root
$\therefore \quad x = -1 \pm 3$	Subtract 1 from both sides
$\therefore \quad x = -4$ or 2	

Check: $x = -4$, $x + 1 = -3$, $(x + 1)^2 = 9$
$x = 2$, $x + 1 = 3$, $(x + 1)^2 = 9$
$x = -4$ or 2 **Answer**

Now $x = -4$ means that $x + 4 = 0$. Similarly $x = 2$ means that $x - 2 = 0$. We can put these two results together to read $(x + 4)(x - 2) = 0$. Hence, the original equation $(x + 1)^2 - 9 = 0$ was equivalent to $(x + 4)(x - 2) = 0$.

Exercise 2.1

Find the roots of the following quadratic equations:

1. $4x^2 - 25 = 0$ 2. $25x^2 - 4 = 0$ 3. $25 - 4x^2 = 0$ 4. $4 - 25x^2 = 0$
5. $6x^2 - 24 = 0$ 6. $49 - x^2 = 0$ 7. $(x + 1)^2 - 16 = 0$
8. $(x + 2)^2 = 16$ 9. $(x - 2)^2 = 9$ 10. $3(x - 1)^2 = 12$

Perfect Squares

Throughout the work in solving the above equations, it will probably have been realised that it is convenient to have x involved only as part of a perfect square, such as x^2, $(x + 1)^2$, $(x - 2)^2$, because we may then obtain x almost immediately by extracting a square root. Thus $(x + 1)^2 = 9$ yields $x + 1 = \pm 3$ and $x = 2$ or -4. Similarly for other examples as in Exercise 2.1.

Expanding the left-hand side of this equation gives a clue for solving the general quadratic equation of the form $ax^2 + bx + c = 0$; that is, the solutions to $(x + 1)^2 = 9$ are also the solutions of

$$(x + 1)(x + 1) = x^2 + 2x + 1 = 9$$

or

$$x^2 + 2x - 8 = 0$$

in which case $a = 1, b = 2, c = -8$ for the general form. We could take this idea further by considering $(x + 1)^2 = 9/4$, which we know has the solution

$$x + 1 = \pm\frac{3}{2}, \text{ i.e. } x = -2\tfrac{1}{2} \text{ or } \tfrac{1}{2}$$

This equation could have been written as $4(x + 1)^2 = 9$ and then developed along the following lines

$$4(x + 1)^2 = 4(x + 1)(x + 1) = 4(x^2 + 2x + 1) = 9$$
$$\therefore 4x^2 + 8x + 4 = 9$$

and the final form is $4x^2 + 8x - 5 = 0$, which, compared with the general quadratic $ax^2 + bx + c = 0$, gives $a = 4$, $b = 8$, $c = -5$. This indicates that a general method for solving the quadratic equation could be made to rely on the creation of perfect squares containing the variable x. To begin with let us see how to recognise perfect squares, so that we can manipulate them into the quadratic equation. Consider the following pattern of expansions:

	Co-efficient of x	*Half the co-efficient of x*	*Constant term*
$(x + 1)^2 = (x + 1)(x + 1) = x^2 + 2x + 1$	2	1	1
$(x + 2)^2 = (x + 2)(x + 2) = x^2 + 4x + 4$	4	2	4
$(x + 3)^2 = (x + 3)(x + 3) = x^2 + 6x + 9$	6	3	9
$(x + 4)^2 = (x + 4)(x + 4) = x^2 + 8x + 16$	8	4	16
$(x + 5)^2 = (x + 5)(x + 5) = x^2 + 10x + ?$	10	5	25
$(x + 7)^2 = (x + 7)(x + 7) = x^2 + ? + 49$	?	?	49
$(x + 8)^2 = (x + 8)(x + 8) = x^2 + ? + 64$	?	?	64
$(x + 10)^2 = (x + 10) \quad ? = x^2 + ? + ?$	?	?	?
$? \quad = \quad ? \quad ? = x^2 + 24x + 144$	24	?	144

(The reader should fill in for ? before continuing.)

Looking through the list of results we see that in each case the constant term is the square of half the coefficient of x. It is this observation that gives us the clue for making perfect squares of the form of those in the above list.

For example, what do we need to add to $x^2 + 20x$ to create a perfect square? We note that half the coefficient of x is 10 and on squaring this we get 100. So the perfect square we are looking for is $x^2 + 20x + 100$, which becomes $(x + 10)^2$. Alternatively, we could think of $x^2 + 20x$ expressed in terms of a perfect square by $(x + 10)^2 - 100$.

As another example consider what needs to be done to $x^2 + 14x + 60$ to make a perfect square of the form in the list above. Again note that half of the coefficient of x is 7, and on squaring this we get 49. So we may write

$$\begin{aligned} x^2 + 14x + 60 &= x^2 + 14x + 49 + 11 \\ &= (x + 7)^2 + 11 \end{aligned}$$

and then we see that the perfect square part was already there.

Looking at both these examples it appears to be possible to insert the perfect square into any quadratic function of x of the form $ax^2 + bx + c$, where $a = 1$.

Exercise 2.2

Express each of the following in terms of a perfect square:

1. $x^2 + 6x$	2. $x^2 + 10x + 1$	3. $x^2 + 18x + 80$
4. $x^2 + x + \frac{1}{4}$	5. $x^2 + 3x + \frac{9}{4}$	6. $x^2 + x + 1$
7. $x^2 + 2x + 2$	8. $x^2 + 12x + 40$	9. $x^2 + 3x + 1$
10. $x^2 + 5x$	11. $2x^2 + 12x + 18$	12. $2x^2 + 40x + 200$
13. $x^2 + bx$	14. $x^2 + bx + 1$	15. $x^2 + bx + c$

Expand the following products:

16. $(x - 1)^2$ 17. $(x - 2)^2$ 18. $(3 - x)^2$

19. $\left(x + \frac{b}{2}\right)^2$ 20. $\left(x - \frac{b}{2}\right)^2$ 21. $\left(x - \frac{b}{2a}\right)^2$

So far we have examined quadratic functions of the type $x^2 + bx + c = 0$, where b is positive. Now consider the following lists of results corresponding to negative values of b:

	Co-efficient of x	*Half the co-efficient of x*	*Constant term*
$(x - 1)^2 = (x - 1)(x - 1) = x^2 - 2x + 1$	-2	-1	1
$(x - 2)^2 = (x - 2)(x - 2) = x^2 - 4x + 4$	-4	-2	4
$(x - 3)^2 = (x - 3)(x - 3) = x^2 - 6x + 9$	-6	-3	9
$(x - 4)^2 = (x - 4)(x - 4) = x^2 - 8x + 16$	-8	-4	16
$(x - 5)^2 = (x - 5)(x - 5) = x^2 - 10x + 25$	-10	-5	25
$(x - 6)^2 = (x - 6)(x - 6) = x^2 - 12x + 36$	-12	-6	36
$(x - 7)^2 = (x - 7)(x - 7) = x^2 - 14x \quad ?$	-14	-7	?
$(x - 8)^2 = (x - 8)(x - 8) = x^2 - 16x \quad ?$	-16	-8	?
$(x - 9)^2 = (x - 9)(x - 9) = x^2 \quad ? \quad + 81$	?	?	81
$(x - 10)^2 = (x - 10) \quad ? \quad = x^2 \quad ? \quad ?$	?	?	?
$? \quad = \quad ? \quad ? \quad = x^2 \quad 24x + 144$	-24	-12	144

(The reader should fill in for ? before continuing.)

As with the previous list we notice the same relation of the constant term being equal to the square of half the coefficient of x. This means that we are now in a position to solve the general quadratic equation by what is known as the method of completing the square.

The Completion of the Square Method

Example 3. Solve the equation $x^2 + 10x + 4 = 0$.

Solution. In order to make the problem more familiar subtract 4 from both sides of the equation and obtain

$$x^2 + 10x = -4$$

Since half the coefficient of x is 5, we add 5^2 to both sides of the equation

$$\therefore \quad x^2 + 10x + 25 = -4 + 25$$
$$(x + 5)(x + 5) = 21$$
$$(x + 5)^2 = 21$$

Extracting the square root of both sides of the equation

$$x + 5 = \pm\sqrt{21} = \pm 4{\cdot}58 \text{ (correct to 2 decimal places)}$$

Subtracting 5 from both sides of the equation we have

$$x = \pm 4{\cdot}58 - 5$$
$$= -9{\cdot}58 \quad \text{or} \quad -0{\cdot}42$$

The roots of the equation $x^2 + 10x + 4 = 0$ are $-9{\cdot}58$ and $-0{\cdot}42$ (correct to 2 decimal places)

Answer

This result enables us to write $x^2 + 10x + 4 \equiv (x + 9{\cdot}58)(x + 0{\cdot}42)$ within the limits of accuracy of our solutions.

Example 4. Find the roots of the equation $x^2 - 7x - 9 = 0$.

Solution. Rewrite the equation as $x^2 - 7x = 9$. Half the coefficient of x is $-7/2$ so we 'complete the square' by adding (49/4) to both sides of the equation

$$\therefore \quad x^2 - 7x + \frac{49}{4} = 9 + \frac{49}{4}$$

$$\therefore \quad \left(x - \frac{7}{2}\right)^2 = \frac{36}{4} + \frac{49}{4} = \frac{85}{4} = 21{\cdot}25$$

Extracting the square root

$$x - \frac{7}{2} = \pm\sqrt{21{\cdot}25} = \pm 4{\cdot}61 \text{ (correct to 2 decimal places)}$$

$$\therefore \quad x = 3{\cdot}5 \pm 4{\cdot}61$$
$$= 8{\cdot}11 \quad \text{or} \quad -1{\cdot}11$$

The roots of the equation $x^2 - 7x - 9 = 0$ are 8·11 and −1·11 (correct to 2 decimal places)

Answer

This result enables us to write $x^2 - 7x - 9 \equiv (x - 8{\cdot}11)(x + 1{\cdot}11)$ within the limits of accuracy of our solution.

Example 5. Find the roots of the equation $3x^2 - 7x - 5 = 0$.

Solution. We first rewrite the equation in a more familiar form after dividing both sides of the equation by 3. This gives

$$x^2 - \frac{7x}{3} = \frac{5}{3}$$

Half the coefficient of x is $-7/6$ so we 'complete the square' by adding (49/36) to both sides of the equation

$$\therefore \quad x^2 - \frac{7x}{6} + \frac{49}{36} = \frac{49}{36} + \frac{5}{3} = \frac{49}{36} + \frac{60}{36} = \frac{109}{36}$$

$$\therefore \quad \left(x - \frac{7}{6}\right)^2 = \frac{109}{36}$$

Extracting the square root

$$x - \frac{7}{6} = \pm\sqrt{\frac{109}{36}} = \pm\frac{\sqrt{109}}{6} = \frac{10{\cdot}44}{6} \text{ (correct to 2 decimal places)}$$

$$\therefore \quad x = \frac{7}{6} \pm \frac{10{\cdot}44}{6} = \frac{17{\cdot}44}{6} \text{ or } -\frac{3{\cdot}44}{6}$$

The roots of the equation $3x^2 - 7x - 5 = 0$ are 2·91 and −0·57 (correct to 2 decimal places)

Answer

This result enables us to write $3x^2 - 7x - 5 \equiv 3(x - 2{\cdot}91)(x + 0{\cdot}57)$ within the limits of accuracy of our solutions.

Having progressed through these three worked examples the reader must be feeling that a formula to represent the solution of the general quadratic equation must be near at hand, since there appears to be very little further variation of the equation left to consider. The solution of general quadratic equations follows the lines of Example 5 above and the reader should compare the solutions very closely.

Solution of the General Quadratic Equation

To solve $ax^2 + bx + c = 0$, $a \neq 0$, first divide through by a and rewrite the equation as follows

$$x^2 + \frac{bx}{a} = -\frac{c}{a}$$

Half the coefficient of x is $\frac{b}{2a}$ so we 'complete the square' by adding $\left(\frac{b^2}{4a^2}\right)$ to both sides of the equation

$$\therefore x^2 + \frac{bx}{a} + \frac{b^2}{4a^2} = \frac{b^2}{4a^2} - \frac{c}{a}$$

$$\therefore \left(x + \frac{b}{2a}\right)^2 = \frac{b^2}{4a^2} - \frac{4ac}{4a^2} = \frac{b^2 - 4ac}{4a^2}$$

Extracting the square root, it follows that

$$x + \frac{b}{2a} = \pm\sqrt{\left(\frac{b^2 - 4ac}{4a^2}\right)} = \pm\frac{\sqrt{(b^2 - 4ac)}}{2a}$$

We finally obtain

$$x = -\frac{b}{2a} \pm \frac{\sqrt{(b^2 - 4ac)}}{2a}$$

i.e. $$x = \frac{-b \pm \sqrt{(b^2 - 4ac)}}{2a}$$ Answer

Comparison with Example 5 suggests $a = 3$, $b = -7$, $c = -5$.

Substitution in the formula that we have just obtained gives

$$x = \frac{-(-7) \pm \sqrt{[(-7)^2 - 4 \times 3 \times (-5)]}}{2 \times 3}$$

$$= \frac{7 \pm \sqrt{(49 + 60)}}{6}$$

$$\therefore x = \frac{7 \pm \sqrt{109}}{6} = 2{\cdot}91 \text{ or } -0{\cdot}57$$ (correct to 2 decimal places as before)

We are now in a position to obtain the roots of any quadratic equation whatsoever.

Exercise 2.3

Solve the following quadratic equations, giving your solutions correct to 2 decimal places. Try questions 1, 2 and 3 from first principles and the rest by using the formula.

1. $x^2 - 6x + 8 = 0$ 2. $2x^2 - 20x + 6 = 0$ 3. $x^2 - 2x - 9 = 0$
4. $3x^2 - 10x - 9 = 0$ 5. $7x^2 + 11x - 3 = 0$ 6. $2x^2 + 2x - 11 = 0$
7. The length of one side of a rectangle is 2 cm longer than the other side. If the area of the rectangle is 13 cm² find the length of its sides.
8. After t seconds a particle is at a distance of s metres from a fixed point O. If the relation between s and t is given by $s = 32t - 16t^2$ find the time taken to get 12 m from O.
9. By using part of the following expressions to form a perfect square, find their least values

 (a) $x^2 - 10x + 21$ (b) $x^2 + 2x + 3$ (c) $x^2 + 8x + 19$

Solution by Factors

When a quadratic function can be expressed as a product of its factors the roots are immediately obvious. For example, the equation given by

$(x-3)(x-4)=0$ is straightaway seen to be satisfied by $x=3$ and $x=4$, i.e. 3 and 4 are the roots of the equation

$$(x-3)(x-4)=0 \text{ (or } x^2-3x-4x+12=x^2-7x+12=0).$$

The following lists of results illustrate the idea relating the factors of the quadratic equation to its roots:

Factors	*Product of the factors*	*Roots*	*Sum of the roots*	*Product of the roots*
$(x-3)(x-2)$	$=x^2-5x+6$	3 and 2	5	6
$(x-4)(x-1)$	$=x^2-5x+4$	4 and 1	5	4
$(x-5)(x)$	$=x^2-5x$	5 and 0	5	0
$(x-6)(x+1)$	$=x^2-5x-6$	6 and -1	5	-6
$(x-7)(x+2)$	$=x^2-5x-14$	7 and -2	5	-14
$(2x-3)(3x-2)$	$=6x^2-13x+6$	$\frac{3}{2}$ and $\frac{2}{3}$	$\frac{13}{6}$	1
$(3x-4)(3x-5)$	$=9x^2-27x+20$	$\frac{4}{3}$ and $\frac{5}{3}$	3	$\frac{20}{9}$
$(4x-3)(5x-3)$	$=20x^2-27x+9$	$\frac{3}{4}$ and $\frac{3}{5}$	$\frac{27}{20}$	$\frac{9}{20}$

In the examples above we started by multiplying the known factors. In general the problems come the other way round; that is, we know the product but we have to find the factors. To get some idea of the nature of the difficulties involved consider the following examples:

$$\begin{aligned}
(5x-4)(2x-3) &= 10x^2-23x+12 & \\
(5x-3)(2x-4) &= 10x^2-26x+12 & \text{(i)}\\
(5x-6)(2x-2) &= 10x^2-22x+12 & \text{(ii)}\\
(5x-2)(2x-6) &= 10x^2-34x+12 & \text{(iii)}\\
(5x-12)(2x-1) &= 10x^2-29x+12 & \\
(5x-1)(2x-12) &= 10x^2-62x+12 & \text{(iv)}\\
(10x-4)(x-3) &= 10x^2-34x+12 & \\
(10x-3)(x-4) &= 10x^2-43x+12 & \\
(10x-6)(x-2) &= 10x^2-26x+12 & \\
(10x-2)(x-6) &= 10x^2-62x+12 & \text{(iii)}\\
(10x-12)(x-1) &= 10x^2-22x+12 & \\
(10x-1)(x-12) &= 10x^2-121x+12 &
\end{aligned}$$

We observe that in each product we have $10x^2$ and $+12$, and only the coefficient of x changes. If we are given $10x^2-?x+12=0$ to factorise then we need to obtain one of the twelve possible pairs of factors. (Really eight since four of them are repeated.) Obtaining the correct pair of factors requires the application of a little skill which takes time—during which one could have solved the equation by the method of completing the square! Again, if the coefficient of x^2 is 1 (unit) the problem of obtaining the factors is much simplified, as we can see by considering the example $x^2-7x+10=0$. Since there is only one x^2 we know that the factors will look like $(x+p)(x+q)=0$ and our task is to find p and q so that when we expand by multiplying the factors we get

$$\begin{aligned}(x+p)(x+q) &= x^2+px+qx+pq\\ &= x^2+(p+q)x+pq = x^2-7x+10\end{aligned}$$

From this result we see that we now need $pq = 10$, $p + q = -7$ or, in words, two numbers which add up to -7 and when multiplied together give 10.

We can get by using $5 \times 2, 2 \times 5, 10 \times 1, 1 \times 10$ or $-5 \times -2, -2 \times -5, -10 \times -1, -1 \times -10$, from which we choose -2 and -5 (or -5 and -2) to give a sum of -7. Hence we can now write $x^2 - 7x + 10 = (x - 2)(x - 5) = 0$ and see immediately that the roots of the equation are 2 and 5.

Example 6. Factorise $x^2 + 9x - 10 = 0$ and hence obtain the roots of the equation.

Solution. We are trying to find p and q so that $(x + p)(x + q) = x^2 + 9x - 10$. That is, $x^2 + (p + q)x + pq = x^2 + 9x - 10$. So we see that we need two numbers which add up to $+9$ and give -10 when multiplied together. 10 and -1 will satisfy the requirements and enable us to write, $x^2 + 9x - 10 = (x + 10)(x - 1) = 0$, and obtain the roots of the equation as 1 and -10. Answer

For the more difficult equation, such as $10x^2 - 43x + 12 = 0$, we need to consider those combinations of the factors of 10 and 12 which might give -43.

Consider the following thoughts:

(i) We know that to obtain $10x^2$ we must have factors which read as $(1x \quad ?)(10x \quad ?)$ or $(10x \quad ?)(1x \quad ?)$ or $(2x \quad ?)(5x \quad ?)$ or $(5x \quad ?)(2x \quad ?)$

(ii) We know that to obtain $+12$ we must have factors which read as $(? \quad \pm 1)(? \quad \pm 12)$ or $(? \quad \pm 2)(? \quad \pm 6)$ or $(? \quad \pm 3)(? \quad \pm 4)$

(iii) To obtain the result $10x^2 - 43x + 12 = 0$ we must combine a pair from (i) with a pair from (ii). For example, $(2x \pm 3)(5x \pm 4)$, which means either

$$(2x + 3)(5x + 4) \text{ which yields } 10x^2 + 23x + 12$$

or

$$(2x - 3)(5x - 4) \text{ which yields } 10x^2 - 23x + 12$$

Unfortunately, neither of these is of the required form.

After a few more trials we obtain $(1x - 4)(10x - 3) = 10x^2 - 43x + 12$, and thereby the solutions $x = 4$ and $x = 0{\cdot}3$ of the equation

$$10x^2 - 43x + 12 = 0.$$

Example 7. Find the solutions to the equation $21x^2 - 34x + 8 = 0$.

Solution. (i) To obtain $21x^2$ we must have factors which read as follows

$(1x \quad ?)(21x \quad ?)$ or $(21x \quad ?)(1x \quad ?)$ or $(3x \quad ?)(7x \quad ?)$ or $(7x \quad ?)(3x \quad ?)$

(ii) To obtain $+8$ we must have factors which read as

$$(? \quad \pm 1)(? \quad \pm 8) \quad \text{or} \quad (? \quad \pm 2)(? \quad \pm 4)$$

(iii) We could obtain the final form by considering all the possible combinations of the factors. Thus

$$(1x \pm 1)(21x \pm 8) = 21x^2 + 29x + 8 \quad \text{or} \quad 21x^2 - 29x + 8$$
$$(1x \pm 2)(21x \pm 4) = 21x^2 + 46x + 8 \quad \text{or} \quad 21x^2 - 46x + 8$$
$$(3x \pm 1)(7x \pm 8) = 21x^2 + 31x + 8 \quad \text{or} \quad 21x^2 - 31x + 8$$
$$(3x \pm 2)(7x \pm 4) = 21x^2 + 26x + 8 \quad \text{or} \quad 21x^2 - 26x + 8$$
$$(3x \pm 4)(7x \pm 2) = 21x^2 + 34x + 8 \quad \text{or} \quad 21x^2 - 34x + 8$$

This last result enables us to write $21x^2 - 34x + 8 = (3x - 4)(7x - 2) = 0$

Hence either $3x - 4 = 0$ or $7x - 2 = 0$, i.e. $x = 1\frac{1}{3}$ or $\frac{2}{7}$. Answer

Exercise 2.4

1. State the roots of the following equations:

(a) $(x - 4)(2x - 5) = 0$ (b) $(2x - 7)(x + 1) = 0$ (c) $(x - 1)(2x + 3) = 0$
(d) $(2x + 1)(3x + 5) = 0$ (e) $(5x + 4)(2x - 1) = 0$ (f) $(3x - 4)(4x - 1) = 0$

2. Factorise the following equations and hence find the solutions:

(a) $3x^2 - x - 2 = 0$ (b) $5x^2 - 36x + 7 = 0$ (c) $11x^2 + 13x + 2 = 0$
(d) $7x^2 - 78x + 11 = 0$ (e) $7x^2 + 4x - 3 = 0$ (f) $2x^2 + 5x - 3 = 0$

(Notice the simplification arising from both the coefficient of x^2 and the constant term being prime numbers.)

3. Factorise the following equations and hence find the solutions:

(a) $2x^2 - 13x + 20 = 0$ (b) $6x^2 + 13x + 5 = 0$ (c) $10x^2 + 3x - 4 = 0$
(d) $6x^2 + 13x + 6 = 0$ (e) $7x^2 + 38x + 15 = 0$ (f) $12x^2 + x - 6 = 0$

Imaginary Roots

It often becomes very tedious to attempt to factorise the quadratic equation $ax^2 + bx + c = 0$ in order to find the solutions, especially on those occasions when easy factors do not exist. However, the formula

$$x = \frac{-b \pm \sqrt{(b^2 - 4ac)}}{2a}$$

will always give the solutions to the equation and its use will often save a great deal of time. The quantity $\sqrt{(b^2 - 4ac)}$ clearly plays an important part in the formula for the following reasons:

(i) If $b^2 > 4ac$ then $b^2 - 4ac > 0$ and we shall have two distinct real roots to the equation as in all the examples worked so far; e.g. $2x^2 + 3x + 1 = 0$ where $a = 2$, $b = 3$, $c = 1$; $b^2 = 9$, $4ac = 8$, $b^2 - 4ac = 1$

(ii) If $b^2 = 4ac$ then $b^2 - 4ac = 0$ and the roots will both be equal to $-\dfrac{b}{2a}$; e.g. $4x^2 + 4x + 1 = 0$, where $a = 4$, $b = 4$, $c = 1$; $b^2 = 16$, $4ac = 16$, $b^2 - 4ac = 0$

(iii) If $b^2 < 4ac$ then $b^2 - 4ac < 0$, and we shall need to express the roots in terms of the square root of a negative quantity; e.g. $x^2 + x + 1 = 0$ where $a = 1$, $b = 1$, $c = 1$; $b^2 = 1$, $4ac = 4$, $b^2 - 4ac = -3$

Example 8. Find the roots of the equation $x^2 + 8x + 25 = 0$.

Solution. Substitute $a = 1$, $b = 8$, $c = 25$ in the formula $x = \dfrac{-b \pm \sqrt{(b^2 - 4ac)}}{2a}$

to get
$$x = \frac{-8 \pm \sqrt{(64 - 100)}}{2a} = \frac{-8 \pm \sqrt{-36}}{2}$$

Now we write $\sqrt{-36} = \sqrt{36} \times \sqrt{-1} = 6\sqrt{-1}$

$\therefore\ x = -4 \pm 3\sqrt{-1}$ (i.e. $-4 - 3\sqrt{-1}$ or $-4 + 3\sqrt{-1}$) Answer

This enables us to write $x^2 + 8x + 25 \equiv (x + 4 + 3\sqrt{-1})(x + 4 - 3\sqrt{-1})$.

The square root of any negative number may be written in the manner shown for $\sqrt{-36}$; that is, we simplify the square root by expressing the number as a product of two factors, one of which is -1.

The following list identifies the simplification:

(a) $\sqrt{-9} = \sqrt{9} \times \sqrt{-1} = 3\sqrt{-1}$
(b) $\sqrt{-25} = \sqrt{25} \times \sqrt{-1} = 5\sqrt{-1}$
(c) $\sqrt{-33} = \sqrt{33} \times \sqrt{-1} = 5{\cdot}75\sqrt{-1}$ (correct to 2 decimal places)
(d) $\sqrt{-52} = 7{\cdot}21\sqrt{-1}$ (correct to 2 decimal places)

Since the square root of any negative number ultimately involves $\sqrt{-1}$ we use the symbol or letter i to represent $\sqrt{-1}$. Using this symbol we will now write $\sqrt{-9}$ as 3i and $\sqrt{-25} = 5\text{i}$ and so on. Reversing these results we have $(3\text{i})^2 = 9\text{i}^2 = -9$, and $(5\text{i})^2 = 25\text{i}^2 = -25$, because $\text{i}^2 = -1$. Of course, i is not a number like 2, 3, 4 or -6, -7 and it is impossible to find a real line which measures i or 2i on which to construct a square of area -1 or -4 respectively, so we call numbers like $\pm\text{i}$, $\pm 2\text{i}$, $\pm 3\text{i}$ or $\frac{2\text{i}}{7}$, $\frac{3\text{i}}{22}$, etc., **imaginary numbers.**

In this book it would not be appropriate to give a definition of a **real number**, and all we can say is that real numbers are either **rational** like $\pm\frac{3}{4}$, $\pm\frac{6}{7}$, ± 8, ± 10, etc., or **irrational** like $\pm\sqrt{2}$, $\pm\sqrt{3}$, $\pm\sqrt{5}$, $\pm\sqrt{\frac{2}{3}}$, $\pm\sqrt{\frac{7}{5}}$.

A **complex number** is a number with two parts, one part real and the other part imaginary. Complex numbers are of the form $3 + 4\text{i}$, or $6 + \frac{19\text{i}}{22}$, $-\frac{2}{7} + \frac{8\text{i}}{9}$, or $\frac{3}{8} - \frac{5\text{i}}{9}$, $\frac{1}{2} + 3\text{i}$ and so on. Thus a complex number is of the form $a + b\text{i}$ where a and b are real numbers. If $a = 0$ we say that the complex number is **pure imaginary**; if $b = 0$ we say that the complex number is **pure real.** (Note: we have described the form taken by complex numbers and shown what a complex number looks like, but we have not defined either real numbers or complex numbers since such definitions are outside the scope of this text.)

Example 9. Find the roots of the equation $3x^2 + 5x + 6 = 0$.

Solution. With $a = 3$, $b = 5$, $c = 6$ we obtain

$$x = \frac{-5 \pm \sqrt{(25 - 72)}}{6} = \frac{-5 \pm \sqrt{-47}}{6}$$

i.e.
$$x = \frac{-5 \pm 6{\cdot}86\text{i}}{2} = -2{\cdot}5 \pm 3{\cdot}43\text{i} \text{ (correct to 2 decimal places)}$$

The roots of the equation are $-2{\cdot}5 + 3{\cdot}43\text{i}$ and $-2{\cdot}5 - 3{\cdot}43\text{i}$ (correct to 2 decimal places)
Answer

This result enables us to write $3x^2 + 5x + 6 \equiv 3(x + 2{\cdot}5 - 3{\cdot}43\text{i})(x + 2{\cdot}5 + 3{\cdot}43\text{i})$

The Sum and Product of the Roots

If we look back to the list on page 19 and examine the two columns giving the sum and product of the roots we should notice a connection between these results and the coefficients in the original quadratic.

Thus $9x^2 - 27x + 20 = 0$ has the roots $\frac{4}{3}$ and $\frac{5}{3}$, which have a sum of $\frac{9}{3} = 3$ and a product of $\frac{20}{9}$. On rewriting the equation in the form

$x^2 - 3x + \frac{20}{9} = 0$ we see that the sum of the roots is equal to the coefficient of x multiplied by -1 (or with its sign changed), in this case 3, and the product of the roots is equal to the constant term. Again, if $20x^2 - 27x + 9 = 0$ is re-written as $x^2 - \frac{27x}{20} + \frac{9}{20} = 0$ then we see that the coefficient of x when multiplied by -1 gives the sum of the roots and the constant term gives the product of the roots.

We can prove that this relation holds for the general quadratic

$$ax^2 + bx + c = 0$$

as follows. Suppose α and β are the two roots of the equation; then we can write

$$ax^2 + bx + c \equiv a(x - \alpha)(x - \beta)$$

If we now expand the right hand side we get

$$\begin{aligned} ax^2 + bx + c &\equiv a\{x^2 - x\beta - x\alpha + \alpha\beta\} \\ &\equiv a\{x^2 - (\alpha + \beta)x + \alpha\beta\} \\ \therefore ax^2 + bx + c &\equiv ax^2 - a(\alpha + \beta)x + a\alpha\beta \end{aligned}$$

Comparing the coefficients, we deduce that

(i) $b = -a(\alpha + \beta)$ i.e. $\alpha + \beta = -\frac{b}{a}$

(ii) $c = a\alpha\beta$ i.e. $\alpha\beta = \frac{c}{a}$

Looking at these results, we see why we consider rewriting $ax^2 + bx + c = 0$ as $x^2 + \frac{bx}{a} + \frac{c}{a} = 0$: by so doing we relate the coefficients directly to the sum and product of the roots.

Example 10. Find the factors and roots of the quadratic equation whose roots have a sum of 14 and a product of 48.

Solution. We are given that $\alpha + \beta = 14$ and $\alpha\beta = 48$

If we consider the equation to be $x^2 + \frac{bx}{a} + \frac{c}{a} = 0$ then

$$-\frac{b}{a} = 14 \quad \text{and} \quad \frac{c}{a} = 48$$

Therefore the equation is

$$x^2 - 14x + 48 = 0$$

which factorises to $(x - 8)(x - 6) = 0$ and has the roots 6 and 8

Check $6 + 8 = 14$; $6 \times 8 = 48$

$\therefore \quad (x - 8)(x - 6) = 0$ Answer

Example 11. Find the quadratic equation whose roots have a sum of -1 and a product of 4.

Solution. We are given that $\alpha + \beta = -1$, $\alpha\beta = 4$. The equation is therefore $x^2 + x + 4 = 0$.

The interesting feature of this example is that the roots of the equation are not real. Using the formula $x = \dfrac{-b \pm \sqrt{(b^2 - 4ac)}}{2a}$ with $a = 1$, $b = 1$, $c = 4$ we get

$$x = \frac{-1 \pm \sqrt{(1-16)}}{2} = \frac{-1 \pm i\sqrt{15}}{2}$$

That is, the roots are $\alpha = \dfrac{-1 + i\sqrt{15}}{2}$ and $\beta = \dfrac{-1 - i\sqrt{15}}{2}$. So the sum and product of the roots will be real numbers even when the roots are complex. We are, of course, always taking a, b and c to be real, i.e. the quadratic has real coefficients.

Exercise 2.5

Find the roots of the equations in questions 1, 2 and 3

1. $x^2 + x + 1 = 0$ 2. $x^2 + 2x + 2 = 0$ 3. $3x^2 + x + 5 = 0$

Find the sum and product of the roots of each of the following equations

4. $2x^2 + 3x + 5 = 0$ 5. $3x^2 + 2x + 11 = 0$ 6. $3x^2 + 4x + 1 = 0$

Find the quadratic equations whose roots α and β have the following sums and products. Also find the factors and roots of these equations

7. $\alpha + \beta = 7$, $\alpha\beta = 10$ 8. $\alpha + \beta = -11$, $\alpha\beta = 18$
9. $\alpha + \beta = 7$, $\alpha\beta = +12$ 10. $\alpha + \beta = -\frac{1}{2}$, $\alpha\beta = -\frac{21}{2}$
11. If $x = 3$, $x = 4$ are the roots of the equation $x^2 + px + q = 0$ what are the values of p and q?
12. Solve the equation $x^2 - 13x + 36 = 0$ and hence solve the equation

$$w^4 - 13w^2 + 36 = 0$$

Harder Problems on the Quadratic Equation

Example 12. Prove that the quadratic equation $x^2 + (m - 1)x + (m - 2) = 0$ always has real roots if m is real. Also find the value of m which gives equal roots.

Solution. The formula for the roots of a quadratic is $x = \dfrac{-b \pm \sqrt{(b^2 - 4ac)}}{2a}$

As we have already seen on page 21 the condition that the roots are real is $b^2 - 4ac \geqslant 0$. With $a = 1$, $b = m - 1$, $c = m - 2$ this means that we require $(m - 1)^2 - 4(m - 2) \geqslant 0$

i.e.

$$m^2 - 2m + 1 - 4m + 8 \geqslant 0$$
$$m^2 - 6m + 9 \geqslant 0$$
$$(m - 3)^2 \geqslant 0$$

Since a perfect square is always positive or zero it follows that $b^2 - 4ac \geqslant 0$, and the roots of this equation are always real. Furthermore, $b^2 - 4ac = 0$ when $m = 3$; that is, this equation has equal roots when $m = 3$, and when this is so $b = 2$. The equal roots are given by $x = \dfrac{-b}{2a} = -1$ Answer

Example 13. Find the values of p which make $x^2 + 2px + p + 2$ a perfect square.

Solution. For the quadratic to be a perfect square the constant $(p + 2)$ must be equal to the square of half the coefficient of x, which is p^2

$$\therefore \quad p^2 = p + 2 \quad \text{or} \quad p^2 - p - 2 = 0 = (p - 2)(p + 1)$$

Therefore the two required values of p are $p = 2$, and $p = -1$.

Check: If $p = 2$, we get $x^2 + 4x + 4 = (x + 2)^2$
If $p = -1$, we get $x^2 - 2x + 1 = (x - 1)^2$
$\therefore \quad p = 2$ or -1 Answer

Example 14. Prove that if one root of the equation $ax^2 + bx + c = 0$ is one third of the other root then $3b^2 = 16ac$. Also find the roots in terms of a and b.

Solution. Let the roots be r and $3r$

$$\therefore \quad 4r = \frac{-b}{a} \qquad 3r^2 = \frac{c}{a}$$

Hence $$16r^2 = \frac{b^2}{a^2} \quad \text{and} \quad 16r^2 = \frac{16c}{3a}$$

$$\therefore \quad \frac{b^2}{a^2} = \frac{16c}{3a} \quad \text{so that} \quad 3b^2 = 16ac$$ Answer

Since $4r = -\frac{b}{a}$ we have $r = -\frac{b}{4a}$ as one root and $3r = -\frac{3b}{4a}$ as the other root

Answer

Example 15. Given that α and β are the roots of the equation $3x^2 + 4x + 6 = 0$ find the equation whose roots are $\frac{1}{\alpha}$ and $\frac{1}{\beta}$.

Solution.
$$\alpha + \beta = -\frac{4}{3} \qquad \text{(i)}$$
$$\alpha\beta = 2 \qquad \text{(ii)}$$

Dividing (i) by (ii) we have $\frac{1}{\beta} + \frac{1}{\alpha} = -\frac{2}{3}$ (which is the sum of the new roots)

If $\alpha\beta = 2$ then $\frac{1}{\alpha\beta} = \frac{1}{2}$ (which is the product of the new roots)

Hence the required equation is $x^2 + \frac{2x}{3} + \frac{1}{2} = 0$

or $$6x^2 + 4x + 3 = 0$$ Answer

Exercise 2.6

1. Prove that the quadratic equation $x^2 + 2mx + m^2 = 0$ has real equal roots for any real value of m.
2. Prove that the equation $x^2 + (m - 2)x + 2 - m = 0$ has real roots if $m \geqslant 2$ or $m \leqslant -2$ but not if $-2 < m < 2$.
3. Find the quadratic equation whose roots are double the roots of the equation $10x^2 + 2x + 3 = 0$.
4. The equation $3x^2 + 5x + 4 = 0$ has the roots α and β. By putting $x = 2/y$ or otherwise find the equation with the roots $2/\alpha$ and $2/\beta$.
5. Show that if the roots of the equation $x^2 + 2mx + n = 0$ are r and $r + 3$ then $4m^2 = 4n + 9$.
6. What is the relation between the roots of $2x^2 + 7x + 3 = 0$ and the roots of $y^2 + 7y + 6 = 0$?
7. By putting $x = y^2$, show that both the roots of the equation $x^2 - 8x + 15 = 0$ are roots of $y^4 - 8y^2 + 15 = 0$. Also find the other two roots of the equation.
8. In a right-angled triangle the sides are of length x, $x + 1$, and $x + 2$. Prove that there is only one such right-angled triangle whose sides are so related.

CHAPTER THREE

DIVISION AND THE REMAINDER THEOREM

Division

We can always divide one number by another number and leave the answer in two parts called the **quotient** and the **remainder**: for example, $59 \div 8 = 7$ with a remainder 3. Having found the result of the division, we are now able to reconstruct the number 59 as

$$59 = 8 \times 7 + 3$$

The 7 is called the quotient and the 3 is called the remainder. Thus we are able to say that the division of 59 by 8 gives a quotient of 7 and a remainder of 3.

As another illustration consider $348 \div 9 = 38$ with remainder 6. This result enables us to reconstruct the number 348 in the form

$$348 = 9 \times 38 + 6$$

In this case the quotient is 38 and the remainder is 6.

The test of one number being a factor of another is that the remainder shall be zero. Similar reasoning applies to finding whether $x - 3$ is a factor of $2x^2 - 2x - 13$. If we divide $2x^2 - 2x - 13$ by $x - 3$ we shall obtain a quotient Q and a remainder R, enabling us to write

$$2x^2 - 2x - 13 \equiv (x - 3)Q + R$$

All we need do now is find out how to perform the division using $x - 3$ as divisor. Division is really a process of successive subtraction. Thus our earlier example of $59 \div 8$ could be thought of as follows (if we do not 'see' the result straightaway): since we know the easy result $40 = 8 \times 5$, we subtract this from 59 to get a remainder of 19. We know the easy result $16 = 8 \times 2$ so we subtract this from 19 to get a remainder of 3

$$\therefore \quad 59 = 8 \times 7 + 3$$

Now let us put this in the form $(5x + 9) \div (x - 2)$. Since we know that $3(x - 2) = 3x - 6$ we could subtract this from $5x + 9$ to get a remainder of $2x + 15$. We can then write

$$5x + 9 \equiv (x - 2)3 + 2x + 15$$

But, have we gone as far as we can? Alternatively, since $4(x - 2) = 4x - 8$ we could subtract this from $5x + 9$ to get a remainder of $x + 17$ so that we can then write

$$5x + 9 \equiv (x - 2)4 + x + 17$$

Again, have we gone as far as we can? With $6(x - 2) = 6x - 12$ we could subtract this from $5x + 9$ to get a remainder of $-x + 21$, so that we can then write

$$5x + 9 \equiv (x - 2)6 - x + 21$$

Truly, there is no end to this process, so in the interests of concluding the division we agree that a remainder will be called *the* remainder when its degree is less than that of the divisor. This means that if we divide by $x - 2$ our remainder will be numerical, i.e. independent of x. If we divide by $x^2 - 3$ then our remainder will be of degree less than 2, and so on. Returning to $(5x + 9) \div (x - 2)$, we now see that we must obtain a remainder which does not contain x. In this case we consider $5(x - 2) = 5x - 10$ subtracted from $5x + 9$, leaving a remainder of 19, so that we can then write

$$5x + 9 \equiv (x - 2)5 + 19$$

and it is this form that we call the result of dividing $5x + 9$ by $x - 2$. Notice that if we put $x = 10$ we have the first part of $59 \div 8$ discussed on page 26.

Example 1. Divide $7x^2 + 3x + 11$ by $x - 5$.

Solution. Since we are dividing by $x - 5$ the remainder must be numerical. This means that we must first eliminate $7x^2$ by working from the left. Since we need to eliminate $7x^2$ we consider $(x - 5)7x = 7x^2 - 35x$. Subtracting this result from $7x^2 + 3x + 11$ we can therefore write

$$7x^2 + 3x + 11 \equiv (x - 5)7x + 38x + 11$$

Our remainder so far is $38x + 11$, which, being of the same degree as the divisor $x - 5$, indicates that we must go further (i.e. it is not yet independent of x). Since we need to eliminate $38x$ we consider $(x - 5)38 = 38x - 190$. Subtracting this result from $38x + 11$ we can therefore write

$$38x + 11 = (x - 5)38 + 201$$

and the final remainder is 201. We may now write

$$\begin{aligned} 7x^2 + 3x + 11 &\equiv (x - 5)7x + (x - 5)38 + 201 \\ &\equiv (x - 5)(7x + 38) + 201 \end{aligned}$$

That is, the quotient is $7x + 38$ and the remainder is 201.

For convenience we usually put the above work in the following form

$$\begin{array}{lll} x - 5)\ 7x^2 + 3x + 11\ (7x + 38 & & \\ \quad\underline{7x^2 - 35x} & & \text{Subtract } 7x^2 - 35x \text{ from } 7x^2 + 3x \\ \qquad\quad 38x + 11 & & \\ \qquad\quad \underline{38x - 190} & & \text{Subtract } 38x - 190 \text{ from } 38x + 11 \\ \qquad\qquad\quad 201 & & \end{array}$$

$\therefore\ 7x^2 + 3x + 11 \equiv (x - 5)(7x + 38) + 201$ Answer

Example 2. Divide $4x^3 + 3x + 9$ by $x + 2$.

Solution. There is no term in x^2 but we consider this written as $0x^2$

$$\begin{array}{lll} x + 2)\ 4x^3 + 0x^2 + 3x + 9\ (4x^2 - 8x + 19 & & \\ \quad\underline{4x^3 + 8x^2} & & \text{Subtract } (x + 2)4x^2 \text{ to eliminate } 4x^3 \\ \qquad -8x^2 + 3x & & \\ \qquad \underline{-8x^2 - 16x} & & \text{Subtract } (x + 2)(-8x) \text{ to eliminate } -8x^2 \\ \qquad\qquad +19x + 9 & & \\ \qquad\qquad \underline{19x + 38} & & \text{Subtract } (x + 2)19 \text{ to eliminate } 19x \\ \qquad\qquad\quad -29 & & \end{array}$$

$\therefore\ 4x^3 + 3x + 9 \equiv (x + 2)(4x^2 - 8x + 19) - 29$ Answer

Exercise 3.1

1. Divide $3x^2 + 4x + 1$ by $x - 1$. Put $x = 10$ and relate the result to dividing 341 by 9.
2. If there is a remainder of 13 when $3x^2 + 4x + m$ is divided by $x - 1$, what is the numerical value of m?
3. When a number n is divided by 7 the quotient is 3 and the remainder is 6. Find: (i) n, (ii) the remainder when $n + 4$ is divided by 7.
4. Divide $f(x) \equiv 3x^3 + x + 1$ by $x + 2$. What number must be added to $f(x)$ in order that the result should be divided by $x + 2$ without remainder?
5. Referring to the worked Example 2 above divide $4x^3 + 3x + 9$ by $4x^2 - 8x + 19$.
6. If $f(x) \equiv x^3 + 2x^2 - x - 5 = (x - 1)(x^2 + 3x + 2) - 3$ what is the remainder when $f(x)$ is divided by $x^2 + 3x + 2$? What is the remainder when $f(x) + 3$ is divided by $x - 1$?

The Remainder Theorem

This theorem enables us to deduce a relation between the divisor and the remainder without the need to perform the division. If we examine some of the examples above we can identify what this relation is.

In one example we had $5x + 9 \equiv (x - 2)5 + 19$. If we put $x = 2$ in $5x + 9$ we get the remainder 19. In another example we had $7x^2 + 3x + 11 \equiv (x - 5)(7x + 38) + 201$. If we put $x = 5$ in $7x^2 + 3x + 11$ we get the remainder $7(25) + 3(5) + 11 = 201$. In the last example $4x^3 + 3x + 9 \equiv (x + 2)(4x^2 - 8x + 19) - 29$. If we put $x = -2$ in $4x^3 + 3x + 9$ we get the remainder $4(-2)^3 + 3(-2) + 9 = -29$.

It appears therefore that if we divide functions of this type by $x - k$ then the remainder is given by substituting $x = k$. In symbols, if $f(x)$ is divided by $x - k$ then the remainder is $f(k)$. (Remember that we are speaking only of functions which consist of adding a finite number of integral powers of x such as $f(x) \equiv px^{10} + qx^8 + rx^5 + sx + m$. Such expressions are called polynomials.)

The Remainder Theorem states that if $f(x)$ is divided by $x - k$ then the remainder is equal to $f(k)$.

We know that if we divide $f(x)$ by $(x - k)$ we will obtain a quotient Q and a remainder R so that we may write $f(x) \equiv (x - k)Q + R$. If we put $x = k$ in this result we have

$$f(k) = 0 + R$$

i.e. the remainder is $f(k)$.

The following example illustrates this result in the case of the general quadratic.

Example 3. Divide $ax^2 + bx + c$ by $x - k$

Solution.

$$
\begin{array}{r}
x - k)\ ax^2 + bx + c\ (ax + (b + ka) \\
\underline{ax^2 - kax \qquad\qquad} \\
(b + ka)x + c \\
\underline{(b + ka)x - k(b + ka)} \\
c + k(b + ka)
\end{array}
$$

$\therefore\ ax^2 + bx + c \equiv (x - k)(ax + b + ka) + c + kb + k^2a$

The remainder is $ak^2 + bk + c$ which is equal to $ax^2 + bx + c$ with k substituted for x.

Factors

If $x - k$ is a factor of $f(x)$ then the remainder after dividing $f(x)$ by $x - k$ will be zero. It follows that since the remainder is $f(k)$, if we find that $f(k) = 0$ we shall know that $x - k$ is a factor of $f(x)$.

Example 4. Prove that $x - 1$, $x + 2$ and $x - 3$ are factors of

$$f(x) \equiv x^4 - x^3 - 7x^2 + x + 6$$

and find the fourth factor.

Solution.

$$\begin{aligned} f(1) &= 1 - 1 - 7 + 1 + 6 = 0 \quad \therefore \quad x - 1 \text{ is a factor of } f(x) \\ f(-2) &= 16 - (-8) - 28 + (-2) + 6 \\ &= 16 + 8 - 28 - 2 + 6 = 0 \quad \therefore \quad x + 2 \text{ is a factor of } f(x) \\ f(3) &= 81 - 27 - 63 + 3 + 6 = 0 \quad \therefore \quad x - 3 \text{ is a factor of } f(x) \end{aligned}$$

So far our results enable us to say that $(x - 1)(x + 2)(x - 3)$ is a factor of $f(x)$. Since $f(x)$ is of degree four in x and $(x - 1)(x + 2)(x - 3)$ is of degree three this suggests that there is another linear factor in $f(x)$ of the form $x - k$. All we know is that $f(k) = 0$, so that we must experiment with one or two values of k in the hope of discovering the fourth factor.

Try $k = 2$. $f(2) = 16 - 8 - 28 + 2 + 6 = -12$ $\therefore$ $x - 2$ is not a factor of $f(x)$
Try $k = -1$. $f(-1) = 1 - (-1) - 7 + (-1) + 6 = 0$ $\therefore$ $x + 1$ is a factor of $f(x)$

Therefore we may write $x^4 - x^3 - 7x^2 + x + 6 \equiv (x - 1)(x + 1)(x + 2)(x - 3)$
Answer

Example 5. Prove that $x - 1$ is always a factor of $f(x) \equiv 2x^3 + x^2 + (p - 3)x - p$ for any value of p. Find the other factors of $f(x)$ when $p = -20$.

Solution. $x - 1$ is a factor of $f(x)$ if $f(1) = 0$. Now $f(1) = 2 + 1 + p - 3 - p = 0$ for any value of p and therefore $x - 1$ is always a factor of $f(x)$. When $p = -20$, $f(x) \equiv 2x^3 + x^2 - 23x + 20$. Dividing $f(x)$ by $x - 1$ we have

$$\begin{array}{l} x - 1)\ 2x^3 + x^2 - 23x + 20\ (2x^2 + 3x - 20 \\ \quad\quad\ \underline{2x^3 - 2x^2} \\ \quad\quad\quad\quad\ 3x^2 - 23x \\ \quad\quad\quad\quad\ \underline{3x^2 - 3x} \\ \quad\quad\quad\quad\quad\quad -20x + 20 \\ \quad\quad\quad\quad\quad\quad \underline{-20x + 20} \\ \quad\quad\quad\quad\quad\quad\quad\quad\quad 0 \end{array}$$

So we may now write $f(x) \equiv (x - 1)(2x^2 + 3x - 20)$. Since

$$2x^2 + 3x - 20 = (2x - 5)(x + 4)$$

then we conclude that $f(x) \equiv (x - 1)(2x - 5)(x + 4)$.
Observe that we could have discovered the factor $x + 4$ by finding $f(-4) = 0$, and the factor $2x - 5$ would have been found from $f(\frac{5}{2}) = 0$.

Exercise 3.2

1. Show that $2x - 3$ is a factor of $f(x) \equiv 2x^3 - x^2 - 7x + 6$ by evaluating $f(\frac{3}{2})$. Obtain the other two factors.
2. By evaluating $f(1), f(-1), f(2)$ express $f(x) \equiv x^3 - 2x^2 - x + 2$ as a product of its factors.
3. The function $f(x)$ is of the third degree in x. If $f(1), f(3), f(-7)$ are each zero and $f(0) = 63$, find $f(x)$.
4. The function $f(x) \equiv x^2 + x + 1$. Explain why it is not possible to find any real number n such that $f(n) = 0$.

5. Two of the factors of $f(x) \equiv x^3 + px^2 + qx + r$ are $x - 1$ and $x + 1$, and $f(x)$ leaves a remainder of 8 when divided by $x - 3$. Find the values of p, q and r.
6. The function $f(x) \equiv mx^3 + nx^2 + x + 1$ is divisible by $x^2 - 1$ without remainder. Find m and n.
7. Invent a quadratic which has the factors $x - 1$ and $x + 2$ and a remainder of 20 when divided by $x - 3$.
8. Prove that $x - 1$ is a factor of $f(x) \equiv m(x^2 - 3x + 2) + n(x^2 + 11x - 12)$ for all values of m and n.
9. Any number with four digits may be considered as $ax^3 + bx^2 + cx + d$ with $x = 10$ (a, b, c, d any digits from 0 to 9, $a \neq 0$). We may also consider dividing this number by 12 in the form $x + 2$, or 13 in the form $x + 3$, and so on. When the digits of the number are reversed we get $dx^3 + cx^2 + bx + a$.
   (i) Add the number and the reversed digit number together and show that the result is always divisible by eleven.
   (ii) Show that the result in (i) does not necessarily hold for a number with three digits.
10. Show that $x - 4$ is a factor of $g(x) \equiv f(x) - f(4)$.

Longer Division

The method of division which was discussed on page 27 is applicable to a divisor of any degree. As we saw in one or two subsequent exercises, if we obtain a result like $f(x) \equiv (x - 7)(x^2 + 3x + 1) + 15$ this really gives us the results of two possible divisions: that is, $x^2 + 3x + 1$ is the quotient when $x - 7$ is the divisor and $x - 7$ is the quotient when $x^2 + 3x + 1$ is the divisor. As already noted, we agree to stop when we have obtained a remainder of lower degree than the divisor, in which case we must start by eliminating the higher powers of x. This in turn suggests that we must always write the expressions in descending powers of x, as illustrated in the next example.

Example 6. Divide $7x^2 - 7x^3 + 6x^4 - 24 + 18x$ by $3x^2 + x - 4$

Solution. $3x^2 + x - 4)\ 6x^4 - 7x^3 + 7x^2 + 18x - 24\ (2x^2 - 3x + 6$

$$\frac{6x^4 + 2x^3 - 8x^2}{-9x^3 + 15x^2 + 18x}$$

At this point we have eliminated x^4 by multiplying the divisor by $2x^2$

$$\frac{-9x^3 - 3x^2 + 12x}{18x^2 + 6x - 24}$$

At this point we have eliminated x^3 by multiplying the divisor by $-3x$

$$\frac{18x^2 + 6x - 24}{0 \quad 0 \quad 0}$$

at this point we have eliminated x^2 by multiplying the divisor by 6. By good fortune this has also yielded a zero remainder and thus enabled us to write

$$6x^4 - 7x^3 + 7x^2 + 18x - 24 \equiv (3x^2 + x - 4)(2x^2 - 3x + 6)$$

Answer

The method is in fact so general that we can apply it to a division involving more than one variable, as in the next example. The actual layout does, however, tend to become a little scattered.

Example 8. Divide $x^2 + 2x + 1 - y^2$ by $x + y + 1$.

Solution. We shall be unable to anticipate the required order of writing down the dividend so the best we can do is to concentrate on one letter and arrange in the order of its descending powers.

$$x + (1 + y))\, x^2 + 2x + 1 - y^2\, (x + 1 - y$$
$$\underline{x^2 + x(y + 1)}$$
$$x(1 - y) + 1 - y^2$$

We have now eliminated x^2. The next step is to eliminate $x(1 - y)$ so multiply the divisor by $1 - y$

$$\underline{x(1 - y) + (1 - y^2)}$$
$$0 \qquad 0$$

(Note: $(1 - y)(1 + y) = 1 - y^2$)

Thus $$x^2 + 2x + 1 - y^2 \equiv (x + 1 + y)(x + 1 - y)$$ Answer

Again, we could have noticed that $x^2 + 2x + 1 = (x + 1)^2$

so that $$x^2 + 2x + 1 - y^2 = (x + 1)^2 - y^2$$

Using the result $A^2 - B^2 = (A - B)(A + B)$ this last line could be modified to read

$$(x + 1)^2 - y^2 = (x + 1 - y)(x + 1 + y)$$

and we have solved the problem by factorising the given expression.

Example 9. Divide $6x^2 + xy - x - y^2 + 7y - 12$ by $2x + y - 3$.

Solution. $$2x + (y - 3))\, 6x^2 + x(y - 1) - 12 + 7y - y^2\, (3x^2 + (4 - y)$$
$$\underline{6x^2 + 3x(y - 3)}$$
$$x(8 - 2y) - 12 + 7y - y^2$$

This first step eliminates x^2

$$\underline{x(8 - 2y) - 12 + 7y - y^2}$$
$$0 \qquad 0 \quad 0 \quad 0$$

Note that $(y - 3)(4 - y) = -12 + 7y - y^2$.

Hence $6x^2 + xy - x - y^2 + 7y - 12 \equiv (2x + y - 3)(3x^2 + 4 - y)$

Answer

Example 10. Show that $x^2 + y^2 + z^2 + 2yz + 2xz + 2xy$ is a perfect square.

Solution. We start with a speculation. To obtain x^2 we must have $(x\ \ ?)(x\ \ ?)$; similarly for y^2 and z^2, so that our first suggestion would be $(x + y + z)(x + y + z)$. Multiplication of the two brackets reveals the result

$$(x + y + z)(x + y + z) = x^2 + y^2 + z^2 + 2yz + 2xz + 2xy$$

Answer

The reader must always be prepared to tackle problems with such a spirit; experimentation with an idea or feeling are very important and vital to the notion of mathematics as an activity.

Exercise 3.3

1. Show that each of the following expressions are perfect squares:
   (i) $x^2 + y^2 + z^2 - 2yz + 2xz - 2xy$
   (ii) $x^2 + y^2 + z^2 + 2yz - 2xz - 2xy$
   (iii) $x^2 + y^2 + z^2 - 2yz - 2xz + 2xy$
   (iv) $x^2 + y^2 + z^2 + w^2 + 2xy + 2xz + 2xw + 2yz + 2yw + 2zw$

2. Divide $a^2 + 2a + 1 - b^2$ by $a + 1 - b$.
3. Divide $6a^2 + ab - a - 12 + 7b - b^2$ by $3a + 4 - b$.
4. Divide the sum of $x^2(y - 1)$, $4x(y - 1)$, $(y - 1)^2$ by $y - 1$.
5. Show that $x - y$ is a factor of $x^3 - y^3$ and find the other factor.
6. Show that $x + y$ is a factor of $x^3 + y^3$ and find the other factor.
7. Using the results in questions 5 and 6 factorise (i) $64a^3 - 27b^3$, (ii) $8p^3 + y^3$.
8. Use the result of $A^2 - B^2 = (A - B)(A + B)$ to factorise

$$(2x + y + 1)^2 - (x - y + 1)^2.$$

9. Divide $4x^4 - 9x^2 - 1 + 6x$ by $2x^2 + 3x - 1$ and use your result to express $4x^4 - 9x^2 - 1 + 6x$ as the difference of two squares, i.e. $A^2 - B^2$.
10. By considering $f(b + c)$ find the factors of

$$f(x) \equiv 2x^2 + xb - xc - 3b^2 - c^2 - 4bc.$$

CHAPTER FOUR

SIMULTANEOUS EQUATIONS OF THE SECOND DEGREE

Quadratic Equations in Two Unknowns

We have seen how to solve any equation of the form $ax^2 + bx + c = 0$ by using the formula $x = \dfrac{-b \pm \sqrt{(b^2 - 4ac)}}{2a}$, observing in the process that if $b^2 - 4ac$ is a perfect square then we might have obtained the factors by inspection. Consider replacing x by $\dfrac{y}{z}$ in the equation $ax^2 + bx + c = 0$ to get $\dfrac{ay^2}{z^2} + \dfrac{by}{z} + c = 0$. Assuming that $z \neq 0$ we may multiply both sides by z to get $ay^2 + byz + cz^2 = 0$, so that we now have an equation of the second degree in the two unknowns y and z. The solution for x will now read $x = \dfrac{y}{z} = \dfrac{-b \pm \sqrt{(b^2 - 4ac)}}{2a}$.

Unfortunately, while this gives two values for x it does not fix the values for y and z. For example, substitution of

$$x = \frac{y}{z} \text{ in } x^2 - 7x + 12 = 0 = (x - 3)(x - 4)$$

gives $y^2 - 7yz + 12z^2 = 0 = (y - 4z)(y - 3z)$, with solutions $y = 4z$ and $y = 3z$.

We may choose as many pairs of values of y and z as we please to satisfy one of these two results, e.g. $y = 4$, $z = 1$; $y = 8$, $z = 2$; $y = 30$, $z = 10$; $y = -6$, $z = -2$; and so on. Clearly, another equation in y and z which has to be satisfied simultaneously with this equation will give a definite number of solutions, but the manipulation will not be as straightforward as in Chapter One.

Simultaneous Equations (One Quadratic, One Linear)

Example 1. Solve the simultaneous equations

$$x^2 - 3xy + 2y^2 = 0 \qquad \text{(i)}$$
$$x + y = 9 \qquad \text{(ii)}$$

Solution. As already discussed, equation (i) can be thought of as a quadratic in $\dfrac{x}{y}$, i.e. $\dfrac{x^2}{y^2} - \dfrac{3x}{y} + 2 = 0$. $y \neq 0$. By inspection we factorise this into

$$\left(\frac{x}{y} - 2\right)\left(\frac{x}{y} - 1\right) = 0,$$

and with multiplication by y^2 we rewrite this result as $(x - 2y)(x - y) = 0$. Consequently the solutions so far are $x = 2y$ and $x = y$.

Substituting $x = 2y$ in (ii) we have $2y + y = 9$, $y = 3$
Substituting $x = y$ in (ii) we have $y + y = 9$, $y = 4{\cdot}5$

The solutions are (a) $x = 6, y = 3$ (b) $x = 4{\cdot}5, y = 4{\cdot}5$ Answer

The general method which we discuss as an alternative is always certain to produce the solution in the event of not 'seeing' the factors of equation (i).

The equation (ii) enables us to write $y = 9 - x$, which we can substitute in equation (i), thereby obtaining an equation in the single unknown x.

Thus $x^2 - 3x(9 - x) + 2(9 - x)^2 = 0 = x^2 - 27x + 3x^2 + 162 - 36x + 2x^2$

$$\therefore \quad 6x^2 - 63x + 162 = 0$$

an equation which we can simplify on division by 3 to give finally

$$2x^2 - 21x + 54 = 0 = (2x - 9)(x - 6)$$

We see that $x = 4{\cdot}5$ or 6, and since $y = 9 - x$ the corresponding results are $y = 4{\cdot}5$ or 3. The solutions are

(a) $x = 6, y = 3$ (b) $x = 4{\cdot}5, y = 4{\cdot}5$ Answer

The above example has a simplicity arising from equation (i) being originally in the form of a quadratic equation in one unknown. An equation such as $x^2 + y^2 = 25$ or $x^2 - 3xy + 2y^2 = 8$ cannot be simplified in this manner and its presence as one of the simultaneous equations requires a little more ingenuity on our part; as the next two examples indicate.

Example 2. Solve the simultaneous equations

$$x^2 - y^2 = 25 \qquad \text{(i)}$$
$$x - y = 1 \qquad \text{(ii)}$$

Solution. We observe that equation (i) may be written $(x - y)(x + y) = 25$, Using equation (ii) in this new result we get $x + y = 25$. We now have two linear equations

$$x - y = 1 \qquad \text{(ii)}$$
$$x + y = 25 \qquad \text{(iii)}$$

From (ii) + (iii) we have $2x = 26$, $x = 13$. By substituting $x = 13$ in (iii) we get $y = 12$.

Check $13^2 - 12^2 = 169 - 144 = 25$. $x = 13, y = 12$ Answer

One further point to notice is that although one equation is of the second degree this pair of equations only has one solution, unlike Example 1 where we obtained two solutions. (In Question 7, Exercise 4.1, we get four solutions!)

The above example was deliberately arranged to give a simple illustration of a method. If we choose an equation of the second degree at random for (i), then we are involved in a more tedious exercise as follows in Example 3.

Example 3. Solve the simultaneous equations

$$x^2 + 3xy + 5y^2 = 19 \qquad \text{(i)}$$
$$x - 2y = 1 \qquad \text{(ii)}$$

Solutions. The equation (i) does not appear to factorise conveniently, so we use (ii) in the form $x = 2y + 1$. Substituting $x = 2y + 1$ in (i) produces the result

$$(2y + 1)^2 + 3(2y + 1)y + 5y^2 = 19$$
$$4y^2 + 4y + 1 + 6y^2 + 3y + 5y^2 = 19$$
$$\therefore \quad 15y^2 + 7y - 18 = 0$$

With $a = 15$, $b = 7$, $c = -18$ we find y by the formula

$$y = \frac{-7 \pm \sqrt{(49 + 1080)}}{30} = \frac{-7 \pm \sqrt{1129}}{30} = \frac{-7 \pm 33{\cdot}6}{30}$$

$$\therefore \quad y = -1{\cdot}35 \quad \text{or} \quad 0{\cdot}89 \text{ (correct to 2 decimal places)}$$

Returning to (ii) $x = -1{\cdot}70$ or $2{\cdot}78$ (correct to 2 decimal places)

$\therefore$ (a) $x = -1{\cdot}70,\ y = -1{\cdot}35$; (b) $x = 2{\cdot}78,\ y = 0{\cdot}89$ Answer

Exercise 4.1

Solve the following pairs of simultaneous equations

1. $x^2 - y^2 = 121$, $x + y = 121$
2. $x^2 - y^2 = 49$, $x - y = 1$
3. $4x^2 + y^2 = 17$, $2x + y = 5$
4. $x^2 + 3xy + 5y^2 = 15$, $x - y = 1$
5. $4x^2 - 4yx - 3y^2 = 20$, $2x - 3y = 10$
6. $x - 3y = 5$, $x^2 + xy + y^2 = 7$
7. $x^2 + y^2 = 41$, $x^2 - y = 11$
8. $x^2 + y^2 = 100$, $x^2 - y^2 = 28$
9. $x^2 - 2xy + y^2 = 16$, $x^2 + 2xy + y^2 = 4$
10. $\frac{1}{x^2} + y = 3$, $\frac{1}{x^2} - \frac{1}{y} = 1$

Simultaneous Equations (Two Quadratic)

The last exercise included pairs of simultaneous equations which produced one, two or four solutions for x and y. At first sight it might appear that we could solve any pair of simultaneous quadratic equations but, unfortunately, the necessary methods are sometimes very devious even for the most innocent pair of equations. Take for example the pair

$$x^2 + y = 11$$
$$x + y^2 = 7$$

There does not appear to be much difference between this suggestion and Questions 7 and 8 in Exercise 4.1, yet the solution is much more elusive even though we can 'see' a solution by inspection, i.e. $x = 3$, $y = 2$.

A piece of mathematical inspiration might be along the following lines.

Example 4. Solve the equation

$$x^2 + y = 11 \qquad \text{(i)}$$
$$x + y^2 = 7 \qquad \text{(ii)}$$

Solution. From (i) $x^2 = 11 - y \quad \therefore \quad x^2 - 9 = 11 - y - 9$

$$(x - 3)(x + 3) = 2 - y \qquad \text{(iii)}$$

From (ii) $x = 7 - y^2$. Since we have just obtained an equation involving $2 - y$ we now rearrange $x = 7 - y^2$ into

$$x - 3 = 4 - y^2 = (2 - y)(2 + y) \qquad \text{(iv)}$$

Thus we have rewritten the original equations in the form

$$(x - 3)(x + 3) = 2 - y \qquad \text{(i) or (iii)}$$
$$x - 3 = (2 - y)(2 + y) \qquad \text{(ii) or (iv)}$$

We see immediately that $x = 3$, $y = 2$ is a solution to both equations simultaneously. The other three roots may be obtained by drawing the graphs of the equations, but the algebraic method is beyond the scope of this text. We are, however, quite safe with problems such as the following example, where we can use the method of solution by substitution.

Example 5. Solve the equation

$$16x^2 + 25y^2 = 544 \qquad \text{(i)}$$
$$xy = 12 \qquad \text{(ii)}$$

Solution. From (ii) $y = \frac{12}{x}$

Substituting for y in (i), we get $16x^2 + 25\left(\frac{144}{x^2}\right) = 544$

Dividing through by 16 $\qquad x^2 + \frac{225}{x^2} = 34$

Multiplication by x^2 gives $\qquad x^4 - 34x^2 + 225 = 0$

We now consider this as a quadratic equation in x^2, which factorises as follows

$$x^4 - 34x^2 + 225 = (x^2 - 25)(x^2 - 9) = 0$$

Hence, either

$$x^2 - 25 = 0 \quad \text{i.e. } x = \pm 5, y = \pm 2{\cdot}4$$

or

$$x^2 - 9 = 0 \quad \text{i.e. } x = \pm 3, y = \pm 4$$

The solutions are $x = 5$, $y = 2{\cdot}4$; $x = -5$, $y = -2{\cdot}4$; $x = 3$, $y = 4$; $x = -3$, $y = -4$ Answer

Graphical Interpretation

In preparation for some of the work to come we remark on the graphical representation of these examples.

For the moment the reader need only accept that each of the equations relating y to x may be represented by a graph, which is either a straight line (like $y = x + 5$) or a curve, as shown in the ensuing figures. We merely wish to relate the solution of the simultaneous equations to the points of intersection of the curves they represent and at this stage we shall not deal with the method of drawing the graphs. The linear equations of Chapter One are represented by straight lines and, since two straight lines intersect in one and only *one* point, it follows that we obtain one and only *one* solution to two linearly independent equations. The quadratic equation of the second degree in one variable is represented by a curve called a **parabola,** and typical examples are given in Fig. 1.

The curve in Fig. 1(*c*) is the same as the curve in (*a*) but has been moved 'up' 3 units. The curve in Fig. 1(*d*) is the same as the curve in (*b*) but has been moved to the right 3 units. Indeed, we may take the comparison further by noticing that the curve in Fig. 1(*b*) is the curve in (*a*) rotated clockwise through 90° about the origin point 0.

The solution to the simultaneous equations

$$y - x^2 = 3 \qquad \text{(i)}$$
$$y - x = 5 \qquad \text{(ii)}$$

is now seen to be given by the intersection of the straight line (ii) and the curve (i) and so we can appreciate why there will be only two solutions. In Fig. 2(*a*) we have drawn the graph of $y = x^2 + 3$ for (i) and the straight line $y = x + 5$ for (ii), and we notice that the intersection of the two graphs gives the solution of the equations as $x = 2$, $y = 7$; $x = -1$, $y = 4$. If, however, the equation (i) had been $y = x^2 + 6$ (i.e. the curve in Fig. 2(*a*) moved 'up' 3 units) then, as we can see in Fig. 2(*b*), there would have been no real points of intersection or solutions; the two solutions would have been imaginary. (See Exercise 4.2, Nos. 9 and 10.)

On the other hand, if we consider the solution to the simultaneous equations

$$y = x^2 + 5\tfrac{1}{4} \qquad \text{(i)}$$
$$y = x + 5 \qquad \text{(ii)}$$

we find by the method of substitution that we get

$$x^2 - x + \tfrac{1}{4} = 0$$
$$\therefore (x - \tfrac{1}{2})(x - \tfrac{1}{2}) = 0$$

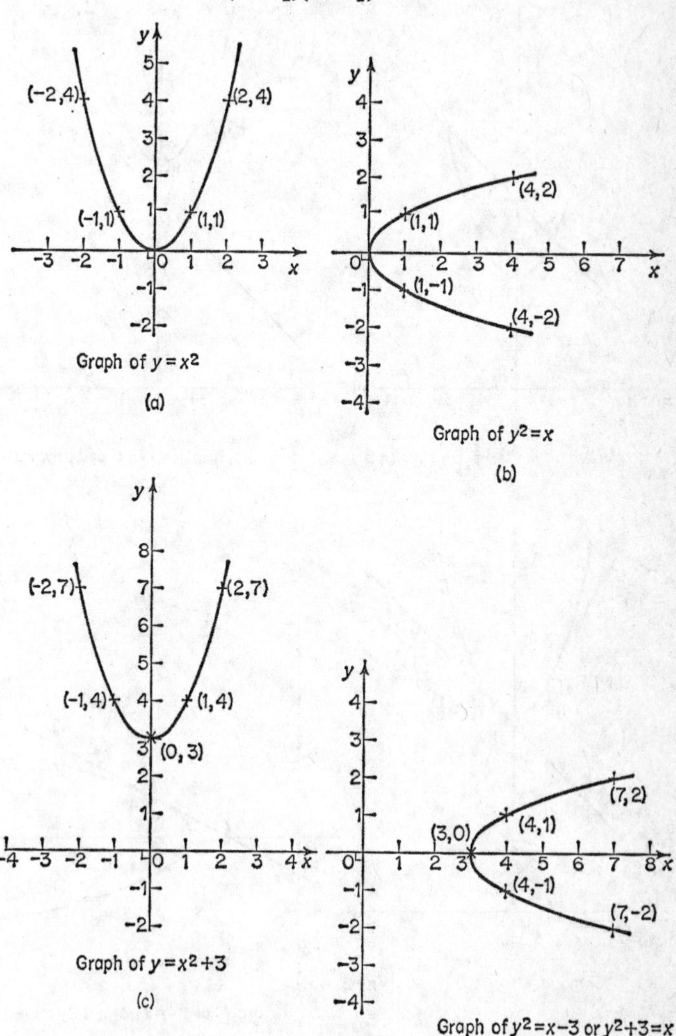

Fig. 1

so that the 'two' solutions are one and the same point, in this case $x = \frac{1}{2}$, $y = 5\frac{1}{2}$. If we now look at Fig. 2(*c*), which represents these two equations, we note that the line is a tangent to the curve, suggesting that we may think of a tangent to a curve as a line which intersects the curve in two points which are coincident.

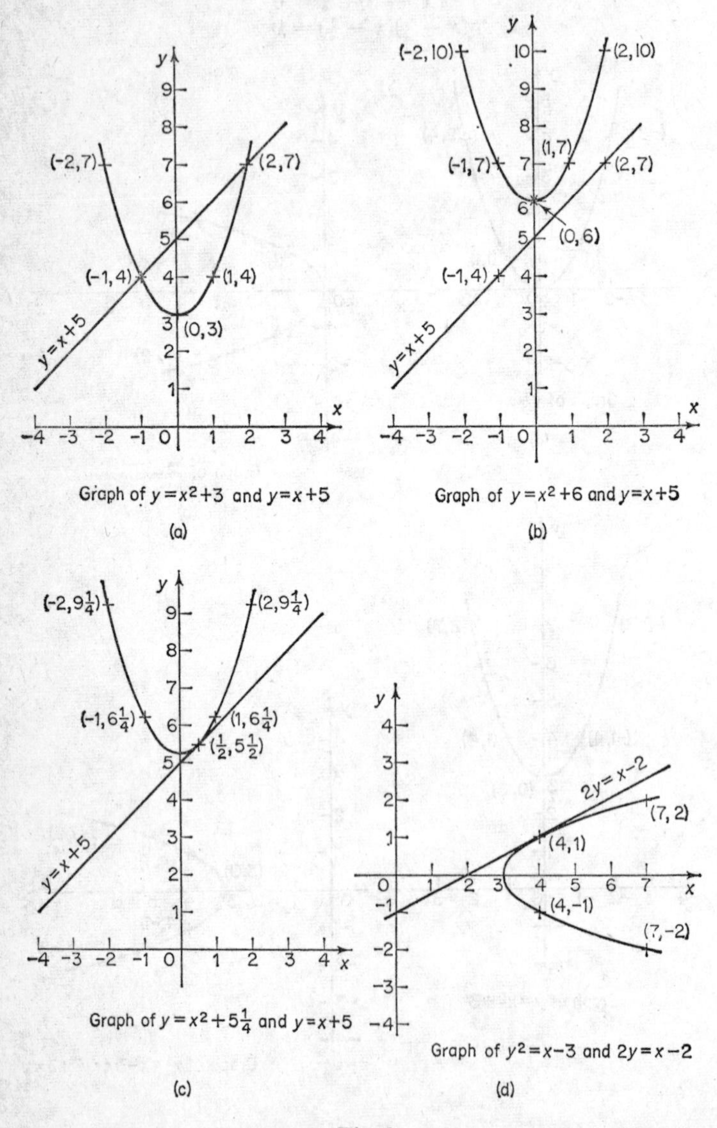

Fig. 2

Similarly, the solution of the simultaneous equations

$$y^2 = x - 3 \qquad \text{(i)}$$
$$2y = x - 2 \qquad \text{(ii)}$$

which is gained by substituting $x = 2y + 2$ in (i) to yield $y^2 = 2y - 1$ or $(y - 1)^2 = 0$, and the final solution $x = 4$, $y = 1$, indicates as shown in Fig. 2(*d*) that the line $2y = x - 2$ is a tangent to the curve $y^2 = x - 3$ at the point $x = 4$, $y = 1$.

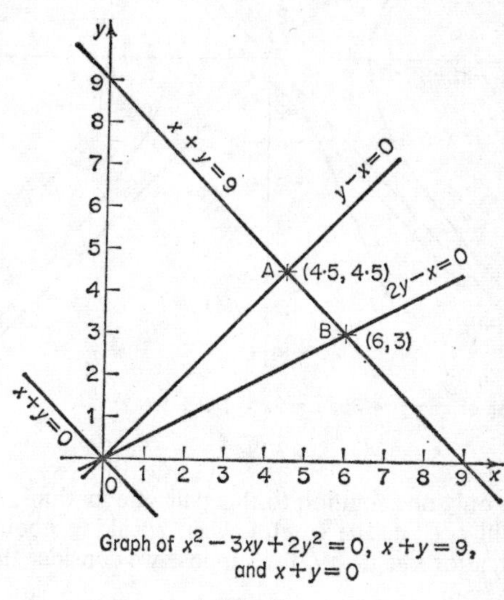

Graph of $x^2 - 3xy + 2y^2 = 0$, $x + y = 9$, and $x + y = 0$

Fig. 3

The two equations in Example 1 (page 33) give the graphs of Fig. 3: that is, the equation $x^2 - 3xy + 2y^2 = (x - 2y)(x - y) = 0$ represents the two lines $x - 2y = 0$ and $x - y = 0$ which intersect at the point $x = 0$, $y = 0$. The solution to the equations

$$x^2 - 3xy + 2y^2 = 0 \qquad \text{(i)}$$
$$x + y = 9 \qquad \text{(ii)}$$

is given by the two points of intersection A and B: at A $x = 4{\cdot}5$, $y = 4{\cdot}5$; at B $x = 6$, $y = 3$.

Note that there will be two and only *two* solutions to the equations. If equation (ii) had been $x + y = 0$, which passes through the point $x = 0$, $y = 0$, then the two solutions would have been the same because the line $x + y = 0$ intersects the other two lines in one point $x = 0$, $y = 0$.

The two equations of Example 2 on page 34 yield the graphs of Fig. 4. The equation $x^2 - y^2 = 25$ gives a two branch curve called a **hyperbola.** In fact, it is called a **rectangular hyperbola** because the curve gets closer to but never quite touches the two perpendicular lines POQ and ROS as we make the numerical value of $\pm x$ greater and greater. Now the line $x - y = 1$ or $y = x - 1$ of equation (ii) in this example is parallel to POQ and so we can

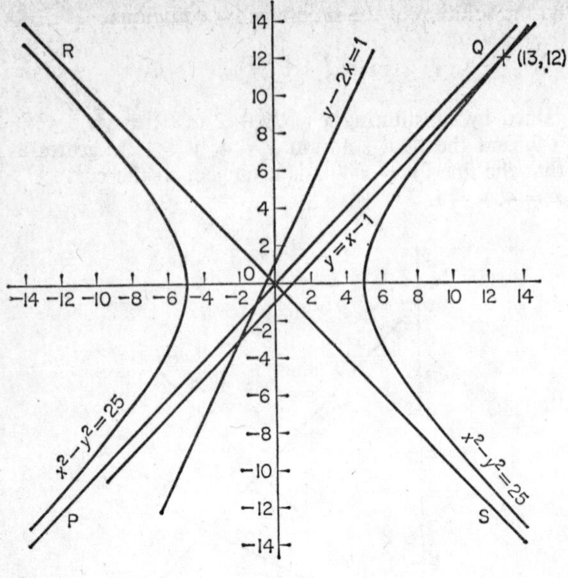

Graph of $x^2-y^2=25$, $y-2x=1$ and $y=x-1$

Fig. 4

see why there is only one solution to this pair of equations, $x = 13$, $y = 12$. However we still need to be careful when speaking about the points of intersection as real or imaginary, for suppose we consider the pair of simultaneous equations

$$x^2 - y^2 = 25 \qquad \text{(i)}$$
$$y - 2x = 1 \qquad \text{(ii)}$$

Using the method of substitution with $y = 2x + 1$ substituted in (i) we get

$$x^2 - (2x + 1)^2 = -3x^2 - 4x - 1 = 25$$

We can rearrange this as

$$3x^2 + 4x + 26 = 0$$

Being unable to factorise this result on sight we now use the formula

$$x = \frac{-4 \pm \sqrt{-296}}{6}$$

This last result reduces to $x = -0{\cdot}67 \pm 2{\cdot}87i$ (correct to 2 decimal places) and we observe at once that we get two imaginary points of intersection. Certainly in Fig. 4 we are unable to indicate such points by the letters A and B as before.

What we must accept is that a straight line will intersect any curve of the second degree in two points which are real or two points which are imaginary (as in the last example). Sometimes these points may coincide, in which case we shall describe them as two coincident points. The equations of Example 3 on page 34 are represented by Fig. 5.

The curve given by $x^2 + 3xy + 5y^2 = 19$ is an ellipse and, as suggested above, we expect to get two intersections of the ellipse with any straight line. Consequently, having found two solutions already, we know that there is no

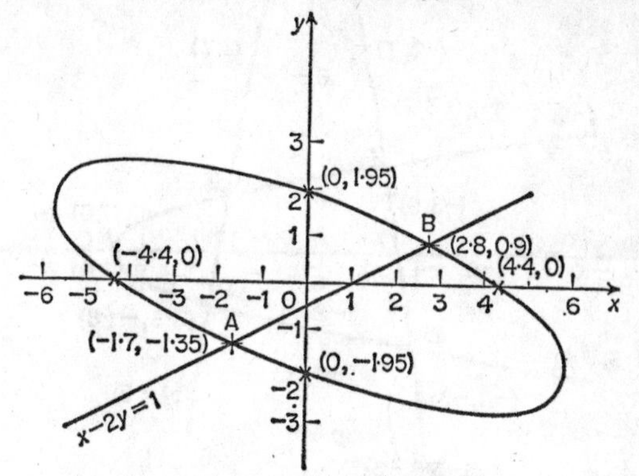

Graph of $x^2 + 3xy + 5y^2 = 19$, and $x - 2y = 1$

Fig. 5

need to look for any more. The two solutions are represented by the points A: $x = -1{\cdot}7$, $y = -1{\cdot}35$, B: $x = 2{\cdot}8$, $y = 0{\cdot}9$.

The example on page 35, to find the solution of the pair of simultaneous equations

$$x^2 + y = 11$$
$$x + y^2 = 7$$

presented a special difficulty because both equations were of the second degree. Here again we can get much information by considering the graphs of the equations in Fig. 6, in this case two parabolas. The solution which we obtained by inspection (or was it inspiration?) is given by A: $x = 3$, $y = 2$. The graph in Fig. 6 shows that there are three other solutions given by the points B, C and D. B: $x = -2{\cdot}8$, $y = 3{\cdot}2$, C: $x = -3{\cdot}8$, $y = -3{\cdot}3$, D: $x = 3{\cdot}6$, $y = -1{\cdot}8$. From this we observe that two simultaneous equations of the second degree will have four solutions. We can now examine the following pair of equations knowing that we should look for four possible solutions.

Example 6. Solve the simultaneous equations

$$x^2 + xy - y^2 = -1 \qquad \text{(i)}$$
$$5x^2 - 16xy + 8y^2 = 5 \qquad \text{(ii)}$$

Solution. Multiply equation (i) by 5 and add the result to (ii) to get

$$10x^2 - 11xy + 3y^2 = 0 = (5x - 3y)(2x - y)$$

Hence $y = 2x$ or $5x = 3y$ are the solutions to this last equation. Substituting $y = 2x$ in equation (i) we have

$$x^2 + 2x^2 - 4x^2 = -1$$
$$x^2 = 1$$
$$x = \pm 1$$

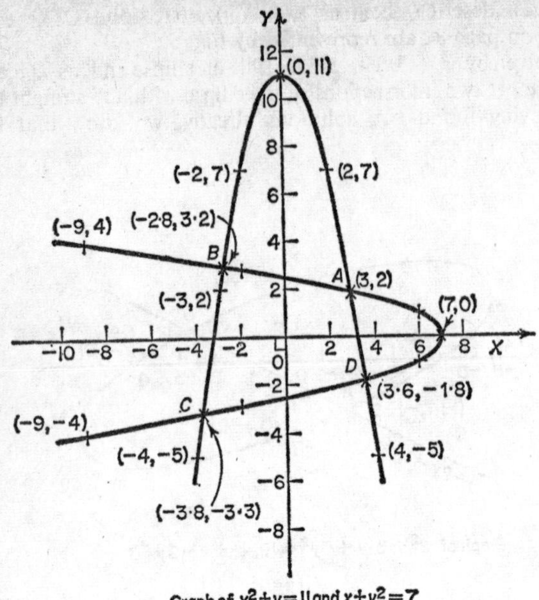

Fig. 6

When $x = \pm 1$ we have $y = \pm 2$. Substituting $x = \frac{3y}{5}$ in (ii) we have

$$\frac{9y^2}{5} - \frac{48y^2}{5} + 8y^2 = 5$$

$$y^2 = 25$$

$$y = \pm 5$$

When $y = \pm 5$, we have $x = \pm 3$. The solutions are $x = 1, y = 2; x = -1, y = -2; x = 3, y = 5; x = -3, y = -5$. Answer

As expected we have obtained four solutions.

Exercise 4.2

Solve the following pairs of simultaneous equations:

1. $x^2 + 9y^2 = 52$, $xy = 8$
2. $x^2 + 2xy + 9y^2 = 68$, $xy = 8$
3. $x^2 - 3xy + y^2 = -1$, $x^2 - 4xy + y^2 = -3$
4. $x^2 + y^2 = 8$, $5x^2 - 2y^2 = 12$
5. $31x^2 - 32xy + 8y^2 = -1$, $5x^2 - 40xy + 16y^2 = -11$
6. $2x^2 + 3xy = -10$, $x^2 + xy - y^2 = -11$
7. $x^2 + 3xy + y^2 = 1$, $x^2 - xy + y^2 = 13$
8. $x^2 + 16y^2 = 64$, $(x - 8)(y - 2) = 0$
9. $y - x^2 = 13$, $y - 2x = 11$
10. $x - y^2 = 10$, $x - 2y = 5$

Common Factors of Numbers

The previous sections of this chapter have involved the addition and subtraction of quadratic equations in order to find their solution. The same procedure may also be used to find the common factors of any two polynomials in the same variable. The principles involved may easily be seen in the following numerical example, where we consider finding the common factor of the two numbers 91 and 208. We start by dividing the larger number by the smaller number

$$\begin{array}{r} 91)208(2 \\ 182 \\ \hline 26 \end{array}$$

In this case we see that we have a quotient of 2 and a remainder of 26, enabling us to reconstruct the larger number as $208 = 91 \times 2 + 26$. Examining this result we feel intuitively that if 208 and 91 have a common factor then it must also be a factor of the remainder 26: that is, $208 - 91 \times 2 = 26$ so a common factor of 208 and 91 is a factor of the left-hand side of the equation and thereby a factor of the right-hand side. This idea reduces the difficulties of the original problem to a simpler search for the common factor of 91 and 26. We now divide 91 by 26 to be able to write $91 = 26 \times 3 + 13$. Again, this result suggests that the common factor of 91 and 26 is also a factor of 13 and clearly 13 is this common factor. Thus $208 = 13 \times 16$ and $91 = 13 \times 7$, from which we see that not only is 13 the common factor, but it is also the highest common factor. This is usually abbreviated to **H.C.F.** or sometimes called the greatest common divisor or **G.C.D.**

Again, consider finding the H.C.F. of the two numbers 2499 and 65637. As before we divide the larger number by the smaller number

$$\begin{array}{r} 2499)65637(26 \\ 4998 \\ \hline 15657 \\ 14994 \\ \hline 663 \end{array}$$

The result so far enables us to write $65637 = 2499 \times 26 + 663$, so we see that any common factor of 65637 and 2499 is also a factor of 663, in which case we reduce the problem to finding the common factor of 663 and 2499

$$\begin{array}{r} 663)2499(3 \\ 1989 \\ \hline 510 \end{array}$$

Hence, $2499 = 663 \times 3 + 510$ so that any common factor of 2499 and 663 is also a factor of 510. We could continue this procedure but since we readily see that $510 = 51 \times 10$ we may as well try 51 (certainly not 10) as the common factor. Since $663 = 51 \times 13$, we deduce that 51 is the H.C.F. of 2499 and 65637.

Common Factors of Algebraic Expressions

Now consider applying the same procedure to finding the common factor of

$$18x^3 + 21x^2 - 49x - 60 \text{ and } 6x^2 - x - 15$$

We start by dividing the higher-degree expression by the lower-degree expression

$$\begin{array}{r} 6x^2 - x - 15)\overline{18x^3 + 21x^2 - 49x - 60}(3x + 4 \\ \underline{18x^3 - 3x^2 - 45x } \\ 24x^2 - 4x - 60 \\ \underline{24x^2 - 4x - 60} \end{array}$$

This result enables us to write

$$18x^3 + 21x^2 - 49x - 60 \equiv (6x^2 - x - 15)(3x + 4)$$

and thereby see that both $6x^2 - x - 15$ and $3x + 4$ are factors of $18x^3 + 21x^2 - 49x - 60$. Because of the higher degree of $6x^2 - x - 15$ we refer to this factor as the highest common factor.

Again, consider finding the highest common factor of

$$4x^3 + 20x^2 + 33x + 20 \text{ and } 4x^2 + 16x + 15.$$

Working as we did before gives

$$\begin{array}{r} 4x^2 + 16x + 15)\overline{4x^3 + 20x^2 + 33x + 20}(x + 1 \\ \underline{4x^3 + 16x^2 + 15x } \\ 4x^2 + 18x + 20 \\ \underline{4x^2 + 16x + 15} \\ 2x + 5 \end{array}$$

This result enables us to write

$$4x^3 + 20x^2 + 33x + 20 \equiv (4x^2 + 16x + 15)(x + 1) + 2x + 5 \quad \text{(i)}$$

and thereby see that if a common factor exists then it will also be a factor of $2x + 5$. Indeed, it may be $2x + 5$.

$$\begin{array}{r} 2x + 5)\overline{4x^2 + 16x + 15}(2x + 3 \\ \underline{4x^2 + 10x } \\ 6x + 15 \\ \underline{6x + 15} \end{array}$$

Therefore

$$4x^2 + 16x + 15 = (2x + 5)(2x + 3)$$

If we return to equation (i) with this last result we now know

$$4x^3 + 20x^2 + 33x + 20 \equiv (2x + 5)(2x + 3)(x + 1) + (2x + 5)$$

which proves that $2x + 5$ is the highest common factor.

Before we consider the theory behind the above work the reader should attempt the following simple examples.

Exercise 4.3

1. Find the H.C.F. of the following pairs of numbers:

(a) 266, 231
(b) 1309, 68
(c) 2431, 1001
(d) 2337, 2261

2. Find the H.C.F. (or the G.C.D.) in each of the following:

(a) $6x^2 + 19x + 15$, $6x^2 + 17x + 12$. (Use the result that their common factor will also be a factor of their difference.)
(b) $6x^2 + 19x + 15$, $12x^2 + 35x + 25$. (Use the result that the common factor will also be a factor of $2(6x^2 + 19x + 15) = 12x^2 + 38x + 30$.)
(c) $5x^2 + 14x - 3$, $2x^2 - x - 21$. (Find the common factor by considering $2(5x^2 + 14x - 3) - 5(2x^2 - x - 21)$.)
(d) Show that $6x^2 + 13x - 28$, $3x^2 - 11x - 4$ do not have a common factor other than 1.
(e) Find the common factor of $3x^2 + 10x + 7$, $6x^2 + 17x + 7$, $9x^2 + 24x + 7$.

3. Find the H.C.F. in each of the following:

(a) $2x^3 + 3x^2 + 3x + 1$, $2x^3 + 4x^2 + 4x + 2$.
(b) $x^3 + x^2 - 10x + 8$, $x^3 + 6x^2 + 10x + 8$. (As an alternative to the division method consider the result of subtracting the two expressions.)
(c) $24x^3 + 14x^2 - 41x - 12$, $8x^2 + 2x - 15$.
(d) $21x^3 + 11x^2 - 146x + 14$, $21x^2 - 52x + 7$.
(e) $24x^3 + 16x^2 - 155x - 72$, $6x^2 + x - 40$.

A Theorem for the Common Factors of Polynomials

This theorem states that if two polynomials have a common factor then the sum or difference of any multiples of the polynomials also has the same factor. (We have already made use of this theorem in Question 2 of Exercise 4.3.)

Proof: Let $f(x)$ and $g(x)$ have the common factor c, and let Q_1 and Q_2 be the quotients when $f(x)$ and $g(x)$ are divided by c. Then because c is a factor of both $f(x)$ and $g(x)$ we may write

$$f(x) = cQ_1, \quad g(x) = cQ_2$$

Note that c may be a function of x such as $3x + 7$ or $x^2 + 10x - 4$, etc. Now, any multiple of $f(x)$ and $g(x)$ is given by $m f(x)$ and $n g(x)$, where m and n may be numbers or algebraic expressions. Hence the sum or difference is given by either $m f(x) + n g(x)$ or $m f(x) - n g(x)$.

But $$mf(x) + ng(x) = mcQ_1 + ncQ_2 = c\{mQ_1 + nQ_2\}$$

so that c is also a factor of $mf(x) + ng(x)$.

Also $$mf(x) - ng(x) = mcQ_1 - ncQ_2 = c\{mQ_1 - nQ_2\}$$

so that c is also a factor of $mf(x) - ng(x)$.

Do note that this theorem is based on the idea that the common factor c exists in the first place.

Again recall the examples of Exercise 4.3. To find the common factor of 266 and 231 consider subtracting the numbers to get $266 - 231 = 35$. Since $35 = 5 \times 7$ then the common factor (if it exists) must be either 5 or 7. Since the numbers 266 and 231 do not end in 0 or 5 it follows that the common factor may be 7. We must divide both numbers by 7 to confirm the result. To find the common factor of $6x^2 + 19x + 15$ and $12x^2 + 35x + 25$, consider (as already suggested) the subtraction

$$2(6x^2 + 19x + 15) - (12x^2 + 35x + 25) = 3x + 5.$$

By the above theorem, if there is a common factor it must be $3x + 5$. We confirm this by seeing that $6x^2 + 19x + 15 = (3x + 5)(2x + 3)$ and $12x^2 + 35x + 25 = (3x + 5)(4x + 5)$.

Another Theorem for the Common Factors of Polynomials

On page 28 we discussed the remainder theorem. The basic result arose from a polynomial $f(x)$ being divided by another polynomial $g(x)$ to give a quotient Q and a remainder R, both of which may be polynomials or constants, and both of a degree which is less than the degree of $f(x)$. Hence $f(x) = g(x)Q + R$. Of course the polynomials in our examples have only been of low degree but the results hold generally. Thus Question 3(*e*) in Exercise 4.3 gave

$$24x^3 + 16x^2 - 155x - 72 = (6x^2 + x - 40)(4x) + 12x^2 + 5x - 72$$

in this example

$$f(x) \equiv 24x^3 + 16x^2 - 155x - 72,\ g(x) = 6x^2 + x - 40,$$
$$Q = 4x,\ R = 12x^2 + 5x - 72.$$

Now if we combine the above theorem for common factors and the remainder theorem we see why the process of successive addition previously discussed produces the H.C.F.

Thus we have $f(x) \equiv g(x)Q + R$ which may be rewritten as

$$f(x) - g(x)Q \equiv R$$

From this we now see that if $f(x)$ and $g(x)$ have a common factor then it must also be a factor of R. Now consider dividing $g(x)$ by R. This time we shall obtain a new quotient Q_1 (say) and a new remainder R_1 (say), enabling us to write

$$g(x) = R.Q_1 + R_1 \text{ or } g(x) - R.Q_1 = R_1$$

from which it follows that the common factor of $g(x)$ and R is also a factor of R_1. By this stage we have reduced the problem to finding the common factor of R and R_1, which are of lower degree than the original $f(x)$ and $g(x)$. As we have already deduced intuitively the final common factor is the highest common factor. As final examples consider the following.

Example 7. Find the H.C.F. of $2x^4 - 5x^3 - 2x + 1$ and $4x^3 + 8x^2 + 5x + 3$.

Solution. The common factor will not be changed by multiplying the first polynomial by 2 before the first division.

$$\begin{array}{r} 4x^3 + 8x^2 + 5x + 3)\ 4x^4 - 10x^3 + 0x^2 - 4x + 2\ (x \\ \underline{4x^4 + \ 8x^3 + 5x^2 + 3x \quad\ } \\ -18x^3 - 5x^2 - 7x + 2 \end{array}$$

If there is a common factor it must also be a factor of the remainder so far,

$$-18x^3 - 5x^2 - 7x + 2$$

Hence, it will also be a factor of $-2(-18x^3 - 5x^2 - 7x + 2)$, so we continue the division as

$$\begin{array}{r} 4x^3 + 8x^2 + 5x + 3)\ 36x^3 + 10x^2 + 14x - 4\ (9 \\ \underline{36x^3 + 72x^2 + 45x + 27} \\ -62x^2 - 31x - 31 \end{array}$$

The common factor must now be a factor of $-62x^2 - 31x - 31$ or (dividing by -31) a factor of $2x^2 + x + 1$, so we re-arrange the division as

$$\begin{array}{r} 2x^2 + x + 1)\ 4x^3 + 8x^2 + 5x + 3\ (2x + 3 \\ \underline{4x^3 + 2x^2 + 2x \qquad} \\ 6x^2 + 3x + 3 \\ \underline{6x^2 + 3x + 3} \end{array}$$

So finally we see that $2x^2 + x + 1$ is the common factor. Answer

Check: $2x^4 - 5x^3 - 2x + 1 = (2x^2 + x + 1)(x^2 - 3x + 1)$

$4x^3 + 8x^2 + 5x + 3 = (2x^2 + x + 1)(2x + 3)$ Answer

Example 8. Find the H.C.F. of $10x^4 + 71x^3 + 149x^2 + 106x + 24$, and $10x^2 + 21x + 9$.

Solution.

$$\begin{array}{r} 10x^2 + 21x + 9)\ 10x^4 + 71x^3 + 149x^2 + 106x + 24\ (x^2 + 5x + 7 \\ \underline{10x^4 + 21x^3 + 9x^2 \qquad\qquad\qquad} \\ 50x^3 + 140x^2 + 106x \\ \underline{50x^3 + 105x^2 + 45x} \\ 35x^2 + 61x + 24 \end{array}$$

The common factor will also be a factor of twice the remainder so far. Such a multiplication avoids awkward fractions in the division.

$$\begin{array}{r} 70x^2 + 122x + 48 \\ \underline{70x^2 + 147x + 63} \\ -25x - 15 = -5\,(5x + 3) \end{array}$$

If there is a common factor it must be $5x + 3$. We confirm this by seeing that

$$10x^2 + 21x + 9 = (5x + 3)(2x + 3) \text{ and}$$

$$10x^4 + 71x^3 + 149x^2 + 106x + 24 = (5x + 3)(2x^3 + 13x^2 + 22x + 8).$$

Answer

If two polynomials do not have a common factor, apart from numerical factors, then we say that the polynomials are relatively prime.

Exercise 4.4

1. Find the H.C.F. of $x^4 + 10x^3 + 35x^2 + 50x + 24$ and $x^3 + 8x^2 + 17x + 10$.
2. Show that $x^4 + 7x^3 + 8x^2 + 19x + 10$ and $x^4 + 7x^3 + 8x^2 + 18x + 11$ are relatively prime.
3. Find the H.C.F. of $25x^4 + 5x^3 - x - 1$ and $20x^4 + x^2 - 1$. (Hint: multiply the first expression by 4.)
4. The result of dividing the polynomial $f(x)$ by $x^2 + x - 6$ is a quotient of $x^2 - 1$ and a remainder of $x - 2$. Find $f(x)$ and show that $x - 2$ is the common factor of $f(x)$ and $x^2 + x - 6$.
5. Suggest two polynomials of degree four which have $x^2 - x + 1$ as their common factor.
6. Suggest three polynomials of degree three of which two have $x^2 + x + 1$ as a common factor, two have $x - 1$ as a common factor but the three polynomials have no common factor.
7. When the polynomial $f(x)$ is divided by $x^2 + 7x + 12$ the remainder is $x + k$. Find the two values of k which enable $f(x)$ and $x^2 + 7x + 12$ to have a common factor.

8. Find the H.C.F. of $x^5 + 3x^4 - 6x^3 - 18x^2 + 5x + 15$ and $x^3 + 5x^2 + 7x + 3$.
9. Find the H.C.F. of $x^5 + 5x^4 + 10x^3 + 10x^2 + 5x + 1$, $x^4 + 4x^3 + 6x^2 + 4x + 1$ and $x^3 + 3x^2 + 3x + 1$.
10. Find the H.C.F. of $x^6 - 1$ and $x^3 + 3x^2 + 3x + 1$.

CHAPTER FIVE

COORDINATE GEOMETRY

Ordered Pairs

Most of the equations we shall be considering represent a relation between the two variables x and y, so that any numerical value for x leads to one or more numerical values of y. Thus, if the relation is given by the equation $y - x = 5$ or $y = x + 5$ then the replacement of x by 2 means that y must be replaced by 7. We can represent this result or solution by writing it out in full as $x = 2$, $y = 7$, or we may adopt an abbreviated form (2, 7) called an **ordered pair**, so that whenever we write (p, q) this means $x = p$ and $y = q$; for example, another solution of the same equation $y = x + 5$ is given by $x = -1, y = 4$, which may be written as the ordered pair $(-1, 4)$. By suggesting more and more values of x we build up a set of solutions, or ordered pairs, as we did on page 2, e.g. (0, 5), (1, 6), (3, 8), (4, 9) will also satisfy the equation $y = x + 5$.

We now seek to give a diagrammatical representation of the complete

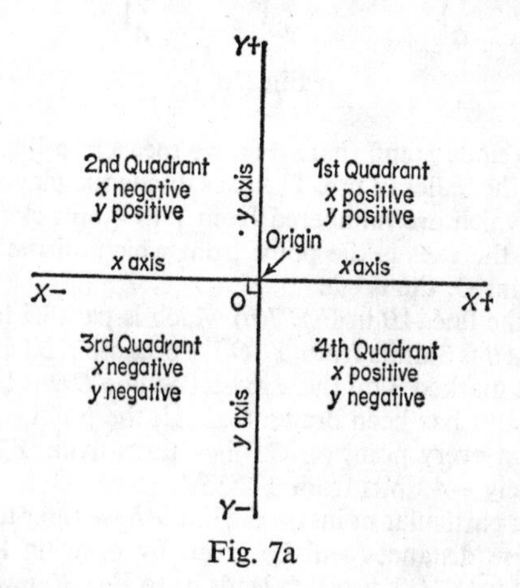

Fig. 7a

relation between x and y, which avoids the need to find so many of the ordered pair solutions. For example, if $y = x + 5$ can be represented by a straight line (and it can), then we shall only need to find two points in order to obtain the whole line which represents the relation. As we saw in Chapter Four with regard to the number and type of solutions of the simultaneous equations, such diagrams will clarify that which algebra tends to obscure.

Rectangular Coordinates

We can associate an ordered pair with a point in a plane, such as a piece of paper, by choosing two perpendicular reference lines in the paper and taking measurements from these lines, with the usual convention that measurements on each side of the reference lines correspond to positive or negative values. Consider Fig. 7(*a*), in which the two perpendicular reference lines are X^-OX^+ and Y^-OY^+. These lines are called the **axes.** X^-OX^+ is the *x* **axis** and Y^-OY^+ is the *y* **axis.**

Graduating the axes to some chosen scale leads to the marked diagram of Fig. 7(*b*). (Throughout coordinate geometry it is assumed that the reader uses A4 size metric graph paper.) Observe that the same scale has been chosen for both axes. This may not always be convenient and, consequently, it is

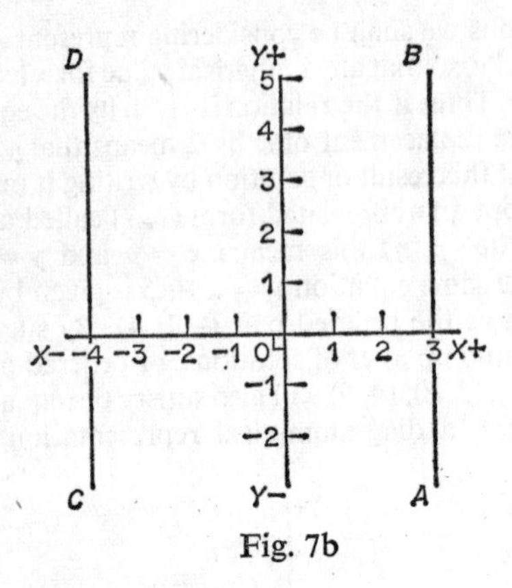

Fig. 7b

very important to understand that when we measure a line on the paper we must be aware of the scales in use. The axes divide the plane into four regions called **quadrants** which are numbered from 1 to 4 anti-clockwise. The point of intersection of the axes is the point from which all the graduations start so we call this point O, the **origin.**

Now consider the line *AB* in Fig. 7(*b*) which is parallel to Y^-OY^+. Every point on the line *AB* is 3 units from Y^-OY^+ because *AB* has also been drawn through the point marked 3 on the *x* axis. The line *CD* is likewise parallel to Y^-OY^+ but, since it has been drawn through the point marked -4 on the *x* axis, we say that every point on *CD* is 4 units from Y^-OY^+ and on the left; or alternatively -4 units from Y^-OY^+.

In order to fix a particular point on the line *AB* we refer to its distance from X^-OX^+ and these distances will be given by drawing lines of rank (i.e. **ordinates**) parallel to X^-OX^+, which leads us to Fig. 7(*c*), where we have two such lines, *PQ* and *RS*, which intersect *AB* and *CD* in the points *K*, *L*, *M* and *N* as shown. Thus every point on *PQ* is 5 units from X^-OX^+ because it has been drawn through the point marked 5 on the *y* axis.

Now the point *L* lies on *PQ* and *AB*, which means that its position may be associated with the ordered pair (3, 5)—that is, 3 units measured in a direction parallel to the *x* axis and 5 units measured in a direction parallel to the *y* axis. Since these measurements are obtained from the ordinate lines (lines of rank) parallel to the two axes we call the two measurements the **coordinates** of the point. Thus the point *L* has an *x* coordinate of 3 and a *y* coordinate of 5. Similarly the coordinates of the point *K* are given by the ordered pair

$(-4, 5)$, i.e. K is the point which is -4 units measured in a direction parallel to the x axis and 5 units measured in a direction parallel to the y axis. The x coordinate of the point K may also be described as 4 units measured in the negative direction parallel to the x axis. Similarly, the point N is $(-4, -2)$ and M is the point $(3, -2)$.

Inequalities

Figs. 7(*b*) and 7(*c*) concentrated on points on a straight line or the point of intersection of two lines, i.e. K, L, M and N. However, the straight line AB, for example, did much more than harness a set of points onto the line; it also divided the plane of the paper. The set of all points on the squared paper is called the **coordinate plane**, and this plane is divided into three sets by the straight line AB: the set of points on the **half plane** on the left; the set of points on the line itself; and the set of points on the **half plane** on the right. Thus all the points to the left of AB have an x coordinate less than 3, i.e. $x < 3$, and all the points to the right of AB have an x coordinate greater than 3, i.e. $x > 3$. We cannot make a similar statement about the y coordinate of these points because this is not restricted by the line AB.

If we now consider the effect on the coordinate plane of the straight line RS we see that the y coordinate of any point on this line is always -2; indeed, we may say that $y = -2$ is the equation of the line RS. However, the line

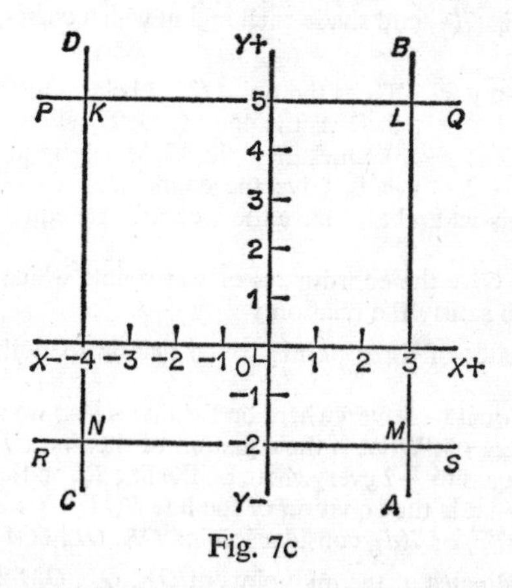

Fig. 7c

also divides the coordinate plane into two other sets of points: namely, those for which $y > -2$, i.e. all points above the line; and those for which $y < -2$, i.e. all points below the line.

If we combine the effects of drawing the lines AB and RS then we see that they partition the coordinate plane into several regions. The shaded region in Fig. 7(*d*) corresponds to $x > 3$ *and* $y > -2$ or, to put it another way, the shaded region corresponds to the set of all points with an x coordinate greater than 3 and a y coordinate greater than -2. Examples of such points are $(6, -1)$, $(10, 7)$, $(3{\cdot}01, -1{\cdot}93)$, etc. Perhaps the reader can now see how the inequalities of page 9 will be dealt with. Similarly the interior region of the square $KLMN$ corresponds to $x > -4$, $x < 3$, $y > -2$, $y < 5$, all four inequalities being applied at the same time. We can shorten this statement by writing $-4 < x < 3$ and $-2 < y < 5$. Typical points which belong to this region are (2, 4), (2·9, −1·9), (−3, −1) and so on.

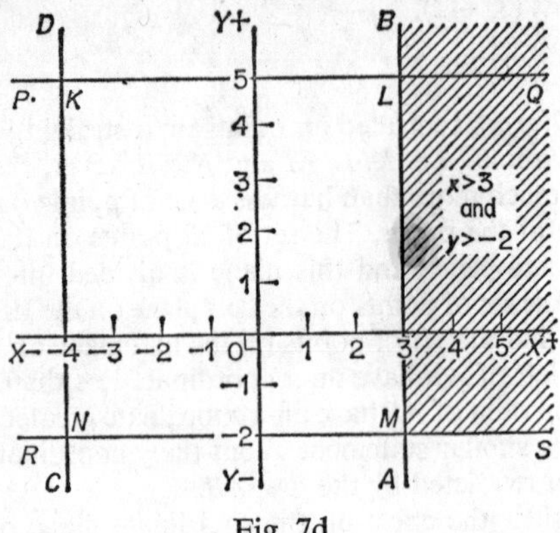

Fig. 7d

Exercise 5.1

1. Make a copy of Fig. 7(*d*) and shade each region which corresponds to the following inequalities:
   (a) $0 < x < 3, 0 < y < 5$. Does the point (2, 5) belong to this set of points?
   (b) $-4 < x < 0, 0 < y < 5$. Does the point (−3, 2) belong to this set of points?
   (c) $0 < x < 3, -2 < y < 0$. Does the point (2, 1) belong to this set of points?
   (d) $-4 < x < 0,\ -2 < y < 0$. Give the coordinates of two points which are members of this set and also have the x coordinate equal to the y coordinate, i.e. $x = y$.
   (e) $x < 0, y < 0$. Give the coordinates of two points which not only lie in this region but also satisfy the relation $y = x + 2$.
2. Give the coordinates of three points which belong to both sets (d) and (e) in Question 1.
3. In Fig. 7(*c*), x is equal to 3 everywhere on the line *AB* so we say that the equation of the line *AB* is $x = 3$. What is the equation of the line *CD*?
4. In Fig. 7(*c*), y is equal to −2 everywhere on the line *RS* so the equation of the line *RS* is $y = -2$. What is the equation of the line *PQ*?
5. Using either Fig. 7(*c*) or 7(*d*), consider joining *OK, OL, OM* and *ON*.
   (a) Find the coordinates of the midpoints of *OK, OL, OM* and *ON*.
   (b) Give the coordinates of any other point on *OK, OL, OM* and *ON* which lies inside the square *KLMN*.
6. In Fig. 7.(*c*), find (a) the coordinates of the midpoints of the sides of the square *KLMN*, (b) the lengths of the sides of the square *KLMN*, (c) the lengths of *OL* and *OM*.

The Distance Between Two Points

We have so far referred to particular points by giving their coordinates, such as (3, 5) or (−2, 1). The general point in the coordinate plane has been referred to as the point (x, y); that is, we make (x, y) represent the general point in the plane and we then make it a particular point by giving numerical values to x and y. There are some occasions when we wish to discuss one, two or more points without giving their coordinates numerical values. We could refer to these points as (a, b); (c, d); (p, q); etc. but, rather than use so many different letters, we choose a subscript notation illustrated by referring to the points as (x_1, y_1), (x_2, y_2), (x_3, y_3) and so on. This way we always know not only which coordinate is being referred to but also which point. It is frequently useful to extend this economy to the point letters themselves so that instead of referring to points $A, B \ldots$ we refer to the point P_1 with coordinates (x_1, y_1), the point P_2 with coordinates (x_2, y_2) and so on. Now it follows that since the position of two points is indicated by their coordinates then the distance between the points should be expressed in terms of these coordinates. Indeed, we may go so far as to *define* the distance d between two points without referring to any diagrams as follows:

Definition. The distance d between two points $P_1(x_1, y_1)$ and $P_2(x_2, y_2)$ is defined by $d = \sqrt{[(x_2 - x_1)^2 + (y_2 - y_1)^2]}$, and this distance will be called the length P_1P_2 or the length of the straight line segment P_1P_2.

Note that this definition applies only when we are using a rectangular coordinate system, i.e. the axes are perpendicular.

Example 1. Find the lengths of the sides of the triangle formed by joining the points $P_1(4, 3)$, $P_2(7, -2)$, $P_3(-1, -5)$.

Solution.

$$\begin{aligned} P_1P_2 &= \sqrt{[(7-4)^2 + (-2-3)^2]} \\ &= \sqrt{(9+25)} = \sqrt{34} \\ &= 5{\cdot}83 \text{ (correct to 2 decimal places)} \\ P_2P_3 &= \sqrt{[(-1-7)^2 + (-5-(-2))^2]} \\ &= \sqrt{(64+9)} = \sqrt{73} \\ &= 8{\cdot}54 \text{ (correct to 2 decimal places)} \\ P_3P_1 &= \sqrt{[(4-(-1))^2 + (3-(-5))^2]} \\ &= \sqrt{(25+64)} = \sqrt{89} \\ &= 9{\cdot}43 \text{ (correct to 2 decimal places)} \end{aligned}$$

The lengths of the sides of the triangle are 5·83, 8·54, 9·43 units (correct to 2 decimal places).

Answer

We again make the observation as we did on page 50 that the distance between two points P_1 and P_2 or the length P_1P_2 is dependent only upon the coordinates of the points P_1 and P_2 and *not* the scales we choose for the x and y axes. This is why we chose to give a definition for distance, before discussing the same result from a diagram as follows.

Consider finding the length of the line segment AB joining the points $A(2, 1)$ and $B(5, 7)$ given in Fig. 8(*a*). We see that the lines through A and B parallel to the axes form a right-angled triangle ACB. Furthermore, C is the point $(5, 1)$.

From Pythagoras's Theorem we have

$$\begin{aligned} AB^2 &= AC^2 + CB^2 \\ \therefore AB^2 &= (5-2)^2 + (7-1)^2 = 9 + 36 \\ \therefore AB &= \sqrt{45} = 6{\cdot}71 \text{ units (correct to 2 decimal places).} \end{aligned}$$

Now consider Fig. 8(*b*) where the scale for the x axis is 6 mm to 1 unit and the scale for the y axis is 6 mm to 2 units. Again we have the result $AC = 3$ units, $CB = 6$ units, that is, CB and AC are *calculated* in units, not mm. In consequence the length of AB is still 6·71 units. Furthermore, it must also be understood that halving the scale along the y axis does not result in halving distances such as AB; indeed, only when both the x and y scales are halved together would a distance like AB be halved. Finally, to find the angle of inclination of the line AB to the x axis, the angle CAB is measured by using

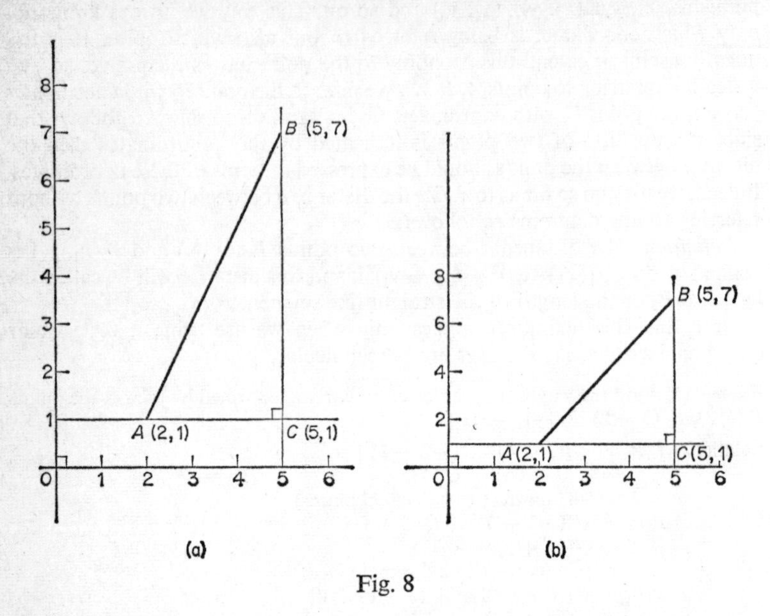

Fig. 8

a protractor only when the x and y scales are the same, otherwise we must work from the result

$$\tan\angle CAB = \frac{7-1}{5-2} = \frac{6}{3} = 2$$

We then consult tangent tables to reveal that $\angle CAB = 63°26'$ and we note that this is the result for *both* Fig. 8(*a*) and 8(*b*).

Observe further that if we write $\tan\angle CAB = CB/AC$ then we mean that CB is the distance 6 units and AC is the distance 3 units, i.e. not the actual measurement in terms of millimetres in Fig. 8(*a*) or 8(*b*). Tan CAB is called the **gradient** of the line AB and we usually denote this by the letter m, i.e. $m = \tan CAB$ for the line AB. We shall say more about gradients on page 57.

The Midpoint of a Straight Line Segment

Again we could define this without the aid of a diagram by saying that the midpoint $M(x, y)$ of the line segment joining $P_1(x_1, y_1)$ and $P_2(x_2, y_2)$ is given by $x = \frac{1}{2}(x_1 + x_2)$, $y = \frac{1}{2}(y_1 + y_2)$. However, since we are now aware of the meaning of length on a graph irrespective of the scales, we shall consider obtaining the required midpoint from the diagram in Fig. 9(*a*).

We first observe the simple cases of the straight line segment being parallel to one or other of the axes. In Fig. 9(*a*) the coordinates of the point C are (4, 2), and therefore the point E has coordinates $(\frac{1}{2}(1 + 4), 2)$ or

$(2\frac{1}{2}, 2)$. Similarly the x coordinate of F is obviously 4 and its y coordinate $\frac{1}{2}(7 + 2) = 4\frac{1}{2}$, i.e. F is the point $(4, 4\frac{1}{2})$.

Now M has the same x coordinate as E and the same y coordinate as F so M is the point $(2\frac{1}{2}, 4\frac{1}{2})$.

Alternatively, suppose M is the point (u, v) then, because E is the midpoint of AC, we have $AE = EC$ in the form $u - 1 = 4 - u$, an equation which has the solution $u = (4 + 1)/2 = 2\frac{1}{2}$. Similarly, because F is the midpoint of CB we have $CF = FB$ in the form $v - 2 = 7 - v$, which has the solution $v = (7 + 2)/2 = 4\frac{1}{2}$.

Finally, M is the point $(2\frac{1}{2}, 4\frac{1}{2})$ as before.

As another numerical example, the midpoint of the line segment joining the points $(-2, 3)$ and $(4, 7)$ is $\left(\frac{-2+4}{2}, \frac{3+7}{2}\right)$, i.e. $(1, 5)$.

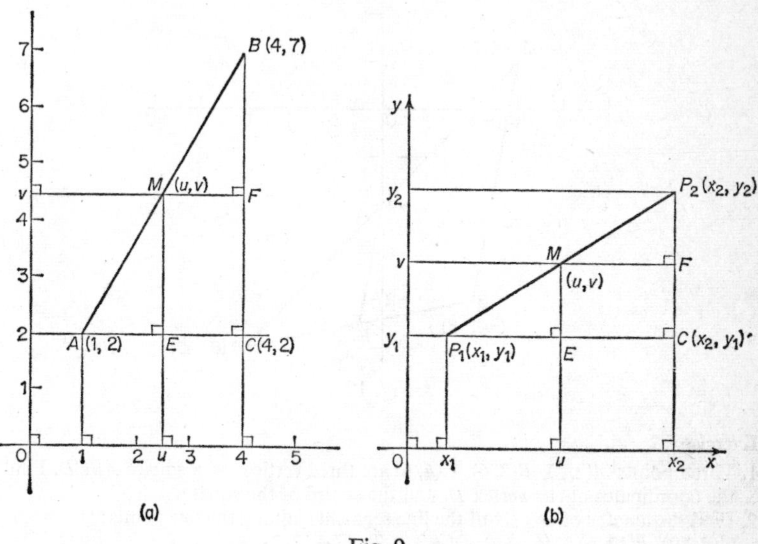

Fig. 9

If we now move to the general example of Fig. 9(*b*), and once again let the point $M(u, v)$ be the midpoint of the line segment joining $P_1(x_1, y_1)$ and $P_2(x_2, y_2)$, then with $C(x_2, y_1)$ and E the midpoint of P_1C we get

$$P_1E = u - x_1 = EC = x_2 - u.$$
$$\therefore u - x_1 = x_2 - u.$$
$$2u = x_1 + x_2$$
$$u = \tfrac{1}{2}(x_1 + x_2)$$

Similarly from $CF = FP_2$ we have $v - y_1 = y_2 - v$

$$2v = y_1 + y_2, \; v = \tfrac{1}{2}(y_1 + y_2)$$

Hence the midpoint of P_1P_2 is the point $(\frac{1}{2}(x_1 + x_2), \frac{1}{2}(y_1 + y_2))$

Example 2. A triangle is formed by joining the points $A(-4, -8)$, $B(-6, 2)$, $C(6, -10)$. E is the midpoint of BC and F is the midpoint of AC. Prove that $EF = \frac{1}{2}AB$.

Solution. Let Fig. 10 represent the problem. The point E has the coordinates $(\frac{1}{2}(-6 + 6), \frac{1}{2}(2 + (-10))$, i.e. E is the point $(0, -4)$. The point F has the coordinates $(\frac{1}{2}(-4 + 6), \frac{1}{2}(-8 + (-10))$, i.e. F is the point $(+1, -9)$.

The length EF is given by $EF = \sqrt{[(1 - 0)^2 + (-9 - (-4))^2]} = \sqrt{26}$.
The length AB is given by

$$AB = \sqrt{[(-6 - (-4))^2 + (2 - (-8))^2]} = \sqrt{104} = 2\sqrt{26}.$$

$$\therefore \quad EF = \tfrac{1}{2}AB \qquad \text{Q.E.D.}$$

Q.E.D: *Quod erat demonstrandum*—which was to be proved.)

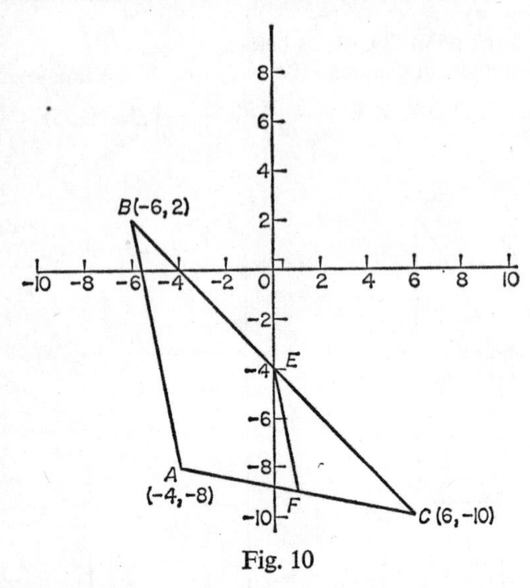

Fig. 10

Exercise 5.2

1. Three points $A(2, 2)$, $B(2, 6)$, $C(6, 6)$ are three vertices of a square $ABCD$. Find the coordinates of the vertex D, and the centre of the square.
2. Find the lengths of any six of the line segments joining the five points: $A(8, 10)$, $B(12, 6)$, $C(6, -4)$, $D(-8, -2)$, $E(-12, 2)$ (Leave your answer as a square root.)
3. $ABCD$ is a square with three of its vertices given by $A(1, 3)$, $B(4, 6)$ $C(7, 3)$. Find the coordinates of the vertex D and the centre of the square. Also show that BD is parallel to the y axis.
4. A circle of radius 5 units is drawn with its centre at the origin. Find the coordinates of the four points of intersection with the axes.
5. The diameter of a circle is the line segment joining $A(-2, 3)$ and $B(6, 7)$. Find the radius and centre of the circle.
6. Find which of the two points $A(2, 11)$, $B(6, 9)$ is closer to the origin and calculate how much closer. Is the midpoint of AB closer to the centre than A or B?
7. A circle of radius 6 units touches both of the positive x and y axes. Find the centre of the circle and the points at which the circle touches the axes. Find also the coordinates of the ends of the diameters of the circle which are parallel to the axes.
8. Show that the points $A(2, 2)$, $B(5, 6)$, $C(9, 3)$, $D(6, -1)$ are the vertices of a square. Find the radius of the inscribed circle and the points at which it touches the sides of the square.

The Gradient of a Straight Line

As we move from one point to another on the line so the x and y coordinates will change, and these changes will in some way indicate the steepness of the line. Consider the straight line which passes through the points $P(1, 2)$ and $Q(4, 6)$ as shown in Fig. 11.

If we consider the x axis as horizontal and the y axis as vertical then a walk from P to Q along the line will rise a vertical distance RQ for a horizontal distance PR and the gradient of the line will be measured by $\frac{RQ}{PR} = \frac{6-2}{4-1} = \frac{4}{3}$

Since the line is straight it does not matter where the points P and Q are chosen. For instance suppose we move from $S(-5, -6)$ to $T(-2, -2)$ on

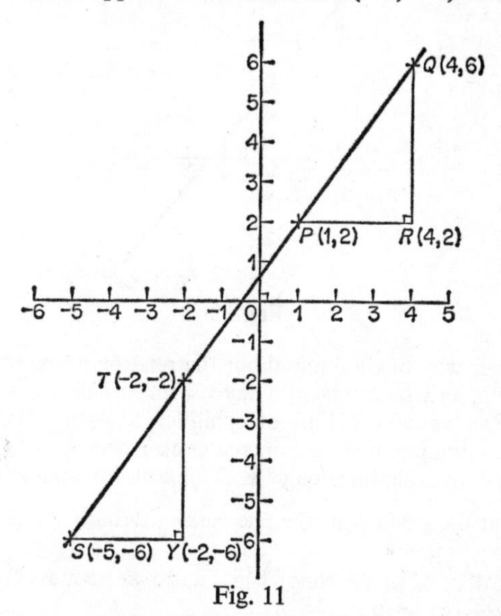

Fig. 11

the same straight line. In this case the gradient will be given by YT/SY, i.e. the increase in y divided by the increase in x as we move from S to T.

$$\therefore \text{ gradient of the line is } \frac{YT}{SY} = \frac{-2-(-6)}{-2-(-5)} = \frac{4}{3} \text{ as before}$$

Alternatively we may use the fact that since the triangles PRQ and SYT are congruent it follows that $\frac{RQ}{PR} = \frac{YT}{SY}$, indicating that the gradient of a straight line is the same everywhere on the line. In coordinate geometry we have to distinguish between lines such as PQ in Fig. 11 and lines such as AB in Fig. 12, and in order to do this we give a formal definition of gradient.

Definition. The gradient of a straight line is

$$\frac{\text{the change in the } y \text{ coordinate}}{\text{the change in the } x \text{ coordinate}}$$

in moving along the line from one point to another.

If we move from $P_1(x_1, y_1)$ to $P_2(x_2, y_2)$ on a straight line, then the gradient is equal to

$$\frac{y_2 - y_1}{x_2 - x_1}$$

Returning to Fig. 12, let us consider moving from A to B. The change in the y coordinate is $-3 - 5 = -8$. The change in the x coordinate is $4 - (-3) = 7$. Therefore the gradient of the line is $-\frac{8}{7}$ or $-1\frac{1}{7}$.

The meaning of a negative gradient is seen by comparing Fig. 11 with

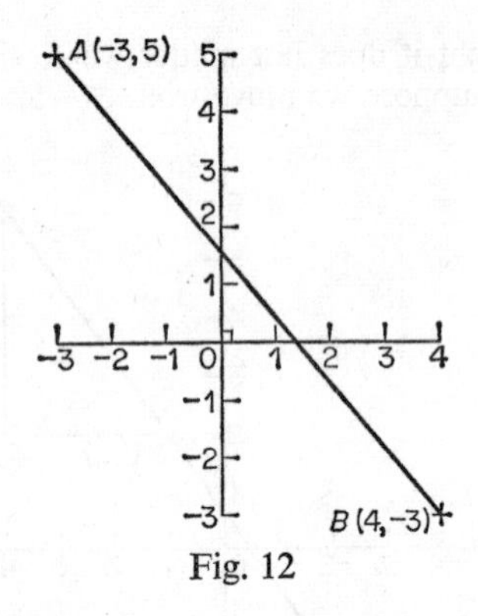

Fig. 12

Fig. 12. Another way of thinking about the relation represented by the line of Fig. 11 is that as x increases so y increases, in which case we say that **y is an increasing function of x.** (Think of uphill left to right.) The line in Fig. 12 represents a relation in which as x increases so y decreases, in which case we say that **y is a decreasing function of x.** (Think of downhill left to right.)

Example 3. Find the gradient of the line passing through the points $A(-5, -6)$ and $B(-2, 7)$.

Solution. We shall consider the changes in x and y as we move from A to B. The gradient of AB is $\dfrac{7 - (-6)}{-2 - (-5)} = \dfrac{13}{3}$

Moving from B to A the gradient is $\dfrac{-6 - 7}{-5 - (-2)} = \dfrac{-13}{-3} = \dfrac{13}{3}$ as before. Answer

Parallel Lines

Different lines which have the same gradient will be parallel. Parallel lines are indicated by the equality of corresponding and alternate angles when the lines are cut by a transversal. This means that we need to associate angles with gradients in order to give a full discussion of parallel lines and later on their intersection by perpendicular lines. Consider the two parallel lines in Fig. 13 to make an angle θ (Greek letter *theta*, pronounced theeta) with the positive direction of the x axis. The gradient of the line AB is given by OB/AO. In the right-angled triangle AOB, $\tan\theta = OB/AO$, so we may also state the gradient of a straight line as being the tangent of the angle which the line makes with the positive direction of the x axis (page 54). Since PQ is parallel to AB, then angle QPR is also θ and the gradient of PQ is also $RQ/PR = \tan\theta$.

The line ST in Fig. 13(*b*), which from previous work we know has a negative gradient, makes an angle α (Greek letter *alpha*) with the positive direction of the x axis. Since α is an obtuse angle the gradient $\tan \alpha$ is negative, a result with which some readers may be familiar from elementary trigonometry. Alternatively, the gradient is $OH/KO = OH/-OK = -\tan \angle OKH = -\tan \beta$ (β is the Greek letter *beta*, pronounced beeta).

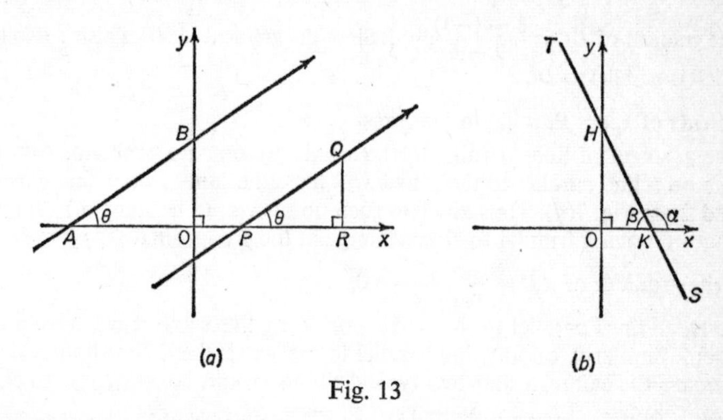

Fig. 13

Perpendicular Lines

Here we wish to discover the special relation between the gradient of two lines which are perpendicular to each other. For this purpose consider the two lines AC and BC in Fig. 14, with the gradient of AC as $\tan \theta$. The gradient of BC is $\tan \alpha = -\tan \beta$ as we saw in Fig. 13 (*b*).

From triangle ABC $\tan \theta = CB/AC$; $\tan \beta = AC/CB$ so that the product of the gradients is $\tan \theta(-\tan \beta) = (CB/AC)(-AC/CB) = -1$.

The usual notation for the gradient of a straight line in coordinate geometry is m. This means that if the gradient of AC is m then the gradient of any line perpendicular to AC is $-1/m$.

Alternatively, if the gradient of two lines are m_1 and m_2 respectively then the condition that these lines should be perpendicular is that $m_1 m_2 = -1$.

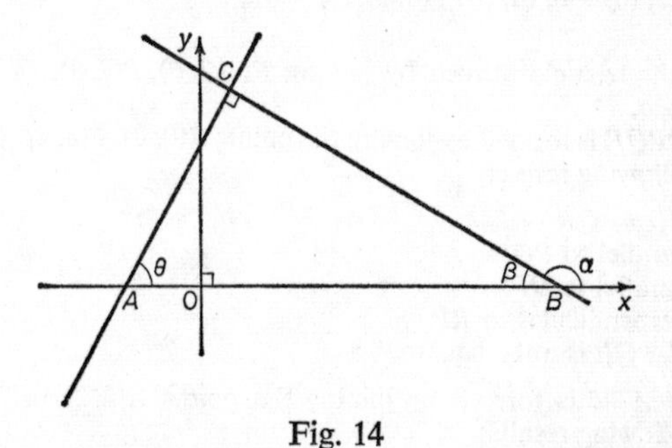

Fig. 14

Example 4. With the four points $A(2, 2)$, $B(5, 6)$, $C(9, 3)$, $D(6, -1)$ prove that: (i) AC is perpendicular to BD; (ii) AB is perpendicular to BC; (iii) AB is parallel to DC.

Solution. The gradient of AC is $\frac{3-2}{9-2} = \frac{1}{7}$. The gradient of BD is $\frac{-1-6}{6-5} = \frac{-7}{1}$. The product of these gradients is $1/7 \times -7 = -1$, hence AC is perpendicular to BD.

The gradient of AB is $\frac{6-2}{5-2} = \frac{4}{3}$. The gradient of BC is $\frac{3-6}{9-5} = -\frac{3}{4}$. The product of these gradients is $\frac{4}{3} \times -\frac{3}{4} = -1$. Hence AB is perpendicular to BC.

The gradient of DC is $\frac{3-(-1)}{9-6} = \frac{4}{3}$. Since the gradient of AB is also $\frac{4}{3}$ it follows that AB is parallel to DC. Q.E.D.

Gradients of Lines Parallel to the Axes

The gradient of lines parallel to the x axis presents no problem. Any two points on a line parallel to the x axis will have the same y coordinate as discussed from Fig. 7(*d*). Thus any two such points would be $A(x_1, y_1)$, $B(x_2, y_1)$ so that in moving from A to B the change in the y coordinate is $y_1 - y_1 = 0$ and the gradient of $AB = \frac{0}{x_2 - x_1} = 0$.

Thus, all lines parallel to the x axis, including the x axis itself, have a zero gradient. Similarly, on any line parallel to the y axis the x coordinates of the points are the same so that two typical points would be $A(x_1, y_1)$, $C(x_1, y_2)$ and the gradient of AC is $\frac{y_2 - y_1}{x_1 - x_1} = \frac{y_2 - y_1}{0}$. Since division by zero is not permitted this result is unusable. We do, however, suggest that all lines parallel to the y axis have a gradient of the form $\frac{1}{0}$ and some readers may presume to handle this by suggesting the gradient is infinity—written ∞—but we shall not deal with the idea of infinity at this stage. Neither shall we consider multiplying the gradient of the axes to gain the result of -1 to indicate perpendicularity, since $0 \times \frac{1}{0} = -1$ is not mathematically acceptable at this stage of our work.

Exercise 5.3

1. Find the gradients of the lines which are perpendicular to the lines with the following gradients

   (a) $\frac{1}{2}$, (b) 0·25, (c) -3, (d) $-\frac{3}{7}$, (e) m, (f) $\frac{1}{m}$, (g) $-\frac{p}{q}$

2. Prove that the triangle formed by joining $R(-2, 0)$, $P(7, 3)$, $Q(8, 0)$ is a right-angled triangle.
3. The figure $OPQR$ is formed by joining the points $R(0, 0)$, $P(6, 2)$, $Q(8, 8)$, $R(2, 6)$. Prove the following results:

   (i) $OP = PQ = QR = RO$
   (ii) OR is parallel to PQ
   (iii) RQ is parallel to OP
   (iv) OQ is perpendicular to RP
   (v) Figure $OPQR$ is not a square

4. The triangle ABC is formed by joining the points $A(2, 2)$, $B(7, 2)$, $C(5, -2)$. Prove the following results:

   (i) Triangle ABC is an isosceles triangle
   (ii) The line joining A to the midpoint of BC (i.e. the median through A) is perpendicular to BC.

5. Find the equation of the straight line which is parallel to the lines $y = 6$ and $y = 2$ and midway between them.
6. Find the equation of the straight line which is parallel to the lines $x = -2$ and $x = 4$ and midway between them.
7. Find the points of intersection of the four lines $x = -2$, $x = 12$, $y = -6$, $y = 8$, and prove that the diagonals of the figure so formed are at right angles.
8. Give the coordinates of two points A and B so that the line joining AB is perpendicular to the line with a gradient of $-\frac{1}{5}$.

CHAPTER SIX

THE EQUATION OF A LOCUS

Locus

When a point moves so as to comply with certain conditions, the path it traces out is called the locus of the point under these conditions. Thus a point which is compelled to move on this page of paper so that it is always 5 cm from the right hand edge of the page can only move on a line which is parallel to the edge of the page and 5 cm away.

The conditions placed upon the point are: (i) the point must be on the page; (ii) the point must always be 5 cm from the right-hand edge. The locus of the point is the line segment AB given in Fig. 15(*a*). We would arrive at the answer by plotting one or two possible positions, then a few more, until observing the answer as the line segment AB.

Another example would be to find the locus of a point which moved on this page so that it was always 5 cm from the centre C of this page. Again we note that the conditions to be complied with are: (i) the point must be on this

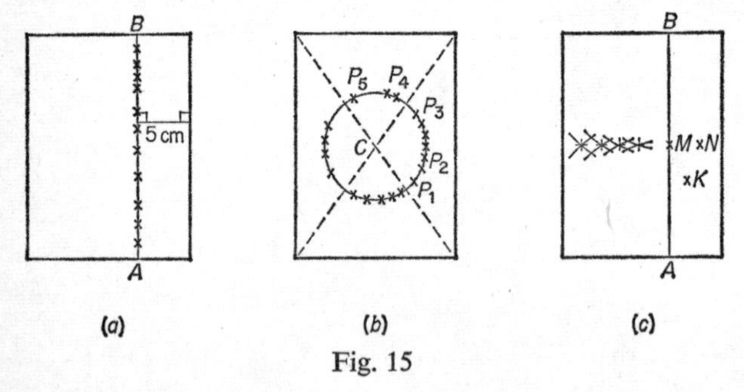

Fig. 15

page; (ii) the point must always be 5 cm from C. We try one or two points and then some more, as shown in Fig. 15(*b*), and as we move from P_1 to P_2 to P_3, etc., we realise that we are plotting points on the circumference of a circle centre C and radius 5 cm, and deduce that the locus is the circumference of this circle, i.e. the set of all points which make up the circumference. If we relax the condition (i) that the point be on this page then we have to use a little imagination and 'see' that the locus is the surface of a sphere of radius 5 cm and centre C. This of course does imply that we always think of a sphere in this manner, i.e. that every point on its surface is the same distance from the centre.

As another example, consider finding the locus of a point which moves on this page so that it is equidistant from the two-points A and B in Fig. 15(*a*). As usual we start by identifying the conditions to be satisfied as: (i) the point must always be on this page; (ii) wherever it is, it must be the same distance from A and B. We begin by plotting one or two trial points: the midpoint M of AB must be one point on the locus (assuming that there are others!) and, as illustrated in Fig. 15(*c*), the point N looks about right—certainly the point K looks wrong because it is nearer to A than B. Recalling the result of Fig. 15(*b*) we try a pair of compasses opened to a radius greater than AM and with centres alternately A and B we draw circles with the same radius and where they intersect will be a point which is the same distance from A as it is from B. Repeating this several times we eventually 'see' that the locus is a straight line through M perpendicular to AB. Thus each time, unless we 'know' the answer, we must build up the locus point by point until we identify its final form.

Exercise 6.1

1. Find the locus of a point which moves on this page so that it is equidistant from the left- and right-hand edges.
2. Find the locus of a point in a plane which is always 6 cm from a fixed point A in the plane.
3. A vertical wheel of radius 30 cm runs along a straight line PQ. What is the locus of the centre of the wheel?
4. Find the locus of the point which moves in the coordinate plane so that it is equidistant from the points $A(3, 1)$ and $B(3, 7)$.
5. A point P moves in the coordinate plane so that it is always 3 units from the line $x = 7$. Find the locus of P.

The Equation of a Locus

We describe any point P in the coordinate plane as $P(x, y)$ but, as soon as we impose conditions on its position, then it will only be able to move according to these conditions in what we have called a 'locus'. Imposing conditions on P means interpreting their influence on x and y. Thus we have seen that if P must move, so that it is only 3 units from the x axis, then the y coordinate of P must always be ± 3 and so the conditions mean $y = 3$ and $y = -3$. The locus is therefore the two lines $y = 3$ and $y = -3$.

Alternatively, we may start with an equation such as $y = 5x + 2$, find all the solutions we can and then interpret each solution as a point on graph paper. Finally we shall describe $y = 5x + 2$ as the equation of the locus, i.e. the equation is the condition that a point $P(x, y)$ will lie on the locus.

Similarly $x^2 + y^2 = 25$ is an equation of a locus which we determine in the usual way by finding as many points as necessary to deduce the complete curve. Thus $(3, 4)$, $(-3, 4)$, $(-3 - 4)$, $(3, -4)$ are four points satisfying the equation but hardly enough to deduce that the locus will be a circle centre the origin and radius 5 units, indeed it will be very tedious to proceed in this way to deduce the locus. However, the reader should appreciate that any equation in x and y which we shall be dealing with can be represented by a curve or a straight line in the coordinate plane. What we need to do is to build a store of standard results which we can subsequently describe as 'well known'.

Consider how we can produce a straight line through a point. Firstly it must have the same gradient everywhere and secondly it will have to pass through the point. The basic problem of producing a particular line is thus expressed by the following example.

Example 1. Find the equation of the straight line which has a gradient of 3 and passes through the point $A(2, 1)$.

Solution. Fig. 16 illustrates the problem. We start by supposing that the point $P(x, y)$ lies on the required line and we then complete the right-angled triangle PAB. This means that B is the point $(x, 1)$.

Finding the gradient in the usual manner we get

$$\frac{BP}{AB} = \frac{y-1}{x-2} = 3$$

Hence $y - 1 = 3(x - 2) \quad \therefore \quad y = 3x - 5$

Now this is an equation which is satisfied by the coordinates of the point (x, y) in complying with the two conditions: (i) the line has a gradient of 3; (ii) the line

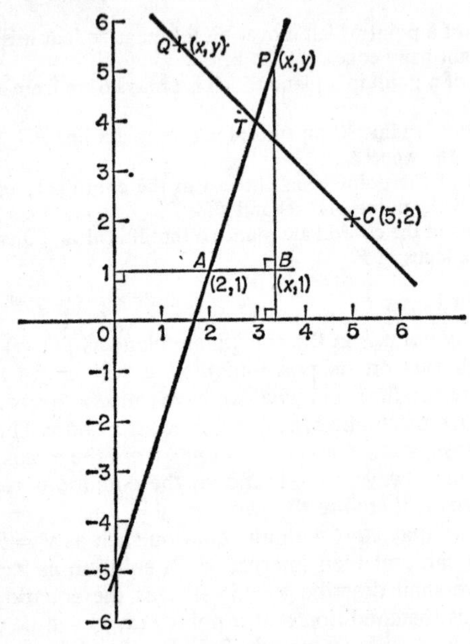

Fig. 16

passes through the point (2, 1). It follows therefore that $y = 3x - 5$ is the equation of the required line. Answer

Observe that the line intersects the y axis at $(0, -5)$ and the x axis at $(\frac{5}{3}, 0)$. These intercepts are significant when we rewrite the equation as $\frac{x}{\frac{5}{3}} + \frac{y}{-5} = 1$

Example 2. Find the equation of the straight line which has a gradient of -1 and passes through the point $C(5, 2)$. Find also the point in which this line intersects the line $y = 3x - 5$.

Solution. We again use Fig. 16, and, as before, we suggest that the point $Q(x, y)$ lies on the line. Although we have not found the line yet, we do know that it slopes in the direction suggested because the gradient is negative.

We now calculate the gradient of the line by moving from $C(5, 2)$ to $Q(x, y)$

$$\therefore \quad \text{the gradient is } \frac{\text{change in } y}{\text{change in } x} = \frac{y-2}{x-5} = -1$$

$$\therefore \quad y - 2 = -1(x - 5)$$

i.e. $y = -x + 7$ is the required equation to the straight line. Answer

Observe that the line intersects the y axis at the point $(0, 7)$ and the x axis at the point $(7, 0)$. As in the previous example we see the significance of these intercepts when the equation is rewritten as $\frac{x}{7} + \frac{y}{7} = 1$.

The intersection T is given by the values of x and y which satisfy the two equations simultaneously

$$y = -x + 7 \qquad \text{(i)}$$

$$y = 3x - 5 \qquad \text{(ii)}$$

$$\therefore \quad 3x - 5 = -x + 7, \quad \text{and} \quad x = 3$$

Substitution in (i) gives $y = -3 + 7 = 4$

The lines intersect at the point $T(3, 4)$ Answer

The Equation of the General Straight Line

The general straight line has a gradient m (excluding lines parallel to the y axis) and intersects the y axis at some point $A(0, c)$ (say). Without drawing a diagram let us obtain the equation to *any* straight line. Again we consider moving from A to a point $P(x, y)$ which also lies on the line. We shall calculate the gradient and call the result m. (Remember that we could write $m = \tan\theta$ where θ is the angle which the line makes with the positive direction of the x axis.) The gradient of AP is

$$\frac{y - c}{x - 0} = m$$

$$\therefore y - c = mx$$

$$\therefore y = mx + c$$

Thus the general straight line may be expressed in the form $y = mx + c$. In Example 1 on page 64 we obtained $y = 3x - 5$, in which case $m = 3$ and $c = -5$. In Example 2 we obtained $y = -x + 7$, in which case $m = -1$, and $c = 7$. Since we now know that the equation of any straight line may be written in the form $y = mx + c$ we could use the following method for a similar example.

Example 3. Find the equation of the straight line passing through the point $(-2, 3)$ and having a gradient of $-\frac{4}{3}$.

Solution. We start with the general equation of a straight line as $y = mx + c$. Since $m = -\frac{4}{3}$ we can modify this to read $y = -\frac{4x}{3} + c$. As the point $(-2, 3)$ will satisfy this equation we can substitute in the equation so far and obtain $3 = +\frac{8}{3} + c$. Hence $c = \frac{1}{3}$.

The required equation is $y = -\frac{4x}{3} + \frac{1}{3}$, which we can finally write as

$$3y = -4x + 1$$

Answer

We may take this idea a stage further now by finding the equation of a straight line passing through two points as in the following example.

Example 4. Find the equation of the straight line joining the points $A(1, 2)$ and $B(-3, 5)$.

Solution. Once again start with the general equation $y = mx + c$ but, instead of working out the gradient of the line AB, substitute the coordinates of the two given points

$$\therefore \quad 2 = m + c \qquad \text{(i)}$$
$$\therefore \quad 5 = -3m + c \qquad \text{(ii)}$$

Subtracting the equations yields $-3 = 4m$, $m = -\frac{3}{4}$
Substituting for m in (i) gives $c = 2\frac{3}{4}$
The equation of the line is therefore $y = -\frac{3}{4}x + 2\frac{3}{4}$
or $\qquad 4y = -3x + 11 \qquad$ Answer

One further observation: since parallel lines have the same gradient it follows that $y = mx + c_1$, $y = mx + c_2$, $y = mx + c_3$ and so on are all parallel lines.

Exercise 6.2

1. Find the equations of the straight lines which join the three points $A(1, 4)$, $B(0, 0)$, $C(7, 5)$.
2. Find the equation of the line which has a gradient of 3 and passes through the point (2, 7).
3. Find the equation of the straight line which is parallel to the line $y = 13x + 2$ and passes through the point (1, 5).
4. Find the points in which the line $\frac{x}{2} + \frac{y}{5} = 1$ intersects the coordinate axes.

   From this result deduce how to write down, on sight, the equations to each of the lines which join the following pairs of points:

   (i) (6, 0), (0, 5) (ii) (7, 0), (0, 8)
   (iii) (0, 3), (−2, 0) (iv) (0, −5), (−4, 0)
   (v) $(a, 0)$, $(0, b)$

   This form of the equation is called 'the intercept form' because it is directly expressed in terms of the intercepts the line makes in the coordinate axes.
5. Find the equation of the line which is perpendicular to the line $y = -2x + 3$ at the point (1, 1).
6. What is the gradient of the line $ax + by + c = 0$? Arrange the following lines in order of steepness starting with the steepest

   (i) $y = x$ (ii) $3y = 4x + 1$ (iii) $4y = 3x - 1$ (iv) $y = \frac{1}{2}x + 20$ (v) $y = 3x - 40$
7. Explain the significance of the sign of c in the equation $y = mx + c$.
8. Find the equation of the perpendicular bisector of the line joining the points $A(2, 3)$ and $B(4, 7)$.
9. From the following selection state which lines are either parallel or perpendicular to each other

   (i) $y = x + 1$ (ii) $y = 2x - 3$ (iii) $2y = 5x - 1$
   (iv) $y - x = 7$ (v) $y + x = 11$ (vi) $5y = -2x + 1$
   (vii) $2y = x - 3$ (viii) $2y = 1 - x$
10. Find the locus of the point which moves so that it is equidistant from the two points $A(0, 0)$ and $B(4, 4)$. Find the point which is equidistant from A, B and $C(4, 10)$.

The Equation of a Circle

Here we recall that we trace a locus, which is a circle, by keeping a moving point $P(x, y)$ a constant distance from a fixed point which is the centre of the circle. We recall also that the distance between two points $A(x_1, y_1)$ and $B(x_2, y_2)$ in rectangular coordinates is given by

$$AB^2 = (x_2 - x_1)^2 + (y_2 - y_1)^2$$

If we now replace $A(x_1, y_1)$ by the point (0, 0) and $B(x_2, y_2)$ by the general point $P(x, y)$ then

$$\begin{aligned} OP^2 &= (x - 0)^2 + (y - 0)^2 \\ &= x^2 + y^2 \end{aligned}$$

If we also suggest that P is always 6 units from O then the final equation is $x^2 + y^2 = 36$, and this we know must be the equation of a circle centre the origin and radius 6 units.

Consider this result in Fig. 17. We are given that OP is always 6 units. With PN perpendicular to the x axis and the point P as (x, y) we have $ON = x$ and $NP = y$. By Pythagoras's Theorem $OP^2 = ON^2 + NP^2$, i.e. $36 = x^2 + y^2$, so that everywhere P moves, its coordinates x and y must satisfy $x^2 + y^2 = 36$. Suppose we now move the circle to a position in which the point $C(20, 4)$ is centre. Again we take $P(x, y)$ to be any point on the circumference and draw PN perpendicular to a line through C which is parallel to the x axis. Therefore N is the point $(x, 4)$, $NP = y - 4$ and $CN = x - 20$.

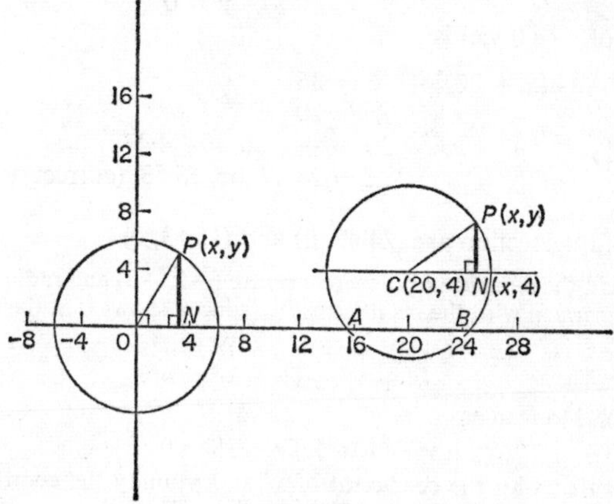

Fig. 17

Applying Pythagoras's Theorem in triangle CNP we have

$$\begin{aligned} CP^2 &= CN^2 + NP^2 \\ \therefore 36 &= (x - 20)^2 + (y - 4)^2 \end{aligned} \qquad \text{(i)}$$

must be the equation of the circle in its new position.

We may expand this form into

$$(x^2 - 40x + 400) + (y^2 - 8y + 16) = 36 \qquad \text{(ii)}$$

and write the final form as

$$x^2 + y^2 - 40x - 8y + 380 = 0 \qquad \text{(iii)}$$

From (i) we see that the equation of a circle appears to consist of the sum of two perfect squares on the left-hand side. In the final form (iii) we note that the equation is of the second degree, that there is no term in xy, that the coefficient of x^2 and y^2 are both unity and there is a connection between the coefficients of x and y and the coordinates of the centre of the circle. On this last remark, if the equation of a circle was $x^2 + y^2 - 10x - 6y - 2 = 0$, we would see that the centre of the circle was at the point (5, 3). The radius requires a little more searching, but we could have recognised the radius straight away had the equation been in the form of (i) above, so we begin by rearranging the equation to get it into this form. Starting with

$$x^2 - 10x + y^2 - 6y = 2$$

we complete the squares (see page 15) to get

$$x^2 - 10x + 25 + y^2 - 6y + 9 = 2 + 25 + 9$$

whence
$$(x - 5)^2 + (y - 3)^2 = 36$$

confirming that the equation is that of a circle of centre (5, 3) and radius 6.

Note that the points A and B where the circle of Fig. 17 intersects the x axis are given by solving the simultaneous equations

$$(x - 20)^2 + (y - 4)^2 = 36$$
$$y = 0$$

Substitution of $y = 0$ yields

$$(x - 20)^2 + 16 = 36$$
$$(x - 20)^2 = 20$$
$$x - 20 = \pm\sqrt{20} = \pm 4{\cdot}47$$
$$\therefore x = 24{\cdot}47 \text{ or } 15{\cdot}53 \text{ (correct to 2 decimal places)}$$

The points of intersection are (24·47, 0) and (15·53, 0)

Example 5. Find the equation of the circle centre (−5, −3) and radius 4 units.

Solution. Comparison with the result already deduced shows that the equation is

$$(x - (-5))^2 + (y - (-3))^2 = 4^2$$

i.e.
$$(x + 5)^2 + (y + 3)^2 = 4^2$$

which we expand to read as

$$x^2 + y^2 + 10x + 6y + 18 = 0 \qquad \text{Answer}$$

Observe again that with the coefficient of x^2 and y^2 unity the coordinates of the centre are given by minus half the coefficients of x and y.

Example 6. Find the centre and radius of the circle given by

$$3x^2 + 3y^2 + 6gx + 6fy + 3c = 0 \quad (g, f \text{ and } c \text{ are constants}).$$

Solution. We divide the equation by 3 in order to reduce the coefficient of x^2 and y^2 to unity. The equation now reads

$$x^2 + y^2 + 2gx + 2fy + c = 0$$

Rearranging to $x^2 + 2gx + y^2 + 2fy = -c$ enables us to see what to do to complete the squares, i.e. adding g^2 and f^2 to both sides.

Thus
$$x^2 + 2gx + g^2 + y^2 + 2fy + y^2 = g^2 + f^2 - c$$

gives the final form

$$(x + g)^2 + (y + f)^2 = g^2 + f^2 - c$$

This represents a circle centre $(-g, -f)$ and radius $\sqrt{(g^2 + f^2 - c)}$ Answer

We observe that the equation of any circle can therefore be written in this form, and if we know g, f and c we can also fix the position and radius of the circle.

Example 7. Find the equation of the circle which passes through the points

$$A(0, -1), B(2, 1), C(6, -3)$$

Solution. We start with the general equation for a circle as

$$x^2 + y^2 + 2gx + 2fy + c = 0$$

Since $A(0, -1)$ lies on this circle we have $0 + 1 + 0 - 2f + c = 0$

i.e. $2f - c = 1$ (i)

Since $B(2, 1)$ lies on this circle we have $4 + 1 + 4g + 2f + c = 0$

i.e. $4g + 2f + c = -5$ (ii)

Since $C(6, -3)$ lies on this circle we have

$$36 + 9 + 12g - 6f + c = 0$$

i.e. $12g - 6f + c = -45$ (iii)

We now have three equations to solve simultaneously for g, f and c.

Adding (i) and (ii) produces $4g + 4f = -4$ or $g + f = -1$

Adding (i) and (iii) produces $12g - 4f = -44$ or $3g - f = -11$

Adding these last two equations gives $4g = -12$, $g = -3$

Substitution of $g = -3$ gives $f = 2$

Substitution of $f = 2$ in (i) gives $c = 3$

The required equation is $x^2 + y^2 - 6x + 4y + 3 = 0$. A circle with centre $(3, -2)$ and radius r given by $r^2 = g^2 + f^2 - c = 9 + 4 - 3 = 10$

$r = \sqrt{10}$ Answer

An alternative method would have been to find the equation to the perpendicular bisectors of AB and AC and then their point of intersection, which is the centre of the circle. We could have used the method of determinants to solve the three equations (see page 103).

Exercise 6.3

1. Find the equation of the circles which satisfy the following requirements:
   (i) centre (2, 6), radius 4. Show that this circle passes through the point (2, 2).
   (ii) centre (−1, −1), radius 3. Show that this circle passes through the point (−1, 2).
   (iii) centre (−5, 5), radius 5. Show that this circle passes through the point (0, 5).
   (iv) passes through the points $A(0, 0)$, $B(2, -2)$, $C(-6, 2)$. Find the centre and radius of this circle.
   (v) centre (0, 4), radius 4. Show that this circle touches the x axis.
   (vi) centre (4, 0), radius 4. Show that this circle touches the y axis.
2. Find the equation of the circle which has for a diameter the line joining $A(2, 2)$ and $B(8, 10)$. (Hint: The midpoint of AB will be the centre and $\frac{1}{2}AB$ the radius.)
3. Find the equation of the circle which passes through the following points, which are the vertices of a square, $A(2, 2)$, $B(8, -6)$, $C(16, 0)$ $D(10, 8)$. (Hint: Consider AC as a diameter of the required circle.)
4. Find the centre and radius of each of the following circles:
   (i) $x^2 + y^2 - 2x - 6y - 90 = 0$
   (ii) $x^2 + y^2 + 2x + 6y - 90 = 0$
   (iii) $x^2 + y^2 + 8x - 10y - 23 = 0$
5. Using the circles given in Question 4 prove the following results:
   (a) (i) intersects the line $x = 1$ in the points (1, 13), (1, −7)
   (b) (ii) intersects the line $y = 5$ in the points (5, 5), (−7, 5)
   (c) (iii) intersects the line $y = x + 1$ in the points (4, 5), (−4, −3)

6. Solve the simultaneous equations: $x^2 + y^2 + 4x - 2y - 44 = 0$ (i)
$y + x = 6$ (ii)
(Hint: Either substitute $y = 6 - x$ in (i) first, or put (i) in perfect square form and then substitute $y = 6 - x$.)
7. Find the locus of the point $P(x, y)$ so that its distances from the two points $O(0, 0)$ and $B(8, 0)$ are such that $PO^2 + BP^2 = 64$. Show that this locus is a circle passing through O and B and find its centre.
8. Prove that the locus of the point $P(x, y)$, which moves so that its distances from two fixed points $O(0, 0)$ and $B(0, 6)$ are related by the equation $PO^2 + PB^2 = 50$, is a circle, of radius 4 and centre (0, 3), which does not pass through either O or B.

The Tangent to a Circle

A tangent to a circle is always perpendicular to the radius through the point of contact. Since we now know how to find the centre and radius of a circle then, by combining this information with the condition for perpendicularity ($m_1m_2 = -1$), we shall be able to find the equation of the tangent as indicated in the following example.

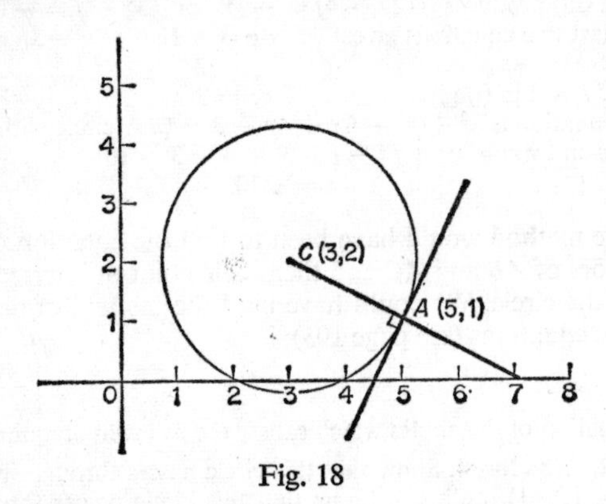

Fig. 18

Example 8. Find the equation of the tangent to the circle $x^2 + y^2 - 6x - 4y + 8 = 0$ at the point $A(5, 1)$. We illustrate the problem with Fig. 18. Note once again that the scales for x and y must be the same if we are to obtain a circle using a pair of compasses.

Solution. The centre of the circle is the point $C(3, 2)$. Hence, the gradient of CA is

$$\frac{1-2}{5-3} = -\frac{1}{2}.$$

Since the tangent at A will be perpendicular to CA then its gradient will be 2.

We now require the equation of a line with this gradient and passing through the point $A(5, 1)$. As we saw before, the line is of the form $\frac{y-1}{x-5} = 2$

i.e. $y - 1 = 2x - 10$

$\therefore\ y = 2x - 9$ is the equation of the required tangent. Answer

The Normal at a Point on a Curve

The line which is perpendicular to a tangent at its point of contact with a curve is called the **normal at that point.** In the case of a circle this is the radius to the point of contact.

For other curves the tangents and normals are not so easy to come by and we have to use other methods to construct them (see page 82). There is, of course, a general principle to be followed in coordinate geometry in order to determine if a straight line is a tangent to a curve. First find the points of intersection of the line with the curve; if they are coincident points then the line is a tangent.

Suppose we have a curve like the one in Fig. 19, which is intersected by the straight line AB in the two points P and Q. Clearly this line is not a tangent to the curve. Now we move the line into position AC and see the two points of intersection come closer together and then even closer in position AD, until they coincide in point T and the line AT is a tangent to the curve. Returning to the case of a tangent to a circle we see the condition that a line be a tangent is that the two points of intersection of the line and circle should coincide.

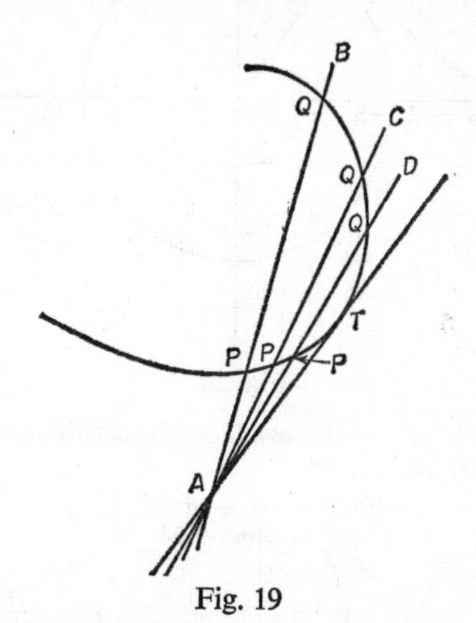

Fig. 19

Example 9. Prove that the line $4y - 3x = 50$ is a tangent to the circle $x^2 + y^2 = 100$.

Solution. We need to prove that the line intersects the circle in two coincident points so we start by solving the two equations simultaneously.

Everywhere on the line $\quad 4y = 3x + 50 \quad$ or $\quad 16y^2 = (3x + 50)^2$
Everywhere on the circle $\quad y^2 = 100 - x^2 \quad$ or $\quad 16y^2 = 1600 - 16x^2$

At a point of intersection we shall have

$$(3x + 50)^2 = 1600 - 16x^2$$
$$\therefore \quad 9x^2 + 300x + 2500 = 1600 - 16x^2$$
$$\therefore \quad 25x^2 + 300x + 900 = 0$$

Dividing by 25.

$$x^2 + 12x + 36 = 0 = (x + 6)(x + 6)$$

i.e. $x = -6$ is the only solution.
Hence the line $4y - 3x = 50$ is a tangent to the circle $x^2 + y^2 = 100$ at the point $(-6, 8)$ as shown in Fig. 20, on page 72. Answer

Example 10. Find the equations of the two tangents which can be drawn from the origin to the circle $x^2 + y^2 - 10y + 16 = 0$. Also find the coordinates of the points of contact.

Solution. Any line through the origin may be written in the form $y = mx$. To find its intersection with the circle substitute $y = mx$ in its equation.

$$\therefore \quad x^2 + m^2x^2 - 10mx + 16 = 0 = x^2(1 + m^2) - 10mx + 16$$

This being a quadratic in x will give the x coordinates for the two points of intersection P and Q as seen in Fig. 21. But we want P and Q to coincide, in which case

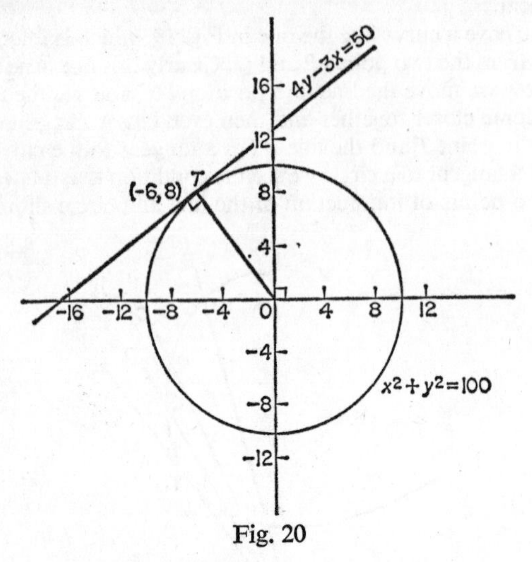

Fig. 20

the quadratic must have two equal roots. Since the condition for equal roots is that '$b^2 = 4ac$' (see page 21), we require

$$(-10m)^2 = 4(1 + m^2)16$$
$$100m^2 = 64m^2 + 64$$
$$36m^2 = 64$$
$$m^2 = \frac{64}{36} = \frac{16}{9} \quad \therefore \quad m = \pm\frac{4}{3} \qquad \text{Answer}$$

Two different values of m indicate the two possible tangents as seen in Fig. 21. Since the tangents pass through the point (0, 0) their equations are $y = 4x/3$ and

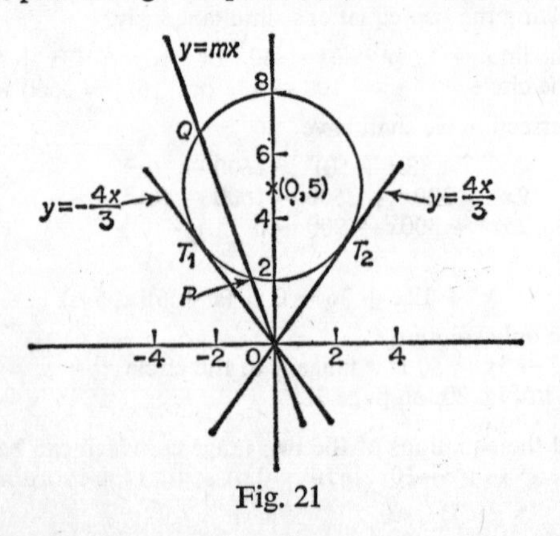

Fig. 21

$y = -4x/3$. To find the points T_1 and T_2 at which the lines touch the circle we continue with the formula

$$x = \frac{-b \pm \sqrt{(b^2 - 4ac)}}{2a} = \frac{+10m}{2(1 + m^2)}$$

For T_2: with $m = \frac{4}{3}$ $\quad x = \frac{40}{6(1 + \frac{16}{9})} = \frac{9 \times 40}{6 \times 25} = 2{\cdot}4$

From $y = \frac{4x}{3}$ and $x = 2{\cdot}4$, we get $y = 3{\cdot}2$

For T_1: with $m = -\frac{4}{3}$ $\quad x = -2{\cdot}4 \quad y = 3{\cdot}2$

The points of contact are $T_1(-2{\cdot}4, 3{\cdot}2)$, $T_2(2{\cdot}4, 3{\cdot}2)$ Answer

Knowing the positions of T_1 and T_2 enables us to calculate the length of the tangents from O. i.e. OT_1 and OT_2.

Exercise 6.4

1. Show that each of the following lines are tangents to the circle $x^2 + y^2 = 100$
   (i) $x = 10$ (ii) $x = -10$ (iii) $y = 10$
   (iv) $y = -10$ (v) $3x + 4y = 50$ (vi) $3x - 4y = 50$
   (Hint: Draw a sketch for parts (i) to (iv) and avoid calculations.)
2. Find the equation of the tangent to the circle $x^2 + y^2 - 6x - 18y + 74 = 0$ at the points (i) (3, 13), (ii) (7, 9). (Hint: Draw a sketch and avoid calculations.)
3. Find the equation of the tangent to the circle $x^2 + y^2 = 169$ at the points $(-5, -12)$ and $(12, 5)$.
4. The line $y = 6$ is a tangent to a circle of radius 4 at the point (8, 6). Find the equation of the two possible circles.
5. Calculate the lengths of OT_1 and OT_2 in the last worked example above, where O is the origin.
6. Find the length of the tangent to the circle $x^2 + y^2 = 64$ from the point $A(0, 11)$, i.e. if the tangent touches the circle at T then the length of the tangent from A is AT. (Hint: use Pythagoras's Theorem.)
7. Prove that the lengths of the tangents from the point $A(0, 12)$ to the circles $(x - 2)^2 + y^2 = 20$ and $(x + 3)^2 + y^2 = 25$, are equal. (Hint: Use the method of question 6.)
8. Find the equations of the tangents to the circle $x^2 + y^2 = 20$ which are parallel to the line $2y + x = 0$.
9. Find the equation of the tangents to the circle $x^2 + y^2 = 4$ at the following points: (i) $(1, \sqrt{3})$, (ii) $(\sqrt{3}, 1)$, (iii) $(\sqrt{2}, \sqrt{2})$.
10. Find the equations of the two tangents which can be drawn from the origin to the circle $x^2 + y^2 - 12x + 18 = 0$. Also find the coordinates of the points of contact.

CHAPTER SEVEN

FURTHER COORDINATE GEOMETRY

Plotting Straight Lines

We have seen on page 63 that any curve consists of a set of points whose x and y coordinates satisfy an equation, consequently described as the equation of the curve. In this chapter we shall only discuss curves of equations of the first and second degree in x and y (see page 2).

Equations of the first degree in x and y produce straight lines; hence their reference name of linear equations. To plot the straight line we need only obtain three of its points: two points to fix the line and the third to check that no error has been made. Two non-parallel straight lines will always intersect in a point which, being common to both the lines, means a pair of x, y values which are common to their equations so that, as already noted, the point of intersection gives the solution to the two simultaneous equations of the lines. We shall now find out more about the use of coordinate geometry for dealing with simultaneous linear inequalities.

The Perpendicular Distance from the Origin to a Line $y = mx + c$

With the equation of a straight line in the form $y = mx + c$ the intercepts on the x and y axis are $-c/m$ and c respectively. Referring to the diagram of Fig. 22(a) we see a line with a negative gradient and a positive intercept on the y axis, which means that the figure represents a choice of line with m negative and c positive. In the right-angled triangle OAB, $AB^2 = c^2 + c^2/m^2 = \frac{c^2}{m^2}(1 + m^2)$. Hence

$$AB = \frac{-c}{m}\sqrt{(1 + m^2)}$$

and
$$\sin\beta = \frac{OB}{AB} = \frac{-m}{\sqrt{(1 + m^2)}}$$

The perpendicular distance p from the origin to the line AB is therefore

$$p = OA\sin\beta = \frac{-c}{m} \times \frac{-m}{\sqrt{(1 + m^2)}} = \frac{c}{\sqrt{(1 + m^2)}}$$

Example 1. Find the distance from the origin to the line $7x + 4y = 10$.

Solution. Arranging the equation in the form $y = mx + c$ we have

$$m = -\frac{7}{4} \quad \text{and} \quad c = \frac{10}{4}$$

$$\therefore \; p = \frac{10}{4\sqrt{(1 + \frac{49}{16})}} = \frac{10}{\sqrt{65}} = \frac{10}{8{\cdot}062} = 1{\cdot}24 \text{ (correct to 2 decimal places)}$$

Answer

The Distance Between Two Parallel Lines

If P is the distance from the origin to one line and p the distance from the origin to the second line then the distance between the two lines is $p - P$, taking the numerical value whether $p - P$ is negative or positive.

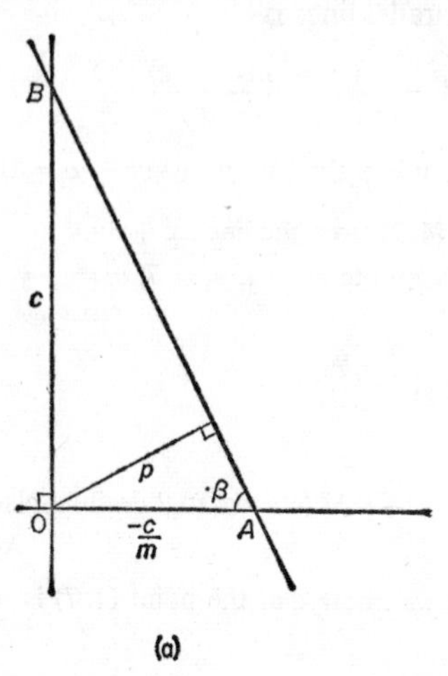

(b)

Fig. 22

Example 2. Find the distance between the two parallel lines $3x + 5y = 10$ and $3x + 5y = -7$

Solution. The distance from the origin to the first line is

$$p = \frac{c}{\sqrt{(1 + m^2)}} \quad \text{where} \quad c = \frac{10}{5}, \quad m = -\frac{3}{5}$$

so that

$$p = \frac{10}{5\sqrt{(1 + \frac{9}{25})}} = \frac{50}{5\sqrt{34}} = \frac{10}{\sqrt{34}}$$

For the second line we get $P = \dfrac{-7}{5\sqrt{(1 + \frac{9}{25})}} = -\dfrac{35}{5\sqrt{34}} = -\dfrac{7}{\sqrt{34}}$

The negative value for P indicates that the line $3x + 5y = -7$ is on the opposite side of the origin to $3x + 5y = 10$.
The distance between the two lines is therefore

$$p - P = \frac{10}{\sqrt{34}} - \left(-\frac{7}{\sqrt{34}}\right) = \frac{17}{\sqrt{34}} = 2{\cdot}92 \text{ (correct to 2 decimal places)}$$

Answer

The Perpendicular Distance From a Point to a Line

Returning to Fig. 22(*b*) we now consider a line through $Q(x_1, y_1)$ parallel to the line $y = mx + c$, which was AB in Fig. 22(*a*). Since LM is parallel to $y = mx + c$ it has a gradient m and an equation $y = mx + d$ (say). Because $Q(x_1, y_1)$ lies on LM we have $y_1 = mx_1 + d$ or $d = y_1 - mx_1$. The equation of this line LM is clearly $y = mx + y_1 - mx_1$, so that $y_1 - mx_1$ is equivalent to the constant c in the equation of the line AB. The distance from the origin to the line LM is therefore

$$P = \frac{y_1 - mx_1}{\sqrt{(1 + m^2)}}$$

so that the distance between the two parallel lines is

$$p - P = \frac{c - y_1 + mx_1}{\sqrt{(1 + m^2)}} = -\frac{(y_1 - mx_1 - c)}{\sqrt{(1 + m^2)}}$$

But this is also the distance from $Q(x_1, y_1)$ to the line $y - mx - c = 0$.

Example 3. Find the distance of the point (4, 7) from the line $2x + 3y = 6$.

Solution. Using the result just proved we substitute $x_1 = 4$, $y_1 = 7$, $m = -\frac{2}{3}$, $c = 2$
The distance of the origin from the line is

$$p = \frac{2}{\sqrt{(1 + \frac{4}{9})}} = \frac{6}{\sqrt{13}}$$

The distance from (4, 7) to the line is

$$d = -\frac{(7 + \frac{8}{3} - 2)}{\sqrt{(1 + \frac{4}{9})}} = -\frac{23}{\sqrt{13}} = -\frac{23}{3{\cdot}61} = -6{\cdot}37 \text{ (correct to 2 decimal places)}$$

Answer

Now since p is positive and d is negative we know that the point (4, 7) is on the opposite side of the line to the origin.

Linear Inequalities

Consider the line $2x + 3y = 24$. To plot the line we find three of its points and tabulate the results in the form of a table.

x	0	12	6
y	8	0	4

We do not choose points at random but reason that $x = 0$ or $y = 0$ are the easiest to start with. These give the intercepts on the axes (page 66).

Now consider the line $2x + 3y = 30$, which is again plotted from three of its points in the following table.

x	0	15	6
y	10	0	6

Both lines are in Fig. 22(c).

We know that the lines are parallel to each other because they have the same gradient of $-\frac{2}{3}$. What we notice is that the increase of the constant term from 24 to 30 moves the line AB into position CD, i.e. further away from the origin. Similarly EF, the line $2x + 3y = 36$, will be parallel to AB and CD but even further away from the origin than CD. Thus the coordinates of any point on the same side of CD as the line EF will make $2x + 3y > 30$, e.g. $P(6, 7)$. Indeed, we may go further and say that the coordinates of any point in the shaded region between CD and EF will make $2x + 3y > 30$ and $2x + 3y < 36$, which we can abbreviate to read $30 < 2x + 3y < 36$. If we include the points **on** the lines CD and EF in the shaded region we can modify this inequality statement to read $30 \leqslant 2x + 3y \leqslant 36$.

Similarly, if we insist that x and y must be positive as well as satisfying the condition that $2x + 3y < 24$, then the shaded interior of the right-angled triangle of Fig. 22(*c*) will contain all the points whose x and y coordinates satisfy simultaneously the three conditions $x > 0$, $y > 0$, $2x + 3y < 24$.

Now let us consider the region bounded by four lines rather than two or three—Fig. 23.

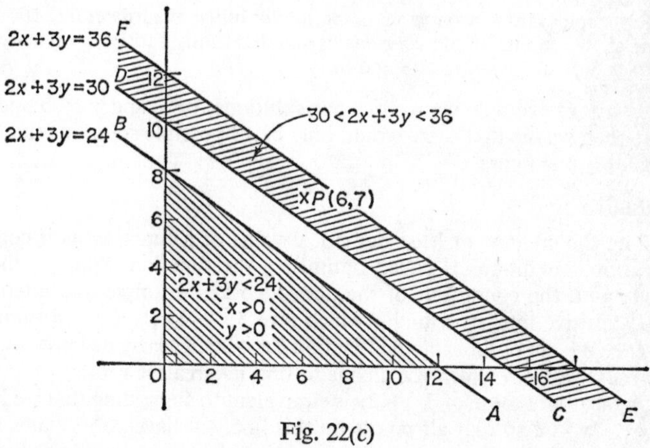

Fig. 22(*c*)

Example 4. Find the region of points on the coordinate plane which gives solutions satisfying simultaneously the inequalities $x > 0$, $y > 0$; $2x + 3y < 18$; $8x + 5y < 40$.

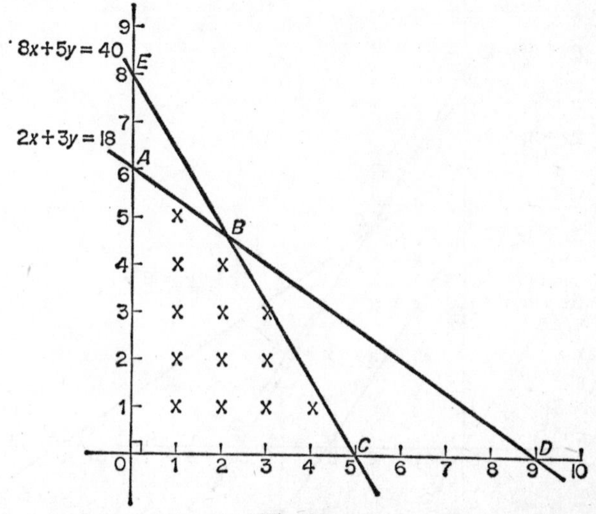

Fig. 23

Solution. We commence by plotting the graphs of $x = 0$ (the y axis), $y = 0$ (the x axis), $2x + 3y = 18$ and $8x + 5y = 40$ in Fig. 23.

x	0	9	4·5
y	6	0	3

x	0	5	2·5
y	8	0	4

We know that everywhere inside the triangular region OAD we have

$$2x + 3y < 18, \quad x > 0, \quad y > 0$$

We know also that everywhere inside the triangular region OEC we have

$$8x + 5y < 40, \quad x > 0, \quad y > 0$$

The intersection of these two regions is the interior of the quadrilateral $OABC$ which contains all the points whose coordinates simultaneously satisfy the requirements that $x > 0$, $y > 0$; $2x + 3y < 18$ and $8x + 5y < 40$. Answer

If in the above example we restrict our solutions for x and y to be positive integers then we see that there would only be 13 possible solutions (indicated by crosses in the figure.)

Optimisation

Finding the greatest or least numerical value of a function as it complies with various conditions is called optimising the function. Suppose that we continue with the conditions of the above worked example and attempt to find the positive integers x and y which give $3x + 2y$ its greatest numerical value. (i.e. We are optimising the function $3x + 2y$. Clearly its least value is 0 so the only problem which remains is to find its greatest value.)

Taking c as any value of $3x + 2y$ is equivalent to suggesting that we have a line $3x + 2y = c$ so that all points on this line will have coordinates which make $3x + 2y = c$. For different values of c this line moves so that it is always parallel to the line $3x + 2y = 0$. We therefore start by drawing the

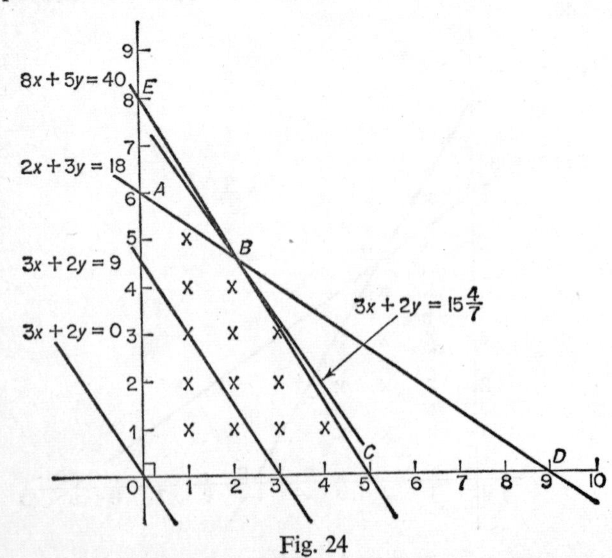

Fig. 24

line $3x + 2y = 0$ as shown in Fig. 24. We have already seen that the coordinates of any point above the line $3x + 2y$ will make $3x + 2y > 0$. What we need to know is what is the furthest distance away from O that we can move a line parallel to $3x + 2y = 0$ yet still get at least one of its points to satisfy the inequalities of the region $OABC$? If we move parallel to $3x + 2y = 0$ we find that the furthest away we can get is the parallel through the point B so that the x, y coordinates of this point must give the greatest value of $3x + 2y$. Unfortunately x, y coordinates of the point B are the solutions of the two simultaneous equations

$$8x + 5y = 40 \quad \text{(i)}$$
$$2x + 3y = 18 \quad \text{(ii)}$$

Multiplication of (ii) by 4 and subtraction from (i) yields

$$-7y = -32$$
$$y = 4\tfrac{4}{7}$$

so clearly B will not give integer solutions for x and y. It follows that we must pull the line back a little till we reach the nearest point with integer coordinates. We can see by inspection (i.e. without redrawing the line) that of the eligible points the point (3, 3) is near (and so is the point (4, 1)).

Substitution of these values gives $3x + 2y = 14$ at (4, 1), $3x + 2y = 15$ at (3, 3). The greatest value of $3x + 2y$ under the given conditions was therefore 15. This type of work is used in the introduction to linear programming and the diagrammatic methods for simple linear inequalities follow this style.

Exercise 7.1

1. Using the inequalities $x > 0$, $y > 0$, $8x + 5y < 40$, $2x + 3y < 18$ of Fig. 24 find the integer values of x and y which give the greatest values of the following expressions:

   (i) $x + y$ (ii) $x + 2y$ (iii) $2x + y$
   (iv) $4x + 3y$ (v) $7x + 5y$ (vi) $5x + 7y$

2. Find the corresponding results if the inequalities in Question 1 had been closed, i.e. $x \geqslant 0$, $y \geqslant 0$; $8x + 5y \leqslant 40$; $2x + 3y \leqslant 18$.
3. A container is packed with two types of cartons, not necessarily exactly. Carton A has a mass of 8 kg and a volume of 2000 cm^3. Carton B has a mass of 5 kg and a volume of 3000 cm^3. There is a profit of 7p on a carton A and 5p on a carton B. The total mass packed into the container must not exceed 40 kg and the total volume of the cartons must not exceed 18 000 cm^3. If x of cartons A and y of cartons B are packed find:

   (i) the total mass packed
   (ii) the total volume of the cartons packed
   (iii) the profit on the container

   By examining Question 2 and Fig. 23 find the greatest profit which can be made on packing one container.
4. Find the greatest value of $5x + 6y$ for integer values of x and y satisfying the conditions $x > 0$; $y > 0$; $4x + 3y \leqslant 24$, $7x + 2y \leqslant 28$.
   (Draw a graph using the first quadrant only and a scale of 1 cm for 1 unit on both axes.)

5. Using the same inequality conditions as in Question 4 find the greatest value of each of the following:
   (i) $2x + y$
   (ii) $3x + y$
6. Find the greatest value and the least value of $x + y$ subject to the conditions $x > 0; y > 0; 4x + 3y \leqslant 24; 7x + 2y \geqslant 28$. (Hint: use your graph for Question 4).
7. Find the distance of the origin from the two lines $4x + 3y = 12$ and $4x + 3y = -12$.
8. Find the distance of the origin from the two lines $4x - 3y = 12$ and $4x - 3y = -12$.
9. Find the distance between the two parallel lines $6x + 8y = 9$ and $6x + 8y = 14$.
10. Find the distance of the point $(-2, 3)$ from the line $12x + 5y = 30$.

The Parabola

We already discussed on page 67 one curve arising from an equation of the second degree: that is the circle, whose general equation is

$$x^2 + y^2 + 2gx + 2fy + c = 0.$$

We recognise from previous experience that this will be a circle because the coefficients of x^2 and y^2 are the same and there is no term in xy. Once we recognise the equation we need only find its centre $(-g, -f)$ and its radius $\sqrt{(g^2 + f^2 - c)}$ and with a pair of compasses draw the circle. There is no need to plot a sequence of points to be joined freehand.

For the parabola, which is given by $y = f(x)$ where $f(x)$ is any quadratic in x, there is no such short cut and we need to plot a sequence of points before discovering the whereabouts and shape of the curve. We have said that $y = ax^2 + bx + c$, $(a \neq 0)$ will give a curve called a **parabola**, and so will the equation $x = ay^2 + by + c$, $(a \neq 0)$, i.e. each variable being expressed as a quadratic function of the other. We shall start with the simplest form of the equation.

The Parabola $y = x^2$

To obtain points for plotting we adopt the usual procedure of suggesting values for the independent variable x and working out the corresponding values for y. We shall restrict the domain of the function to $8 \leqslant x \leqslant 8$. Our first observation is that $\pm x$ gives the same value of y so our table of points is as follows:

x	0	± 1	± 2	± 3	± 4	± 5	± 6	± 7	± 8
$y = x^2$	0	1	4	9	16	25	36	49	64

For the average piece of graph paper we shall need a different scale for the x and y axes. We have chosen (Fig. 25) the following scales: x, 1 cm to 2 units and y, 1 cm to 5 units (here scaled down somewhat because of printing requirements).

The curve obtained by 'joining' the points in Fig. 25 is called a parabola. The points of the table are indicated by X, and if there had been any doubt as to where the curve was in between some of these points then we would have calculated the coordinates of an intermediate point, e.g. $x = \pm 5\frac{1}{2}$, $y = 30\frac{1}{4}$. Having drawn a graph, we should always examine it for any particular features such as the following:

(1) Since $y = x^2$ we have $x = \pm\sqrt{y}$ so that the curve may be used to find square roots as well as squares. The point A, for example, has coordinates $x = 7{\cdot}4$ and $y = 55$, and within the limits of accuracy of the graph drawn we deduce that $\sqrt{55} = 7{\cdot}4$.

(2) We observe that the curve passes through the origin and that the x axis (i.e. the line $y = 0$) is a tangent to the curve at the origin.

(3) The least value of y is 0.

(4) The curve is symmetrical about the y axis (i.e. the line $x = 0$). The point where the parabola intersects its axis of symmetry is called the **vertex** of the parabola. This means that if we fold the paper along the y axis then the curve in the first quadrant coincides with that part in the second quadrant.

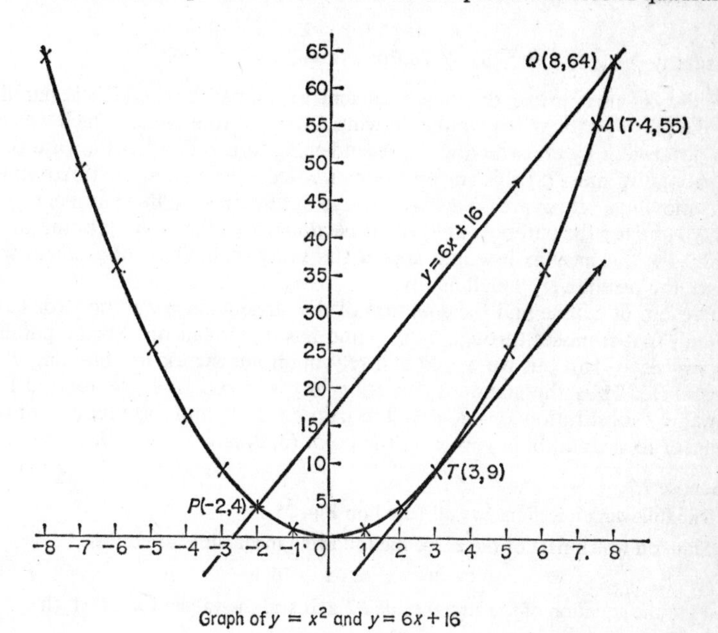

Graph of $y = x^2$ and $y = 6x + 16$

Fig. 25

Algebraically this means that for each point (x, y) on the curve there is a point $(-x, y)$ on the curve. This result may be expressed by saying that if $y = f(x)$ is the equation of the curve and if $y = f(-x)$ (i.e. changing the sign of x makes no difference to the value of y), then the curve is symmetrical about the y axis. Similarly, if $x = f(y)$ and $x = f(-y)$ then this curve would be symmetrical about the x axis, e.g. $y^2 = 4ax$.

Again with the graph of $y = x^2$, consider the points P and Q which are the intersections of the straight line $y = 6x + 16$ with the curve $y = x^2$. We may think of the points P and Q as giving the solutions of the two simultaneous equations

$$y = x^2 \qquad \text{(i)}$$
$$y = 6x + 16 \qquad \text{(ii)}$$

Substitution for y in (i) gives

$$x^2 - 6x - 16 = 0 = (x - 8)(x + 2)$$
$$\therefore x = 8 \text{ or } -2, y = 64 \text{ or } 4$$

giving $P(-2, 4)$ and $Q(8, 64)$ as the solutions.

Further observations relating the equation of the line PQ to that of the curve are the inequalities $x^2 < 6x + 16$ and $x^2 > 6x + 16$. (We know that $x^2 = 6x + 16$ at P and Q.) If we choose any value of x between P and Q then, since the curve is below the line PQ, it follows that $y = x^2$ is less than $y = 6x + 16$ for the same value of x. Hence, in between P and Q, $x^2 < 16x + 16$

i.e. $x^2 < 6x + 16$ for $-2 < x < 8$

Similarly $x^2 > 6x + 16$ for $x < -2$ and $x > 8$

Finally, let us consider drawing a tangent to the parabola which is parallel to the line $y = 6x + 16$. Unlike drawing a tangent to a circle, which we can do accurately by constructing a perpendicular to a radius, in the case of a parabola we must (at this stage of our knowledge) use an approximation by eye, moving a set square on a straight edge until we have a line parallel to PQ which touches the parabola at T. By inspection we see that T is the point (3, 9). This is not as easy as it would appear since any slight error of gradient will place the results very much astray.

We can of course find the equation of the tangent since we know (or have found) that it passes through $T(3, 9)$ and has a gradient of 6 (being parallel to $y = 6x + 16$), but we would still rely upon guesswork in obtaining T at this stage. Thus the standard form $y = mx + c$ becomes $y = 6x + c$ followed by substitution $x = 3$, $y = 9$ to gain $c = -9$, and the equation of the tangent to the parabola $y = x^2$ at the point (3, 9) is $y = 6x - 9$.

Exercise 7.2

The following questions are all based on Fig. 25.

1. State on which part of the curve the following inequalities are satisfied
$$9 < x^2 < 6x + 16$$
2. Find the equation of the straight line PT and find the value of x for which
(i) $x^2 < x + 6$
(ii) $x^2 > x + 6$
3. Find the equation of the tangent to the curve $y = x^2$ at the point $(-3, 9)$.
4. Find the angle which the tangent at T makes with the x axis.
5. Draw the tangent at the point $(-4, 16)$. Find the gradient of this tangent and its equation.
6. Find the following square roots: (i) $\sqrt{20}$, (ii) $\sqrt{30}$, (iii) $\sqrt{60}$.
7. Comment on the position of the graphs of the following equations in relation to the graph of $y = x^2$
(i) $y = x^2 + 3$ (ii) $y = x^2 - 10$ (iii) $y = -x^2$
(iv) $x = y^2$ (v) $x = -y^2$ (iv) $x = y^2 + 3$

The General Parabola $y^2 = 4ax$ ($a > 0$)

Here we regard a as a constant, for any one parabola. Thus in one case we may have $a = 2$ and in another case $a = 10$, thereby giving the two parabolas $y^2 = 8x$ and $y^2 = 40x$. We may tabulate the points on the graph in the usual way but here, for convenience, instead of choosing $x = 1, 2, 3,$ etc., we choose $x = a, 4a,$ etc., $a > 0$, as shown on the following table.

x	0	a	$4a$	$9a$	$16a$	$12a$
$y^2 = 4ax$	0	$4a^2$	$16a^2$	$36a^2$	$64a^2$	$48a^2$
y	0	$\pm 2a$	$\pm 4a$	$\pm 6a$	$\pm 8a$	$\pm 6{\cdot}9a$

Choosing a scale of x: 1 cm to $1a$ units, y: 1 cm to $1a$ units, we obtain the graph of Fig. 26. We would usually know the value of a but here we show the more awkward situation when a is not known.

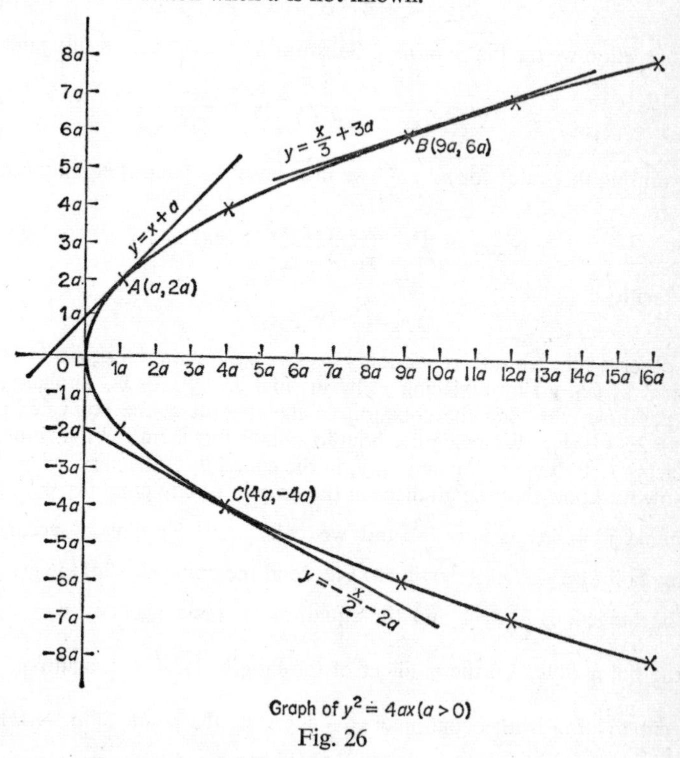

Graph of $y^2 = 4ax\ (a > 0)$

Fig. 26

The Equation of a Tangent to the Parabola $y^2 = 4ax$

We wish to find the equation of the tangent to $y^2 = 4ax$ at the point (x_1, y_1) by considering the intersections with the line $y = mx + c$. At the points of intersection we must have

$$4ax = y^2 \quad \text{(i)}$$

$$mx = y - c \quad \text{(ii)}$$

From m(i) $-4a$ (ii) we get $my^2 - 4ay + 4ac = 0$, which gives the two values of y for the two points of intersection. These points are coincident if $(-4a)^2 = 4m(4ac)$ (page 21)

$$\therefore\ a = cm, \quad c = \frac{a}{m}$$

Hence the equation of a tangent to $y^2 = 4ax$ must be of the form $y = mx + \frac{a}{m}$. $(a = cm)$

We now apply the condition that this line must be a tangent at the point (x_1, y_1).

The point (x_1, y_1) is on the line

$$\therefore \quad y_1 = mx_1 + c \qquad \text{(iii)}$$

The point (x_1, y_1) is on the parabola

$$\therefore \quad y_1^2 = 4\,cmx_1 \quad (a = cm) \qquad \text{(iv)}$$

By inspection we see that $c = mx_1$. Substitution of $c = mx_1$ in (iii) gains

$$m = \frac{y_1}{2x_1} = \frac{2ay_1}{4ax_1} = \frac{2ay_1}{y_1^2} = \frac{2a}{y_1}$$

Substituting this value for $m = \frac{2a}{y_1}$ we now have the tangent equations as

$$y = mx + \frac{a}{m} = \frac{2ax}{y_1} + \frac{2ax_1}{y_1} = \frac{2a}{y_1}(x + x_1)$$

and finally

$$yy_1 = 2a(x + x_1)$$

We observe that from the equation $y^2 = 4ax$ we obtain the equation of the tangent at (x_1, y_1) by replacing y^2 by yy_1 and $2x$ by $x + x_1$. Similarly, for the parabola $x^2 = 4ay$ the equation of the tangent at the point (x_1, y_1) is $xx_1 = 2a(y + y_1)$. We recognise how to obtain this form for the future by replacing x^2 by xx_1 and $2y$ by $y + y_1$ in the equation of the curve $x^2 = 4ay$.

Now we know that the gradient of the tangent at any point (x_1, y_1) of the parabola $y^2 = 4ax$ is $m = \frac{2a}{y_1}$ and we can draw the tangent accurately at any point we please. Referring to Fig. 26 at the point $A(a, 2a)$ the gradient of the tangent is $\frac{2a}{2a} = 1$ and the equation of this tangent is $y = x + a$. At the point $B(9a, 6a)$ the gradient of the tangent is $\frac{2a}{6a} = \frac{1}{3}$, again an easy tangent to draw with equation $y = \frac{x}{3} + 3a$. At the point $C(4a, -4a)$ the tangent has a gradient $-\frac{2a}{4a} = -\frac{1}{2}$ and its equation is $y = -\frac{x}{2} - 2a$.

Example 5. Find the equation of the tangent at the point (25, 15) on the curve

$$y^2 = 9x$$

Solution. Comparing $y^2 = 9x$ with $y^2 = 4ax$ we see that $a = 2\frac{1}{4}$. Substituting $x_1 = 25$, $y_1 = 15$, in the result $yy_1 = 2a(x + x_1)$ the equation of the tangent is therefore

$$15y = 4\tfrac{1}{2}(x + 25)$$

or

$$30y = 9(x + 25) \qquad \text{Answer}$$

Alternatively, write $y^2 = \frac{9}{2}(2x)$. Replace y^2 by yy_1 and $2x$ by $x + x_1$ to obtain the equation of the tangent as $yy_1 = \frac{9}{2}(x + x_1)$.

Returning briefly to three previous questions, consider first of all Fig. 25 on page 81—the graph of $y = x^2$. If we compare this with $x^2 = 4ay$ we see that after taking $a = \frac{1}{4}$ we get the equation of the tangent at the point $T(3, 9)$ as $xx_1 = 2a(y + y_1)$; to become

$$3x = \tfrac{1}{2}(y + 9)$$

i.e.

$$y = 6x - 9 \text{ (page 82) as before}$$

Secondly, consider Question 3, Exercise 7.2 on page 82. Here the parabola was $y = x^2$, i.e. $a = \frac{1}{4}$, and the equation of the tangent $xx_1 = 2a(y + y_1)$. At $(-3, 9)$ this becomes $-3x = \frac{1}{2}(y + 9)$, i.e. $y = -6x - 9$.

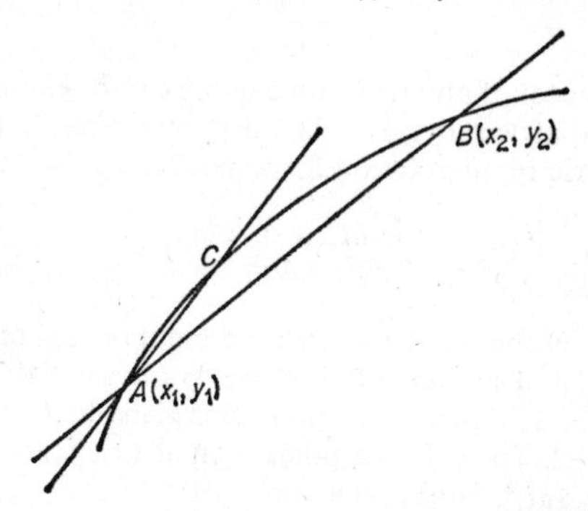

Fig. 27

Thirdly, consider Question 5, Exercise 7.2, where $xx_1 = 2a(y + y_1)$ at $(-4, 16)$ becomes $-4x = \frac{1}{2}(y + 16)$, i.e. $y = -8x - 16$ as before.

An alternative method for finding the equation of the tangent at the point (x_1, y_1) on $y^2 = 4ax$ uses a more general idea which we shall discuss in the section on differentiation.

Consider the part of the parabola in Fig. 27 on which two points $A(x_1, y_1)$, $B(x_2, y_2)$ have been taken. The equation of the line AB may be found from the two expressions for the gradient of the line, namely

$$\frac{y - y_1}{x - x_1} = \frac{y_2 - y_1}{x_2 - x_1} \qquad \text{(i)}$$

where (x, y) is a general point on AB (see again page 89).
Since this is the parabola $y^2 = 4ax$ then $y_2{}^2 = 4ax_2$ and $y_1{}^2 = 4ax_1$

$$\therefore y_2{}^2 - y_1{}^2 = (y_2 - y_1)(y_2 + y_1) = 4a(x_2 - x_1)$$

Substituting this result for $x_2 - x_1$ in (i) we get

$$\frac{y - y_1}{x - x_1} = \frac{(y_2 - y_1)4a}{(y_2 - y_1)(y_2 + y_1)}$$

$$\therefore \quad y - y_1 = \frac{4a(x - x_1)}{y_2 + y_1} \qquad \text{(ii)}$$

Now move B to position C closer to A. Clearly the line AC is nearer to becoming the tangent at A the closer we take C to A. We consider that we have made AB the tangent at A when we have made B the same point as A (i.e.

the tangent at A being a line which intersects the curve in two coincident points at A). As B tends to A so y_2 tends to y_1 and thereby (ii) above represents the tangent at A when $y_2 = y_1$

$$\therefore y - y_1 = \frac{2a}{y_1}(x - x_1) \text{ is the required tangent at } (x_1, y_1)$$

Rearrangement yields $yy_1 - y_1^2 = 2ax - 2ax_1$, and since $y_1^2 = 4ax_1$ we get $yy_1 = 2a(x + x_1)$ as before (page 84). Note again that the gradient of the tangent is $2a/y_1$.

Parameters

When the coordinates of any point on a curve can be expressed in terms of one variable, then that variable is called a parameter. The usual and most convenient parametric representation for a parabola $y^2 = 4ax$ is

$$x = at^2, y = 2at$$

i.e.

$$y^2 = (2at)^2 = 4a^2t^2 = 4ax$$

Thus we may speak of the point $t = 1$ to represent $(a, 2a)$ or the point $t = 3$ to represent $(9a, 6a)$ and so on. In Fig. 26 we would say that the point A was $t = 1$ or given by $t = 1$, similarly the point B is given by $t = 3$, and the point C is given by $t = -2$. To replace a point such as (x_1, y_1) we usually suggest the point $t = t_1$, i.e. $(at_1^2, 2at_1)$ and so on.

The Equation of the Tangent in Parametric Form

Using the parametric form above we recall that the gradient of the tangent at the point (x_1, y_1), being $m = \frac{2a}{y_1}$, may now be replaced by $\frac{2a}{2at_1} = \frac{1}{t_1}$, which means that the equation of the tangent at (x_1, y_1), $y = mx + \frac{a}{m}$ may be rewritten in the form $y = \frac{x}{t_1} + at_1$. If we multiply this last equation by t_1 we get $yt_1 = x + at_1^2$.

The next example reveals a very interesting property of the parabola.

Example 6. Find the equation of the normal at the point $(at^2, 2at)$ on the parabola $y^2 = 4ax$. Find also the point of intersection of the normal with the x axis.

Solution. Let the problem be illustrated by Fig. 28. The equation of the tangent at $P(at^2, 2at)$ is $y = x/t + at$, with gradient $1/t$. The normal at P, being perpendicular to the tangent, must have a gradient of $-t$. The equation of the normal therefore is $y = -tx + c$, where c is to be found by making this line pass through $P(at^2, 2at)$

$$\therefore \quad 2at = -tat^2 + c, \text{ hence } c = at^3 + 2at$$

Hence the normal at $P(at^2, 2at)$ is given by $y = -tx + 2at + at^3$. This line intersects the x axis at Q where $y = 0$, $x = 2a + at^2$; i.e. the point $Q(2a + at^2, 0)$.

Now the coordinates of N, the foot of the perpendicular from P to the x axis, are $x = at^2, y = 0$; consequently $NQ = OQ - ON = 2a$, which is a constant distance no matter where we take the point P.

The distance NQ is called the **subnormal** and the parabola is known for this property of having a constant subnormal.

Parameters for the Circle

Using the result $\cos^2\theta + \sin^2\theta = 1$ for any angle θ, consider the equation of a circle $x^2 + y^2 = a^2$ having the origin as centre and radius a. By putting $x = a\cos\theta$ and $y = a\sin\theta$ we have $a^2\cos^2\theta + a^2\sin^2\theta = a^2$ to indicate that θ may be used as a parameter for any point on the circle.

The equation of the tangent to the circle $x^2 + y^2 = a^2$ at the point (x_1, y_1) was $xx_1 + yy_1 = a^2$ (page 73 and page 280, Example 6.4, No. 9).

With $x_1 = a\cos\theta_1$ and $y_1 = a\sin\theta_1$ this equation now becomes $x\cos\theta_1 + y\sin\theta_1 = a$. However, the use of a parameter to represent points is not always a gain and it would be foolish to consider plotting curves only if a

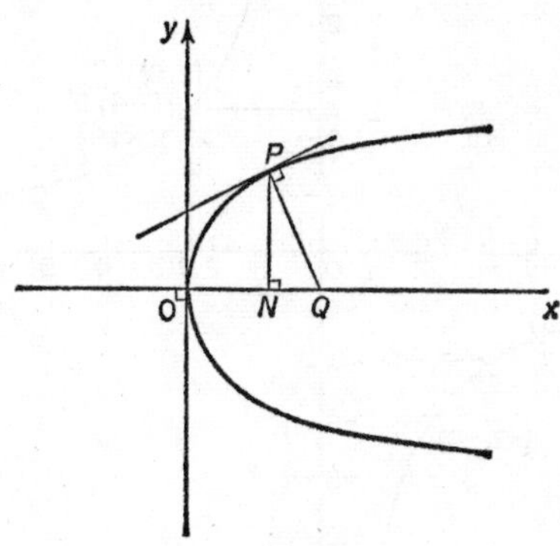

Fig. 28

parameter could be found. For example, the rectangular hyperbola given by $xy = k^2$ allows a parameter t with $x = kt$, $y = k/t$ to represent the points, but $y = k^2/x$ is quite adequate for finding anything we need at this stage.

The Rectangular Hyperbola $xy = k^2$

Just as the straight line $y = mx$ through the origin expresses the fact that y is directly proportional to x so the rectangular hyperbola $y = k^2/x$ expresses the fact that is y directly proportional to $1/x$, or in other words y is inversely proportional to x. We shall plot the graph of $xy = 12$ or $y = 12/x$ for $-8 \leqslant x \leqslant 8$, using the following table of results in Fig. 29. Thus $k^2 = 12$.

x	± 1	± 2	± 3	± 4	± 6	± 8
$y = 12/x$	± 12	± 6	± 4	± 3	± 2	$\pm 1{\cdot}5$

(1) We notice that the graph has two branches: one in the first quadrant and one in the third.

(2) The graph is symmetrical with respect to the origin; that is, a point on one branch has a mirror image on the other branch when reflected in the origin, alternatively from $y = f(x)$; $-y = f(-x)$, so that if a point (h, k) lies on the graph then so does a point $(-h, -k)$. Alternatively, consider the points in which the line $y = mx$ intersects the hyperbola. These points are $(+\sqrt{(12/m)}, +\sqrt{12m})$ and $(-\sqrt{(12/m)}, -\sqrt{12m})$.

(3) If the curve was continued in all four directions it would get closer to,

but never quite touch, the x and y axes. (The curve is said to touch the axes at infinity, and the axes are called the **asymptotes** of the hyperbola.) If we choose any point $P(x, y)$ on the curve and then draw PA and PB perpendicular to the axes as shown in Fig. 29, we have $AP = x$, $BP = y$ and the area of the rectangle $OAPB$ is $AP.BP = xy = 12$, so that no matter where P is chosen on the curve the rectangle $OAPB$ will always have an area of 12 square units.

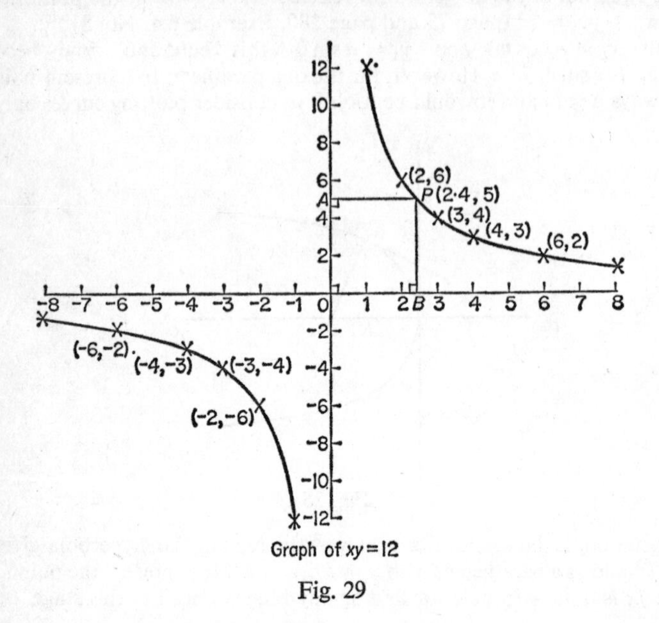

Fig. 29

(4) The curve $xy = k^2$ is symmetrical about the line $y = x$. Such symmetry in a locus is revealed by each point $Q(u, v)$ of the locus, having an image point $R(v, u)$ on the locus. For example, the locus $xy = 12$ of Fig. 29 contains the two points (3, 4), (4, 3) and the two points (2, 6), (6, 2) and the general points $(x, 12/x)$, $(12/x, x)$. If the line $y = x$ be drawn on Fig. 29 the curve will not look symmetrical about the line because the x and y scales are not the same. The symmetry is apparent if $y = x$ is drawn on Fig. 30.

(5) The curve $xy = k^2$ is also symmetrical about the line $y = -x$. Such symmetry in a locus is revealed by each point $D(m, n)$ of the locus having an image point $E(-m, -n)$ on the locus. For example, the locus $xy = 12$ of Fig. 29 contains the two points (3, 4), (−3, −4) and the two points (2, 6), (−2, −6) and the general points $(x, 12/x)$, $(-x, -12/x)$. If the line $y = -x$ be drawn on Fig. 29 the curve will not look symmetrical about the line because the x and y scales are not the same. The symmetry is apparent if $y = -x$ is drawn on Fig. 30.

Equation of a Tangent to the Rectangular Hyperbola $xy = k^2$

We shall use the same method to obtain the equation of the tangent as for the parabola. First consider the equation of the line joining two points on the hyperbola, $L(x_1, y_1)$ and $M(x_2, y_2)$, by writing down two expressions for its gradient, using (x, y) as a general point on the line, i.e.

$$\frac{y - y_1}{x - x_1} = \frac{y_2 - y_1}{x_2 - x_1}$$

But $$y_2 - y_1 = \frac{k^2}{x_2} - \frac{k^2}{x_1} = k^2\left(\frac{1}{x_2} - \frac{1}{x_1}\right) = -\frac{k^2}{x_1x_2}(x_2 - x_1)$$

$$\therefore \frac{y - y_1}{x - x_1} = -\frac{k^2(x_2 - x_1)}{x_1x_2(x_2 - x_1)} = -\frac{k^2}{x_1x_2}$$

For LM to become a tangent at L we make M move along the curve to coincide with L, and put $x_2 = x_1$, $k^2 = x_1y_1$ in the above equation.

$$\therefore \frac{y - y_1}{x - x_1} = -\frac{x_1y_1}{x_1^2} = -\frac{y_1}{x_1}$$

$$\therefore x_1y - x_1y_1 = -y_1x + y_1x_1$$

i.e. $$x_1y + xy_1 = 2x_1y_1 = 2k^2$$

is the equation of the tangent at the point (x_1, y_1).

Example 7. Find the solution of the two simultaneous equations

$$x^2 + y^2 = 13 \quad \text{(i)}$$
$$xy = 6 \quad \text{(ii)}$$

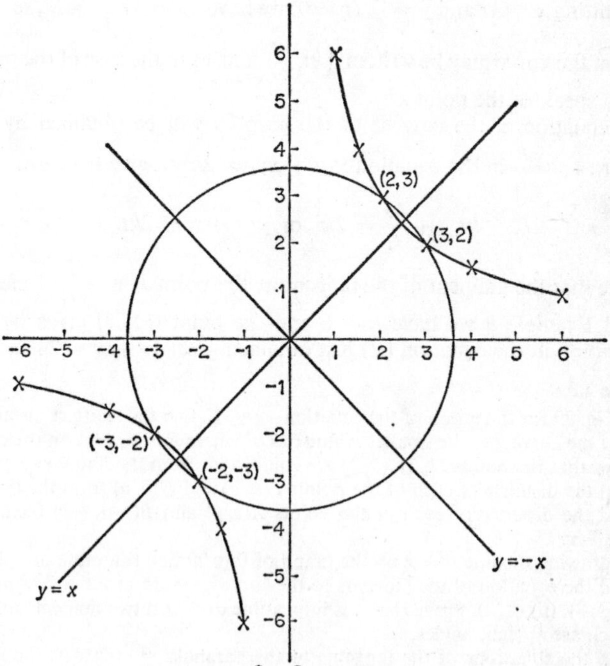

Graph of $x^2 + y^2 = 13$ and $xy = 6$

Fig. 30

Solution. The solutions will be given by the points of intersection of the circle (i) and the rectangular hyperbola (ii) as seen in Fig. 30. The circle is drawn not by attempting a radius of $\sqrt{13}$ but by opening the compasses from (0, 0) to (2, 3).

The solutions are $x = 2$, $y = 3$; $x = 3$, $y = 2$; $x = -2$, $y = -3$; $x = -3$, $y = -2$. Answer

Example 8. Prove that the tangents to $yx = k^2$ at the points (x_1, y_1) and $(-x_1, -y_1)$ are parallel.

Solution. Using the formula we have just obtained the tangents are

$$x_1y + xy_1 = 2k^2 \quad \text{(i)}$$

and

$$-x_1y - xy_1 = 2k^2 \quad \text{(ii)}$$

Both tangents have a gradient of $-\frac{y_1}{x_1}$ and are therefore parallel.

Example 9. Find the equations of the tangent and normal at the point (12, 3) on the rectangular hyperbola $xy = 36$.

Solution. The equation of the tangent at (x_1, y_1) is $x_1y + xy_1 = 2 \times 36$. At (12, 3) this becomes $12y + 3x = 72$ or $4y + x = 24$. Answer

The normal is perpendicular to the tangent and therefore has a gradient of 4. The equation of the normal at (12, 3) is given by $\frac{y-3}{x-12} = 4$

i.e. $$y - 3 = 4x - 48 \quad \text{or} \quad y = 4x - 45.$$ Answer

Parameters for the Rectangular Hyperbola $xy = k^2$

By putting $x = kt$ and $y = \frac{k}{t}$, $(t \neq 0)$ we have $xy = kt.\frac{k}{t} = k^2$, so that any point on the curve may be written $\left(kt, \frac{k}{t}\right)$ and, as in the case of the parabola, we may speak of the point t.

The equation of the tangent at the point t_1 will be obtained by writing $x_1 = kt_1$, $y_1 = \frac{k}{t_1}$ in the equation $x_1y + xy_1 = 2k^2$ to give the form

$$kt_1y + \frac{kx}{t_1} = 2k^2 \quad \text{or} \quad yt_1^2 + x = 2kt_1 \quad (A)$$

Observe that the gradient of the tangent at the point t_1 is $-\frac{1}{t_1^2}$. Referring to worked Example 9 we have $k = 6$ and the point (12, 3) given by $t_1 = 2$. Substitution in the equation (A) just obtained gives $4y + x = 24$ as before.

Exercise 7.3

1. In Fig. 29 for the graph of the equation $xy = 12$ find the shortest distance from O to the curve, i.e. the smallest value of OP where P is a point on the curve.
2. Show that the points (2, 3), (3, 2) are equidistant from the line $y = x$.
3. Find the distance of each of the points $P(a, b)$ and $Q(b, a)$ from the line $y = x$.
4. Find the distance of each of the points $A(a, b)$ and $B(-a, -b)$ from the line $y = -x$.
5. By drawing the line $y = x$ on the graph of Fig. 29 find the value of $\sqrt{12}$.
6. Find the equation of the tangents to the curve $xy = 16$ at the following points: (i) (2, 8), (ii) (8, 2). Prove that it is impossible to obtain two tangents to $xy = 16$ which are at right angles.
7. Find the equations of the tangents to the parabola $y^2 = 20x$ at the following points: (i) (5, 10), (ii) (5, −10). Show that these tangents are at right angles and intersect at the point (−5, 0).

8. Find the equations of the tangent and normal at the point $t = 3$, $(kt, k/t)$ on the rectangular hyperbola $xy = 25$.
9. Find the equations of the two tangents to the hyperbola $xy = 1$ which are parallel to the line $y + x = 0$. Also find the points at which the tangents touch $xy = 1$.
10. Prove that any point on the parabola $y^2 = 4ax$ is the same distance from the point (9, 0) as it is from the line $x = -a$.

CHAPTER EIGHT

LINEAR EQUATIONS IN THREE OR FOUR VARIABLES

Unique Solutions

We have considered systems of linear equations in only two variables x and y in Chapter One and have extended our ideas to systems of equations of the second degree, first algebraically and then diagrammatically, in order to clarify the source and number of our solutions. We now return to linear equations but, instead of recruiting coordinate geometry to help us, we shall turn to a new idea of determinants. An equation in three variables x, y and z still permits a representation in three-dimensional geometry. However, when dealing with four or more variables we shall have to resort to the purely algebraic means of solution, which depend on the methods of substitution and comparison encountered in Chapter One.

For the moment we shall consider the sets of equations to be **determinate**, i.e. to have a unique solution. To recall the earlier method consider the next example.

Example 1. Solve the following equations:

$$3x + 3y - 2z - w = 40 \qquad \text{(i)}$$
$$2x + 3y - z - w = 35 \qquad \text{(ii)}$$
$$3x + y + 4z - 2w = 15 \qquad \text{(iii)}$$
$$-x + y + z + w = 35 \qquad \text{(iv)}$$

Solution. The general idea is to combine various multiples of the equations in order to eliminate the variables one by one until we are left with just two simultaneous equations, in two unknowns, which we can solve in the usual manner. So we start by comparing the equations and notice that subtracting (ii) from (i) will remove y and w from the result. We pursue this method of elimination with a view to producing equations in fewer unknowns.

(i) − (ii) gives $\quad x - z = 5 \qquad$ (v)
(ii) + (iv) gives $\quad x + 4y = 70 \qquad$ (vi)
2(i) − (iii) gives $\quad 3x + 5y - 8z = 65 \qquad$ (vii)

We now have three equations from which w is missing, or has been eliminated.

(vii) − 8(v) gives $\quad -5x + 5y = 25$ or $-x + y = 5 \qquad$ (viii)

We now have two equations in the two variables x and y, namely (vi) and (viii)

(vi) + (viii) gives $\quad 5y = 75$ and $y = 15$

Substitution of $y = 15$ in (vi) gives $x + 60 = 70$ and $x = 10$
Substitution of $x = 10$ in (v) gives $z = 5$
Substitution for x, y and z in (iv) gives $-10 + 15 + 5 + w = 35$ and $w = 25$

$$x = 10, y = 15, z = 5, w = 25 \qquad \text{Answer}$$

Example 2. When two motors of type x and one motor of type y are working against one motor of type z the net power generated is 10 units. When one motor of type x

and three motors of type y are working against one motor of type z the net power generated is 30 units. But, when three motors of type x and three motors of type y are working against two motors of type z the net power generated is 25 units. What is the power generated by each type of motor?

Solution. The three above conditions may be expressed in the following three equations, where x, y and z are the powers of the respective motors

$$2x + y - z = 10 \quad \text{(i)}$$
$$x + 3y - z = 30 \quad \text{(ii)}$$
$$3x + 3y - 2z = 25 \quad \text{(iii)}$$

(i) − (ii) gives $x - 2y = -20$ (iv)
2(ii) − (iii) gives $-x + 3y = 35$ (v)
(iv) + (v) gives $y = 15$

Substitution of $y = 15$ in (iv) gives $x = 30 - 20 = 10$
Substitution for x and y in (i) gives $20 + 15 - z = 10 \quad \therefore \quad z = 25$

$$x = 10, y = 15, z = 25 \text{ units of power}$$ Answer

We have now solved two sets of equations. Four equations in four variables followed by three equations in three variables and each time we have obtained a unique (only one) solution.

Consistency and Dependence

We have seen before (page 6) that a unique solution may not always be possible, for two reasons 1. the equations may not be consistent and 2. some of the equations may be linearly dependent.

An illustration of case 1 is the following set of equations.

$$x + y + z = 10 \quad \text{(i)}$$
$$x + 2y + z = 12 \quad \text{(ii)}$$
$$3x + 3y + 3z = 14 \quad \text{(iii)}$$

Here, (iii) is inconsistent with (i), for if (i) is true then (iii) should read $3x + 3y + 3z = 30$. We say that this pair of equations is inconsistent.

An illustration of case 2 is the following set of equations.

$$x + 2y + 3z = 14 \quad \text{(i)}$$
$$2x - 3y + 5z = 11 \quad \text{(ii)}$$
$$4x + y + 11z = 39 \quad \text{(iii)}$$

Here, (iii) is a linear combination of 2(i) + (ii) and therefore gives no new information after the first two equations. So we really have two equations in three unknowns. If we set about solving them in the usual way we get the following results

2(i) − (ii) gives $7y + z = 17 \quad z = 17 - 7y$
Substituting this result in (i) $x + 2y + 51 - 21y = 14$
$$x = 19y - 37$$

Putting $y = t$ and using t as a parameter for the solutions we have

$$x = 19t - 37, y = t, z = 17 - 7t$$

Since any value of t gives a fresh set of solutions it follows that the two equations in three unknowns have an unlimited number of solutions.

Exercise 8.1

1. Three different types of unknown resistances are given. The only meter available will read large numbers only so that a combination of the resistances has to be read. The experimenter takes the following readings:

$$x + y + z = 10$$
$$x + 2y + z = 13$$
$$x + y + 2z = 15$$

Find each resistance.

Solve the following sets of simultaneous equations:

2.
$$x + 2y - z + 2w = -6 \quad \text{(i)}$$
$$2x + y + 3z - 4w = 15 \quad \text{(ii)}$$
$$x + 3y + 2z + 3w = -10 \quad \text{(iii)}$$
$$-3x + 4y + 3z + 2w = -16 \quad \text{(iv)}$$

3.
$$3x + y - z + w = 1 \quad \text{(i)}$$
$$4x + y + z - 2w = 1 \quad \text{(ii)}$$
$$2x - y - 3z + w = 2 \quad \text{(iii)}$$
$$x + y + 3z - 2w = 0 \quad \text{(iv)}$$

4.
$$2x + 3y + z + w = 5 \quad \text{(i)}$$
$$x + 2y + 2z + 3w = 7 \quad \text{(ii)}$$
$$3x + 4y - z + w = -4 \quad \text{(iii)}$$
$$x + 4y + 8z + 12w = 26 \quad \text{(iv)}$$

5.
$$2x - 3y - z - w = 1 \quad \text{(i)}$$
$$4x + 2z + 3w = 5 \quad \text{(ii)}$$
$$2x - 6y + 3z + 4w = -4 \quad \text{(iii)}$$
$$3x + 6y + z + w = 3 \quad \text{(iv)}$$

6.
$$x - y + z + w = 13 \quad \text{(i)}$$
$$2x + y + 2z + w = 1 \quad \text{(ii)}$$
$$-x + y + 3z + w = -2 \quad \text{(iii)}$$
$$3x + 2y + 5z + 3w = 5 \quad \text{(iv)}$$

7. Show that the second equation in the following set is a linear combination of the other two equations. Obtain the solutions of the equations as functions of a parameter t by putting $z = t$:

$$2x + 4y - z = 75 \quad \text{(i)}$$
$$x + 4y = 70 \quad \text{(ii)}$$
$$3x + 4y - 2z = 80 \quad \text{(iii)}$$

8. Show that the following four equations are not linearly independent. (Hint: Examine the relationship between (i), (iii) and (iv).) Solve the independent equations using the parameter $y = t$ and find the solution when $t = 1$:

$$x + 2y - 3z + w = 10 \quad \text{(i)}$$
$$2x + 5y - z - 7w = -33 \quad \text{(ii)}$$
$$x + 4y - 7z + 7w = 42 \quad \text{(iii)}$$
$$x + 3y - 5z + 4w = 26 \quad \text{(iv)}$$

9. Show that the following set of equations are inconsistent:

$$2x + y - z = 10 \quad \text{(i)}$$
$$x + 3y - z = 30 \quad \text{(ii)}$$
$$-3x - 4y + 2z = 20 \quad \text{(iii)}$$

10. Three motors acting together give a total power of 6 units. Two of them exactly counterbalance the third. Is there only one answer to the power rating? What if it is known that each has an integer for its power rating?

Geometric Interpretation

Just as a straight line depicts all possible solutions of a linear equation in two variables so a flat plane depicts all possible solutions of a linear equation in three variables. We use a two-dimensional system of rectangular coordinates for working with two variables and a three-dimensional system of rectangular coordinates for working with three variables. In the latter case the x, y and z axes are three lines mutually perpendicular at their point of intersection, the origin.

The graphical identification of the notions of inconsistency and independence of equations in three variables is so simple and illuminating that it is worth discussing it at this stage without indulging in the study of three-dimensional coordinate geometry wherein a coordinate representation of individual points is essential. All we need to remember at this moment is that:

(i) Any two straight lines intersect in a point. Two parallel planes will be said to intersect in a point at infinity and we shall be unable to represent their point of intersection with finite numbers.

(ii) Any two planes intersect in a straight line. Two parallel planes will be said to intersect in a line at infinity and we shall be unable to obtain a finite and meaningful representation of this line at our stage.

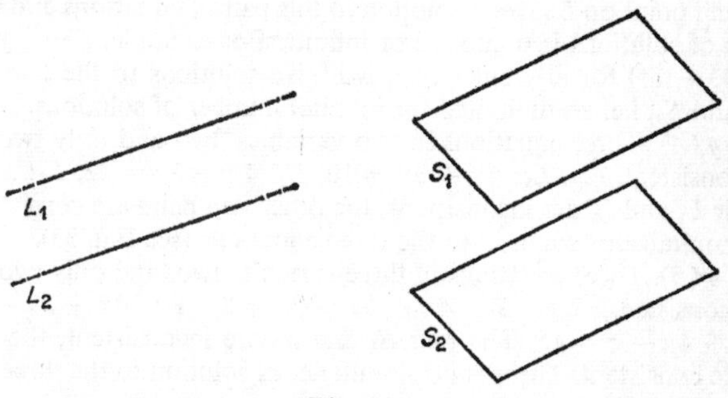

Fig. 31

It follows that three planes will intersect in a point, such as the vertex of a tetrahedron; or in three parallel lines, e.g. the edges of a prism.

Consider the following cases:

Case 1 (*A*). Two dependent or equivalent linear equations in two variables, e.g. $4x + 3y = 10$ and $12x + 9y = 30$, are represented by all the points on the same line.

Case 1 (*B*). Two dependent or equivalent linear equations in three variables, e.g. $4x + 5y - 2z = 8$ and $-8x - 10y + 4z = -16$, are represented by all the points on the same plane.

Case 2 (*A*). Two inconsistent linear equations in two variables, e.g. L_1: $4x + 3y = 10$ and L_2: $4x + 3y = 22$, are represented by a pair of parallel straight lines, L_1 and L_2. There is no solution for the system (see Fig. 31).

Case 2 (*B*). Two inconsistent linear equations in three variables, e.g. S_1: $4x + 5y - 2z = 8$ and S_2: $4x + 5y - 2z = 20$, are represented by a pair of parallel planes, S_1 and S_2. There is no solution for the system (see Fig. 31).

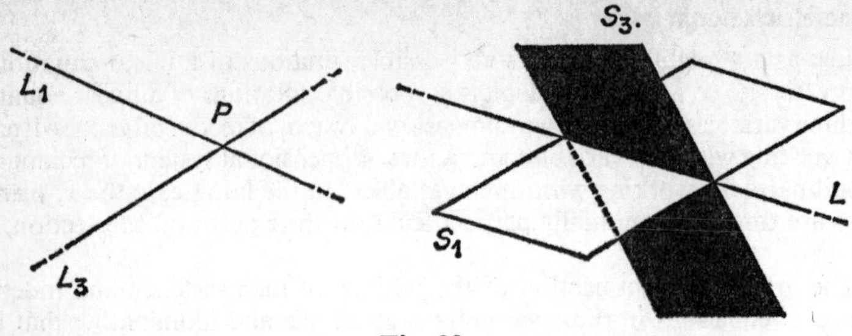

Fig. 32

In both cases, the fact that a system of two inconsistent linear equations has no common solution corresponds to the geometric fact of saying that their parallel graphs have no point in common.

Case 3 (A). Two consistent and independent linear equations in two variables, e.g. L_1: $4x + 3y = 10$ and L_3: $x + y = 9$, are represented by two lines which intersect in one, and only one point (see Fig. 32).

Case 3 (B). Two consistent and independent linear equations in three variables, e.g. S_1: $4x + 5y - 2z = 8$ and S_3: $4x + 4y - z = 15$, are represented by two planes which intersect in one, and only one, line L (see Fig. 32). Thus each point on L gives a solution to this pair of equations and clearly the number of solutions is unlimited or infinite. For example, $z = t$, $y = t - 7$, $x = \frac{1}{4}(43 - 3t)$ for all values of t, will give solutions to the two equations for S_1 and S_3, i.e. an unlimited (or infinite) number of solutions.

Case 4 (A). Three equations in two variables, two and only two, of which are inconsistent, e.g. L_1: $4x + 3y = 10$, L_2: $4x + 3y = 22$, L_3: $x + y = 9$. The pair L_1 and L_2 are inconsistent, the other two pairs are consistent. There is no simultaneous solution to the three equations (see Fig. 33).

Case 4 (B). Three equations in three variables two, and only two, of which are inconsistent, e.g. S_1: $4x + 5y - 2z = 8$, S_2: $4x + 5y - 2z = 20$, S_3: $4x + 4y - z = 15$. The pair S_1 and S_2 are inconsistent, the other two pairs are consistent. There is no simultaneous solution to the three equations (see Fig. 33).

Case 5 (A). Three consistent equations in two variables, one of which is a linear combination of the other two, have a unique solution, e.g.

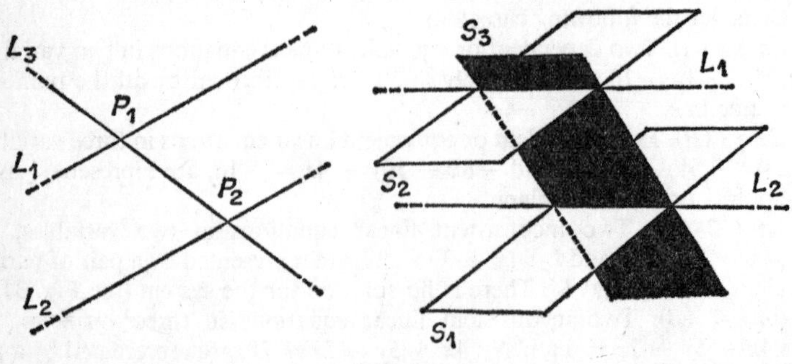

Fig. 33

L_1: $4x + 3y = 10$, L_3: $x + y = 9$, L_4: $-2x + y = 60$ which all intersect at the point $(-17, 26)$. We can obtain the equation of L_4 from $-3(4x + 3y = 10) + 10(x + y = 9)$ (see Fig. 34).

Case 5 (*B*). Three consistent equations in three variables one of which is a linear combination of the other two, e.g. S_1: $4x + 5y - 2z = 8$, S_3: $4x + 4y - z = 15$, S_5: $8x + 7y - z = 37$ (i.e. the equation of S_5 is given by $3(4x + 4y - z = 15) - (4x + 5y - 2z = 8)$ have an unlimited number of solutions given by all the points on the common line L (see Fig. 34).

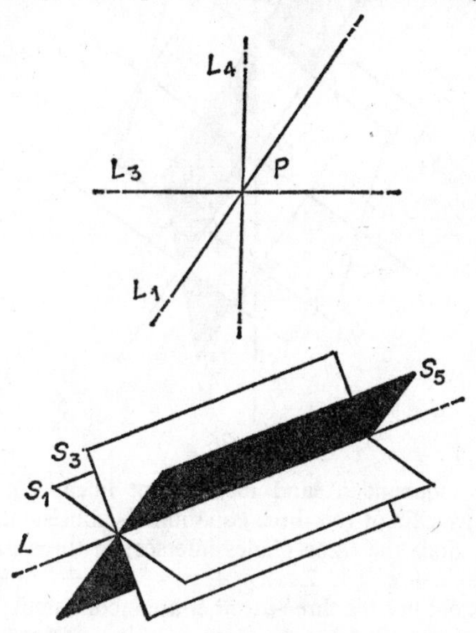

Fig. 34

Case 6 (*A*). Three consistent and independent linear equations in two variables have no simultaneous solution, being represented by three straight lines which form the sides of a triangle, e.g. L_1: $4x + 3y = 10$, L_3: $x + y = 9$, L_5: $2x - y = 6$, which intersect in three points $P_1(-17, 26)$, $P_2(5, 4)$, $P_3(2\frac{4}{5}, -\frac{2}{5})$ (see Fig. 35).

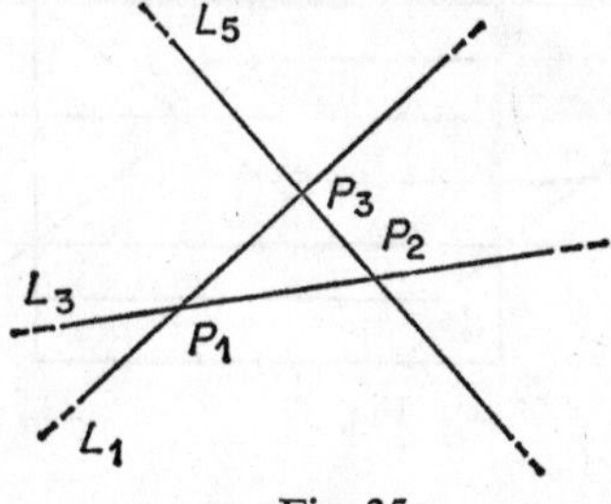

Fig. 35

Case 6 (*B*). Three consistent and independent linear equations in three variables are represented by three planes which intersect in a single point which gives the unique solution, e.g. S_1: $4x + 5y - 2z = 8$, S_3: $4x + 4y - z = 15$, S_4: $x + y - 2z = 16$, which all intersect in the 'point' given by $x = 16$, $y = -14$, $z = -7$ (see Fig. 36).

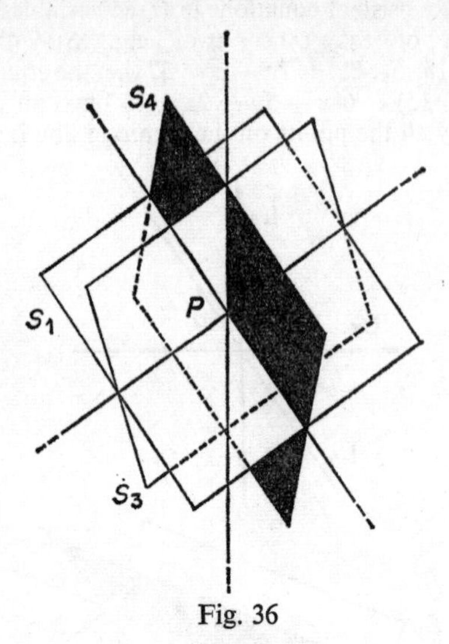

Fig. 36

Case 7. Three independent and inconsistent linear equations in three variables where any pair of the three equations is consistent have no simultaneous solution, since the three planes intersect in three parallel lines (see Fig. 37), e.g. $x + y + z = 1$, $2x + y + z = 2$, $3x + 2y + 2z = 5$, where each pair is consistent but the three together are inconsistent; i.e. add the first two and subtract the result from the third equation, to get $0 = 2$!

Solution of the first pair of equations is given by $x = 1$, $y = t$, $z = -t$, where t is a parameter. Substitution in the third equation gives $3 + 2t - 2t = 5$, i.e. $3 = 5$!

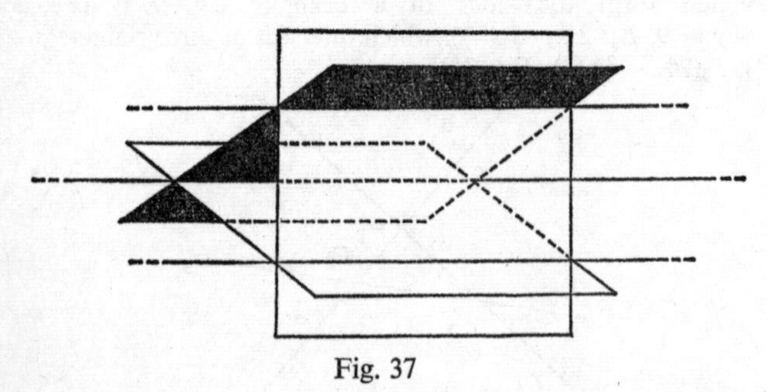
Fig. 37

Summary

The concepts of independence and consistency for systems of two equations may be generalised as the concepts of linear independence and consistency for systems of three or more equations. To be determinate, a physical problem leading to linear equations in n variables must specify n, and in general only n, conditions which are both consistent and linearly independent. The problem then has a unique solution.

The graphs of linear equations in three variables are (flat) planes, which are identical for any two dependent equations and parallel for any two inconsistent equations.

The three graph planes of a system of three linear equations in three variables intersect: (1) in a single point when the equations are linearly independent and consistent; (2) in a straight line when the equations are consistent but also linearly dependent; (3) in two parallel lines when one pair of equations is separately inconsistent but the other two pairs are separately consistent and independent; and (4) in three parallel lines when the equations are linearly independent, but also inconsistent with each pair being consistent.

These four possibilities correspond respectively to the cases in which the system of equations is: (1) determinate with a unique solution; (2) indeterminate with an unlimited (infinite) number of possible solutions; and (3 or 4) indeterminate with no possible solution (just as when each pair of equations is separately inconsistent and their parallel graph-planes do not intersect at all).

Exercise 8.2

From the following set of equations choose the selection requested in questions 1 to 4 and state which cases on pages 95–98 apply.

(i) $4x + 3y + 2z = 14$ (ii) $2x + y + z = 8$ (iii) $x + 2y + z = 10$
(iv) $3x + 3y + 2z = 12$ (v) $2x + 2y + z = 6$ (vi) $6x + 6y + 4z = 23$

1. Two inconsistent equations.
2. Three inconsistent equations each pair being consistent.
3. One equation which is a linear combination of two others, i.e. 3 linearly dependent equations.
4. Three linearly independent and consistent equations.
5. Find the solution to the simultaneous equations (ii), (iii), (v).
6. What is represented by the following three equations?

$$x + y + z = 3$$
$$x + y + z = 4$$
$$x + y + z = 7$$

7. What is represented by the following three equations?

$$x + y + z = 3$$
$$x + y + 2z = 7$$
$$2x + 2y + 3z = 12$$

CHAPTER NINE

DETERMINANTS

Second-order Determinants (Cramer's Rule)

Having derived a general formula for solving quadratic equations in one variable, we shall now derive general formulae for solving systems of linear equations in two or more variables.

Any system of two linear equations in two variables can be written in the form

$$a_1x + b_1y = k_1 \quad \text{(i)}$$
$$a_2x + b_2y = k_2 \quad \text{(ii)}$$

where x and y are the two variables and a_1, a_2, b_1, etc., are constants; for instance, two particular equations of this type are

$$x - 2y = -8$$
$$x + y = 7$$

and here the values of the constants are

$$a_1 = 1, b_1 = -2, k_1 = -8$$
$$a_2 = 1, b_2 = 1, \quad k_2 = 7$$

Applying the standard method to solve the above equations, we get

$$\text{(i)} \times b_2: \quad a_1b_2x + b_1b_2y = k_1b_2 \quad \text{(iii)}$$
$$\text{(ii)} \times b_1: \quad a_2b_1x + b_1b_2y = k_2b_1 \quad \text{(iv)}$$
$$\text{(iii)} - \text{(iv)}: \quad (a_1b_2 - a_2b_1)x = k_1b_2 - k_2b_1$$
$$\text{(i)} \times a_2: \quad a_1a_2x + a_2b_1y = a_2k_1 \quad \text{(v)}$$
$$\text{(ii)} \times a_1: \quad a_1a_2x + a_1b_2y = a_1k_2 \quad \text{(vi)}$$
$$\text{(vi)} - \text{(v)}: \quad (a_1b_2 - a_2b_1)y = a_1k_2 - a_2k_1$$

$$x = \frac{k_1b_2 - k_2b_1}{a_1b_2 - a_2b_1}$$
$$y = \frac{a_1k_2 - a_2k_1}{a_1b_2 - a_2b_1}$$

These solutions express the value of the variable x and y in terms of the constant coefficients, a_1, b_1, etc., in the preceding typical equations. Hence they are solution formulae for any pair of equations in these typical forms.

If we wish, we can solve any system of two linear equations in two variables simply by substituting the appropriate values for the constants in these solution formulae. For instance, the system of two equations in the example above may be solved by the above substitutions as follows

$$x = \frac{(-8)(1) - (-2)(7)}{(1)(1) - (-2)(1)} = \frac{-8 + 14}{1 + 2} = \frac{6}{3} = 2$$
$$y = \frac{(1)(7) - (-8)(1)}{3} = \frac{7 + 8}{3} = \frac{15}{3} = 5$$

(Note here that the denominator in the two formulae, being the same, needs to be evaluated only the first time.)

In the above forms, of course, these solution formulae are not easy to remember, but suppose we rewrite their common denominator in the form at the left side of the following formula of definition

$$\begin{vmatrix} a_1 & b_1 \\ a_2 & b_2 \end{vmatrix} = a_1b_2 - b_1a_2$$

This sort of square array of quantities between two vertical lines is called a **determinant**. The quantities, a_1, b_1, etc., in the determinant are called its **elements.** Since, in this case, there are two elements in each row and column, the determinant is called a determinant of the second order: i.e, $a_1 \quad b_1$ is the first row, $a_2 \quad b_2$ is the second row; $a_1 \quad a_2$ make the first column and $b_1 \quad b_2$ make the second column. The defined value of the determinant is given by the right-hand side of the above formula, which is called the **determinant's expansion.** The formula simply states the following rule for expanding (evaluating) a determinant of the second order: take the product of the elements on the diagonal from the upper left-hand corner to the lower right-hand corner, and subtract from it the product of the elements on the diagonal from the upper right-hand corner to the lower left-hand corner, thus:

$$\overset{(+)\qquad(-)}{\begin{vmatrix} a_1 & b_1 \\ a_2 & b_2 \end{vmatrix}} = a_1b_2 - b_1a_2, \text{ or } a_1b_2 - a_2b_1$$

For instance, substituting for a_1, b_1, etc., gives

$$\begin{vmatrix} 1 & -2 \\ 1 & 1 \end{vmatrix} = 1(1) - (-2)1 = 1 + 2 = 3$$

Moreover, when we define a second-order determinant by this expansion rule we may also write the numerators of the above solution formulae in the same form. Then the formulae become

$$x = \frac{\begin{vmatrix} k_1 & b_1 \\ k_2 & b_2 \end{vmatrix}}{\begin{vmatrix} a_1 & b_1 \\ a_2 & b_2 \end{vmatrix}}, \quad y = \frac{\begin{vmatrix} a_1 & k_1 \\ a_2 & k_2 \end{vmatrix}}{\begin{vmatrix} a_1 & b_1 \\ a_2 & b_2 \end{vmatrix}}$$

This is a special case of a more general formula known as Cramer's Rule (page 105) after the mathematician, Gabriel Cramer, who first stated it in 1760. The Rule's pattern is simple to remember if you note, after writing the equations in the form $ax + by = k$, that: (1) the common denominator determinant—called the determinant of the system or the determinant of the detached coefficients—has the same arrangement of a's and b's as the original equations; and (2) the numerator determinants have the same arrangement except that k's are substituted for the a-coefficients of x in the x-formula, and for the b-coefficients of y in the y-formula.

Since pairs of linear equations in two variables are simple to solve in any case, the greatest value of Cramer's Rule lies in its extension to systems of linear equations in more variables.

Example 1. Solve the following equations by the method of determinants.

$$17x - 13y = 11 \quad \text{(i)}$$
$$5x + 4y = 10 \quad \text{(ii)}$$

Solution.

$$x = \frac{\begin{vmatrix} 11 & -13 \\ 10 & 4 \end{vmatrix}}{\begin{vmatrix} 17 & -13 \\ 5 & 4 \end{vmatrix}} = \frac{44 - (-130)}{68 - (-65)} = \frac{174}{133}$$

$$y = \frac{\begin{vmatrix} 17 & 11 \\ 5 & 10 \end{vmatrix}}{133} = \frac{170 - 55}{133} = \frac{115}{133}$$

$$x = \frac{174}{133}, \quad y = \frac{115}{133} \qquad \text{Answer}$$

Example 2. Find the point of intersection of the pair of straight lines

$$y - 10x = 19 \quad \text{(i)}$$
$$2y - 4x = 8 \quad \text{(ii)}$$

Solution. Notice the interchange of places for y and x.

$$y = \frac{\begin{vmatrix} 19 & -10 \\ 8 & -4 \end{vmatrix}}{\begin{vmatrix} 1 & -10 \\ 2 & -4 \end{vmatrix}} = \frac{-76 - (-80)}{-4 - (-20)} = \frac{4}{16}$$

$$x = \frac{\begin{vmatrix} 1 & 19 \\ 2 & 8 \end{vmatrix}}{16} = \frac{8 - 38}{16} = \frac{-30}{16}$$

$$x = -\tfrac{15}{8}, \; y = \tfrac{1}{4} \qquad \text{Answer}$$

Observe that we do not have to consider how to combine the equations in order to eliminate either x or y. The same determinant method applies regardless of the different coefficients.

Exercise 9.1

Using the method of determinants solve the following equations

1. $7x + 4y = 13$
   $3x + 5y = 8$
2. $13x + 19y = -7$
   $7x + 11y = 3$
3. $19x + 17y = -1$
   $3x + 4y = 2$
4. $23x - 11y = 14$
   $5x - y = 9$
5. $4x - 8y = 4$
   $-9x + 6y = 3$
6. $3x + y + 1 = 0$
   $5x - 2y - 3 = 0$
7. What happens if the method of determinants is used in an attempt to solve the two inconsistent equations $x + 2y = 7$, $x + 2y = 1$?

Determinants of Higher Order

A system of three linear equations in three variables can be written in the form

$$\begin{aligned} a_1x + b_1y + c_1z &= k_1 \qquad \text{(i)} \\ a_2x + b_2y + c_2z &= k_2 \qquad \text{(ii)} \\ a_3x + b_3y + c_3z &= k_3 \qquad \text{(iii)} \end{aligned}$$

Solving by the usual method of eliminating each variable in turn we have

$$\begin{aligned} &\text{(i) } c_3 - \text{(iii) } c_1\text{: } x(a_1c_3 - a_3c_1) + y(b_1c_3 - b_3c_1) = k_1c_3 - k_3c_1 \qquad \text{(iv)} \\ &\text{(ii) } c_3 - \text{(iii) } c_2\text{: } x(a_2c_3 - a_3c_2) + y(b_2c_3 - b_3c_2) = k_2c_3 - k_3c_2 \qquad \text{(v)} \end{aligned}$$

We now eliminate y by finding

$$\text{(iv) } (b_2c_3 - b_3c_2) - \text{(v) } (b_1c_3 - b_3c_1)$$

which leaves

$$\begin{aligned} x\{(a_1c_3 - a_3c_1)(b_2c_3 - b_3c_2) - (a_2c_3 - a_3c_2)(b_1c_3 - b_3c_1)\} \\ = (k_1c_3 - k_3c_1)(b_2c_3 - b_3c_2) - (k_2c_3 - k_3c_2)(b_1c_3 - b_3c_1) \end{aligned}$$

This result reduces to

$$x = \frac{k_1b_2c_3 - k_1c_2b_3 - b_1k_2c_3 + b_1c_2k_3 + c_1k_2b_3 - c_1b_2k_3}{a_1b_2c_3 - a_1c_2b_3 - b_1a_2c_3 + b_1c_2a_3 + c_1a_2b_3 - c_1b_2a_3}$$

with similar formulae for y and z.

The first thing to notice is that the denominator consists of the detached coefficients of x, y and z in the three equations. Furthermore, it is the total number of positive and negative permutations of a, b and c. Thus, taking abc as positive one interchange of adjacent letters yields another permutation and a change in sign, $-bac$ say. A further interchange of adjacent letters gives another permutation and another change in sign. There are six permutations in all so that the sequence of changes reads as follows

$$abc, \ -bac, \ +bca, \ -cba, \ +cab, \ -acb$$

The 1, 2, 3 order of the suffixes is preserved throughout, hence the denominator consists of the sum of all the different permutations (both negative and positive) of the letters a, b and c. The second thing to notice is that the numerator is the denominator with a_1 replaced by k_1, a_2 replaced by k_2 and a_3 replaced by k_3.

Here again we can solve any system of equations in the above form simply by substituting the appropriate values for the constants in the solution formulae. We now rewrite their common denominator in the simpler form at the left side of the following definition formula

$$\begin{vmatrix} a_1 & b_1 & c_1 \\ a_2 & b_2 & c_2 \\ a_3 & b_3 & c_3 \end{vmatrix} = a_1b_2c_3 - a_1c_2b_3 - b_1a_2c_3 + b_1c_2a_3 + c_1a_2b_3 - c_1b_2a_3$$

Since the left side of this formula is also a square array of elements between two vertical lines, it is by definition a determinant and, as there are three elements in each row and column, the determinant is called a **determinant of the third order.**

Obviously, the expansion rule of a determinant of the third order is much more complicated than that for a second-order determinant. We may state it most simply if we first supplement the above determinant with its first two columns repeated as below, drawing through it the indicated diagonal lines. With reference to such a diagram, the expansion rule then becomes

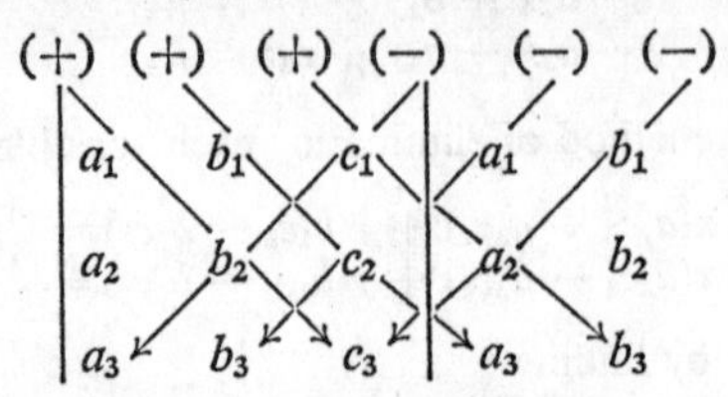

Add the products of the three elements on each of the diagonals sloping down to the right, and subtract from this sum the products of the three elements on each of the diagonals sloping down to the left.

The preceding solution formula may now be restated in determinant form as

$$x = \frac{\begin{vmatrix} k_1 & b_1 & c_1 \\ k_2 & b_2 & c_2 \\ k_3 & b_3 & c_3 \end{vmatrix}}{\begin{vmatrix} a_1 & b_1 & c_1 \\ a_2 & b_2 & c_2 \\ a_3 & b_3 & c_3 \end{vmatrix}} = \frac{|k \quad b \quad c|}{|a \quad b \quad c|}$$

The symbols in the numerator and denominator of the second fraction here are simply briefer ways of writing the corresponding typical determinants in the first fraction. This notational device cannot be used, of course, when the elements of a determinant have specific numerical values, but it is very convenient for writing general formulae for determinants expressed in subscript letters as above.

For instance, using this abbreviated notation, we can now write Cramer's Rule for the solution of a system of two linear equations in two variables as

$$x = |k \quad b| / |a \quad b|$$
$$y = |a \quad k| / |a \quad b|$$

or we can write the statement of Cramer's Rule for a system of three linear equations in three variables as

$$x = |k \quad b \quad c| / |a \quad b \quad c|$$
$$y = |a \quad k \quad c| / |a \quad b \quad c|$$
$$z = |a \quad b \quad k| / |a \quad b \quad c|$$

and, since it is shown in higher mathematics that this rule is general for a system of n linear equations in n variables, corresponding statements can be made of solution formulae for $n = 4, 5, 6$, etc. For instance, when $n = 4$

$$x = |k \quad b \quad c \quad d| / |a \quad b \quad c \quad d|$$
$$y = |a \quad k \quad c \quad d| / |a \quad b \quad c \quad d|$$
$$z = |a \quad b \quad k \quad d| / |a \quad b \quad c \quad d|$$
$$w = |a \quad b \quad c \quad k| / |a \quad b \quad c \quad d|$$

The expansion rule for a determinant of the third order is, of course, somewhat complex to be practical in an equation-solving routine. And the expansion rules of determinants of still higher orders are successively more complicated. Consequently, if determinants could be evaluated only by expansion rules they would have little practical value.

It happens, however, that there are other ways of evaluating determinants which are both simpler and quicker. We shall therefore discuss such methods first before applying determinants to the solutions of systems of linear equations by Cramer's Rules when more than two variables are involved.

Expansion by Minors

The minor of an element in a determinant is defined as the determinant of next lower order which is obtained by striking out the row and column in which the given element occurs. In the case of a third-order determinant, for instance, the minor of the element a_1 is

$$\begin{vmatrix} a_1 & b_1 & c_1 \\ a_2 & b_2 & c_2 \\ a_3 & b_3 & c_3 \end{vmatrix} = \begin{vmatrix} b_2 & c_2 \\ b_3 & c_3 \end{vmatrix}$$

The minor of the element b_2 is

$$\begin{vmatrix} a_1 & b_1 & c_1 \\ a_2 & b_2 & c_2 \\ a_3 & b_3 & c_3 \end{vmatrix} = \begin{vmatrix} a_1 & c_1 \\ a_3 & c_3 \end{vmatrix}$$

And the minor of the element c_2 is

$$\begin{vmatrix} a_1 & b_1 & c_1 \\ a_2 & b_2 & c_2 \\ a_3 & b_3 & c_3 \end{vmatrix} = \begin{vmatrix} a_1 & b_1 \\ a_3 & b_3 \end{vmatrix}$$

The reason for so defining the minor of an element is this important general expansion rule: a determinant is equal to the sum of the products formed by multiplying each element in the Nth row or column by its minor, provided the first such product is given the sign of $(-1)^{N+1}$ and each product thereafter is given an alternate sign.

Thus, if $N = 1$ or any other odd number, then $(-1)^{N+1} = 1$, and the expansion by minors begins with a plus sign. For instance, if we expand a third-order determinant by minors of the first column then $N = 1$ and $(-1)^{N+1} = (-1)^{1+1} = (-1)^2 = 1$. Hence the expansion in terms of the minors of the first column begins with a plus sign as in the formula

$$|a \quad b \quad c| = \begin{vmatrix} a_1 & b_1 & c_1 \\ a_2 & b_2 & c_2 \\ a_3 & b_3 & c_3 \end{vmatrix} = a_1 \begin{vmatrix} b_2 & c_2 \\ b_3 & c_3 \end{vmatrix} - a_2 \begin{vmatrix} b_1 & c_1 \\ b_3 & c_3 \end{vmatrix} + a_3 \begin{vmatrix} b_1 & c_1 \\ b_2 & c_2 \end{vmatrix}$$

The expansion in terms of the minors of the first row is given by

$$|a \quad b \quad c| = a_1 \begin{vmatrix} b_2 & c_2 \\ b_3 & c_3 \end{vmatrix} - b_1 \begin{vmatrix} a_2 & c_2 \\ a_3 & c_3 \end{vmatrix} + c_1 \begin{vmatrix} a_2 & b_2 \\ a_3 & b_3 \end{vmatrix}$$

But if $N = 2$ or any other even number, then $(-1)^{N+1} = -1$, and the expansion by minors begins with a minus sign. For instance, if we expand the same third-order determinant by the minors of the second row, then $N = 2$ and $(-1)^{N+1} = (-1)^{2+1} = (-1)^3 = -1$. Hence the expansion begins with a minus sign, as in the formula

$$|a \quad b \quad c| = -a_2 \begin{vmatrix} b_1 & c_1 \\ b_3 & c_3 \end{vmatrix} + b_2 \begin{vmatrix} a_1 & c_1 \\ a_3 & c_3 \end{vmatrix} - c_2 \begin{vmatrix} a_1 & b_1 \\ a_3 & b_3 \end{vmatrix}$$

We must hasten to add at this point that complete general proof for all basic determinant formulae requires complicated reasoning which is ordinarily detailed only in higher mathematics beyond the scope or intention of this book.

The question of which row or column to select for expansion depends upon the pattern of the numerical values of the elements. In general, the choice of a row or column with most zeros or small numbers will shorten the work of calculation.

Example 3. Expand the following determinant by minors of the first column, and then by minors of the second column

$$D = \begin{vmatrix} 7 & 0 & 9 \\ 32 & 1 & 9 \\ 3 & 0 & 6 \end{vmatrix}$$

Solution. By minors of the first column

$$D = 7\begin{vmatrix} 1 & 9 \\ 0 & 6 \end{vmatrix} - 32\begin{vmatrix} 0 & 9 \\ 0 & 6 \end{vmatrix} + 3\begin{vmatrix} 0 & 9 \\ 1 & 9 \end{vmatrix}$$
$$= 7(6 - 0) - 32(0 - 0) + 3(0 - 9)$$
$$= 42 - 0 - 27 = 15 \qquad \text{Answer}$$

By minors of the second column

$$D = -0 + (1)\begin{vmatrix} 7 & 9 \\ 3 & 6 \end{vmatrix} - 0$$
$$= 42 - 27 = 15 \qquad \text{Answer}$$

The procedure in the second case is obviously shorter than the first. This is because of the more judicious choice of a column by which to expand.

Example 4. Expand the following determinant by minors of the first row and also by minors of the fourth column

$$D = \begin{vmatrix} 6 & 9 & 3 & 1 \\ 4 & 2 & 7 & 0 \\ 6 & 4 & 2 & 0 \\ 9 & 1 & 9 & 1 \end{vmatrix}$$

Solution. When we delete the row and column in which an element occurs the remaining minor is a determinate of order 3

$$D = 6\begin{vmatrix} 2 & 7 & 0 \\ 4 & 2 & 0 \\ 1 & 9 & 1 \end{vmatrix} - 9\begin{vmatrix} 4 & 7 & 0 \\ 6 & 2 & 0 \\ 9 & 9 & 1 \end{vmatrix} + 3\begin{vmatrix} 4 & 2 & 0 \\ 6 & 4 & 0 \\ 9 & 1 & 1 \end{vmatrix} - 1\begin{vmatrix} 4 & 2 & 7 \\ 6 & 4 & 2 \\ 9 & 1 & 9 \end{vmatrix}$$
$$= 6\left(2\begin{vmatrix} 2 & 0 \\ 9 & 1 \end{vmatrix} - 7\begin{vmatrix} 4 & 0 \\ 1 & 1 \end{vmatrix} + 0\right)$$
$$-9\left(4\begin{vmatrix} 2 & 0 \\ 9 & 1 \end{vmatrix} - 7\begin{vmatrix} 6 & 0 \\ 9 & 1 \end{vmatrix} + 0\right)$$
$$+3\left(4\begin{vmatrix} 4 & 0 \\ 1 & 1 \end{vmatrix} - 2\begin{vmatrix} 6 & 0 \\ 9 & 1 \end{vmatrix} + 0\right)$$

$$-1\left(4\begin{vmatrix}4 & 2\\1 & 9\end{vmatrix}-2\begin{vmatrix}6 & 2\\9 & 9\end{vmatrix}+7\begin{vmatrix}6 & 4\\9 & 1\end{vmatrix}\right)$$

$$\begin{aligned}&= 6(2(2-0)-7(4-0))\\&\quad -9(4(2-0)-7(6-0))\\&\quad +3(4(4-0)-2(6-0))\\&\quad -(4(36-2)-2(54-18)+7(6-36))\\&= 6(-24)-9(-34)+3(4)-(-146)\\&= 464-144=320\end{aligned}$$

Answer

By minors of the fourth column

$$D = -\begin{vmatrix}4 & 2 & 7\\6 & 4 & 2\\9 & 1 & 9\end{vmatrix}+0-0+\begin{vmatrix}6 & 9 & 3\\4 & 2 & 7\\6 & 4 & 2\end{vmatrix}$$

$$= -\left(4\begin{vmatrix}4 & 2\\1 & 9\end{vmatrix}-2\begin{vmatrix}6 & 2\\9 & 9\end{vmatrix}+7\begin{vmatrix}6 & 4\\9 & 1\end{vmatrix}\right)$$

$$+\left(6\begin{vmatrix}2 & 7\\4 & 2\end{vmatrix}-9\begin{vmatrix}4 & 7\\6 & 2\end{vmatrix}+3\begin{vmatrix}4 & 2\\6 & 4\end{vmatrix}\right)$$

$$\begin{aligned}&= -[4(36-2)-2(54-18)+7(6-36)]\\&\quad +[6(4-28)-9(8-42)+3(16-12)]\\&= -(-146)+174=320\end{aligned}$$

Answer

Exercise 9.2

Expand the following determinants by minors of a judiciously selected row or column

1. $\begin{vmatrix}1 & 1 & 0\\3 & 3 & 1\\-2 & 3 & 1\end{vmatrix}$

2. $\begin{vmatrix}0 & 1 & -3 & 0\\5 & -2 & 2 & 4\\0 & 1 & 0 & -1\\3 & -1 & 0 & 5\end{vmatrix}$

3. $\begin{vmatrix}0 & 0 & 3 & -3\\1 & 2 & 0 & 1\\0 & 0 & 1 & 4\\1 & 1 & 5 & 2\end{vmatrix}$

4. $\begin{vmatrix}a & 0 & 0\\0 & b & 0\\0 & 0 & c\end{vmatrix}$

5. $\begin{vmatrix}0 & 0 & 1 & 1\\-1 & 3 & 5 & 0\\2 & 0 & -2 & -1\\1 & -1 & 3 & 0\end{vmatrix}$

6. $\begin{vmatrix}1-x & 2 & 0\\3 & 1 & 2+x\\0 & 4 & 0\end{vmatrix}$

Further Properties of Determinants

The calculation of a determinant's value can usually be shortened if we change its form before we expand it. We shall discuss the following five theorems for third-order determinants only, but the results are true for determinants of any order.

Theorem 1. Changing all columns to rows or all rows to columns, in the same order, leaves the value of a determinant unchanged

$$D = \begin{vmatrix}a_1 & b_1 & c_1\\a_2 & b_2 & c_2\\a_3 & b_3 & c_3\end{vmatrix} = \begin{vmatrix}a_1 & a_2 & a_3\\b_1 & b_2 & b_3\\c_1 & c_2 & c_3\end{vmatrix}$$

According to the formula for the expansion of the determinant, it makes no difference whether we expand a determinant by minors of the kth row or kth column.

This means that any theorem concerning the rows of a determinant applies equally to its columns, and vice versa.

Theorem 2. Multiplying all the elements of any row (or column) of a determinant by the same quantity multiplies the value of the determinant by that quantity, i.e.

$$N|a \quad b \quad c| = |Na \quad b \quad c| = |a \quad Nb \quad c| = |a \quad b \quad Nc|$$

Proof

$$N|a \quad b \quad c| = N\left(a_1\begin{vmatrix} b_2 & c_2 \\ b_3 & c_3 \end{vmatrix} - a_2\begin{vmatrix} b_1 & c_1 \\ b_3 & c_3 \end{vmatrix} + a_3\begin{vmatrix} b_1 & c_1 \\ b_2 & c_2 \end{vmatrix}\right)$$

$$= Na_1\begin{vmatrix} b_2 & c_2 \\ b_3 & c_3 \end{vmatrix} - Na_2\begin{vmatrix} b_1 & c_1 \\ b_3 & c_3 \end{vmatrix} + Na_3\begin{vmatrix} b_1 & c_1 \\ b_2 & c_2 \end{vmatrix}$$

$$= |Na \quad b \quad c|$$

From this theorem it follows that if we multiply all the elements in any row (or column) of a determinant by the same quantity, and then divide the entire determinant by this quantity, or vice versa, we leave the value of the determinant unchanged. The result is as follows

$$\begin{vmatrix} a & b & c \end{vmatrix} = \frac{1}{N}\begin{vmatrix} Na & b & c \end{vmatrix} = \frac{1}{N}\begin{vmatrix} a & Nb & c \end{vmatrix} = \frac{1}{N}\begin{vmatrix} a & b & Nc \end{vmatrix}$$

$$= N\begin{vmatrix} \frac{a}{N} & b & c \end{vmatrix} = N\begin{vmatrix} a & \frac{b}{N} & c \end{vmatrix} = N\begin{vmatrix} a & b & \frac{c}{N} \end{vmatrix}$$

These theorems are useful in simplifying the evaluation of determinants, for they can be applied to remove fractions or larger numbers from the rows or columns before expansion.

Example 5. Evaluate the following determinant

$$D = \begin{vmatrix} \frac{1}{4} & 2 & 3 \\ \frac{3}{4} & 0 & 7 \\ \frac{1}{2} & 0 & 5 \end{vmatrix}$$

Solution. Simplify column 1 by multiplying it by 4, and offset this operation by dividing the entire determinant by 4. Then expand by minors of the second column

$$D = \frac{1}{4}\begin{vmatrix} (4)\frac{1}{4} & 2 & 3 \\ (4)\frac{3}{4} & 0 & 7 \\ (4)\frac{1}{2} & 0 & 5 \end{vmatrix} = \frac{1}{4}\begin{vmatrix} 1 & 2 & 3 \\ 3 & 0 & 7 \\ 2 & 0 & 5 \end{vmatrix}$$

This is equivalent to extracting the factor of $\frac{1}{4}$ from the first column

$$= \frac{1}{4}\left(-2\begin{vmatrix} 3 & 7 \\ 2 & 5 \end{vmatrix}\right)$$

$$= -\tfrac{1}{2}(15 - 14) = -\tfrac{1}{2}$$

Answer

Example 6. Evaluate

$$D = \begin{vmatrix} 7 & 3 & 5 \\ 7 & 0 & 0 \\ 16 & 48 & 32 \end{vmatrix}$$

Solution. Simplify row 3 by dividing it by 16, and offset this operation by multiplying the entire determinant by 16. Then expand by minors of the second row

$$D = 16\begin{vmatrix} 7 & 3 & 5 \\ 7 & 0 & 0 \\ 1 & 3 & 2 \end{vmatrix}$$

$$= 16\left(-7\begin{vmatrix} 3 & 5 \\ 3 & 2 \end{vmatrix}\right)$$

$$= -112(6 - 15) = 1008$$

Answer

Exercise 9.3

Simplify and evaluate the following determinants

1. $\begin{vmatrix} 3 & 1 & -1 \\ 6 & 2 & 5 \\ 9 & 3 & 7 \end{vmatrix}$ 2. $\begin{vmatrix} 4 & 6 & 1 & 3 \\ 2 & 3 & 0 & -5 \\ 6 & 9 & 1 & 4 \\ 8 & 12 & 2 & 2 \end{vmatrix}$ 3. $\begin{vmatrix} 1 & 7 & 3 & 9 & -2 \\ 0 & 3 & 1 & 6 & 5 \\ 0 & \frac{3}{2} & \frac{1}{2} & 3 & \frac{5}{2} \\ 2 & 0 & 4 & -6 & 8 \\ 1 & 0 & 3 & 5 & -9 \end{vmatrix}$

4. Show that

$$\begin{vmatrix} 1 & a & -3 \\ -5 & b & 0 \\ 2 & c & -4 \end{vmatrix} + \begin{vmatrix} 1 & d & -3 \\ -5 & e & 0 \\ 2 & f & 4 \end{vmatrix} = \begin{vmatrix} 1 & a+d & -3 \\ -5 & b+e & 0 \\ 2 & c+f & 4 \end{vmatrix}$$

5. Show that

$$\begin{vmatrix} 1-x & 2 & 0 \\ 3 & 3 & 2+x \\ 0 & 4 & 0 \end{vmatrix} = 0$$

if $x = 1$ or $x = -2$

6. Show that if we multiply $\begin{vmatrix} a & b \\ c & d \end{vmatrix}$ by $\begin{vmatrix} e & f \\ g & h \end{vmatrix}$

the correct answer is given by

$$\begin{vmatrix} ae+bg & af+bh \\ ce+dg & cf+dh \end{vmatrix}$$

Theorem 3. If the elements in any two columns (or rows) of a determinant are proportional, the determinant has the value 0

i.e. $$|a \quad b \quad Nb| = 0$$

Proof

$$|a \quad b \quad Nb| = N|a \quad b \quad b| \qquad \text{(from Theorem 2)}$$

$$= N\left(a_1\begin{vmatrix} b_2 & b_2 \\ b_3 & b_3 \end{vmatrix} - a_2\begin{vmatrix} b_1 & b_1 \\ b_3 & b_3 \end{vmatrix} + a_3\begin{vmatrix} b_1 & b_1 \\ b_2 & b_2 \end{vmatrix}\right)$$

$$= N[a_1(0) - a_2(0) + a_3(0)] = 0$$

Theorem 4. If each of the elements in a column (or row) of a determinant can be expressed as the sum of two quantities the determinant itself can be expressed as the sum of two determinants.

i.e. $$|m+n \quad b \quad c| = |m \quad b \quad c| + |n \quad b \quad c|$$

Proof follows directly from expansion of the above determinants by minors of the first column of each.

This theorem, together with Theorem 3, enables us to establish the more important Theorem 5.

Theorem 5. If all the elements in any column (or row) of a determinant are multiplied by the same constant and the resulting products are added to the corresponding elements in another column (or row) the value of the determinant is unchanged

i.e. $$|a \quad b \quad c+Na| = |a \quad b \quad c|$$

Proof

$$|a \quad b \quad c+Na| = |a \quad b \quad c| + |a \quad b \quad Na| \qquad \text{(Theorem 4)}$$
$$= |a \quad b \quad c| + 0 = |a \quad b \quad c| \qquad \text{(Theorem 3)}$$

The great practical importance of this theorem is that it may be applied repeatedly to reduce all, or all but one, of the elements in a selected row or column to 0. Thus it may greatly expedite the labour of calculating the determinant's value.

Example 7. Apply Theorem 5 to evaluate the determinant

$$D = \begin{vmatrix} 6 & 9 & 3 & 1 \\ 4 & 2 & 7 & 0 \\ 6 & 4 & 2 & 0 \\ 9 & 1 & 9 & 1 \end{vmatrix}$$

Solution. Subtracting row 1 from row 4 we have

$$\begin{vmatrix} 6 & 9 & 3 & 1 \\ 4 & 2 & 7 & 0 \\ 6 & 4 & 2 & 0 \\ 3 & -8 & 6 & 0 \end{vmatrix}$$

Expanding in terms of the minors of the last column

$$= -\begin{vmatrix} 4 & 2 & 7 \\ 6 & 4 & 2 \\ 3 & -8 & 6 \end{vmatrix} + 0 - 0 + 0$$

Adding 2 times row 2 to row 3 and -2 times row 1 to row 2

$$= -\begin{vmatrix} 4 & 2 & 7 \\ -2 & 0 & -12 \\ 15 & 0 & 10 \end{vmatrix}$$

$$= -(-2)\begin{vmatrix} -2 & -12 \\ 15 & 10 \end{vmatrix}$$

$$= 2(-20 + 180) = 320$$ Answer

Example 8. Show without expanding the determinant that

$$D = \begin{vmatrix} b & c & 1 \\ \frac{a}{2} & 0 & 1 \\ \frac{a+b}{3} & \frac{c}{3} & 1 \end{vmatrix} = 0$$

Solution. Adding $-\frac{1}{3}$ row 1 and $-\frac{2}{3}$ row 2 to row 3

$$D = \begin{vmatrix} b & c & 1 \\ \frac{a}{2} & 0 & 1 \\ 0 & 0 & 0 \end{vmatrix}$$

$$= 0 - 0 + 0 = 0$$ Answer

Exercise 9.4

Evaluate the following determinants, expediting each step by means of Theorem 5.

1. $\begin{vmatrix} 1 & 3 & 2 \\ 5 & 7 & 10 \\ 7 & 9 & 14 \end{vmatrix}$

2. $\begin{vmatrix} 1 & 6 & 1 & 2 \\ -3 & -5 & 2 & -2 \\ 5 & 13 & -3 & 6 \\ -2 & -6 & 7 & -1 \end{vmatrix}$

Show, without expanding that each of the following determinants = 0

3. $\begin{vmatrix} 3 & -3 & 2 & 4 \\ 6 & -6 & 4 & 8 \\ -2 & 1 & 3 & 6 \\ 6 & -5 & 1 & 2 \end{vmatrix}$

4. $\begin{vmatrix} 1 & 5 & 3 & 4 & 1 \\ 3 & 12 & 6 & 9 & -2 \\ 5 & 8 & -2 & 3 & 1 \\ -2 & 1 & 3 & -2 & 7 \\ 2 & -1 & -5 & 3 & 14 \end{vmatrix}$

5. $\begin{vmatrix} 1 & a & b+c \\ 1 & b & a+c \\ 1 & c & a+b \end{vmatrix}$

6. $\begin{vmatrix} 1 & 5 & 3 & 2 \\ 3 & -2 & -8 & 7 \\ -2 & 6 & 10 & 1 \\ 2 & 1 & -3 & 6 \end{vmatrix}$

Applications of Cramer's Rule

Having introduced Cramer's Rule earlier to show how and why determinants are defined (pages 101 and 105), and having already applied the rule to the solution of determinant systems of linear equations in two variables (page 101, we may now use the preceding theorems to extend the application of Cramer's Rule to the solution of determinant systems of linear equations in three or more variables.

One of the many conveniences of solving systems of linear equations by determinants is that the methods can be applied, when appropriate, to find the value of only one variable without the need of solving the entire system for all variables as when more elementary methods are used.

Example 9. Solve the following system of equations for z only

$$\begin{aligned} 3x + 3y - 2z - w &= 40 \\ 2x + 3y - z - w &= 35 \\ 3x + y + 4z - 2w &= 15 \\ -x + y + z + w &= 35 \end{aligned}$$

Solution.

$$|\,a\,b\,c\,d\,| = \begin{vmatrix} 3 & 3 & -2 & -1 \\ 2 & 3 & -1 & -1 \\ 3 & 1 & 4 & -2 \\ -1 & 1 & 1 & 1 \end{vmatrix}$$

Adding column 1 to columns 2, 3 and 4

$$= \begin{vmatrix} 3 & 6 & 1 & 2 \\ 2 & 5 & 1 & 1 \\ 3 & 4 & 7 & 1 \\ -1 & 0 & 0 & 0 \end{vmatrix}$$

Expand in terms of the minors of row 4

$$= -(-1) \begin{vmatrix} 6 & 1 & 2 \\ 5 & 1 & 1 \\ 4 & 7 & 1 \end{vmatrix}$$

Adding -1 times row 3 to row 2 and -2 times row 3 to row 1

$$= \begin{vmatrix} -2 & -13 & 0 \\ 1 & -6 & 0 \\ 4 & 7 & 1 \end{vmatrix}$$

$$= \begin{vmatrix} -2 & -13 \\ 1 & -6 \end{vmatrix} = 12-(-13) = 25$$

Divide column 3 by 5 and multiply the determinant by 5 according to Theorem 5

$$|abkd| = 5\begin{vmatrix} 3 & 3 & 8 & -1 \\ 2 & 3 & 7 & -1 \\ 3 & 1 & 3 & -2 \\ -1 & 1 & 7 & 1 \end{vmatrix}$$

Adding row 4 to rows 1 and 2, and twice row 4 to row 3

$$= 5\begin{vmatrix} 2 & 4 & 15 & 0 \\ 1 & 4 & 14 & 0 \\ 1 & 3 & 17 & 0 \\ -1 & 1 & 7 & 1 \end{vmatrix}$$

Expand in terms of the minors of column 4

$$= 5\begin{vmatrix} 2 & 4 & 15 \\ 1 & 4 & 14 \\ 1 & 3 & 17 \end{vmatrix}$$

Adding -1 times row 3 to row 2, and -2 times row 3 to row 1

$$= 15\begin{vmatrix} 0 & -2 & -19 \\ 0 & 1 & -3 \\ 1 & 3 & 17 \end{vmatrix}$$

Expand in terms of the minors of column 1

$$= 5\begin{vmatrix} -2 & -19 \\ 1 & -3 \end{vmatrix}$$

$$= 5(6 + 19) = 125$$

$$z = |abkd|/|abcd| \quad \text{Cramer's Rule}$$

$$= 125/25 = 5$$ Answer

Equation of the Straight Line Joining Two Points

The equation of the straight line joining $A(x_1, y_1)$ and $B(x_2, y_2)$ is given by the determinant $D = 0$ where

$$D = \begin{vmatrix} x & y & 1 \\ x_1 & y_1 & 1 \\ x_2 & y_2 & 1 \end{vmatrix}$$

If we subtract the first row from the other two rows in turn the determinant reduces to

$$D = \begin{vmatrix} x & y & 1 \\ x_1 - x & y_1 - y & 0 \\ x_2 - x & y_1 - y & 0 \end{vmatrix}$$

Putting $x = x_1$ and $y = y_1$ makes $D = 0$ since the second row is a row of zeros. Putting $x = x_2$ and $y = y_2$ again makes $D = 0$ since the third row becomes a row of zeros. These results mean that the linear equation given by $D = 0$ is the equation of the line AB.

Alternatively we suggest that the equation is $y = mx + c$ (i)

Because $A(x_1, y_1)$ is on this line, $y_1 = mx_1 + c$ (ii)

Because $B(x_2, y_2)$ is on this line, $y_2 = mx_2 + c$ (iii)

From (ii) – (iii) $\dfrac{y_1 - y_2}{x_1 - x_2} = m$

From x_2 (ii) $- x_1$ (iii) $$c = \frac{x_1y_2 - x_2y_1}{x_1 - x_2}$$

Substituting these results in (i) we get

$$y = \frac{(y_1 - y_2)x}{x_1 - x_2} + \frac{(x_1y_2 - x_2y_1)}{x_1 - x_2}$$

i.e. $$x(y_1 - y_2) - y(x_1 - x_2) + (x_1y_2 - x_2y_1) = 0$$

i.e. $$x\begin{vmatrix} y_1 & 1 \\ y_2 & 1 \end{vmatrix} - y\begin{vmatrix} x_1 & 1 \\ x_2 & 1 \end{vmatrix} + 1\begin{vmatrix} x_1 & y_1 \\ x_2 & y_2 \end{vmatrix} = 0$$

i.e. $$\begin{vmatrix} x & y & 1 \\ x_1 & y_1 & 1 \\ x_2 & y_2 & 1 \end{vmatrix} = 0$$

Example 10. Find the equation of the line joining the points $A(3, -7)$, $B(-2, 8)$.
Solution. The equation of the line is

$$\begin{vmatrix} x & y & 1 \\ 3 & -7 & 1 \\ -2 & 8 & 1 \end{vmatrix} = 0$$

i.e. $$x(-7 - 8) - y(3 - (-2)) + 24 - 14 = 0$$
$$-15x - 5y + 10 = 0$$
$$\text{or} \quad 3x + y = 2$$ Answer

Example 11. Show that the three points $A(8, 9)$, $B(-1, 6)$, $C(-10, 3)$ are collinear.
Solution. The equation of the line AB is

$$\begin{vmatrix} x & y & 1 \\ 8 & 9 & 1 \\ -1 & 6 & 1 \end{vmatrix} = 0$$

If this line passes through the point C then

$$\begin{vmatrix} -10 & 3 & 1 \\ 8 & 9 & 1 \\ -1 & 6 & 1 \end{vmatrix}$$

must be zero. Subtracting the first row from the other two in turn reduces the determinant to

$$\begin{vmatrix} -10 & 3 & 1 \\ 18 & 6 & 0 \\ 9 & 3 & 0 \end{vmatrix} = \begin{vmatrix} 18 & 6 \\ 9 & 3 \end{vmatrix} = 0$$

Hence the three points are collinear.

Exercise 9.5

1. Solve the following system of equations.

$$\begin{aligned} x + y + z + w &= -2 \\ 2x + y - z - w &= 1 \\ 3x - 2y + z + 2w &= 2 \\ x + 2y - z - 3w &= 6 \end{aligned}$$

2. You are told that from a system of equations, the solution for x was exactly the following

$$x = \frac{\begin{vmatrix} 1 & 2 & -1 & 1 \\ 5 & 3 & 2 & 5 \\ 7 & 4 & 3 & -6 \\ 2 & 6 & -2 & 7 \end{vmatrix}}{\begin{vmatrix} 1 & 2 & -1 & 1 \\ 3 & 3 & 2 & 5 \\ 5 & 4 & 3 & -6 \\ -7 & 6 & -2 & 7 \end{vmatrix}}$$

write the solutions for the other unknowns, and write down what the original equations must have been.

3. Evaluate the determinants

$$\begin{vmatrix} 1 & 1 & 1 \\ 1 & 1+a & 1 \\ 1 & 1 & 1+b \end{vmatrix} \text{ and } \begin{vmatrix} 1 & 1 & 1 & 1 \\ 1 & 1+a & 1 & 1 \\ 1 & 1 & 1+b & 1 \\ 1 & 1 & 1 & 1+c \end{vmatrix}$$

4. Evaluate the determinant

$$\begin{vmatrix} a & b & c \\ b & c & a \\ c & a & b \end{vmatrix}$$

5. Writing $f(x) = \begin{vmatrix} x^2 & x & 1 \\ 4 & 2 & 1 \\ 1 & 1 & 1 \end{vmatrix}$

use the remainder theorem to prove that $x - 1$ and $x - 2$ are factors of the determinant. Evaluate the determinant to confirm this result.

6. Apply the idea in Question 5 to show that $x - b$ is a factor of the determinant

$$D = f(x) = \begin{vmatrix} x^2 & x & 1 \\ b^2 & b & 1 \\ a^2 & a & 1 \end{vmatrix}$$

Find the other factors of D and evaluate the determinant.

7. Find the equation of the line joining the points $A(2, 3)$, $B(3, 1)$ and show that the point $C(0, 7)$ also lies on this line.

8. Prove that each of the following sets of three points are collinear:

(a) (7, 2), (4, 0), (10, 4)
(b) (0, 5), (7, 0), (−7, 10)
(c) (0, −6), (16, 0), (8, −3)

Summary

1. When a determinant has two identical rows (or columns) its value is zero.

2. When two rows (or columns) are interchanged the value of the determinant remains numerically the same but changes sign.

3. When the elements of a row (or column) are multiplied by the same non-zero number the value of the determinant is also multiplied by that number.

4. A positive or negative multiple of any row (or column) may be added to any other row (or column) without changing the value of the determinant.

CHAPTER TEN

INDICES AND LOGARITHMS

The Fundamental Law of Indices

This fundamental law of indices states that if m and n are positive integers then $a^m \times a^n = a^{m+n}$: the quantity a^m being the product of m factors each one being a. We can always extend this result to the product $a^m \times a^n \times a^p$, where m, n and p are each positive integers. Thus

$$a^m \times (a^n \times a^p) = a^m \times a^{n+p} = a^{m+n+p}$$

or

$$(a^m \times a^n) \times a^p = a^{m+n} \times a^p = a^{m+n+p},$$

the same result being obtained whether we first associate a^n with a^p or a^m with a^n. All we have done here is use the associative law for multiplication. From the fundamental law we deduce the following results for positive integers m and n

(i) $$a^m \div a^n = a^{m-n} \text{ if } m > n$$

(ii) $$a^m \div a^n = \frac{1}{a^{n-m}} \text{ if } m < n$$

(iii) $$(a^m)^n = (a^n)^m = a^{mn}$$

We can prove these results by using the fundamental law as follows:

(i) Since $m - n$ is positive, $a^{m-n} \times a^n = a^{m-n+n} = a^m$. Now divide both sides of the equation by a^n to get $\dfrac{a^{m-n} \times a^n}{a^n} = \dfrac{a^m}{a^n}$

i.e.

$$a^{m-n} = a^m \div a^n \qquad \text{Q.E.D.}$$

(ii) Since $n - m$ is positive then $\dfrac{1}{a^{n-m} \times a^m} = \dfrac{1}{a^{n-m+m}} = \dfrac{1}{a^n}$

Now multiply both sides of the equation by a^m to get

$$\frac{a^m}{a^{n-m} \times a^m} = \frac{a^m}{a^n},$$

i.e.

$$a^m \div a^n = \frac{1}{a^{n-m}} \qquad \text{Q.E.D.}$$

(iii) $(a^m)^n = a^m \times a^m \times a^m \times a^m \ldots \times a^m$, a product of n factors, each one being a^m,

$$\begin{aligned} &= a^{2m} \times a^m \times a^m \ldots \times a^m \\ &= a^{2m} \times (\text{a product of } n - 1 \text{ factors, each one being } a^m) \\ &= a^{3m} \times (\text{a product of } n - 2 \text{ factors, each one being } a^m) \\ &= a^{4m} \times (\text{a product of } n - 3 \text{ factors, each one being } a^m) \\ &= a^{nm} \end{aligned}$$

Similarly

$$(a^n)^m = a^{mn} = a^{nm}$$

Two further results which we need to be reminded of are

(iv) $(ab)^m = a^m b^m$ and (v) $\left(\frac{a}{b}\right)^m = \frac{a^m}{b^m}$

Having identified these laws for positive indices m and n we **assume** that they are true for all rational numbers, i.e. any number of the form p/q where p and q are integers (positive or negative) and $q \neq 0$. All that remains to be done is to find a meaning for some of the new expressions which are consistent with these laws.

Example 1. To obtain a meaning for $a^{1/n}$, where n is a positive integer.

Solution. We start with some numerical examples: $a^{\frac{1}{2}} \times a^{\frac{1}{2}} = a^1 = a$, but since we know that $\sqrt{a} \times \sqrt{a} = a$ it follows that $a^{\frac{1}{2}} = \sqrt{a}$, e.g. $9^{\frac{1}{2}} = 3$, $25^{\frac{1}{2}} = 5$.

Next consider $a^{\frac{1}{3}} \times a^{\frac{1}{3}} \times a = a^{\frac{1}{3}+\frac{1}{3}+\frac{1}{3}} = a^1 = a$, but since we know that $\sqrt[3]{a} \times \sqrt[3]{a} \times \sqrt[3]{a} = a$ it follows that $a^{\frac{1}{3}} = \sqrt[3]{a}$, i.e. a cube root of a (e.g. $8^{\frac{1}{3}} = 2$, $125^{\frac{1}{3}} = 5$).

Now consider $(a^{1/n})^n = a^{n/n} = a$ from (iii) above. Hence $a^{1/n} = \sqrt[n]{a}$, i.e. an nth root of a (e.g. $32^{\frac{1}{5}} = 2$, $81^{\frac{1}{4}} = 3$, $128^{\frac{1}{7}} = 2$).

Answer

For cases such as $17^{\frac{1}{5}}$ we are unable to simplify the results as we did for $32^{\frac{1}{5}}$ although we can find the fifth root of 17 by using logarithms as the reader possibly already knows. For a number such as $96^{\frac{1}{5}}$ we can go part of the way towards simplification by realising that since $96 = 32 \times 3$ the result (iv) above enables us to write

$$96^{\frac{1}{5}} = (32 \times 3)^{\frac{1}{5}} = 32^{\frac{1}{5}} \times 3^{\frac{1}{5}} = 2 \times 3^{\frac{1}{5}} \text{ or } 2\sqrt[5]{3}$$

Similarly

$$250^{\frac{1}{3}} = (125 \times 2)^{\frac{1}{3}} = 125^{\frac{1}{3}} \times 2^{\frac{1}{3}} = 5 \times 2^{\frac{1}{3}} = 5\sqrt[3]{2}$$

Example 2. To obtain a meaning for $a^{m/n}$, m and n being positive integers.

Solution. Now since $a^{m/n} = (a^{1/n})^m$ we may interpret $a^{m/n}$ as an (nth root of a) taken or raised to the power of m, i.e. $(\sqrt[n]{a})^m$. For example, $32^{\frac{3}{5}} = (32^{\frac{1}{5}})^3 = (2)^3 = 8$.

Another interpretation giving the same result is to consider $a^{m/n} = (a^m)^{1/n}$, in which case we now find an nth root of (a taken to the power m)—that is $\sqrt[n]{a^m}$. Numerically this becomes $32^{\frac{3}{5}} = (32^3)^{\frac{1}{5}} = (32768)^{\frac{1}{5}} = 8$, as before. Clearly the first method of evaluation is the easier.

Example 3. To obtain a meaning for a^0, $a \neq 0$.

Solution. Consider the result $a^0 \times a^n = a^n \times a^0 = a^{n+0} = a^n$. Since multiplication by 1 is the only way to leave a^n unchanged, it follows that $a^0 = 1$.

Example 4. To obtain a meaning for a^{-n} when n is a positive integer.

Solution. Consider $a^n \times a^{-n} = a^{n-n} = a^0 = 1$

then

$$\frac{a^n \times a^{-n}}{a^n} = a^{-n} = \frac{1}{a^n}$$

e.g. $2^{-3} = \frac{1}{2^3} = \frac{1}{8}$. In words, a^{-n} is the reciprocal of a^n, thus $a^{-1} = \frac{1}{a}$

Finally we consider $a^{-1/n}$ where n is a positive integer. We have

$$(a^{-1/n})^n = a^{-n/n} = a^{-1} = 1/a.$$

Taking the nth roots of both sides we get

$$(a^{-1/n}) = \left(\frac{1}{a}\right)^{1/n} = \frac{1}{a^{1/n}}$$

e.g. $$8^{-\frac{1}{3}} = \frac{1}{8^{\frac{1}{3}}} = \frac{1}{2}$$

Exercise 10.1

Write down the following in their simplest form:

1. $36^{\frac{1}{2}}$ 2. $8^{\frac{2}{3}}$ 3. $64^{\frac{1}{6}}$ 4. $64^{-\frac{1}{6}}$ 5. 2^{-3}
6. $(\frac{1}{2})^{-3}$ 7. $4^{\frac{3}{2}}$ 8. $125^{\frac{2}{3}}$ 9. $125^{-\frac{2}{3}}$ 10. $9^{-\frac{3}{2}}$
11. $4^{\frac{1}{2}}$ 12. $(\frac{1}{2})^4$ 13. 2^{-4} 14. 4^{-2} 15. $(-2)^4$
16. $x^{\frac{1}{2}} \times x^{\frac{2}{3}}$ 17. $x^4 \div x^{\frac{1}{2}}$ 18. $x^{-\frac{1}{2}} \times x^{\frac{3}{2}}$ 19. $(8^{\frac{2}{3}})^3$ 20. $(8^3)^{\frac{2}{3}}$
21. $(x^{\frac{1}{2}} + y^{\frac{1}{2}})^2$ 22. $(x^{\frac{1}{2}} + y^{\frac{1}{2}})(x^{\frac{1}{2}} - y^{\frac{1}{2}})$ 23. $32^{\frac{1}{5}}$ 24. $63^{\frac{1}{2}}$ 25. $(108)^{-\frac{1}{2}}$

One problem arises from our discussion of the meaning of $a^{1/n}$ where you probably notice that we referred to $a^{1/n}$ as *an* nth root of a so that we wrote $a^{1/n} = \sqrt{a}$. Consider $a^{\frac{1}{2}} = \sqrt{a}$, based on the result that $\sqrt{a} \times \sqrt{a} = a$. There is another answer for this result, namely $(-\sqrt{a}) \times (-\sqrt{a}) = a$. Now when we write $\sqrt{a}$ we mean $+\sqrt{a}$, thus $\sqrt{49} = 7$. If we require the negative root then we state $-\sqrt{49} = -7$. Thus, $a^{\frac{1}{2}}$ is either $+\sqrt{a}$ or $-\sqrt{a}$. This means that $9^{\frac{1}{2}} = \pm 3$ but $\sqrt{9} = 3$. This leads to a complication because $(9^{\frac{1}{2}})^2$ represents only the one value, 9, but, $(9^2)^{\frac{1}{2}}$ represents two values $\pm\sqrt{(9^2)} = \pm 9$ so that we are not being consistent with the law that $(a^m)^n = (a^n)^m = a^{mn}$ when m is a fraction. In order to be consistent with this law we shall in this book take $\sqrt[n]{a}$ as *one* of the n values of $a^{1/n}$, given thus, $a^{\frac{1}{2}} = \sqrt{a}$, i.e. $9^{\frac{1}{2}} = 3$. If we require $9^{\frac{1}{2}} = -3$ etc., we shall say so.

Rationalisation

A number such as $\sqrt{3}$, $\sqrt{2}$, $\sqrt{8}$ is called an irrational number because it cannot be expressed as a quotient of two integers like $7{\cdot}5 = \frac{15}{2}$ and $-2{\cdot}3 = \frac{-7}{3}$ or $\frac{7}{-3}$.

Now suppose we have to simplify $\sqrt{\frac{7}{9}}$, i.e. $\left(\frac{7}{9}\right)^{\frac{1}{2}}$, since this is the same as $\frac{7^{\frac{1}{2}}}{9^{\frac{1}{2}}}$ we obtain the result $\frac{(7)^{\frac{1}{2}}}{3}$, and $\sqrt{7}$ will have to be either left as it is or written as 2·65, after consulting tables of square roots. Thus $\frac{(7)^{\frac{1}{2}}}{3}$ is the simplification and $\frac{2{\cdot}65}{3} = 0{\cdot}88$ (2 decimal places) is the evaluated form. On the other hand, if we attempt to simplify $\left(\frac{9}{7}\right)^{\frac{1}{2}}$ we obtain $\frac{3}{\sqrt{7}}$, and we consider that a further simplification is necessary by rationalising the denominator so as to make

$$\frac{3}{\sqrt{7}} = \frac{3 \times \sqrt{7}}{\sqrt{7} \times \sqrt{7}} = \frac{3\sqrt{7}}{7}$$

Using the result $x^2 - y^2 = (x - y)(x + y)$, so that $\frac{1}{(x + y)} = \frac{x - y}{x^2 - y^2}$ or

$\frac{1}{x-y} = \frac{x+y}{x^2-y^2}$, we now consider rationalising the denominator in the expression such as $\frac{1}{\sqrt{2}-1}$ by writing

$$\frac{1}{(\sqrt{2}-1)}\frac{(\sqrt{2}+1)}{(\sqrt{2}+1)} = \frac{\sqrt{2}+1}{2-1} = \frac{\sqrt{2}+1}{1}$$

Similarly

$$\frac{1}{\sqrt{2}+1} = \sqrt{2}-1$$

Example 5. Simplify

$$\frac{1}{x+\sqrt{(x^2+1)}}$$

Solution. Rationalise the denominator as follows

$$\frac{1}{x+\sqrt{(x^2+1)}} = \frac{1}{x+\sqrt{(x^2+1)}} \times \frac{x-\sqrt{(x^2+1)}}{x-\sqrt{(x^2+1)}} = \frac{x-\sqrt{(x^2+1)}}{x^2-(x^2+1)}$$

$$= \sqrt{(x^2+1)} - x \qquad \text{Answer}$$

Example 6. Given that $\frac{a}{b}$ is a good approximation to $\sqrt{2}$ prove that $\frac{a+2b}{a+b}$ is a better approximation.

Solution. We are being asked to prove that $\left(\frac{a+2b}{a+b}\right)^2$ is closer to 2 than $\frac{a^2}{b^2}$. We can illustrate this numerically by taking $a = 7$, $b = 5$, in which case $\frac{a^2}{b^2} = \frac{49}{25}$, so that $2 - \frac{a^2}{b^2} = 2 - \frac{49}{25} = \frac{1}{25}$

$\frac{a+2b}{a+b} = \frac{17}{12}$ and $2 - \left(\frac{a+2b}{a+b}\right)^2 = 2 - \frac{289}{144} = -\frac{1}{144}$ so we see that $\frac{17}{12}$ is closer to $\sqrt{2}$ than $\frac{7}{5}$ because $\frac{1}{144}$ is smaller than $\frac{1}{25}$. The closeness of $\frac{a^2}{b^2}$ to 2 is given by $2 - \frac{a^2}{b^2} = \frac{2b^2-a^2}{b^2}$. The closeness of $\left(\frac{a+2b}{a+b}\right)^2$ to 2 is given by

$$2 - \frac{(a+2b)^2}{(a+b)^2} = 2 - \frac{a^2+4ab+4b^2}{a^2+2ab+b^2}$$

$$= \frac{2a^2+4ab+2b^2-(a^2+4ab+4b^2)}{a^2+2ab+b^2} = \frac{a^2-2b^2}{(a+b)^2}$$

Now $(a+b)^2 > b^2$ for $a > 0, b > 0$, so that $\frac{a^2-2b^2}{(a+b)^2}$ is less than $\frac{2b^2-a^2}{b^2}$ in magnitude although of different sign. This means that if the approximation $\frac{a}{b}$ is less than $\sqrt{2}$ then the approximation $\frac{a+2b}{a+b}$ is greater than $\sqrt{2}$ (but still closer to $\sqrt{2}$).

If we return to the numerical case suggested after 7/5 we produced a closer or better approximation to $\sqrt{2}$ in 17/12. Now let us start again but this time with 17/12, i.e. $a = 17$, $b = 12$. Thus

$$\frac{a + 2b}{a + b} = \frac{17 + 24}{29} = \frac{41}{29}$$

so that

$$\left(\frac{a + 2b}{a + b}\right)^2 = \left(\frac{41}{29}\right)^2 = \frac{1681}{841}$$

The closeness of this approximation is given by

$$2 - \frac{1681}{841} = \frac{1}{841}$$

and since $\frac{1}{841}$ is less than $\frac{1}{144}$ we have $\frac{41}{29}$ as a better approximation to $\sqrt{2}$ than $\frac{17}{12}$.

Note that the errors in the approximations are $\frac{1}{25} = \frac{1}{5^2}$, $\frac{1}{144} = \frac{1}{12^2}$, $\frac{1}{841} = \frac{1}{29^2}$ so that a continuation of this process produces closer and closer approximations. This procedure produces a sequence of rational numbers which get as close as we please to (i.e. tend to) the irrational number $\sqrt{2}$.

Exercise 10.2

Write down the following in their simplest form with rational denominators:

1. $\left(\frac{27}{8}\right)^{\frac{1}{3}}$
2. $\left(\frac{27}{8}\right)^{-\frac{1}{3}}$
3. $\left(\frac{27}{8}\right)^{\frac{2}{3}}$
4. $\left(\frac{27}{8}\right)^{-\frac{3}{4}}$
5. $\left(\frac{4}{9}\right)^{\frac{3}{2}}$
6. $\left(\frac{4}{9}\right)^{-\frac{3}{2}}$
7. $\left(\frac{2}{3}\right)^{\frac{1}{2}}$
8. $\left(\frac{16}{27}\right)^{\frac{1}{2}}$
9. $\left(\frac{11}{13}\right)^{0}$
10. $\left(-\frac{11}{13}\right)^{0}$
11. $\left(\frac{16}{27}\right)^{\frac{1}{4}}$
12. $\left(\frac{125}{16}\right)^{-\frac{3}{8}}$
13. $\frac{1}{\sqrt{3} - 1}$
14. $\frac{1}{\sqrt{3} + 1}$
15. $\frac{1}{\sqrt{3} + \sqrt{2}}$
16. $\frac{1}{\sqrt{3} - \sqrt{2}}$
17. $\frac{1}{\sqrt{x} + 2}$
18. $\frac{1}{\sqrt{x} + \sqrt{y}}$
19. $\frac{1}{\sqrt{y} - \sqrt{x}}$
20. $\frac{1}{x - \sqrt{(x^2 - 1)}}$
21. Starting with $\frac{17}{10}$ as a first approximation to $\sqrt{3}$ show that with $a = 17, b = 10$ then $\frac{a + 3b}{a + b}$ is a closer approximation to $\sqrt{3}$.
22. Examine the worked example for approximations to $\sqrt{2}$ and also Question 21 and suggest a formula for getting better approximations to $\sqrt{5}$, and starting with $\frac{11}{5}$ find the next approximation.
23. A sequence of approximations which gets closer to $\sqrt{2}$ is given by the worked example above, namely

$$\frac{7}{5}, \frac{17}{12}, \frac{41}{29}, \cdots$$

Give the next three approximations in the sequence.

24. Give the first five terms in the sequence of approximations for $\sqrt{3}$ in Question 21.

Logarithms (Base 10)

It is assumed that the reader is familiar with the use of logarithms for multiplication and division and is able to use the tables for logarithms and antilogarithms at the end of this book. We recall that a logarithm is an index and that the tables enable us to express any positive number as a power of 10. For example, the tables inform us that the logarithm of 2·213 is 0·3450, which we write as log 2·213 = 0·3450. This result enables us to write $2{\cdot}213 = 10^{0 \cdot 3450}$.

Because $22{\cdot}13 = 2{\cdot}213 \times 10 = 2{\cdot}213 \times 10^1$ we can use the law of indices to find the logarithm of 22·13 as follows:

$$2{\cdot}213 = 10^{0 \cdot 3450} \therefore 22{\cdot}13 = 10^{0 \cdot 3450} \times 10^1$$

Hence $22{\cdot}13 = 10^{1 \cdot 3450}$ and we can write log 22·13 = 1·3450. Similarly

$$221{\cdot}3 = 2{\cdot}213 \times 10^2 = 10^{0 \cdot 3450} \times 10^2 = 10^{2 \cdot 3450}$$

Again

$$0{\cdot}2213 = 2{\cdot}213 \times \frac{1}{10} = 2{\cdot}213 \times 10^{-1} = 10^{0 \cdot 3450} \times 10^{-1} = 10^{\bar{1} \cdot 3450}$$

The last step is to be noted. We do not combine 0·3450 − 1 to give −0·6550, instead we leave the index in two parts: the whole number part (**characteristic**) as −1 written $\bar{1}$, and the fraction part (**mantissa**) ·3450

e.g. $\quad 10^{\bar{1} \cdot 2675}$ means $10^{-1} \times 10^{0 \cdot 2675} = 10^{-1} \times 1{\cdot}851 = 0{\cdot}1851$

By so doing we may restrict all the numbers to the table of positive logarithms, as otherwise we would have to print another set with negative logarithms.

Because these logarithms are based on powers of 10 we sometimes need to write the results as $\log_{10} 22{\cdot}13 = 1{\cdot}3450$ to avoid confusion with logarithms based on powers of a number different from 10.

One final point is to notice that since $1 = 10^0$ then $\log_{10} 1 = 0$.

Logarithms (Bases other than 10)

Since $221{\cdot}3 = 10^{2 \cdot 345}$ we are able to write $\log_{10} 221{\cdot}3 = 2{\cdot}345$

Similarly since $\quad 8 = 2^3$ we are able to write $\log_2 8 = 3$

" " $\quad 81 = 3^4$ we are able to write $\log_3 81 = 4$

" " $\quad 1024 = 4^5$ we are able to write $\log_4 1024 = 5$

" " $\quad 15625 = 5^6$ we are able to write $\log_5 15625 = 6$

Again, considering negative indices:

since $0{\cdot}2213 = 10^{\bar{1} \cdot 3450}$ we are able to write $\log_{10} 0{\cdot}2213 = \bar{1}{\cdot}3450$.

Similarly since $\frac{1}{8} = 2^{-3}$ we are able to write $\log_2 (\frac{1}{8}) = -3$ or $\bar{3}{\cdot}0$

" " $\frac{1}{81} = 3^{-4}$ we are able to write $\log_3 (\frac{1}{81}) = -4$ or $\bar{4}{\cdot}0$ and so on.

Observe that $\log \frac{1}{8} = \log (1 \div 8) = \log 1 - \log 8 = 0 - \log 8$

$\log \frac{1}{8} = -\log 8$ and so on, in any base.

Exercise 10.3

1. Write down the logarithms to base 10 of the following numbers:

(i) 17·16 (ii) 221·6 (iii) 2862·0 (iv) 36970·0
(v) 477400·0 (vi) 0·1716 (vii) 0·02216 (viii) 0·002862
(ix) 0·0003697 (x) 0·00004774

2. Write down the following logarithms:

   (i) $\log_2 8$ (ii) $\log_7 49$ (iii) $\log_{10} 10$ (iv) $\log_2 16$
   (v) $\log_4 64$ (vi) $\log_{10} 12$ (vii) $\log_5 1$ (viii) $\log_2 1$
   (ix) $\log_3 1$ (x) $\log_5 125$

3. Write down the following logarithms:

   (i) $\log_2 \frac{1}{8}$ (ii) $\log_4 \frac{1}{16}$ (iii) $\log_9 81$ (iv) $\log_9 \frac{1}{81}$
   (v) $\log_4 \frac{1}{64}$ (vi) $\log_7 \frac{1}{49}$ (vii) $\log_{11} \frac{1}{121}$ (viii) $\log_{\frac{2}{3}} \frac{8}{27}$

4. Express the following results in the notation of indices:

   (i) $\log_{10} 100 = 2$ (ii) $\log_{10} 1000 = 3$ (iii) $\log_3 81 = 4$ (iv) $\log_7 49 = 2$
   (v) $\log_2 128 = 7$ (vi) $\log_7 343 = 3$ (vii) $\log_2 x = y$ (viii) $\log_3 x = y$
   (ix) $\log_b a = c$

5. Find x in each of the following (the logarithms are to base 10 unless indicated otherwise):

   (i) $\log x = 1$ (ii) $\log 1 = x$ (iii) $\log_2 x = 3$ (iv) $\log x = 0{\cdot}7076$
   (v) $\log_3 x = 3$ (vi) $\log_x 16 = 4$ (vii) $\log_x 9 = 2$ (viii) $\log_3 x = 0$
   (ix) $\log 4^x = 3 \log 4$ (x) $\log_3 (9^x) = 1$

6. Simplify the following by expressing the result as a single logarithm:

   (i) $\log 5 + \log 2$ (ii) $\log 3 + \log 5 - \log 15$
   (iii) $\log 27 + \log 4 - 2 \log 6$ (iv) $6 \log 3 - 3 \log 6$
   (v) $2 \log 5 - \log 10$ (vi) $2 \log 6 - \log 12$

Finding the *n*th Roots by Logarithms

One of the laws of indices enables us to write $(a^{1/n})^n = a$, $(a \neq 0)$, where n is any integer (not zero). Taking the logarithm of both sides of this expression when $a > 0$ we get

$$\log a = \log (a^{1/n}) + \log (a^{1/n}) + \log (a^{1/n}) + \ldots + \log (a^{1/n}) \text{ for } n \text{ terms}$$

$$\therefore \log a = n \log (a^{1/n})$$

and we have proved that $\log (a^{1/n}) = 1/n \log a$, enabling us to obtain the nth root of any positive number we please.

Example 7. Find one of the fifth roots of 3.

Solution. We require $3^{\frac{1}{5}}$. $\log 3^{\frac{1}{5}} = \frac{1}{5} \log 3 = \frac{1}{5} (0{\cdot}4771) = 0{\cdot}09542$.
Since $\log 3^{\frac{1}{5}} = 0{\cdot}09542$ it follows by taking antilogarithms that $3^{\frac{1}{5}} = 1{\cdot}25$ (correct to 2 decimal places). The other four roots involve complex numbers.

Example 8. Find $(0{\cdot}754)^{\frac{1}{4}}$

Solution. $\log (0{\cdot}754)^{\frac{1}{4}} = \frac{1}{4} \log 0{\cdot}754 = \frac{1}{4}(\bar{1}{\cdot}8774)$. Now recall that this means $\frac{1}{4}(-1 + 0{\cdot}8774)$, and in order to use the given logarithm tables we must keep the negative characteristic and the positive mantissa separate. So we consider the expression as $\frac{1}{4}(-4 + 3 + 0{\cdot}8774) = \frac{1}{4}(-4 + 3{\cdot}8774)$ (pause and think!), which becomes

$$= -1 + 0{\cdot}9694 \text{ (to 4 decimal places)}$$
$$= \bar{1}{\cdot}9694$$

Consequently $0{\cdot}754^{\frac{1}{4}} = 0{\cdot}9319$ (antilogarithm tables) Answer

Observe that this is only one of the four roots, e.g. $-0{\cdot}9319$ is another root.

The Graph of $y = 2^x$

So far whenever the base of the logarithms has been different from 10 (notice that all bases are positive) we have deliberately arranged straightforward questions so that the answers may be derived from equally straightforward numerical results. Thus we all know the result $8^{\frac{1}{3}} = 2$, so that $\log_2 8 = 3$ is merely a rephrasing of that knowledge in the context of logarithms, but it is a more difficult problem to be asked for $\log_2 7{\cdot}3$ or $\log_3 8$. By drawing the graph of $y = 2^x$ we can obtain some of these results for logarithms to base 2 since $\log_2 y = \log_2 (2^x) = x \log_2 2 = x$.

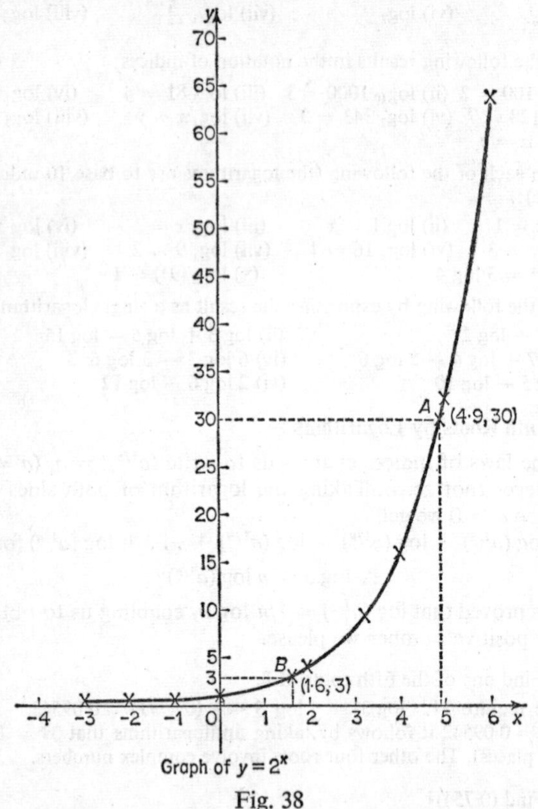

Fig. 38

Similarly if we need to work with logarithms to base 3 then we could draw the graph of $y = 3^x$ to get $\log_3 y = x$. But, even with a number as small as 3, when we calculate y for $x = 4$ we have a result of $y = 3^4 = 81$ and then $3^5 = 243$ so that we rapidly approach large values for y. In order to keep the values of y within reasonable bounds for the size of Fig. 38 we consider $y = 2^x$ only. As usual we start with the table of values as follows:

x	6	5	4	3	2	1	0	-1	-2	-3
$y = 2^x$	64	32	16	8	4	2	1	0·5	0·25	0·125

Using this graph and the relation $\log_2 y = x$ we can now find results like the following:

(i) At A, $\log_2 30 = 4{\cdot}9$ (reading taken from the graph)
(ii) At B, $\log_2 3 = 1{\cdot}6$ (reading taken from the graph)

Take careful note of these two results: for $\log_{10} 3 = 0{\cdot}4771$ and $\log_{10} 30 = 1{\cdot}4771$, but we *do not get* $\log_2 30 = 2{\cdot}6$ if $\log_2 3 = 1{\cdot}6$! This is the advantage of using base 10 logarithms when the numbers are denary.

Logarithmic Plotting

Suppose we now consider the graph of $y = kx^m$ where k and m are constants. For example, $y = 6x^4$ or $y = 2x^3$.

We know from our experience already that the graphs will be curves and not straight lines, as illustrated in Fig. 39. By taking logarithms we can convert such graphs into straight lines.

$$\text{Thus } \log y = \log kx^m = \log k + \log x^m = m \log x + \log k$$

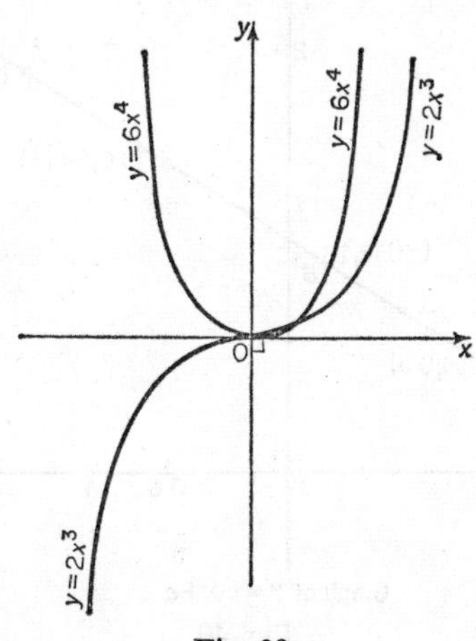

Fig. 39

Now put $\log y = Y$, $\log x = X$ and $\log k = c$ so that the new equation reads $Y = mX + c$, which we recognise as a straight line if we plot Y against X. Furthermore the gradient of the straight line is m and the intercept on the y axis is c. The whole object of this exercise is that we sometimes have experiments in which we suspect that the relation between y and x is of the form $y = kx^m$ (or we hope it is) but we need too many results to find a good approximation to curves such as those in Fig. 39. However, comparatively few results would be enough to obtain the best straight line fit, so we look for the expected result in the form of $\log y = m \log x + c$ as the next example shows.

Example 9. Assuming the existence of the relation $y = kx^m$, $(k > 0)$ find k and m and a and b from the following table of results by plotting a suitable graph.

x	4	40	20	b
y	29·87	118·9	a	10

Solution. Taking logarithms we get $\log y = m \log x + \log k$.
Substitute $\log y = Y$, $\log x = X$, $\log k = c$ and plot the line $Y = mX + c$. We now rewrite the table of given information.

x	4	40	20	b
$X = \log x$	0·602	1·602	1·301	?
y	29·87	118·9	a	10
$Y = \log y$	1·4752	2·0753	?	1

Choosing a scale of 4 cm to 1 unit on each axis we plot the two points $P(0{\cdot}6, 1{\cdot}475)$ and $Q(1{\cdot}6, 2{\cdot}075)$ to obtain the straight line in Fig. 40. The equation of this line is $Y = mX + c$.

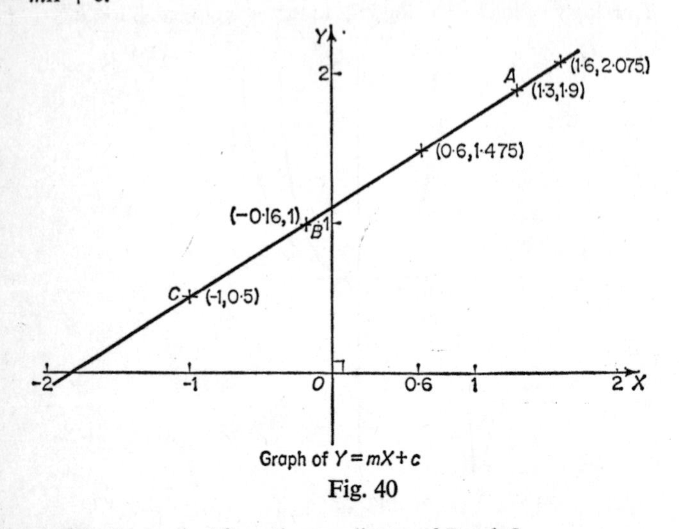

Graph of $Y = mX + c$

Fig. 40

We calculate the gradient from the coordinates of P and Q

$$\therefore \quad m = \frac{2{\cdot}075 - 1{\cdot}475}{1{\cdot}6 - 0{\cdot}6} = 0{\cdot}6$$

The value of c is given by putting $X = 0$, i.e. the intercept on the Y axis. From the graph $c = 1{\cdot}1$. We now recall that $c = \log k$ so that $\log k = 1{\cdot}1$; we then use anti-logarithm tables to get $k = 12{\cdot}59$. Hence $y = 12{\cdot}59\, x^{0{\cdot}6}$ was the original relation according to the graph.

In fact, the results were calculated from the relation $y = 13x^{0{\cdot}6}$. With a larger scale we may have been able to get $c = 1{\cdot}114$ to give $k = 13$, but under the circumstances the result is fair. A moment's thought suggests that we are being optimistic in quoting the result as $12{\cdot}59x^{0{\cdot}6}$, i.e. four significant figures. Taking $12{\cdot}6x^{0{\cdot}6}$ or $13{\cdot}00x^{0{\cdot}6}$ would be justified.

To find a. At A; $X = 1{\cdot}301$ gives $Y = 1{\cdot}9$ on the graph. $\log y = 1{\cdot}9$ and $y = 79{\cdot}43$. Again $y = 80$ would be more realistic. (The calculated value from $y = 13x^{0{\cdot}6}$ is 78·42.)

To find b. At B; $X = \log x = -0{\cdot}16 = -1 + 0{\cdot}84 = \bar{1}{\cdot}84$

$$\therefore \quad x = 0{\cdot}6918 = 0{\cdot}69 \text{ (realistically)}$$ Answer

Again if we substitute $x = 0{\cdot}6918$ in $y = 13x^{0{\cdot}6}$ we shall obtain $y = 10{\cdot}42$ and not 10, due to the limitations of reading from the graph.

Exercise 10.4

1. Find the following roots:

   (i) $2^{\frac{1}{5}}$ (ii) $8^{\frac{1}{4}}$ (iii) $10^{\frac{1}{5}}$ (iv) $1{\cdot}234^{\frac{1}{3}}$ (v) $0{\cdot}3^{\frac{1}{3}}$ (vi) $0{\cdot}8^{\frac{1}{4}}$ (vii) $0{\cdot}1^{\frac{1}{5}}$ (viii) $0{\cdot}4^{-\frac{1}{2}}$
   (ix) $0{\cdot}6^{-\frac{1}{3}}$ (x) $0{\cdot}8^{\frac{2}{5}}$

2. From the graph of Fig. 38 find the a in each of the following:

   (i) $\log_2 a = 3{\cdot}8$ (ii) $\log_2 a = 2{\cdot}5$ (iii) $\log_2 50 = a$ (iv) $\log_2 10 = a$

3. Use the graph of Fig. 40 to solve the following:

   (i) Find x when $y = 60$ (ii) Find y when $x = 10$

4. The following table gives some values of x and y which are known to be related by the equation $y = kx^m$. By drawing a graph of $\log y$ against $\log x$ find the values of k, m, a and b.

x	6·31	6·13	10	b
y	14·78	105·0	a	80

5. Using the graph of Fig. 38 for $y = 2^x$ solve the following pairs of simultaneous equations:

   (a) $y - 2^x = 0$, $y + x = 0$ (b) $y - 2^x = 0$, $y - 12x = 0$

6. Draw a graph of $y = 3^x$; $-2 \leqslant x \leqslant 4$ and use it to find the following:

   (i) $\log_3 50$ (ii) $\log_3 1$ (iii) $\log_3 0{\cdot}5$
   (iv) The solution to the simultaneous equations

   (a) $y - 3^x = 0$, $y + x = 0$ (b) $y - 3^x = 0$, $y - 9x = 0$

7. Given that $\log x = y$ and $\log a = b$ find the numerical value of the following in terms of y and b.

   (i) $\log 10x$ (ii) $\log ax$ (iii) $\log\left(\frac{a}{x}\right)$ (iv) $\log (a^2)$

   (v) $\log\left(\frac{\sqrt{x}}{a}\right)$ (vi) $\log\left(\frac{a}{10}\right)$ (vii) $\log\left(\frac{a}{10}\right)^{-1}$ (viii) $\log \sqrt{ax}$

CHAPTER ELEVEN

SEQUENCES AND SERIES

Sequences

A sequence of numbers is an ordered set. To be placed or written in sequence means that a set of numbers has been written down in some order, which enables each number to be referred to by its position in the sequence. The easiest sequence is that of the counting numbers 1, 2, 3, 4, 5, 6, 7, 8, . . . 30. The most useful feature of such an arrangement is the fact that it is predictable, so much so that after the eighth term we regard the pattern to be so well established that we discontinue writing the individual terms in favour of a dribble of dots. By mathematical convention we indicate our intention to stop at 30 by adding this term at the end and closing the sequence by a full stop.

We could argue that giving the first eight terms to establish the pattern or interrelation between consecutive terms is unnecessary, since the first four terms were adequate for this purpose, but there is no standard convention in this matter. We know, however, that the counting numbers are unlimited, so that whereas we have just given a sequence of only a finite number of them, which we now call a **finite sequence,** it is possible to think of arranging an unlimited or infinite number of the counting numbers, which we call an **infinite sequence,** i.e. a sequence without end or, in other words, a sequence with no last term. This idea is represented by appropriately not writing a last term so that 1, 2, 3, 4, . . . is to be taken as representing an **infinite sequence.**

The question of predictability is very important in mathematics, for we need to be able to speak of terms of a sequence both in particular and general. In the example we have here we are able to quote any term we please. For example, the twentieth term is 20, the eighty-fourth term is 84 and so on. From this example we can produce another sequence such as

$$1, -2, 3, -4, 5, \ldots \qquad \text{(A)}$$

As written it indicates an infinite sequence having 1 as its first term. Because the signs are alternately positive and negative we describe this sequence as an **infinite, alternating sequence**; that is, the word alternating is used with respect to the signs of the consecutive terms and nothing else. An example of another alternating sequence is

$$-11, \tfrac{1}{12}, -13, \tfrac{1}{14}, -15, \tfrac{1}{16}, \ldots \qquad \text{(B)}$$

which is just as predictable as the two already mentioned. Observe that the first term of a sequence does not have to be 1—indeed, it seldom is. Again concentrating on predictability, in sequence (A) there is no difficulty in seeing that the twentieth term is -20 and the thirty-third term is 33. Similarly, in the sequence (B) the twentieth term is $\frac{1}{30}$ and the thirty-third term is -43; observations which depend upon discovering the pattern of writing, the fact that the minus sign is only attached to the even-positioned terms in (A), just as the even-positioned terms of (B) are all positive and in reciprocal form.

We now need to turn from the particular term to the general one, which in mathematics is always spoken of as the *n*th term. Furthermore, in this branch of work we refer to the terms as $u_1, u_2, u_3, u_4, \ldots$ and employ the abbreviation $\{u_n\}$ to represent the sequence. This means that we write

$$\{u_n\} = u_1, u_2, u_3, \ldots$$

to represent an infinite sequence with the added information that whatever is written within the curly brackets will represent the *n*th term of the sequence, thus n always takes the values 1, 2, 3, ... in that order.

As an example of the use of this notation we have the following

$$\{n\} = 1, 2, 3, 4, 5, \ldots \text{ (i.e. the } n\text{th term } u_n = n)$$
$$\{2^n\} = 2, 2^2, 2^3, 2^4 \ldots$$

or

$$\{2^n\} = 2, 4, 8, 16, \ldots \text{ (i.e. the } n\text{th term } u_n = 2^n)$$
$$\{(-1)^n\} = -1, 1, -1, 1, -1, \ldots \text{ (i.e. the } n\text{th term } u_n = (-1)^n)$$

To obtain the twenty-third term of the sequence $\{(-1)^n\}$ we put $n = 23$ to get $u_{23} = (-1)^{23} = -1$.

Similarly, to obtain the twenty-third term of the sequence $\{n^2\}$ we put $n = 23$ to get $u_{23} = 23^2$.

Exercise 11.1

Write out the first four terms and find u_{10} in each of the following sequences

1. $\left\{\frac{1}{n}\right\}$
2. $\{2n\}$
3. $\{2n + 1\}$
4. $\{2n - 1\}$
5. $\{n^3\}$
6. $\{(-1)^{n+1}\}$
7. $\{a + n - 1\}$
8. $\{a + (n - 1)d\}$
9. $\{r^{n-1}\}$
10. $\{ar^{n-1}\}$

Write down the next two terms of the following sequences and also give the *n*th term:

11. 1, 3, 5, 7, ...
12. −1, 3, −5, 7, ...
13. $1, \frac{1}{2}, \frac{1}{3}, \frac{1}{4}, \ldots$
14. $\frac{1}{5}, \frac{1}{6}, \frac{1}{7}, \frac{1}{8}, \ldots$
15. 3, 6, 12, 24, 48, ...
16. $\frac{1}{2}, \frac{3}{4}, \frac{5}{6}, \frac{7}{8}, \ldots$

Write down the next two terms of the following sequences:

17. $\frac{7}{5}, \frac{17}{12}, \frac{41}{29}, \ldots$
18. $\frac{13}{7}, \frac{34}{20}, \frac{94}{54}, \ldots$

Limits

Looking at the first five examples in the previous exercise we can immediately see an essential difference between the first sequence and the next four. The terms of the first sequence $\{1/n\}$ decrease as n increases, a process which we might have described as 'the terms are getting gradually smaller' if it was not for the fact that we are unable to give a clear mathematical meaning to the word 'gradually'. The greater that n becomes the smaller that $1/n$ becomes, indeed by choosing n large enough we can get $1/n$ as close to 0 as we please. The terms of the sequence $\{2n\}$ clearly increase as n increases, so that each term is greater than the preceding term. The greater that n becomes the greater that $2n$ becomes, and we can make $2n$ as large as we please merely by choosing the appropriate n.

For example, is it possible to get $1/n < 1/10\,000\,000$? The answer is yes, by taking $n = 10\,000\,001$ so that all succeeding terms after this will also be less than $1/10\,000\,000$.

Is it possible to find a term in $\{2n\}$ which is greater than $A = 60\,000\,000$? Again the answer is yes; by taking $n = 30\,000\,001$, so that all succeeding terms after this will also be greater than 60 000 000.

We have here two different mathematical ideas. The first sequence has terms which are converging to a particular number, in this case 0—that is, converging in the sense that by choosing n large enough we can get $1/n$ as close to 0 as we please. The second sequence has terms which are diverging in the sense that given any number A whatever, we can always find a value for n such that after this value all the terms will be greater than A.

Not all sequences of increasing terms will behave like $\{2n\}$, for consider the sequence $\{3 - 1/n\} = 2, 2\frac{1}{2}, 2\frac{2}{3}, 2\frac{3}{4}, \ldots$, wherein each term is greater than the preceding term and yet no term will ever be equal to or greater than 3. There will, of course, always be terms greater than any number which is less than 3; for example, 2·999. Writing $2{\cdot}999 = 3 - 0{\cdot}001 = 3 - 1/1000$, we recognise this as the thousandth term of the sequence, so all the terms after this one will be greater than 2·999, but still, of course, less than 3. So we are able to get terms of the sequence which are as close to 3 as we please, and furthermore, having got one term as close as we please, all the succeeding terms will be even closer. The number 3 is called the **limit** of the sequence $\{3 - 1/n\}$.

In slightly more formal language we say that u_n tends to 3 as n tends to infinity (i.e. there is no stopping the value of n). A simple abbreviation of this statement is written $u_n \to 3$ as $n \to \infty$. So the idea of a limit is associated with an infinite sequence whereby n must be allowed to tend to infinity, i.e. to remain unchecked, or unbounded.

The important point to note is that we shall find the limit of a sequence by inspection and deduce the result from experience because the formal method of finding a limit is beyond the scope of this work. In the interests of gaining the necessary experience consider the following examples.

Example 1. The limit of the sequence $\{1/n\}$ is 0.

Discussion. Writing the first six terms will give some feel for the pattern of the sequence

$$\therefore \quad \left\{\frac{1}{n}\right\} = 1, \tfrac{1}{2}, \tfrac{1}{3}, \tfrac{1}{4}, \tfrac{1}{5}, \tfrac{1}{6}, \ldots$$

We note that (i) each term is positive, (ii) the terms are decreasing, (iii) the greater we make n the smaller $1/n$ becomes, (iv) the greater we make n the closer $1/n$ gets to 0, (v) we can get $1/n$ as close to 0 as we please by taking n large enough. The limit of the sequence $\{1/n\}$ is 0.

Example 2. The limit of the sequence $\{5 + 1/n\}$ is 5.

Discussion. Again we write down the first few terms

$$\therefore \quad \left\{5 + \frac{1}{n}\right\} = 6, 5\tfrac{1}{2}, 5\tfrac{1}{3}, 5\tfrac{1}{4}, 5\tfrac{1}{5}, 5\tfrac{1}{6}, \ldots$$

We note that (i) each term is positive; (ii) the terms are decreasing; (iii) the greater n becomes the smaller $1/n$ becomes and the nearer the terms get to 5, because we can get $1/n$ as close to 0 as we please by taking n large enough then we can get $5 + 1/n$ as close to 5 as we please by the same process of increasing n. The limit of the sequence $\{5 + 1/n\}$ is 5 or, in other words, the sequence $\{5 + 1/n\}$ converges to 5 as n tends to infinity.

Example 3. The limit of the sequence $\{(-1)^n/n\}$ is 0.

Discussion. In this case we have

$$\left\{\frac{(-1)^n}{n}\right\} = -1, \tfrac{1}{2}, -\tfrac{1}{3}, \tfrac{1}{4}, -\tfrac{1}{5}, \tfrac{1}{6}, -\tfrac{1}{7}$$

We note that (i) the terms alternate in sign, (ii) the absolute values of the terms are decreasing in the manner of $\{1/n\}$, (iii) the greater we make n the nearer the terms get to 0. The limit of the sequence is 0.

Example 4. The limit of the sequence $\{r^n\}$ is 0 if $-1 < r < 1$.

Discussion. Taking a positive value for r we get a sequence of positive terms only. If we take $-1 < r < 0$ we have a sequence of terms which alternate in sign. Two numerical examples to illustrate these points are

$$(A)\ r = \tfrac{3}{4}.\ \{r^n\} = \tfrac{3}{4}, \tfrac{9}{16}, \tfrac{27}{64}, \tfrac{81}{256}, \tfrac{243}{1024}, \tfrac{729}{4096}, \ldots$$

$$(B)\ r = -\tfrac{1}{2}.\ \{r^n\} = -\tfrac{1}{2}, \tfrac{1}{4}, -\tfrac{1}{8}, +\tfrac{1}{16}, -\tfrac{1}{32}, +\tfrac{1}{64}, -\tfrac{1}{128}, \ldots$$

We note that (i) in both the cases (A) and (B) the absolute values of the terms are decreasing, and furthermore decreasing more rapidly than the corresponding terms of the sequence $\{1/n\}$; (ii) as long as $-1 < r < 1$ the terms of the sequence will decrease in absolute value; and (iii) the greater we make n the closer the terms approach 0—indeed we may get as close as we please to 0 by taking n great enough.

Clearly if $r = 1$ then $\{r^n\}$ becomes 1, 1, 1, 1, 1, 1, . . . so the limit of the sequence is 1. If $r = -1$ then $\{r^n\}$ becomes $-1, +1, -1, +1, -1, +1, -1, +1, \ldots$ Since there is no *single* number which r^n approaches as n tends to infinity then we say that this sequence, i.e. $\{(-1)^n\}$ does not have a limit. We do, however, refer to the sequence as having an upper limit of 1 and a lower limit of -1, but the finer points of this analysis are not appropriate at this stage.

Just to try to ensure that the notion of a limit is understood consider two more numerical illustrations.

Example 5. The sequence $\left\{\dfrac{n}{n+1}\right\}$ converges to 1.

Discussion. We have

$$\left\{\frac{n}{n+1}\right) = \tfrac{1}{2}, \tfrac{2}{3}, \tfrac{3}{4}, \tfrac{4}{5}, \ldots$$

The terms are positive and increasing but never greater than 1. The nth term $u_n = \left\{\dfrac{n}{n+1}\right\}$ may be rewritten $u_n = \left\{\dfrac{1}{1 + 1/n}\right\}$

Since we know that

$$\frac{1}{n} \to 0 \text{ as } n \to \infty \text{ then we deduce (correctly) that } \frac{1}{1 + 1/n} \to \frac{1}{1 + 0} = 1.$$

The sequence $\left\{\dfrac{n}{n+1}\right\}$ converges to 1.

Example 6. The sequence 0·3, 0·33, 0·333, 0·3333, 0·33333 converges to $\frac{1}{3}$.

Discussion. The sequence represents a recurring decimal which would normally be written as $0{\cdot}\dot{3}$. We could represent the sequence in the more general manner of $\left\{\frac{1}{3}\left(1-\frac{1}{10^n}\right)\right\}$. (The reader should check this by substituting $n = 1, 2, 3, \ldots$). We know from Example 4 that with $r = \frac{1}{10}$, $r^n \to 0$ as $n \to \infty$, so we use this result here to deduce that $\frac{1}{3}\left(1-\frac{1}{10^n}\right) \to \frac{1}{3}(1-0)$ as $n \to \infty$. The sequence therefore converges to $\frac{1}{3}$.

Exercise 11.2

Find the limits of the following sequences as n tends to infinity:

1. $\left\{\frac{1}{n^2}\right\}$ 2. $\left\{1+\frac{1}{n^2}\right\}$ 3. $\left\{\frac{1}{6^n}\right\}$ 4. $\left\{\left(\frac{5}{6}\right)^n\right\}$ 5. $\left\{\frac{n+1}{n}\right\}$

6. $\left\{\frac{1-n}{n}\right\}$ 7. $\left\{\frac{n+2}{n+1}\right\}$ 8. $\left\{\frac{n^2+1}{2n^2+3}\right\}$ 9. $\left\{\frac{n^2+2n}{3n^2}\right\}$ 10. $\left\{\left(\frac{1}{5}\right)^n\right\}$

11. Using the result in Example 6 give the limits of the following sequences:
    (i) 0·1, 0·11, 0·111, 0·1111, . . . (ii) 2·6, 2·66, 2·666, 2·6666, . . .

12. Write down the first five terms of the following sequences and find their respective limits.
    (i) $\left\{1-\frac{1}{10^n}\right\}$ (ii) $\left\{\frac{n}{2^n}\right\}$ (iii) $\{(-1)^n + (-1)^n\}$

Series

A series is a sum of an ordered set of terms; that is, a sequence whose terms are added to each other. Examples are:

(i) $1 + 2 + 3 + 4 + 5 + 6 + \ldots + 79 + 80$

(ii) $\frac{1}{2} + \frac{1}{2^2} + \frac{1}{2^3} + \frac{1}{2^4} + \ldots$

(iii) $\frac{1}{6} + \frac{1}{7} + \frac{1}{8} + \ldots$

(iv) $9 + (-3) + 1 + \left(-\frac{1}{3}\right) + \frac{1}{9} + \left(-\frac{1}{27}\right) \ldots$

The series in (i) is a finite series and would be described as the sum of the first eighty natural numbers. The other three series are infinite series. In each case the terms of the series are predictable because we can obtain a formula for the nth term of the series. The nth terms are

(i) $u_n = n$ (ii) $u_n = \frac{1}{2^n}$

(iii) $u_n = \frac{1}{5+n}$ (iv) $u_n = 9\left(-\frac{1}{3}\right)^{n-1}$

The last series is also described as an alternating series because the terms alternate in sign. It may also be written as $9 - 3 + 1 - \frac{1}{3} + \frac{1}{9} - \frac{1}{27} \ldots$, which does not contradict the definition of a series as being a *sum* of the terms of a sequence because the terms of a sequence are not restricted to positive terms only. The interesting point about the infinite series is that we can find a sum at all: considering the number of terms involved, series (iii) is impossible to sum, but (ii) and (v) are quite easy, as we shall shortly see.

The notation for the sum of a series is Σ (**sigma**), and we indicate the number and starting point of the terms concerned by writing something like

$$\sum_{n=1}^{20} u_n$$

to mean

$$u_1 + u_2 + u_3 + \ldots + u_{19} + u_{20}$$

Alternatively we write S_{20} to represent the sum of the *first* twenty terms of the series concerned.

Thus $\sum_{n=1}^{20} u_n$ means the same as S_{20}. The advantage of the sigma notation arises when the starting point is not the first term of the series, for we can then indicate this by writing

$$\sum_{n=7}^{20} u_n$$

which means

$$u_7 + u_8 + u_9 + \ldots + u_{19} + u_{20}$$

The sum of the infinite series is indicated by either $\sum_{n=1}^{\infty} u_n$ or S_∞ but it must be observed that we have a different meaning for the word sum in this context.

The Arithmetical Progression

This type of series (abbreviated to A.P.) is one in with the difference between consecutive terms is always constant. Typical examples are as follows

(i) $1 + 3 + 5 + 7 + 9 \ldots$
(ii) $55 + 50 + 45 + 40 + \ldots$

In each case the **common difference** between the terms may be found by subtracting any term from the following term so that in (i) $5 - 3$ or $9 - 7$ or $7 - 5$ all give the common difference of 2, and in (ii) $50 - 55$, or $40 - 45$, etc., give the common difference of -5.

Having obtained the common difference relation between consecutive terms we are now able to predict any term we please. Suppose we require the twentieth term of each series. We shall need to find the pattern of the terms, and our thoughts might proceed along the following lines:
'$u_1 = 1$, $u_2 = 3$, which is $1 + 2$; $u_3 = 5$, which is $1 + 2 + 2$; $u_4 = 7$, which is $1 + 2 + 2 + 2$ or $1 + 3(2)$; $u_5 = 9$, which is $1 + 4(2)$; this suggests $u_6 = 1 + 5(2)$ (Check: $u_6 = 11$); then $u_7 = 1 + 6(2)$, the pattern of results suggests that $u_{20} = 1 + 19(2) = 39$'.

Flushed with success, it is easy to see that $u_n = 1 + (n - 1)2 = 2n - 1$. Incidentally, this series would be described as the sum of the first twenty odd numbers. Now that we know the first and last terms we can find the sum of twenty terms as follows

$$S_{20} = 1 + 3 + 5 + \ldots + 37 + 39$$

The series in reverse order is

$$S_{20} = 39 + 37 + 35 + \ldots 3 + 1$$

Adding the two expressions we get

$$2S_{20} = 40 + 40 + 40 + \ldots + 40 + 40 = 40 \times 20$$

$$\therefore S_{20} = \frac{40 \times 20}{2} = 400 = 10\,\{\text{first} + \text{last terms}\} = 10\,\{u_1 + u_{20}\}$$

Similarly $S_{50} = 25$ {first + last terms} $= 25\,\{u_1 + u_{50}\}$, and since the fiftieth term is given by $u_{50} = 1 + 49(2)$ the result follows.

For the second series again we need to find the pattern of the terms and again we piece it together as '$u_1 = 55$, $u_2 = 50$, which is $55 - 5$; $u_3 = 55 - 5 - 5$; $u_4 = 55 - 3(5)$; $u_5 = 55 - 4(5)$; so $u_{20} = 55 - 19(5) = -40$'.

Again we overflow into $u_n = 55 - (n - 1)5 = 60 - 5n$.

Hence $\quad S_{20} = 55 + 50 + 45 + \ldots - 35 - 40$
In reverse $\quad S_{20} = -40 - 35 - 30 - \ldots + 50 + 55$

$$\therefore 2S_{20} = 15 + 15 + 15 + \ldots + 15 + 15 = 15 \times 20$$
$$\therefore S_{20} = 150 = 10\,\{\text{first} + \text{last terms}\} = 10\,\{u_1 + u_{20}\}$$

The general notation for an A.P. is to represent the first term by a and the common difference by d. The letter l is sometimes used to represent the last term to be summed.

Example 7. Find the nth term and the sum of the first n terms of the series

$$a + (a + d) + (a + 2d) + (a + 3d) + \ldots$$

Solution. Here we have $u_1 = a$, $u_2 = a + d$, $u_3 = a + 2d$ and hence $u_n = a + (n-1)d$

$S_n = a + (a + d) + (a + 2d) + \ldots + (a + (n - 2)d) + (a + (n - 1)d)$

In reverse

$S_n = (a + (n - 1)d) + (a + (n - 2)d) + (a + (n - 3)d) + \ldots + (a + d) + a$

$2S_n = [2a + (n - 1)d] + [2a + (n - 1)d] + \ldots$
$\qquad + [2a + (n - 1)d] + [2a + (n - 1)d]$

$2S_n = n[2a + (n - 1)d]$

Hence $S_n = \frac{n}{2}[2a + (n - 1)d]$ Answer

Observe that $S_n = \frac{n}{2}\Big[a + a + (n - 1)d\Big] = \frac{n}{2}\Big[\text{first} + \text{last terms}\Big]$

as previously found. To check the result for (ii) above we had $n = 20$, $a = 55$, $d = -5$. Substituting in the formula we have just obtained gives

$$S_{20} = \frac{20}{2}\Big[110 + 19(-5)\Big] = 10\Big[110 - 95\Big] = 150 \text{ (as before).}$$

Exercise 11.3

1. Find the arithmetical progressions in the following examples:

(i) $2 + 4 + 6 + 8 + \ldots$
(ii) $2 + 4 + 8 + 16 + \ldots$
(iii) $100 + 50 + 25 + \ldots$
(iv) $100 + 50 + 0 - 50 \ldots$
(v) $10 + 9 + 8 + 7 + \ldots$
(vi) $\frac{1}{2} + \frac{1}{3} + \frac{1}{4} + \frac{1}{5} + \ldots$
(vii) $0{\cdot}1 + 0{\cdot}2 + 0{\cdot}3 + 0{\cdot}4 + \ldots$
(viii) $k + 3k + 5k + 7k + \ldots$
(ix) $13 - 11 + 9 - 7 + 5 \ldots$
(x) $x + (x + 3) + (x + 6) + (x + 9) \ldots$

2. Write down the next two terms and the nth term in each of the following A.P.'s:

(i) $6 + 8 + 10 + \ldots$
(ii) $10 + 8 + 6 + \ldots$
(iii) $10 + 20 + 30 + \ldots$
(iv) $30 + 20 + 10 + \ldots$
(v) $0 + d + 2d + 3d + \ldots$
(vi) $0 - k - 2k - 3k - \ldots$
(vii) $a + (a + d) + (a + 2d) + \ldots$
(viii) $a + (a - 3) + (a - 6) + (a - 9) + \ldots$
(ix) $1 + 1\frac{1}{4} + 1\frac{1}{2} + \ldots$
(x) $3 + 2\frac{7}{8} + 2\frac{3}{4} + \ldots$

3. Write down the first, second and tenth terms together with the common difference of the series with the following nth terms:

(i) $3 + 7(n - 1)$ (ii) $5 + 2(n - 1)$ (iii) $2n + 3$
(iv) $6 - 2n$ (v) $8 - 7n$ (vi) $\frac{1}{5}(8n + 3)$

4. Find the following sums:

(i) S_{20} for $1 + 2 + 3 + 4 + \ldots$
(ii) S_n for $1 + 2 + 3 + 4 + \ldots$ (i.e. the sum of the first n terms)
(iii) S_{15} for $30 + 28 + 26 + \ldots$
(iv) S_{31} for $1 + 3 + 5 + \ldots$
(v) S_{101} for $4 + 8 + 12 + 16 + \ldots$

5. A woman's salary starts at £1500 per annum and rises by annual increments of £100 at the beginning of each year. What would be her salary at the end of 10 years and what is the total salary to be earned in those ten years?
6. Find the sum of the first n terms of the A.P. $1 + 3 + 5 + 7 + 9 \ldots$ and show that the sum of the first n odd numbers is always a perfect square. Express 25 and 49 as sums of ordered odd numbers.
7. The sum of the first thousand even numbers starting with 2 is 10 001 000. Use this result, and this result only, to obtain (i) the sum of the first thousand odd numbers starting with 1, and (ii) the sum of the first thousand natural numbers starting with 1.
8. The eleventh term of an A.P. is 34 and the twenty-first term is 64. Find (i) the first term, (ii) the common difference and (iii) the sum of the first fifteen terms of the series.

The Geometrical Progression

This type of series (abbreviated to G.P.) is one in which the ratio of consecutive terms is constant. Typical examples are

(i) $3 + 6 + 12 + 24 + 48 + \ldots$
(ii) $1000 + 200 + 40 + 8 + 1{\cdot}6 + \ldots$
(iii) $4 + (-12) + 36 + (-108) + \ldots$
or $4 - 12 + 36 - 108 + - + - \ldots$
(iv) $66 - 33 + 16\frac{1}{2} - 8\frac{1}{4} + 4\frac{1}{8} - + - + \ldots$
or $66 + (-33) + 16\frac{1}{2} + (-8\frac{1}{4}) + 4\frac{1}{8} + \ldots$

One immediate difference between an A.P. and a G.P. is that an alternating series may be a G.P. but never an A.P., as we see in cases (iii) and (iv) above. In each case the **common ratio** is obtained by dividing any term by the preceding term, so that in (i) $6 \div 3$, $12 \div 6$ or $48 \div 24$ each give the common ratio of 2, and in (ii) $200 \div 1000$, $8 \div 40$, etc., each give the common ratio of $\frac{1}{5}$ or 0·2. In (iii) the common ratio of -3 is given by $-12 \div 4$, or $36 \div (-12)$, etc., and finally in (iv) the common ratio is $-\frac{1}{2}$ as found from $-33 \div 66$ or $15\frac{1}{2} \div (-7\frac{3}{4})$ and so on.

As in the case of the A.P. we need to find only two things in order to be able to generate the complete progression. In the case of the A.P. it was the first term and the common difference, in the case of the G.P. we require only the first term and the common ratio.

As before we need to be able to predict any term of the series we please and this depends upon finding the pattern of the terms. Taking the G.P. given in (i) above we may continue the series with ease as 96, 192, 384, etc., but what is not so obvious is how to obtain the twentieth term (say) without having the previous nineteen terms available. Our thoughts might proceed along the following lines for (i)

'$u_1 = 3$; $u_2 = 6$, which is $3 + 3$ or 3×2; $u_3 = 12$, which is $3 + 3 + 6$ or $3 \times 2 \times 2$; $u_4 = 24$, which is $3 \times 2 \times 2 \times 2$; i.e. we see the pattern emerging in terms of multiplication rather than addition. $u_5 = 48 = 3(2)^4$; $u_6 = 3(2)^5$; so $u_n = 3(2)^{n-1}$. If we require the tenth term, then $u_{10} = 3(2)^9 = 1536$.

In the case of (iv) above we think possibly along the lines '$u_1 = 66$; $u_2 = -33$, which is $66 \div (-2)$ or $66(-\frac{1}{2})$;

$$u_3 = 15\tfrac{1}{2} = -33(-\tfrac{1}{2}) = 66(-\tfrac{1}{2})(-\tfrac{1}{2}) \text{ or } 66(-\tfrac{1}{2})^2;$$

$u_4 = 66(-\frac{1}{2})^3 = -\frac{66}{8} = -8\frac{1}{4}$; so $u_n = 66(-\frac{1}{2})^{n-1}$.' If we require the sixth term then $u_6 = 66(-\frac{1}{2})^5 = -\frac{66}{32} = -\frac{33}{16}$.

Summation

The standard method of summation of a number of terms is very straightforward once the first and last terms are known. Consider finding the sum of the first fifteen terms of the series

$$3 + 6 + 12 + 24 + 48 + \ldots$$

The fifteenth and last term is $3(2)^{14}$, so we have

$$S_{15} = 3 + 6 + 12 + 24 + \ldots + 3(2)^{13} + 3(2)^{14}$$

Multiplying this result by the common ratio 2 we obtain

$$2S_{15} = 6 + 12 + 24 + \ldots + 3(2)^{13} + 3(2)^{14} + 3(2)^{15}$$

Subtracting the two results gives

$$S_{15} - 2S_{15} = 3 - 3(2)^{15}$$

i.e.

$$(1 - 2)S_{15} = 3(1 - (2)^{15})$$

$$\therefore S_{15} = \frac{3(1 - (2)^{15})}{1 - 2} = 3(2^{15} - 1) = 98304$$

It is always a surprise to see so few terms, which start from such small beginnings, amount to so much.

Example 8. Find the tenth term and the sum of the first ten terms of the G.P.

$$7 - 21 + 63 - \ldots$$

Solution. The common ratio is $-21 \div 7 = -3$ or $63 \div (-21) = -3$. Since the first term is 7, the tenth term is $7(-3)^9$, which we shall leave in this form for the moment.

$\therefore \quad S_{10} = 7 - 21 + 63 - 189 \ldots +7(-3)^8 + 7(-3)^9$

or

$$S_{10} = 7 + (-21) + 63 + (-189) + \ldots +7(-3)^8 + 7(-3)^9$$

Multiply this result by the common ratio -3

$\therefore \quad -3S_{10} = (-21) + 63 + (-189) + \ldots +7(-3)^9 + 7(-3)^{10}$

Subtraction of these two lines gives

$$S_{10} - (-3)S_{10} = 7 + 0 + 0 + 0 + \ldots + 0 - 7(-3)^{10}$$

$$\therefore \quad (1 - (-3))S_{10} = 7 - 7(-3)^{10} = 7(1 - (-3)^{10})$$

$$\therefore \quad S_{10} = \frac{7(1 - (-3)^{10})}{1 - (-3)} = \frac{7}{4}(1 - (-3)^{10}) = -7 \times \frac{59048}{4} = -103334 \quad \text{Answer}$$

Example 9. Find the sum of the first n terms of the general G.P. given by

$$a + ar + ar^2 + ar^3 + ar^4 \ldots$$

Solution. This is the standard notation for a G.P. having a as the first term and r as the common ratio. The nth term is ar^{n-1}

$$\therefore \quad S_n = a + ar + ar^2 + ar^3 + \ldots + ar^{n-3} + ar^{n-2} + ar^{n-1}$$

Multiply this result by the common ratio r

$$\therefore \quad rS_n = ar + ar^2 + ar^3 + \ldots + ar^{n-3} + ar^{n-2} + ar^{n-1} + ar^n$$

Subtraction of these two results now gives

$$S_n - rS_n = a + 0 + 0 + 0 + \ldots + 0 + 0 + 0 - ar^n$$

$$\therefore \quad (1 - r)S_n = a - ar^n = a(1 - r^n)$$

Hence

$$S_n = \frac{a(1 - r^n)}{1 - r} \quad \text{Answer}$$

Since this gives the sum of the first n terms of any G.P. we can check back on Example 8 above where $a = 7$ and $r = -3$, $n = 10$ so that

$$S_{10} = \frac{7(1 - (-3)^{10})}{1 - (-3)} = \frac{7}{4}(1 - (-3)^{10})$$

as before.

The Convergence of an Infinite G.P.

Back on page 129 we discussed what happens to r^n when $-1 < r < 1$, as n tends to infinity. We saw that $r^n \to 0$ as $n \to \infty$ when $-1 < r < 1$, and the inclusion of this result into the above expression for the sum

$$S_n = \frac{a(1 - r^n)}{1 - r}$$

is now relevant because we see that if $-1 < r < 1$ we are able to say that as n tends to infinity and r^n tends to 0, so S_n tends to

$$\frac{a(1 - 0)}{1 - r} = \frac{a}{1 - r}.$$

Now do note that S_n *tends to* $\frac{a}{1 - r}$—that is we can get S_n as close as we please to $\frac{a}{1 - r}$ merely by taking n large enough. In consequence we say that the sum of the infinite series is $\frac{a}{1-r}$ and we agree that even with this different meaning of the word *sum* we shall continue to use the equality sign and write

$$S_\infty = \frac{a}{1-r}.$$

To repeat, then, we agree that $S_\infty = \frac{a}{1-r}$ is to be accepted as meaning that $S_n \to \frac{a}{1-r}$ as $n \to \infty$.

Example 10. Find the sum of the infinite G.P. whose first three terms are given by

$$1 + \tfrac{1}{3} + \tfrac{1}{9} + \ldots$$

Solution. In the standard notation we have $a = 1$, $r = \frac{1}{3}$ so that $S_n = \frac{1(1 - (\frac{1}{3})^n)}{1 - \frac{1}{3}}$
The sum of the infinite series is found by letting n tend to infinity in the result for S_n. We know that $(\frac{1}{3})^n \to 0$ as $n \to \infty$ so we are left with

$$S^n \to \frac{1(1-0)}{1-\frac{1}{3}} = \frac{3}{2} \quad \text{as} \quad n \to \infty$$

Therefore the sum of the infinite series is $1\frac{1}{2}$. An alternative phrase would be that the series converges to the sum $1\frac{1}{2}$. Answer

Exercise 11.4

1. State which of the following series are A.P.'s or G.P.'s and in each case find the common ratio or difference and the 11th term:

   (i) $1 + \frac{1}{4} + \frac{1}{16} + \frac{1}{64} + \ldots$
   (ii) $2 + 4 + 8 + 16 + \ldots$
   (iii) $3 + 5 + 7 + 9 + \ldots$
   (iv) $1 + 5 + 25 + 125 + \ldots$
   (v) $1 - 2 + 4 - 8 + \ldots$
   (vi) $243 - 81 + 27 - + - + \ldots$

2. Find the sum of the following series:

   (i) $1 + x + x^2 + x^3 + x^4 + \ldots + x^{12}$
   (ii) $1 - x + x^2 - x^3 + x^4 \ldots + x^{12}$
   (iii) $1 + x + x^2 + x^3 + \ldots x^9$
   (iv) $1 + 2x + 4x^2 + 8x^3 + \ldots$ to seven terms

3. Find the sums of the following infinite series:

   (i) $1 + \frac{1}{2} + \frac{1}{4} + \frac{1}{8} + \ldots$
   (ii) $3 - 1 + \frac{1}{3} - \frac{1}{9} + \ldots$
   (iii) $1 + x + x^2 + x^3 + \ldots; -1 < x < 1$
   (iv) $1 - x + x^2 - x^3 + \ldots; -1 < x < 1$

4. Given that $S_{10} = 1 + 2x + 3x^2 + 4x^3 + 5x^4 + 6x^5 + 7x^6 + 8x^7 + 9x^8 + 10x^9$ find xS_{10}, and by combining the two results obtain the sum

$$S_{10} = \frac{1 - 11x^{10} + 10x^{11}}{(1-x)^2}$$

5. Find the sum of the infinite series $1 + \frac{1}{x} + \frac{1}{x^2} + \frac{1}{x^3} + \frac{1}{x^4} + \ldots$ where $x > 1$

6. The third and fifth terms of a G.P. of positive terms are 50 and 1250 respectively. Find the first term, the common ratio and the sum of the first six terms.

Arithmetic and Geometric Means

We define these respective means as follows:

1. The arithmetic mean of n quantities is (the sum of these quantities) divided by n. Thus if a, b, c and d are the quantities then $\frac{1}{4}(a+b+c+d)$ is their arithmetic mean.

2. The geometric mean of n quantities is the nth root of their product. Thus if a, b, c, d and e are the quantities then their geometric mean is $\sqrt[5]{(abcde)}$.

For two positive quantities we can easily prove that the A.M. $\geqslant$ G.M., for if we suppose a and b are the two quantities then

$$\text{the arithmetic mean } A = \tfrac{1}{2}(a+b), \text{ the geometric mean } G = \sqrt{ab}$$

$$\therefore \text{A.M.} - \text{G.M.} = \tfrac{1}{2}\{a - 2\sqrt{ab} + b\} = \tfrac{1}{2}\{\sqrt{a} - \sqrt{b}\}^2$$

The right-hand side, being a perfect square, is therefore positive or zero.

$$\therefore \text{A.M.} > \text{G.M. if } a \neq b \text{ and A.M.} = \text{G.M. if } a = b$$

It follows that A.M. $\geqslant$ G.M. The need to take roots and deal with real numbers only does require reservations about the signs of the quantities concerned, e.g. the geometric mean of -9 and 4 is not real.

Three consecutive terms of an A.P. are

$$a + (n-1)d,\ a + nd,\ a + (n+1)d,$$

and the arithmetical mean of the two outer terms is

$$\tfrac{1}{2}\{a + (n-1)d + a + (n+1)d\} = \tfrac{1}{2}\{2a + 2nd\} = a + nd,$$

which is the middle term. Thus any term of an A.P. (other than the first term) is the arithmetic mean of the two adjacent terms.

Similarly, if we take three consecutive terms of a G.P., we get

$$ar^{n-1},\ ar^n,\ ar^{n+1}$$

and the geometric mean of the two outer terms is

$$\sqrt{(ar^{n-1})(ar^{n+1})} = \sqrt{a^2r^{2n}} = ar^n,$$

which is the middle term. Thus any term of a G.P. (except the first term) is the geometric mean of the two adjacent terms.

Example 11. Prove that a square has a greater area than any rectangle with the same perimeter.

Solution. Let the sides of the rectangle be a and b so that the perimeter is $2(a+b) = p$. The area A of the square with perimeter $p = 2(a+b)$ is $\dfrac{(a+b)^2}{4}$.

The area of the rectangle is ab. From A.M. $\geqslant$ G.M. we have $\dfrac{a+b}{2} \geqslant \sqrt{ab}$ which gives $\dfrac{(a+b)^2}{4} \geqslant ab$ (squaring both sides).

Hence $A \geqslant ab$.

Example 12. How many terms must be taken of the infinite series

$$1 + \tfrac{1}{4} + \tfrac{1}{16} + \tfrac{1}{64} + \ldots$$

in order that their sum shall be within 0·01 of the sum of the infinite series?

Solution. The series is a G.P. with $a = 1$, $r = \frac{1}{4}$ so that the sum to infinity is

$$S = \frac{1}{1 - \dfrac{1}{4}} = \frac{4}{3}$$

The sum

$$S_n = \frac{1\left(1-\frac{1}{4^n}\right)}{1-\frac{1}{4}} = \frac{4}{3}\left(1-\frac{1}{4^n}\right) = \frac{4}{3} - \frac{1}{3 \times 4^{n-1}}$$

We require

$$S - S_n = \frac{4}{3} - \left\{\frac{4}{3} - \frac{1}{3 \times 4^{n-1}}\right\} = \frac{1}{3 \times 4^{n-1}} < \frac{1}{100}$$

This means that we need n so that $3 \times 4^{n-1} > 100$.

With a problem as elementary as this we may find n by trial substitution since $n = 4$ gives $3 \times 4^{n-1} = 3 \times 4^3 = 3 \times 64 > 100$, then $n = 4$ will do but possibly $n = 3$ might be better. $3 \times 4^{3-1} = 48$ so clearly we need four or more terms to get within 1/100 of the sum to infinity.

If we modify the problem to require a closeness of 1/1 000 000 the solution is not so easily obtained. We would require in this case to have $3 \times 4^{n-1} > 1\,000\,000$. Taking logarithms means that we need

$$\log 3 + \log 4^{n-1} = \log 3 + (n-1)\log 4 > \log 1\,000\,000 = 6.$$

From tables this becomes $(n - 1)0{\cdot}6021 > 6 - 0{\cdot}4771 = 5{\cdot}5229$

$$\therefore \quad n - 1 > \frac{5{\cdot}5229}{0{\cdot}6021}$$

The right-hand side of the inequality lies between 9 and 10, so we satisfy the inequality by taking $n = 11$

Check

$$S_{11} = \frac{1\left(1-\frac{1}{4^{11}}\right)}{1-\frac{1}{4}} = \frac{4}{3}\left(1-\frac{1}{4^{11}}\right) \quad \therefore \quad S - S_{11} = \frac{1}{3 \times 4^{10}}$$

$$\log(3 \times 4^{10}) = \log 3 + 10 \log 4 = 0{\cdot}4771 + 6{\cdot}021 = 6{\cdot}4981$$

$$\therefore \quad 3 \times 4^{10} = 3\,149\,000{\cdot}0$$

which is greater than 1 000 000 and the result above is correct. Answer

Exercise 11.5

1. Find the A.M.s and G.M.s of the following sets of numbers and confirm that A.M. > G.M. in each case:
   (i) 9, 4
   (ii) 2, 9, 12
   (iii) 2, 4, 8, 64
2. Find the number of terms of the series $1 + 1/10 + 1/100 + 1/1000 + \ldots$ which must be taken in order to be within 1/1 000 000 of the sum to infinity.
3. £1000 is invested at a compound interest rate of 10% per annum paid at the end of each year. Show that at the end of the fifth year the investment amounts to £1000 × $(1{\cdot}1)^5$ and find how long it takes to double the investment to the nearest year.
4. A series of perpendicular lines *AB*, *BC*, *CD*, *DE*, *EF*, . . . are drawn in an anti-clockwise order such that each line is perpendicular to the previous one in the series. The relative lengths of the lines are

   $$AB = BC = 2CD = 2DE = 4EF = 4FG = 8GH = 8HI = 16IJ, \text{ etc.,}$$

   and in the final picture *A*, *C*, *E*, *G*, *I*, . . . lie on a straight line. With $AB = \sqrt{8}$ find the distance of *A* from the end of (i) the second line, (ii) the sixth line, (iii) the 4*n*th line. Also find the limit of your answer (iii) as *n* tends to infinity.

5. An equilateral triangle shape is made from a set of equal spheres which fill the triangle, as in the game of snooker. How many spheres are required to make a triangle with (i) 3 and (ii) 4 spheres along each side?
6. How many equal spheres are needed to make a complete pyramid of five layers in the form of a regular tetrahedron shape (each face an equilateral triangle)?
7. A ball when dropped from a height of x cm onto a horizontal table rebounds to a height of $\frac{3}{4}$ cm. Find the total distance the ball will move through when it is dropped from a height of 120 cm until it comes to rest on the table.
8. A clock strikes the hour and once on the half hour. How many times does the clock strike in 24 hours?
9. Novice gamblers frequently attempt to win by 'doubling up' each time they lose. If such a gambler loses £1, then £2, then £4 by this means what are his total losses if he loses ten bets in a row?

CHAPTER TWELVE

THE BINOMIAL EXPANSION: THE METHOD OF INDUCTION

Pascal's Triangle

Being able to see the pattern in a set of given terms of a series is a first step towards finding either the sum of the series or the limit of its terms. Associated with the binomial series we shall be discussing is a pattern of coefficients which form what is called Pascal's triangle. The top five rows of this triangle are given below. Study the arrangement and you will see that once the two top rows are written down the rest will follow in the manner indicated

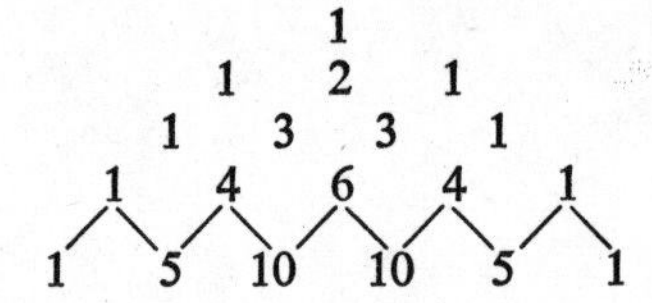

By the connecting lines between the fourth and fifth rows, the next row would be

$$1 \quad 6 \quad 15 \quad 20 \quad 15 \quad 6 \quad 1 \quad \text{and so on}$$

We could mix in a little algebra and the pattern would remain just as evident

$$\begin{array}{ccccccccccc} & & & & & 1 & & & & & \\ & & & 1 & & 2x & & 1x^2 & & & \\ & & 1 & & 3x & & 3x^2 & & 1x^3 & & \\ & 1 & & 4x & & 6x^2 & & 4x^3 & & 1x^4 & \\ 1 & & 5x & & 10x^2 & & 10x^3 & & 5x^4 & & 1x^5 \end{array}$$

but here the pattern is in the coefficients of the powers of x. We have merely taken the first pattern and then attached powers of x in ascending order along each row from left to right. Inserting some more algebra to balance the powers of x, we have

$$\begin{array}{ccccccccccc} & & & & & 1 & & & & & \\ & & & 1a^2 & & 2ax & & 1x^2 & & & \\ & & 1a^3 & & 3a^2x & & 3ax^2 & & 1x^3 & & \\ & 1a^4 & & 4a^3x & & 6a^2x^2 & & 4ax^3 & & 1x^4 & \\ 1a^5 & & 5a^4x & & 10a^3x^2 & & 10a^2x^3 & & 5ax^4 & & 1x^5 \end{array}$$

and the extensions of the triangle are again easy to produce.

Now consider the expansions of powers of positive integral powers of $a + x$; that is, $(a + x)^2$, $(a + x)^3$, and so on, meaning by the expansion the result of evaluating the powers of $a + x$ in the form of a series.

$$(a + x)^2 = (a + x)(a + x) = a^2 + 2ax + x^2 \quad \text{(2nd row of Pascal's triangle)}$$

$$\begin{aligned}(a + x)^3 = (a + x)^2(a + x) = {} & a^2 + 2ax + x^2 \\ & a + x \\ \hline & a^3 + 2a^2x + ax^2 \\ & \quad + a^2x + 2ax^2 + x^3 \\ \hline & a^3 + 3a^2x + 3ax^2 + x^3 \quad \text{(3rd row of Pascal's triangle)}\end{aligned}$$

$$\begin{aligned}(a + x)^4 = (a + x)^3(a + x) = {} & a^3 + 3a^2x + 3ax^2 + x^3 \\ & a + x \\ \hline & a^4 + 3a^3x + 3a^2x^2 + ax^3 \\ & \quad a^3x + 3a^2x^2 + 3ax^3 + x^4 \\ \hline & a^4 + 4a^3x + 6a^2x^2 + 4ax^3 + x^4 \quad \text{(4th row of Pascal's triangle)}\end{aligned}$$

Similarly we may obtain the expansion of $(a + x)^5$, $(a + x)^6$ by using the relevant rows of the triangle. For instance, $(a + x)^{11}$ will have its coefficients of the powers of a and x given by the row which starts 1, 11. We describe the expansions in a little more detail by referring to them as expansions in ascending powers of x or descending powers of a.

Example 1. Expand $(-3y + 1)^4$ in ascending powers of y.

Solution. To be consistent with the order of the expansion above, we rewrite this as $(1 - 3y)^4$. With $x = -3y$ we have

$$(1 + x)^4 = 1 + 4x + 6x^2 + 4x^3 + x^4$$

Now replace the x with $-3y$ and we have

$$1 - 12y + 54y^2 - 108y^3 + 81y^4 \qquad \text{Answer}$$

Example 2. Expand $(1 + x)^6$ and by putting $x = 0{\cdot}003$ find the value of $(1{\cdot}003)$ correct to eight places of decimals.

Solution. $(1 + x)^6 = 1 + 6x + 15x^2 + 20x^3 + 15x^4 + 6x^5 + x^6$ by using the sixth row of the Pascal triangle. Substituting $x = 0{\cdot}003$ involves finding x^2, x^3, etc., and this we do separately

$$\begin{aligned} x &= 0{\cdot}003 & 6x &= 0{\cdot}018 \\ x^2 &= 0{\cdot}003^2 = 0{\cdot}000\,009 & 15x^2 &= 0{\cdot}000\,135 \\ x^3 &= 0{\cdot}000\,000\,027 & 20x^3 &= 0{\cdot}000\,000\,54 \\ x^4 &= 0{\cdot}000\,000\,000\,081 & 15x^4 &= 0{\cdot}000\,000\,001\,215 \end{aligned}$$

There is nothing to be gained by going any further. Hence by adding the results so far

$$(1{\cdot}003)^6 = 1{\cdot}018\,135\,54 \text{ (correct to eight decimal places)} \qquad \text{Answer}$$

An interesting point is that $6 \log 1{\cdot}003 = 6 \times 0{\cdot}0012 = 0{\cdot}0072$. Hence by logarithms, $(1{\cdot}003)^6 = 1{\cdot}016$—an incorrect result which arises from the fact that we have used only four-figure logarithms whose error in approximation to the four figures has been multiplied by 6.

Exercise 12.1

1. Add the sixth and seventh lines to the Pascal triangle on page 140 and write down the expansion of $(a + x)^6$ and $(a + x)^7$.

2. Add all the terms of each of the rows of the numerical Pascal triangle and deduce the sum of the terms in (i) the 10th row and (ii) the nth row.
3. In any row—e.g. the seventh row of the Pascal triangle 1, 7, 21, 35, 35, 21, 7, 1—the even numbered terms are u_2, u_4, u_6, u_8 and the odd numbered terms are u_1, u_3, u_5, u_7. Show that the sum of the odd numbered terms is equal to the sum of the even numbered terms in (i) the 7th row and (ii) the 10th row. Using the result in Question 2 (ii) find the sum of the odd numbered terms in the nth row.
4. Expand $(a + x)^6$ in ascending powers of a.
5. Expand $(3x + 2y)^4$ in descending powers of y.
6. Expand $(x - y)^5$ in descending powers of x.
7. Evaluate $(1{\cdot}01)^6$ correct to 4 decimal places, by finding the expansion of $(1 + x)^6$ and then putting $x = 0{\cdot}01$.
8. Evaluate $(0{\cdot}99)^6$ correct to 4 decimal places by finding the expansion of $(1 + x)^6$ and then putting $x = -0{\cdot}01$.
9. Find the values of the following without the use of tables.

   (i) $(1 + \sqrt{5})^3 - (1 - \sqrt{5})^3$
   (ii) $(\sqrt{2} + \sqrt{5})^3 - (\sqrt{2} - \sqrt{5})^3$
   (iii) $(3 + \sqrt{2})^4 - (3 - \sqrt{2})^4$

10. It is a simple matter to arrange the following quantities in ascending order: $53^{\frac{1}{12}}$, $904^{\frac{1}{12}}$, and $1000^{\frac{1}{12}}$, because the index is the same in each case and the magnitude of the result is dependent only upon the base numbers 53, 904 and 1000. Had $1000^{\frac{1}{12}}$ been expressed as $(10^3)^{\frac{1}{12}} = 10^{\frac{1}{4}}$ we could have effected the comparison merely by converting all the numbers to the same power. Thus to compare the magnitudes of $4^{\frac{1}{3}}$ and $3^{\frac{1}{2}}$ we express each as $(4^2)^{\frac{1}{6}}$ and $(3^3)^{\frac{1}{6}}$, i.e. $16^{\frac{1}{6}}$ and $27^{\frac{1}{6}}$, to see that $3^{\frac{1}{2}} > 4^{\frac{1}{3}}$. Working from the method of this example compare the following to arrange in ascending order.

    (i) $5^{\frac{1}{2}}, 8^{\frac{1}{3}}$ (iii) $2^{\frac{1}{2}}, 3^{\frac{1}{3}}, 4^{\frac{1}{4}}$
    (ii) $5^{\frac{1}{2}}, 24^{\frac{1}{4}}$ (iv) $5^{\frac{2}{3}}, 7^{\frac{1}{2}}$

The Binomial Expansion

The sum of any two algebraic terms such as $a + x$ is called a binomial; consequently the expansion of $(a + x)^n$ is called a binomial expansion. The Pascal triangle was used to obtain the expansion of $(a + x)^n$ where n was a positive integer and did not exceed 10 in the examples that were discussed. To obtain the expansion of $(a + x)^{20}$ by the same means would be an extremely tedious procedure but since the binomial expansion has such a wide application there is a standard notation for representing the coefficients. In this text we shall only deal with the expansion for an index n which is a positive integer.

Combinations and Permutations

We add a brief reminder of the idea and notation of combinations, which are arrangements of a given number of quantities without regard to their order. For example, the number of different combinations of three articles a, b, c taken two at a time is given by ab, bc, ca.

ab and ba being the same combination because no regard is paid to the order of the letters, similarly with bc and cb.

To represent this numerical result we write $C_2^3 = 3$, i.e. the number of different combinations of three things taken two at a time is 3.

The number of different combinations of five things taken three at a time would be written as C_3^5 and given by *abc, abd, acd, bcd, ebc, ebd, ecd, aec, aed, abe*, so that $C_3^5 = 10$. We now need to gain a connection between the

5 and 3 in C_3^5 and the result 10 or the n and r in C_r^n and whatever the result might be. To find this connection we now consider permutations.

Permutations of a set of articles are arrangements in which the order is important, i.e. important in the sense that *abc*, *acb*, *bac*, *bca*, *cba*, *cab* are the six different permutations of the three articles or letters a, b and c. If we have n different letters to be permutated three at a time (i.e. P_3^n) then we may think along the following lines.

To take the first position we may choose n letters, but having filled it there are only $n - 1$ letters to take the second position, and having filled these two positions there will be $n - 2$ letters to fill the third position. The total number of different permutations will then be $n(n - 1)(n - 2)$. Had we required permutations taking four letters at a time the total number of different permutations would have been $n(n - 1)(n - 2)(n - 3)$. Sometimes the word 'different' is left to be implied and we speak merely of the number of combinations and permutations.

Example 3. Find the number of permutations and the number of combinations of 10 different letters taken four at a time without repetition.

Solution. To find the number of permutations. The first letter may be taken in 10 different ways and having done this the second letter may be taken in 9 different ways. This means that there are 90 permutations of two letters from 10, i.e. each time a first letter is chosen there are 9 second choices to go with it and since we may choose 10 different first choices then there are 90 permutations of two letters from ten. Each one of these 90 permutations may be followed by eight different letters so that the number of permutations taking three at a time from the ten letters is $10 \times 9 \times 8$. Similarly the number of permutations taking four at a time is $10 \times 9 \times 8 \times 7 = 5040$.

Answer

Reminder. *abcd*, *bacd*, *cbda*, etc., are the same combination of the four letters a, b, c, d. Now any one combination of four different letters would give $4 \times 3 \times 2 \times 1$ permutations so that all the possible permutations will be given by all the combinations multiplied by $4 \times 3 \times 2 \times 1$

$$\therefore (4 \times 3 \times 2 \times 1)\, C_4^{10} = P_4^{10}$$

$$\therefore C_4^{10} = \frac{P_4^{10}}{4 \times 3 \times 2 \times 1} = \frac{10 \times 9 \times 8 \times 7}{4 \times 3 \times 2 \times 1}$$

Answer

It is at this stage that we ought to begin to see the connection between the 10 and 4 and the final result. We now seek to improve the notation and make the results easier to recognise by introducing what are called **factorials**.

Definition. For any positive integer n, factorial n, written $n!$, is defined by $n! = n(n - 1)(n - 2)(n - 3)(n - 4) \ldots (4)(3)(2)(1)$. We define $0! = 1$. Thus $4! = 4 \times 3 \times 2 \times 1$, $10! = 10 \times 9 \times 8 \times 7 \times 6 \times 5 \times 4 \times 3 \times 2 \times 1$ and $6! = 6 \times 5 \times 4 \times 3 \times 2 \times 1$. Combining these results, we see that

$$\frac{10!}{4!\,6!} = \frac{10 \times 9 \times 8 \times 7}{4 \times 3 \times 2 \times 1}$$

Therefore $$C_4^{10} = \frac{10 \times 9 \times 8 \times 7}{4 \times 3 \times 2 \times 1} = \frac{10!}{4!\,6!}$$

Observation of the pattern of this result leads to

$$C_t^n = \frac{n!}{t!(n-t)!}$$

that is, the number of combinations of n different letters taken t at a time is C_t^n. Alternatively, we may reason that each combination of t letters allows $t!$ permutations so that the number of permutations of n letters taken t at a time is $(t!)C_t^n$. But, for these permutations the first letter may be taken in n different ways and, having made the first choice, it leaves $n - 1$ different ways of taking the second letter and thereafter $n - 2$ ways. Hence

$$P_3^n = n(n-1)(n-2) = n(n-1)(n-3+1)$$
$$P_4^n = n(n-1)(n-2)(n-3) = n(n-1)(n-2)(n-4+1)$$

and similarly

$$P_t^n = n(n-1)(n-2)(n-3)\ldots(n-t+1)$$
$$\therefore (t!)C_t^n = P_t^n$$
$$\therefore C_t^n = \frac{P_t^n}{t!} = \frac{n(n-1)(n-2)(n-3)\ldots(n-t+1)}{t!}.$$
$$\therefore C_t^n = \frac{\{n(n-1)(n-2)\ldots(n-t+1)\}}{t!}\frac{(n-t)(n-t-1)\ldots 3.2.1}{(n-t)!}$$
$$= \frac{n!}{t!(n-t)!}$$

Example 4. Find the number of different combinations of twelve things taken seven at a time.

Solution. We require $C_7^{12} = \frac{12!}{7!5!} = \frac{12.11.10.9.8.7.6.5.4.3.2.1}{(7.6.5.4.3.2.1)(5.4.3.2.1)}$

(using multiplication dots rather than crosses)

$$\therefore \quad C_7^{12} = \frac{12.11.10.9.8}{5.4.3.2.1} = 11.9.8$$

$$C_7^{12} = 792$$ Answer

Example 5. Show that $C_t^n = C^n_{n-t}$

Solution. $C_t^n = \frac{n!}{t!(n-t)!}$, For C^n_{n-t} we now write $n - t$ where we had t in C_t^n.

$$\therefore \quad C^n_{n-t} = \frac{n!}{(n-t)!(n-(n-t))!} = \frac{n!}{(n-t)!t!} = C_t^n$$ Answer

As an example of this last result we have $C_7^{12} = C_5^{12}$ or $C_8^{10} = C_2^{10}$ and so on.

Exercise 12.2

Evaluate the following:

1. C_4^6 2. C_4^7 3. C_2^6 4. C_3^7
5. C_{10}^{10} 6. C_5^8 7. C_3^8 8. C_2^{14}
9. Show that $C_4^{10} + C_3^{10} = C_4^{11}$ and try to suggest a general formula of this kind starting with C_t^n.
10. Find the number of different combinations of the letters a, b, c, d, e taken
    (i) four at a time
    (ii) four at a time and containing ab

11. Find how many different numbers can be formed from the six digits 1, 2, 3, 4, 5, 6 taken four at a time without repetition. How many of these numbers are divisible by (i) 5, (ii) 2?
12. How many numbers in between 1000 and 2000 are divisible by 5? (Repetition allowed, i.e. 1115, etc.)

The Binomial Expansion (*continued*)

Considering $(a + x)^6$ as $(a + x)(a + x)(a + x)(a + x)(a + x)(a + x)$ and the expansion in ascending powers of x, we obtain the various coefficients as follows:

(i) a^6 can only be obtained by selecting an a from each bracket.
(ii) Terms in a^5x arise from selecting x from one bracket and a^5 from the remaining five brackets. From six we can choose one at a time C_1^6 ways.
(iii) Terms in a^4x^2 arise from selecting x^2 from two of the six brackets and we can do this C_2^6 ways, and a^4 from the remaining four brackets.
(iv) Terms in a^3x^3 arise from selecting x^3 from three of the brackets and we can do this C_3^6 ways, and a^3 from the remaining three brackets.

Similarly for a^2x^4, ax^5, x^6 until we have

$$(a + x)^6 = a^6 + C_1^6a^5x + C_2^6a^4x^2 + C_3^6a^3x^3 + C_4^6a^2x^4 + C_5^6ax^5 + x^6.$$

The Binomial Theorem for a Positive Integral Index *n*

This theorem states that

$$(a + x)^n = a^n + C_1^na^{n-1}x + C_2^na^{n-2}x^2 + C_3^na^{n-3}x^3 + \dots + C_t^na^{n-t}x^t + \dots C^n_{n-1}ax^{n-1} + x^n$$

As in the case of $(a + x)^6$ we obtain the coefficients (called the binomial coefficients) of the terms by considering the manner in which the term is selected from the n brackets of the expansion.

(i) Terms in $a^{n-1}x$ arise out of selecting x from one bracket and a^{n-1} from the remaining $n - 1$ brackets, and this may be done in C_1^n ways.
(ii) Terms in $a^{n-2}x^2$ arise out of selecting x^2 from the multiplication of two brackets and a^{n-2} from the rest; selecting two brackets from n brackets may be done in C_2^n ways.
(iii) The general term of the expansion represented by terms $a^{n-t}x^t$ arises from obtaining x^t from multiplying t brackets and a^{n-t} from the remaining $n - t$ brackets. We can select t brackets from n brackets in C_t^n ways so that the general term is given by $C_t^na^{n-t}x^t$. There are of course $n + 1$ terms in all and $C_t^na^{n-t}x^t$ is the $(t + 1)$th term. If the binomial coefficients are written out, the early terms of the expansion become

$$(a + x)^n = a^n + na^{n-1}x + \frac{n(n-1)}{2!}a^{n-2}x^2 + \frac{n(n-1)(n-2)}{3!}a^{n-3}x^3 + \dots$$

a form worth remembering for those occasions when only the first few terms are required, as in the following example.

Example 6. Find the value of $(1{\cdot}007)^{20}$ correct to three decimal places.

Solution. This is similar to the earlier problems but Pascal's triangle would be much too tedious to employ so we now use the Binomial theorem.

$$(1 + x)^{20} = 1 + 20x + \frac{20.19.x^2}{2} + \frac{20.19.18.x^3}{2.3} + \ldots$$

Putting $x = 0{\cdot}007$ we get

$$(1{\cdot}007)^{20} = 1 + 20(0{\cdot}007) + 190(0{\cdot}007)^2 + 1140(0{\cdot}007)^3 + \ldots$$
$$\therefore\ (1{\cdot}007)^{20} = 1 + 0{\cdot}14 + 0{\cdot}009\,31 + 0{\cdot}000\,391$$
$$= 1{\cdot}150 \text{ (correct to 3 decimal places)}$$ Answer

By logarithms

$$20 \log 1{\cdot}007 = 20 \times 0{\cdot}0029 = 0{\cdot}058$$
$$\therefore\ (1{\cdot}007)^{20} = 1{\cdot}143$$

This gives some idea of the way in which the approximation of logarithms to four significant figures, although sometimes exaggerated, is nevertheless very close.

An alternative method for obtaining the expansion would be to consider

$$\left(1 + \frac{7x}{10}\right)^{20} = 1 + 20.\left(\frac{7x}{10}\right) + \frac{20.19}{2}.\left(\frac{7x}{10}\right)^2 + \frac{20.19.18}{2.3}\left(\frac{7x}{10}\right)^3 + \frac{20.19.18.17}{2.3.4}\left(\frac{7x}{10}\right)^4 + \ldots$$
$$= 1 + 14x + 93{\cdot}1x^2 + 391x^3$$

Now put $x = 0{\cdot}01$ to make $(1 + 7x/10)^{20} = (1 + 0{\cdot}007)^{20}$ and the result follows.

Example 7. Find the coefficient of x^5 in the expansion of $(2 - x)^9$.

Solution. From the binomial theorem the term with x^5 in $(a - x)^9$ is $C_5{}^9a^4(-x)^5$. The coefficient of x^5 is therefore

$$C_5{}^9(2)^4(-1)^5 = \frac{9.\overset{2}{\not{8}}.7.\not{6}}{\not{4}.\not{3}.\not{2}.1}(-16) = -2016$$ Answer

Example 8. Find the coefficient of x^7 in the expansion of $(3 - x/3)^{11}$.

Solution. The required coefficient is

$$C_7{}^{11}\,(3)^4\left(-\frac{1}{3}\right)^7 = \frac{11.10.9.8}{4.3.2.1}\left(-\frac{1}{3}\right)^3 = -\frac{110}{9}$$ Answer

Example 9. Expand $(1 - 2x - x^2)^{10}$ in ascending powers of x as far as x^3.

Solution. We put this expression into binomial form by writing

$$(1 - 2x - x^2)^{10} = (1 - [2x + x^2])^{10}$$

and expand binomially as

$$1 - 10\,[2x + x^2] + \frac{10.9}{2}\,[2x + x^2]^2 - \frac{10.9.8}{2.3}\,[2x + x^2]^3 + - + - + \ldots$$

the unwritten terms involving x^4 and higher powers of x. We now have

$$(1 - 2x - x^2)^{10} = 1 - 20x - 10x^2 + 45\{4x^2 + 4x^3 + x^4\} - 120\{8x^3 + 12x^4 + 6x^5 + x^6\} + \ldots$$
$$= 1 - 20x - 10x^2 + 180x^2 + 180x^3 - 960x^3 + \text{(terms in } x^4 \text{ and higher powers of } x)$$
$$= 1 - 20x + 170x^2 - 780x^3 \ldots$$

Alternatively consider

$$[(1 - 2x) - x^2]^{10} = (1 - 2x)^{10} - 10(1 - 2x)^9x^2 + 45(1 - 2x)^8x^4 - + - +$$

We need only consider the first two terms here because thereafter x will be to powers of 4 and above.

$$\therefore \quad [(1-2x)-x^2]^{10} = (1-20x+180x^2-960x^3+\text{terms in } x^4 \text{ and above})$$
$$-10x^2(1-18x+\text{terms in } x^2 \text{ and above})$$
$$= 1-20x+180x^2-960x^3-10x^2+180x^3\ldots$$
$$= 1-20x+170x^2-780x^3\ldots \text{ as before}$$

Exercise 12.3

1. Write down the binomial expansions of the following:

   (i) $(1+x)^8$ (ii) $(1-x)^8$ (iii) $(x-1)^8$
   (iv) $(2+x)^7$ (v) $(2-x)^7$ (vi) $(x-2)^7$
   (vii) $(3+2x)^6$ (viii) $(3-2x)^6$ (ix) $(2x-3)^6$

2. Find the coefficients of x^9 in the expansion of:

   (i) $(1+x)^{11}$ (ii) $(1-x)^{11}$ (iii) $(2+x/2)^{11}$

3. Write down the first four terms of the expansion of the following in ascending powers of x:

   (i) $(2+x)^{10}$ (ii) $(3+x)^{10}$

   and find $(2{\cdot}0001)^{10}$, $(3{\cdot}000\,01)^{10}$ correct to 2 decimal places.

4. Expand each of the following as far as the term in x^3:

   (i) $(1+x+x^2)^{12}$ (ii) $(1+x-x^2)^{12}$ (iii) $(1+x+2x^2)^{12}$

5. Write down the $t+1$ term in the binomial expansion of $(x+1/x)^{10}$ and find the term which is independent of x.

Mathematical Induction

It is not always possible or convenient to prove a theorem by a direct argument, especially on those occasions when the theorem is introduced at a time or place in our studies when the underlying mathematical ideas have not been fully developed. But, whenever a theorem can be proved by taking particular cases in succession then the method of proof by mathematical induction becomes very useful. In broad detail it means that if by putting $n = 1, 2, 3$, etc., we are able to select and prove particular cases of the theorem then we can prove the truth of the theorem by assuming it to be true for $n = k$ and demonstrating that this implies that the theorem is true for $n = k+1$ and consequently for all values of n. If this statement appears somewhat vague consider the following examples as first illustrations.

Example 10. Prove that the sum S_n of the first n natural numbers is given by $S_n = \frac{1}{2}n(n+1)$, i.e.

$$S_n = 1+2+3+4+\ldots+(n-1)+n = \tfrac{1}{2}n(n+1)$$

Solution. We verify the suggested result in the simplest case when $n = 1$. Thus $S_1 = 1 = \frac{1}{2}(1)(1+1) = 1$, and the formula is true for $n = 1$.

Now we assume that the formula is true when $n = k$: that is, we are assuming that

$$S_k = 1+2+3+4+\ldots+(k-1)+k = \tfrac{1}{2}k(k+1)$$

On this assumption

$$S_{k+1} = 1+2+3+4+\ldots+(k-1)+k+(k+1) = S_k+(k+1)$$
$$= \tfrac{1}{2}k(k+1)+(k+1) = (k+1)(\tfrac{1}{2}k+1) = \tfrac{1}{2}(k+1)[(k+1)+1]$$

This means that on the assumption that the formula was true for S_k we obtain the result $S_{k+1} = \frac{1}{2}(k+1)(k+2)$, which is exactly the same result we get by substituting $k+1$ for n in the formula $S_n = \frac{1}{2}n(n+1)$.

This last step proves that whenever the formula is true for $n = k$ then it is also true for $n = k+1$. Now we do not have to assume that it is true for $n = 1$ because we know that it is true. But, since it is true for $n = 1$ then it must be true for $n + 1 = 2$ and since it is true for $n = 2$ then it must be true for $n + 1 = 3$.

By continuing this argument we see that the formula must be true for all positive integer values of n.

Example 11. Prove by induction that the sum of n terms of the series

$$1.2 + 2.3 + 3.4 + 4.5 + \ldots + n(n + 1) = \tfrac{1}{3}n(n + 1)(n + 2)$$

Solution. First we verify that the formula is true when $n = 1$.
If $S_n = \frac{1}{3}n(n + 1)(n + 2)$ then $S_1 = \frac{1}{3}(2)(3) = 2$ which is true since the first term is 1.2. The sum of the first $k + 1$ terms is

$$\begin{aligned} S_{k+1} &= 1.2 + 2.3 + 3.4 + 4.5 + \ldots k(k + 1) + (k + 1)(k + 2) \\ &= S_k + (k + 1)(k + 2) \end{aligned}$$

If we assume that the formula is true for S_n when $n = k$ we may write

$$S_k = \tfrac{1}{3}k(k + 1)(k + 2)$$

and, substituting this assumed result above, we have

$$\begin{aligned} S_{k+1} &= \tfrac{1}{3}k(k + 1)(k + 2) + (k + 1)(k + 2) \\ &= (\tfrac{1}{3}k + 1)(k + 1)(k + 2) \\ &= \tfrac{1}{3}(k + 3)(k + 1)(k + 2) = \tfrac{1}{3}(k + 1)(k + 2)(k + 3) \\ &= \tfrac{1}{3}(k + 1)[(k + 1) + 1][(k + 1) + 2] \end{aligned}$$

Now this is exactly the result we would get by putting $n = k + 1$ in the formula $S_n = \frac{1}{3}n(n + 1)(n + 2)$, which means that if the formula is true for $n = k$ it is also true for $n = k + 1$. But it really is true for $n = 1$, as already confirmed. Hence it must be true for $n = 1 + 1 = 2$, and being true for $n = 2$ it must also be true for $n = 2 + 1 = 3$, and so on. Consequently it is true for *all* positive integer values of n.

Example 12. Prove that $3^{2n} - 1$ is always divisible by 8 [n a positive integer].

Solution. When $n = 1$, $f(n) = 3^{2n} - 1 = 3^2 - 1 = 8$ so that the result is true for $n = 1$. Assume the result is true when $n = k$ then

$$\begin{aligned} f(k + 1) &= 3^{2k+2} - 1 = 3^2(3^{2k} - 1) + 8 = 9(3^{2k} - 1) + 8 \\ \therefore\ f(k + 1) &= 9f(k) + 8 \end{aligned}$$

so, if $f(k)$ is divisible by 8 then so is $f(k + 1)$. But, $f(1)$ is divisible by 8 so, $f(2)$ is divisible by 8 from this last result. Since $f(2)$ is divisible by 8 then so is $f(3)$ and so on. Hence $f(n) = 3^{2n} - 1$ is divisible by 8. (See Exercise 12.4, No. 6.)

Example 13. Prove that

$$\frac{1}{1.2} + \frac{1}{2.3} + \frac{1}{3.4} + \ldots + \frac{1}{n(n + 1)} = \frac{n}{n + 1}$$

Solution. We are being asked to prove that the sum of n terms of the series is given by $S_n = n/(n + 1)$. When $n = 1$, we have $S_1 = \frac{1}{2}$, which is correct since the first term is $1/(1.2)$. Assuming that the formula is true when $n = k$, i.e. $S_k = k/(k + 1)$, consider finding S_{k+1} by adding the $k + 1$th term of the series to S_k to get

$$S_{k+1} = S_k + \frac{1}{(k + 1)(k + 2)}$$

$$\therefore\ S_{k+1} = \frac{k}{k + 1} + \frac{1}{(k + 1)(k + 2)}, \text{ since } S_k = \frac{k}{k + 1}$$

$$\therefore\ S_{k+1} = \frac{k(k + 2)}{(k + 1)(k + 2)} + \frac{1}{(k + 1)(k + 2)} = \frac{(k + 1)^2}{(k + 1)(k + 2)} = \frac{k + 1}{k + 2}$$

$$= \frac{k + 1}{(k + 1) + 1}$$

which is exactly the result we would have obtained by substituting $n = k + 1$ in the formula $s_n = \frac{n}{n+1}$. Hence, if the formula is true for $n = k$ it is also true for $n = k + 1$. But, the formula is true for $n = 1$, hence it must be true for $n = 2$.

Hence $S_n = \frac{n}{n+1}$ for all positive integers n. Answer

Notice that with the last example we have the opportunity to sum the series to infinity; that is, to find S_n as $n \to \infty$. But $S_n = \frac{n}{n+1} = \frac{1}{1 + 1/n}$ and as $n \to \infty, \frac{1}{n} \to 0$

$$\therefore S_n \to \frac{1}{1+0} = 1$$

Hence the infinite series $\frac{1}{1.2} + \frac{1}{2.3} + \frac{1}{3.4} + \frac{1}{4.5} + \ldots$ converges to a sum of 1.

Exercise 12.4

Use the method of induction to prove the following results:

1. $1 + 3 + 5 + 7 + \ldots + (2n - 1) = n^2$
2. $2 + 4 + 6 + 8 + \ldots + 2n = n(n + 1)$
3. $1^3 + 2^3 + 3^3 + 4^3 + \ldots + n^3 = \frac{1}{4}n^2(n + 1)^2$
4. $6^n - 5n + 4$ is always divisible by 5. (Hint: find $f(k + 1) - f(k)$)
5. $\frac{1}{1.2.3} + \frac{1}{2.3.4} + \frac{1}{3.4.5} + \frac{1}{4.5.6} + \ldots \frac{1}{n(n+1)(n+2)}$ (i.e. n terms) $= \frac{n(n+3)}{4(n+1)(n+2)}$

   Also find the sum of the infinite series.
6. $9^n - 1$ is divisible by 8.
7. $9^n - 8n - 1$ is divisible by 64 for $n \geqslant 2$. (Hint: find $f(k + 1) - f(k)$)
8. $1^2 + 3^2 + 5^2 + \ldots + (2n - 1)^2 = \frac{n}{3}(2n - 1)(2n + 1)$
9. $n(n + 1)$ is divisible by 2.
10. $n^3 - n$ is divisible by 6 for $n \geqslant 2$. (Hint: find $f(k + 1) - f(k)$.)

CHAPTER THIRTEEN

TRIGONOMETRIC FUNCTIONS AND EQUATIONS

Sin, Cos and Tan

From elementary mathematics the reader should already be familiar with the basic trigonometric functions—sin θ, cos θ, tan θ defined for angles from 0° to 90°. We shall need to extend these definitions for angles which are greater than 90°, but before doing so consider a brief revision of the relationship between sin θ, cos θ and tan θ when θ is an acute angle. Consider Fig. 41(*a*) to represent a unit circle (i.e. a circle having a radius of 1 unit) centre O with a tangent drawn at A. The point P is on the circumference of the first quadrant and OP produced intersects the tangent in the point B. NP is perpendicular to OA.

With the angle PON called θ we have $\tan\theta = \dfrac{AB}{OA} = \dfrac{AB}{1} = AB$

$$\sin\theta = \frac{NP}{OP} = \frac{NP}{1} = NP$$

$$\cos\theta = \frac{ON}{OP} = \frac{ON}{1} = ON$$

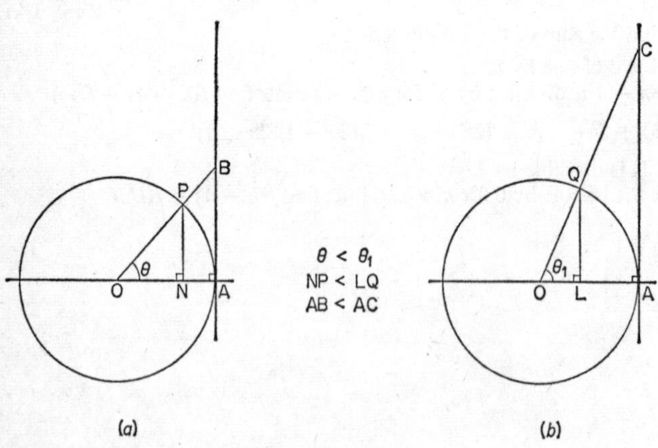

Fig. 41

The single construction of Fig. 41(*a*) will therefore give us a means of comparing all of the trigonometric functions of an angle. Since AB is clearly greater than NP we see that $\tan\theta > \sin\theta$ for $0 < \theta < 90°$. We can obtain another comparison from the following inequality:

Area of triangle OAP < Area of sector OAP < Area of triangle OAB

The area of the sector is $\theta/360$ of the area of the circle, which is $\pi 1^2$, and this gives the area of the sector as

$$\frac{\pi\theta}{360}$$

Since the area of triangle OAP is $\frac{1}{2}.OA.NP = \frac{1}{2}\sin\theta$ and the area of triangle OAB is $\frac{1}{2}.OA.AB = \frac{1}{2}\tan\theta$, the inequality reads as

$$\tfrac{1}{2}\sin\theta < \frac{\pi\theta}{360} < \tfrac{1}{2}\tan\theta$$

or

$$\sin\theta < \frac{\pi\theta}{180} < \tan\theta$$

One further observation from Fig. 41(*a*) and (*b*); as θ increases within $0 < \theta < 90°$, AB and PN increase. In other words, both $\sin\theta$ and $\tan\theta$ are increasing functions of θ in the first quadrant. Likewise, as θ increases so ON decreases in Fig. 41, so we can say that as θ increases so $\cos\theta$ decreases. Again, in other words, $\cos\theta$ is a decreasing function of θ in the first quadrant.

One very important point to remember is that the relation between $NP = \sin\theta$ (or $AB = \tan\theta$) and θ is quite different from the relation between $\overset{\frown}{AP}$ and θ. If we double the size of θ then the length of $\overset{\frown}{AP}$ is also doubled; indeed, a multiplication of θ by any positive number k will multiply the length of $\overset{\frown}{AP}$ by the same number k, which means that the length $\overset{\frown}{AP}$ is directly proportional to θ. But there is no such relationship between $\sin\theta$ and θ—for example, $\sin 15° = 0{\cdot}2588$ and $\sin 30° = 0{\cdot}5000$, so that multiplication of θ by 2 does not multiply $\sin\theta$ by 2, i.e. in general $\sin 2\theta \neq 2\sin\theta$. The same result applies to $\tan\theta$ not being directly proportional to θ—for example, $\tan 15° = 0{\cdot}2679$ and $\tan 45° = 1$, so that in general $\tan 3\theta \neq 3\tan\theta$. Similarly for $\cos\theta$.

Pythagoras's Theorem

Elementary trigonometry provides a very elegant proof of this theorem, which states that if in any right-angled triangle, c is the length of the hypotenuse and a and b the length of the other two sides then $c^2 = a^2 + b^2$. (There are other descriptions of the theorem in terms of area, but our statement is all we need.)

Proof: From Fig. 42, which is drawn in the usual notation of the side of length c opposite angle C and the side of length a opposite angle A, etc., we have

$$AN = b\cos\theta \text{ and } NB = a\sin\theta$$

Since $$AB = c = NB + AN \text{ then } c = a\sin\theta + b\cos\theta \qquad \text{(i)}$$

In the triangle ABC we have $\sin\theta = a/c$ and $\cos\theta = b/c$ and these results, when substituted in (i), give

$$c = \frac{a^2}{c} + \frac{b^2}{c}$$

and finally $$c^2 = a^2 + b^2 \qquad \text{Q.E.D.}$$

This result may be written as

$$1 = \frac{a^2}{c^2} + \frac{b^2}{c^2} = \sin^2\theta + \cos^2\theta$$

from which $\cos^2\theta = 1 - \sin^2\theta$ and $\cos\theta = \pm\sqrt{(1 - \sin^2\theta)}$, and we see that if we know $\sin\theta$ we can always find $\cos\theta$ without using tables. The sign to be chosen from $\pm$ will depend upon the size of θ. For example, if θ is an acute angle then $\cos\theta$ is positive but if θ is an obtuse angle then $\cos\theta$ is negative.

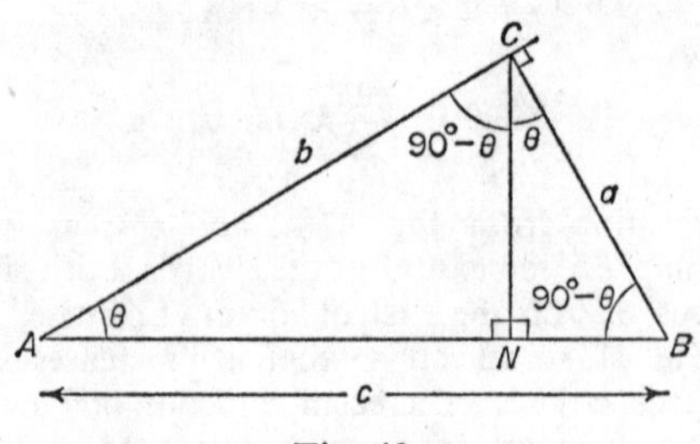

Fig. 42

The relation between $\sin\theta$, $\cos\theta$ and $\tan\theta$ is much closer than the above results suggest for $0 < \theta < 90°$. Using Fig. 42 it is easily seen that with $\angle ABC = 90° - \theta$ we have, $\sin(90° - \theta) = \frac{b}{c} = \cos\theta$.

With $\tan\theta = \frac{a}{b} = \frac{a}{c} \cdot \frac{b}{c} = \frac{\sin\theta}{\cos\theta}$ it is clear that we could use the tables for $\sin\theta$ to do the work of the cosine and tangent tables because

$$\cos\theta = \sin(90° - \theta)$$

$$\tan\theta = \frac{\sin\theta}{\cos\theta} = \frac{\sin\theta}{\sin(90° - \theta)}$$

but convenience dictates that we employ all three sets of tables.

Again note that $(\tan\theta)\tan(90° - \theta) = \frac{a}{b} \times \frac{b}{a} = 1.$

Examination of the result $\sin^2\theta + \cos^2\theta = 1$ shows that since the minimum value of a perfect square is zero then the maximum value of $\sin\theta$ or $\cos\theta$ is 1, and we know from previous work that $\sin 0 = 0$, $\cos 0 = 1$, $\sin 90° = 1$, $\cos 90° = 0$. It is at this stage that we see the problems of discussing $\tan\theta$ as θ tends to $90°$, because with $\tan\theta = \frac{\sin\theta}{\cos\theta}$ and $\cos\theta \to 0$ as $\theta \to 90°$ and $\sin\theta \to 1$ as $\theta \to 90°$ it follows that $\tan\theta$ becomes greater and greater as $\theta \to 90°$, indeed we see that $\tan\theta$ tends to infinity, i.e. its value is unbounded as θ tends to $90°$. We shall see more about the comparisons of the three trigonometric functions on page 159.

Trigonometric Equations

Trigonometric functions are useful in stating many basic scientific principles and occur quite frequently in equations. When only constant values of the trigonometric functions appear in an equation they may be treated like any other constant. For instance, the constant $\frac{1}{2}$ may be written as $\cos 60°$ so that $x + y \cos 60° = 3$ is equivalent to $x + \frac{1}{2}y = 3$.

When variable trigonometric functions appear in an equation the equality is of a basically different type. Then it is called a **trigonometric equation** because it cannot be fully solved by algebraic methods alone. Thus, $2y = \sin x$ is a trigonometric equation. Its solution, however, is more difficult than its apparent simplicity might suggest. Most of the elementary trigonometric equations reduce to a quadratic form in either sin, cos or tan and thereby enable us to employ algebraic methods of solution, as we can see in the next examples.

Example 1. Solve the equation $2 \tan x = 3$, for x.

Solution. If we put $t = \tan x$ the equation takes the form $2t = 3$ with the solution $t = 1{\cdot}5$. This means that we have reduced the equation to $\tan x = 1{\cdot}5$ so that all we need to do now is consult the tangent tables to find that $x = 56° \, 19'$ to the nearest minute.

(*Note:* This is a solution for x an acute angle. We do not know yet whether there are other values of x which might give other solutions to $\tan x = 1{\cdot}5$, but we shall see very soon that there are infinitely many solutions to this result.)

Example 2. Solve the equation $2 \cos^2 x + \cos x - 1 = 0$ for x. (Recall that $\cos^2 x$ means $(\cos x)(\cos x)$ and not $\cos x^2$, which we shall write as $\cos (x^2)$ for clarity; $\cos^2 x$ is read as 'cos squared x' and not 'cos x squared'.)

Solution. Putting $c = \cos x$ the equation becomes

$$2c^2 + c - 1 = 0 = (2c - 1)(c + 1)$$

The solutions are

$$c = \tfrac{1}{2} \text{ or } -1$$

$$\therefore \quad \cos x = \tfrac{1}{2} \text{ or } \cos x = -1$$

Consulting cosine tables we find that $x = 60°$ for $\cos x = \frac{1}{2}$ but nowhere can we find a solution for $\cos x = -1$ (in fact this means $x = 180°$) so that even though we have obtained a solution we do not yet know whether we have obtained the *complete* solution to the equation.

Example 3. Solve the equation $\sin x + \cos x = 1{\cdot}5$, for x.

Solution. With equations like this one we need to combine the result $\sin^2 x + \cos^2 x = 1$ in the sense that wherever we have $\sin^2 x$ we may substitute $1 - \cos^2 x$ or alternatively, wherever we have $\sin x$ we may substitute $\sqrt{(1 - \cos^2 x)}$ or $-\sqrt{(1 - \cos^2 x)}$. Returning to the original equation, we have $\sin x = 1{\cdot}5 - \cos x$, which becomes $\sin^2 x = (1{\cdot}5 - \cos x)^2$ on squaring both sides of the equation. Finally we get the form

$$1 - \cos^2 x = (1{\cdot}5 - \cos x)^2 = 2{\cdot}25 - 3 \cos x + \cos^2 x$$

Rearrangement gives

$$2 \cos^2 x - 3 \cos x - 1{\cdot}25 = 0$$

By putting $c = \cos x$ this becomes a recognisable quadratic equation

$$2c^2 - 3c - 1{\cdot}25 = 0$$

The solution is therefore

$$c = \frac{3 \pm \sqrt{(3^2 - 4(2)(-1{\cdot}25))}}{4} = \frac{3 \pm \sqrt{19}}{4} = \frac{3 \pm 4{\cdot}36}{4} = 1{\cdot}84 \text{ or } -0{\cdot}34$$

$$\therefore \quad \cos x = 1{\cdot}84 \text{ or } -0{\cdot}34 \text{ (correct to 2 decimal places)} \qquad \text{Answer}$$

We know that cos x is never greater than 1 so the result cos $x = 1{\cdot}84$ will not give a real solution to our equation. We also know that there is no acute angle which has a cosine equal to $-0{\cdot}34$. Again we see a need to extend the definition of the trigonometric functions, but before doing so, consider the following exercise.

Exercise 13.1

1. Using the tables find the following:

   (i) $\sin 40°$ and $\cos 50°$
   (ii) $\tan 40°$, $\tan 50°$, $(\tan 40°)(\tan 50°)$
   (iii) $\sin 30°$, $\cos 60°$
   (iv) $\tan 30°$, $\tan 60°$, $(\tan 30°)(\tan 60°)$

2. Find x in each of the following:

   (i) $\sin x = 0{\cdot}5$
   (ii) $\cos x = 0{\cdot}5$
   (iii) $\tan x = 1{\cdot}7321$
   (iv) $\sin x = 0{\cdot}3456$

3. Prove the following results by using Pythagoras's Theorem in a 90°, 60°, 30° triangle or a 90°, 45°, 45° triangle.

   (i) $\sin 60° = \frac{1}{2}\sqrt{3}$ (ii) $\sin 30° = 0{\cdot}5$ (iii) $\tan 60° = \sqrt{3}$
   (iv) $\tan 30° = \frac{1}{3}\sqrt{3}$ (v) $\tan 45° = 1$ (vi) $\sin 45° = \cos 45° = \frac{1}{2}\sqrt{2}$

4. Solve the following equations for x, an acute angle:

   (i) $2\sin^2 x - 5\sin x + 2 = 0$ (ii) $2\cos^2 x - 5\cos x + 2 = 0$
   (iii) $\sin x + \cos x = 1$ (iv) $\sin x - \cos x = 1$
   (v) $3\tan^2 x - 4\tan x + 1 = 0$

Inverse Trigonometric Functions

For a trigonometric function like $y = \sin x$ the inverse trigonometric function, stated in words, reads

'x is equal to the angle whose sine is y'

The customary way of writing this is $x = \sin^{-1} y$ or $x = \text{arc}\sin y$. It is important not to mistake $x = \sin^{-1} y$ for the index notation $x = (\sin y)^{-1}$, which gives $x = 1/\sin y$.

Examples of this notation are as follows

$$0{\cdot}5 = \sin 30° \qquad 30° = \sin^{-1}(0{\cdot}5)$$
$$1 = \tan 45° \qquad 45° = \tan^{-1}(1)$$
$$\tfrac{1}{2}\sqrt{3} = \cos 30° \qquad 30° = \cos^{-1}(\tfrac{1}{2}\sqrt{3})$$

The inverse trigonometric function is being used every time we consulted the tables in order to find the angle. For example, the answers to Question 2 in the previous Exercise 13.1 were all instances of finding a solution to the inverse function, e.g. for $\sin x = 0{\cdot}3456$ we thought $x = \sin^{-1} 0{\cdot}3456$ which was found to be $20°13'$ so that $20°13' = \sin^{-1} 0{\cdot}3456$.

Angles of any Magnitude

Only positive angles are usually considered in elementary mathematics, and these only from 0° to 360° but, in advanced mathematics, this limited concept of the amount of opening between two lines may be generalised to include angles of any real magnitude, positive or negative. For this purpose, it is customary to take the positive x axis of an x, y rectangular coordinate system as the standard reference line.

Imagine a wheel in the plane of such a system of coordinates, with its hub at the origin O, and with its outer rim intersecting the positive x axis at the fixed point Q. If this wheel is set so that a given spoke OP lies along the x axis in its positive direction, the point P at the outer end of the spoke will coincide with the fixed point Q on the x axis. Then, since there is no opening between the two lines OQ and OP, we may say that the angle $QOP = 0°$.

If, now, our wheel is rotated about the origin so that the spoke OP moves in an anti-clockwise direction—like the hands of a clock moving backwards—the generated angle QOP is regarded as positive. But if the wheel is rotated in the opposite, clockwise direction—like the hands of a clock moving normally—the generated angle QOP is regarded as negative. Thus, the angle

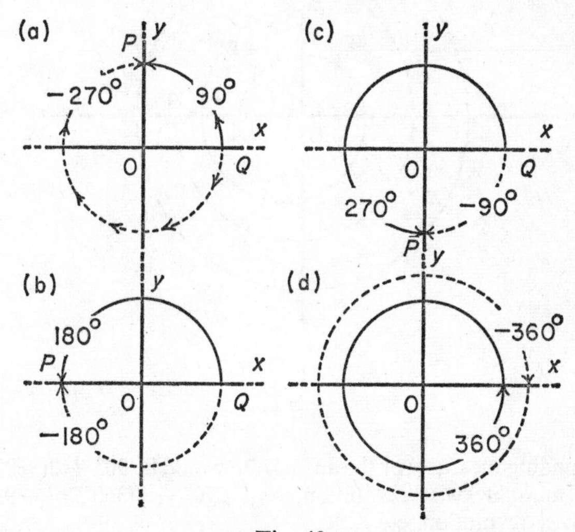

Fig. 43

between the positive x axis and the positive y axis is $(+90°)$ measured anti-clockwise, but is $-270°$ measured clockwise, as in Fig. 43(*a*). And the angle between the positive x axis and itself, which we previously noted to be $0°$, is also $360°$ measured through one complete revolution about the origin in an anti-clockwise direction, but is $-360°$ measured through one complete revolution about the origin in a clockwise direction, as in Fig. 43(*d*).

As our imaginary wheel makes more than one complete rotation about its hub at the origin, anti-clockwise or clockwise, the spoke OP will, of course duplicate its positions the first time round. Suppose, then, that the wheel comes to rest on its last partial rotation so that the positive x axis and the spoke OP form the positive angle θ, or the negative angle $\alpha = \theta - 360°$.

If now our mathematical purposes require us to keep track of the fact that this is a second, third or nth revolution of OP about O, we may do so by defining the angle QOP as $= \theta + 360°$ for a second anti-clockwise revolution, or as $=\theta + (n - 1)360°$ for an nth anti-clockwise revolution, and as $\alpha - 360°$ for a second clockwise revolution, or as $=\alpha - (n - 1)360°$ for an nth clockwise revolution.

In Fig. 44, for example, OP is shown coinciding with the positive y axis on its second revolution so that $\theta = 90°$ and $\alpha = 90° - 360° = -270°$. Hence the angle $AOP = 90° + 360° = 450°$ measured anti-clockwise, or $AOP = -270° - 360° = -630°$ measured clockwise. The next time around

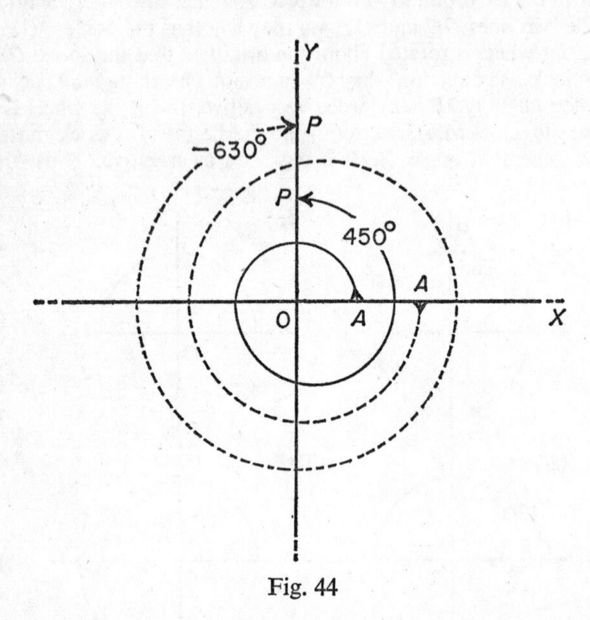

Fig. 44

the corresponding measures of the angle AOP would be $90° + 2(360°) = 810°$ for a third anti-clockwise revolution, or $-270° - 2(360°) = -990°$ for a third clockwise revolution, etc.

Exercise 13.2

1. Solve the following equations for an acute angle x:

(i) $x = \sin^{-1} 0{\cdot}7193$ (ii) $x = \sin^{-1} 0{\cdot}6293$
(iii) $x = \tan^{-1} 0{\cdot}7813$ (iv) $x = \cos^{-1} 0{\cdot}9703$
(v) $2x = \cos^{-1} 0{\cdot}8660$ (vi) $3x = \tan^{-1} 1$

What are all the possible values from $-1000°$ to $+1000°$ of the following angles

2. between the positive and negative directions of the x axis?
3. between the positive x axis and the negative y axis?
4. between the positive x axis and itself?
5. between the positive x axis and the graph of $y = x$ in the first quadrant?
6. between the positive x axis and the graph of $y = x$ in the third quadrant?

Radian Measure of Angles

While discussing the magnitude of angles it is convenient to introduce a different unit of measure—that of radians rather than of degrees. This unit is essential in the calculus and is therefore often used in other branches of mathematics as well.

Suppose the imaginary wheel described above to have a radius $OP = r$, and suppose the wheel to be turned so that the arc $QP = r$, as in Fig. 45. An angle of one **radian** is then defined as an angle equal to angle POQ. In more formal language, meaning the same thing: an angle of one **radian**

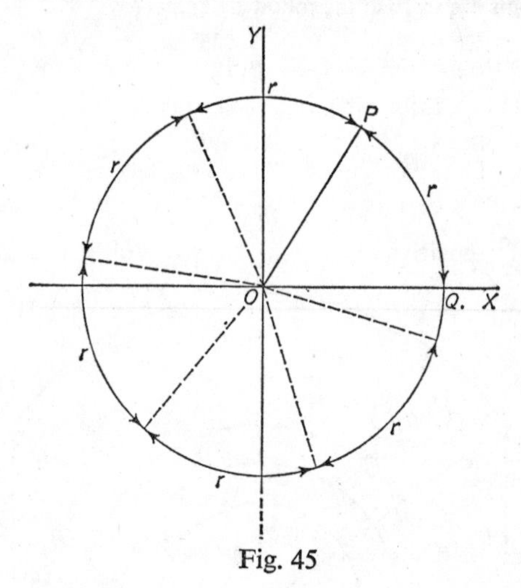

Fig. 45

is an angle subtended at the centre by an arc of a circle equal in length to the radius of the circle.

Since the circumference of a circle is $2\pi r = 2(3{\cdot}14\ldots)r = (6{\cdot}28\ldots)r$ (as shown in Fig. 45) we can measure off $6{\cdot}28\ldots$ arcs of length r on the circumference of a circle of radius r.
Therefore 2π radians $= 6{\cdot}28\ldots$ radians $= 360°$.
Dividing by 2π,

1 radian $= 360°/2\pi =$ **57·3°**, approximately, or **57° 18′** approximately

Or, dividing by 360

1° $= 2\pi/360 =$ **0·017453** radians, approximately

Normally, however, angles measured in radians are expressed in terms of π, thus

$$\begin{aligned} \pm\pi \text{ radians} &= \pm 180° \\ \pm\pi/2 \text{ radians} &= \pm 90° \\ \pm\pi/4 \text{ radians} &= \pm 45° \\ \pm 4\pi \text{ radians} &= \pm 720° \text{ etc.} \end{aligned}$$

Consequently, when the measure of an angle is given as $n\pi$ this is understood to mean $n\pi$ radians. The measure of an angle will be implied by the statement in which it occurs. For example, if we write $\theta_2 = 180° - \theta_1$ then the presence of 180° implies that both θ_2 and θ_1 are measured in degrees. If we write $\theta_2 = \pi - \theta_1$ then the presence of π implies that both θ_2 and θ_1 are measured in radians. If we write $\theta_2 = 180° - \theta_1 = \pi - \theta_1$ we are making two statements; first that $\theta_2 = 180° - \theta_1$ when the angles are measured in degrees, and second that $\theta_2 = \pi - \theta_1$ when the angles are measured in radians. The abbreviation for radians is **rad.** Thus π rad $= 180°$.

Exercise 13.3

How many radians are there in the following angles?

1. 30°	2. −60°	3. 235°	4. −225°
5. 450°	6. −720°	7. 10°	8. −2°

How many degrees are there in the following angles?

9. $\pi/4$	10. $-\pi/3$	11. $\pi/6$	12. -3π
13. $4/\pi$	14. -9π	15. 0·1 radians	16. −0·2 radians

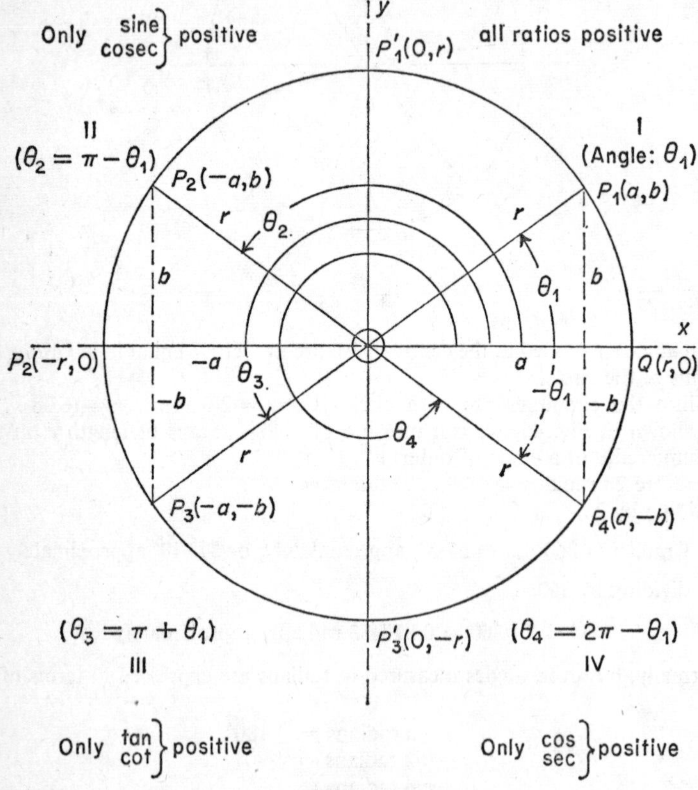

Fig. 46

Trigonometric Functions of any Angle

Although trigonometric functions are usually defined in elementary mathematics only for positive angles from 0° to 90°, in advanced mathematics these definitions also are extended. When generalised, they apply to angles of any magnitude and either sign.

Returning to the imaginary wheel described above, let its radius again be r so that Q is the fixed point (r, O) on the positive x axis (page 158). Let the point P at the outer end of the spoke OP be designated by coordinates as $P(x, y)$ so that $r = \sqrt{(x^2 + y^2)}$ and let the angle QOP again be designated θ. Then, regardless of the magnitude or sign of θ, the generalised definitions for the six basic trigonometric functions of θ are

$$\sin\theta = y/r \qquad \operatorname{cosec}\theta = r/y$$
$$\cos\theta = x/r \qquad \sec\theta = r/x$$
$$\tan\theta = y/x \qquad \cot\theta = x/y$$

Suppose $P(x, y)$ to assume any first quadrant position $P_1(a, b)$ so that θ is the acute angle θ_1 shown in the first quadrant of Fig. 46. The above definitions are then equivalent to the familiar elementary ones for functions of an acute angle in a right-angled triangle. For here r is the length of the hypotenuse; x, equal to a, is the length of the side adjacent to the acute angle θ_1; and y, equal to b, is the length of the side opposite θ_1
Hence, substituting $x = a$, $y = b$, we get

$$\sin\theta_1 = y/r = b/r$$
$$\cos\theta_1 = x/r = a/r$$

The other trigonometric functions of θ_1 have their usual values, listed in the Quadrant I column of Table 1 below.

Table 1. Typical Values of Trigonometric Functions

Quadrant		I		II		III		IV	
θ	0	θ_1	$\pi/2$	θ_2	π	θ_2	$3\pi/2$	θ_4	2π
P	Q	P_1	P_1'	P_2	P_2	P_3	P_3'	P_4	Q
(x, y)	$(r, 0)$	(a, b)	$(0, r)$	$(-a, b)$	$(-r, 0)$	$(-a, -b)$	$(0, -r)$	$(a, -b)$	$(r, 0)$
$\sin\theta$	0	b/r	1	b/r	0	$-(b/r)$	-1	$-(b/r)$	0
$\cos\theta$	1	a/r	0	$-(a/r)$	-1	$-(a/r)$	0	a/r	1
$\tan\theta$	0	b/a	$\pm\infty$	$-(b/a)$	0	b/a	$\pm\infty$	$-(b/a)$	0
$\cot\theta$	$\mp\infty$	a/b	0	$-(a/b)$	$\mp\infty$	a/b	0	$-(a/b)$	$\mp\infty$
$\sec\theta$	1	r/a	$\pm\infty$	$-(r/a)$	-1	$-(r/a)$	$\mp\infty$	r/a	1
$\operatorname{cosec}\theta$	$\mp\infty$	r/b	1	r/b	$\pm\infty$	$-(r/b)$	-1	$-(r/b)$	$\mp\infty$

But suppose $P(x, y)$ to assume the position $P_2(-a, b)$, so that θ is the corresponding second quadrant angle

$$\theta_2 = 180° - \theta_1 = \pi - \theta_1$$

or the position $P_3(-a, -b)$, so that θ is the corresponding third quadrant angle

$$\theta_3 = 180° + \theta_1 = \pi + \theta_1$$

or the position $P_4(a, -b)$, so that θ is the corresponding fourth quadrant angle

$$\theta_4 = 360° - \theta_1 = 2\pi - \theta_1$$

Now we no longer have a right-angled triangle of which θ_2, θ_3 or θ_4 is an acute angle. Nevertheless, the generalised definitions above still give us the same absolute values—numerical values without regard to sign—for the trigonometric functions of these angles. But in each quadrant the signs of certain functions change as the signs of x and y change while r, being the length of the line, always remains positive.

In the second quadrant, for instance, by substituting $y = b$, $x = -a$ we find

$$\sin \theta_2 = y/r = b/r = \sin \theta_1$$
$$\cos \theta_2 = x/r = -a/r = -\cos \theta_1$$
$$\tan \theta_2 = y/x = b/-a = -b/a = -\tan \theta_1$$

or in the third quadrant, by substituting $x = -a$, $y = -b$, we find

$$\sin \theta_3 = y/r = -b/r = -\sin \theta_1$$
$$\cos \theta_3 = x/r = -a/r = -\cos \theta_1$$
$$\tan \theta_3 = y/x = -b/-a = b/a = \tan \theta_1$$

All other values and changes of sign for the trigonometric functions of angles in different quadrants are shown in the four *Quadrant* columns of Table 1 and may be verified.

Suppose $P(x, y)$ next to assume any of the positions on the coordinate axes: $Q(r, 0)$ so that $\theta = 0°$, or P_1' $(0, r)$ so that $\theta = 90° = \pi/2$, or $P_2(-r, 0)$ so that $\theta = 180° = \pi$, etc., as also shown in Fig. 46. All entries for the sin and cos functions in the appropriate (alternate) columns of the table, and most entries for the other functions in these columns, may still be verified in the same way. For instance, by appropriate substitutions for x and y again, we find

$$\sin 0° = y/r = 0/r = 0$$
$$\cos 0° = x/r = r/r = 1$$
$$\cos 90° = x/r = 0/r = 0$$
$$\cos 180° = x/r = -r/r = -1 = -\cos 0°$$
$$\cot 270° = x/y = 0/-r = 0 = \cot 90°$$

But when we make similar substitutions for tan 90°, sec 90°, cot 180°, etc., we get

$$\tan 90° = y/x = r/0 = (?)$$
$$\sec 90° = r/x = r/0 = (?)$$
$$\cot 180° = x/y = -r/0 = (?)$$

Here arises a complication with which we have now become familiar. Routine substitutions give definite values of all the basic trigonometric functions of θ for most values of θ but, in a few special cases such as those now being considered, the same substitutions produce a result in the form $N/0$ to which we can usefully assign no definite values. With strict correctness, therefore, we must say that 90° has no defined tangent and no defined secant. Nevertheless, in some standard tables of natural trigonometric functions you will find on the first line the '∞' for tan 90° and sec 90°. This, of course does not imply an equality, 'tan 90° = ∞, etc.' to be read: 'tan 90° equals infinity, etc.' What such an entry does imply may be seen by glancing farther down the same columns of a table of natural trigonometric functions. There you will find that for values of $\theta = 87°$, $88°$, etc., getting closer and closer to 90° in the first quadrant, both the tangent and secant of θ become larger and larger. The final entries for tan 90° and sec 90° in such a table are therefore simply shorthand—if sometimes misunderstood—ways of recording the fact that we can find values of either of these functions as large as we like, provided that we only take values of θ sufficiently close to 90° but still less than 90°.

Recall, moreover, from Fig. 46 and Table 1, that for a second quadrant angle $\theta_2 = 180° - \theta_1$

$$\tan \theta_2 = -b/a = -\tan \theta_1$$

and

$$\sec \theta_2 = -r/a = -\sec \theta_1$$

This means that for values of $\theta = 93°$, $92°$, etc., getting closer and closer to 90° from the second quadrant side, the tan and sec of θ have similar sets of larger and larger values, but prefixed by a minus sign. In other words, we can find values of either function less than any pre-assigned quantity, provided that we only take values of θ sufficiently close to 90° but still more than 90°.

When tabulating typical values of trigonometric functions as in Table 1 on p. 159, therefore, it is likewise customary to indicate this fact by the entry '$\pm\infty$' or '$\mp\infty$'. This of course means only that $\tan \theta$ and $\sec \theta$ become indefinitely larger and larger with a plus sign as θ gets closer and closer to 90° from the first quadrant side, but indefinitely larger and larger with a minus sign as θ gets closer and closer to 90° from the second quadrant side.

When this sequence of signs is reversed from one quadrant to the next the order of signs in the entry is reversed, as in the case of '$\mp\infty$' for cot 180° or sec 270°, etc. Thus, for example, the entry '$\mp\infty$' for sec 270° records more than the fact that sec 270° 'does not exist' in the sense of 'has no definite numerical value'. The entry also tells us that the secant function is one which goes out of the 'negative door' of 'definite numerical existence' at the end of the third quadrant, but comes back in at the 'positive door' at the beginning of the fourth quadrant. Note, finally, that the values of the generalised trigonometric functions summarised in Table 1 on page 159 depend only upon the final position of the side OP of an angle $QOP = \theta$. For this reason they apply equally well to negative angles and to positive or negative angles greater than 360° in absolute value. Typical examples are

$$\sin(-\theta_1) = y/r = -b/r = -\sin \theta_1$$

and

$$\sin(\theta_1 \pm 2\pi n) = y/r = b/r = \sin \theta_1$$

In view of these generalised definitions, the fact that tables of trigonometric functions range only from 0° to 90° may at first seem puzzling. The explanation is, however, that the absolute values of these functions are repeated in each quadrant somewhat, although not exactly, as the values of common logarithmic mantissas are repeated between each pair of consecutive integral powers of 10. Hence a table of natural trigonometric functions contains only first-quadrant entries for much the same reason that a table of common logarithms contains mantissas only for numbers from 10 to 100 or from 100 to 1000. Anyone who uses a trigonometric table is expected to be able to apply it to angles in other quadrants in the same way that anyone who uses a logarithm table is expected to be able to apply it to numbers with other decimal places. The **steps for finding a trigonometric function of any angle θ** are as follows:

1. Find the corresponding first quadrant angle θ_1.
2. Find the required function of θ.
3. Determine whether the required function of θ has the same sign or a minus sign.

In Step 1, if θ is a negative angle or an angle greater than 360° we can always find the equivalent first-, second-, third- or fourth-quadrant angle θ_1, θ_2, θ_3 or θ_4 respectively—as previously explained (p. 159). If θ or its equivalent angle is then a first-quadrant angle this is θ_1, and there is no further problem of Step 1. But if θ or its equivalent angle is in the second, third or fourth quadrant we can always find θ_1 by substitution in the above formula for corresponding angles (p. 159), transposed and combined as follows

$$\theta_1 = 180° - \theta_2 = \theta_3 - 180° = 360° - \theta_4$$

Step 2 follows as in elementary mathematics.

Step 3 may be concluded simply by inspection of Fig. 46, or of its sketched or mentally pictured equivalent. Although standard trigonometry texts give many rules and formulae for this purpose, all are easy to confuse and are derived in the first place from diagrams like that in Fig. 46 anyway. Rather than try to learn these rules by rote and, therefore, run the risks of mis-recollection, you will do better to sketch out such a diagram whenever necessary, until its plan becomes so firmly fixed in your mind's eye that you can mentally formulate the textbook rules for yourself.

Example 4. Find $\sin\theta$, $\cos\theta$ and $\tan\theta$ when $\theta = -5\pi/6$ rad.

Solution. $5\pi/6$ radians $= 5\pi/6 \times 360/2\pi$ degrees $= 150°$.

Step 1: The positive angle equivalent to θ is

$$360° + \theta = 360° - 150° = 210° = \theta_3 \quad \text{(page 159)}$$

Hence the corresponding first quadrant angle is

$$\theta_1 = \theta_3 - 180° = 210° - 180° = 30° \quad \text{(formula above)}$$

Step 2: From a table of natural logarithms

$$\sin 30° = 0{\cdot}5000$$
$$\cos 30° = 0{\cdot}8660$$
$$\tan 30° = 0{\cdot}5774$$

Step 3: But from Fig. 46 and Table 1

$$\sin\theta_3 = -b/r = -\sin\theta_1 \quad \text{(page 159)}$$
$$\cos\theta_3 = -a/r = -\cos\theta_1$$
$$\tan\theta_3 = -b/-a = b/a = \tan\theta_1$$

Hence our answer is

$$\sin -150° = -0{\cdot}5000$$
$$\cos -150° = -0{\cdot}8660 \quad \text{(substitution)}$$
$$\tan -150° = 0{\cdot}5774$$

Answer

Example 5. Find $\sin\theta$, $\cos\theta$, $\tan\theta$ when $\theta = 510°$.

Solution. Step 1: The angle less than 360° equivalent to θ is

$$\theta - 360° = 510° - 360° = 150° = \theta_2 \quad \text{(page 158)}$$

Hence the corresponding first quadrant angle is

$$\theta_1 = 180° - \theta_2 = 180° - 150° = 30° \quad \text{(formula above)}$$

Step 2 is now the same as in Example 1.

Step 3: In this second quadrant case, however, we find from Fig. 46 and Table 1 that

$$\sin\theta_2 = b/r = \sin\theta_1$$
$$\cos\theta_2 = -a/r = -\cos\theta_1$$
$$\tan\theta_2 = b/-a = -b/a = -\tan\theta_1$$

Hence this time our answer is

$$\sin 510° = 0{\cdot}5000$$
$$\cos 510° = -0{\cdot}8660$$
$$\tan 510° = -0{\cdot}5774 \qquad \text{Answer}$$

When θ is an exact multiple of 90° of course, Step 2 is unnecessary and the required function values can be read directly from the diagram as in Table 1.

Example 6. Find the six basic trigonometric functions of $\theta = 990°$.

Solutions. Step 1: the angle less than 360° equivalent to θ is

$$\theta - 2(360°) = 990° - 720° = 270°$$

Steps 2 and 3: hence the trigonometric functions of θ are those listed in the $3\pi/2$ column of Table 1.

$$\sin 990° = -1, \text{ etc.} \qquad \text{Answer}$$

Exercise 13.4

1. Complete the following table of signs for the trigonometric functions of angles in the several quadrants

Quadrant	I	II	III	IV
sin	+	+	−	
cos	+	−		
tan	+			
cot				
sec				
cosec				

2. Referring only to Fig. 46, write formulae like the one below for $\tan\theta_1$.

$$\sin\theta_1 = \sin(\pi - \theta_1) = -\sin(\pi + \theta_1) = -\sin(2\pi - \theta_1)$$

3. Find the values of the following trigonometric functions:

(i) $\sin 120°$ (ii) $\tan 135°$ (iii) $\cos 765°$ (iv) $\sin -330°$
(v) $\sin -240°$ (vi) $\sin 210°$ (vii) $\tan 330°$ (viii) $\tan 1140°$

4. Find u in each of the following examples where $0 < u < 360°$:

(i) $u = \sin^{-1} 0{\cdot}5$ (ii) $u = \cos^{-1} 0{\cdot}5$ (iii) $u = \tan^{-1}(1)$
(iv) $u = \cos^{-1}(-0{\cdot}5)$ (v) $u = \sin^{-1}(-0{\cdot}5)$ (vi) $u = \tan^{-1}(-1)$

Complete Solution of Trigonometric Equations

At the beginning of this chapter we were able to find only first-quadrant roots of trigonometric equations but, now that we have seen how and why the trigonometric functions are defined for angles of any magnitude, we must realise that a determinate trigonometric equation has an infinite (indefinite) number of roots. The roots with values from 0° to 360° are called the **principal**

roots because the **complete solution** of a trigonometric equation can be expressed as its roots in the four basic quadrants, plus exact multiples of 360° or 2π; namely, $n(360°)$ or $2\pi n$, where $n = 0, 1, 2, 3$, etc.

Example 7. Completely solve the equation in Example 1, page 153

$$2 \tan x - 3 = 0$$

Solution. We have already found its first quadrant trigonometric solution to be $x = \tan^{-1} 1{\cdot}5 = 56° 19'$. But since

$$\tan x = \tan (180° + x) \quad \text{(page 160)}$$

it follows that another (third quadrant) root is

$$x = \tan^{-1} 1{\cdot}5 = 180° + 56° 19' = 236° 19'$$

Hence the principal roots of the equation are

$$56° 19' \text{ and } 236° 19'$$

And since

$$\tan x = \tan [x \pm n(360°)] \quad \text{(page 161)}$$

the complete solution of the equation is

$$x = 56° 19' \pm n(360°), 236° 19' \pm n(360°)$$

for $n = 0, 1, 2, 3, 4$, etc., hereafter understood.

Check $\quad 2 \tan 56° 19' - 3 = 0$ (substitution)
$\quad 2 \tan 236° 19' - 3 = 0$ Answer

For the same reason that it is possible to express the complete solution of a trigonometric equation in terms of the principal roots, it is, of course, only necessary to check the principal roots, as above.

Example 8. Solve completely the equation in Example 2, page 153.

$$2 \cos^2 x + \cos x - 1 = 0$$

Solution. We have already found the first-quadrant trigonometric solution for the first of these simplified equations to be

$$x = \cos^{-1} \tfrac{1}{2} = 60° \quad \text{(page 153)}$$

But since

$$\cos x = \cos (360° - x) \quad \text{(page 161)}$$

it follows that another (fourth-quadrant) root is

$$x = \cos^{-1} \tfrac{1}{2} = 360° - 60° = 300°$$

Moreover, since

$$\cos 180° = -1 \text{ (Table 1, page 159)}$$

still another root is

$$x = \cos^{-1} (-1) = 180°$$

Hence the three principal roots of the equation are

$$x = 60°, 180°, 300°$$

and the complete solution is

$$\begin{aligned} x = {} & 60° \pm n(360°) \\ & 180° \pm n(360°) \\ & 300° \pm n(360°) \end{aligned}$$

Answer

Check:

$2 \cos^2 60° + \cos 60° - 1 = 2(\tfrac{1}{2})^2 + \tfrac{1}{2} - 1 = 1 - 1 = 0$ (substitution)
$2 \cos^2 180° + \cos 180° - 1 = 2(-1)^2 - 1 - 1 = 2 - 2 = 0$ etc.

In the above examples all results which previously checked algebraically have been found to lead to roots which also check trigonometrically. In other instances, however, results which check at the preliminary algebraic stage of solution may be found to lead to inverse trigonometric equations for which there are no possible trigonometric solutions.

Example 9. Solve completely the equation $2 \sin x + 2 \operatorname{cosec} x - 5 = 0$ (Exercise 13.1, No. 4 (i).

Solution. With $\operatorname{cosec} x = 1/\sin x$ multiply the equation by $\sin x$ and rearrange to get

$$2 \sin^2 x - 5 \sin x + 2 = 0 = (2 \sin x - 1)(\sin x - 2)$$
$$\therefore \quad \sin x = 0{\cdot}5 \quad \text{or} \quad \sin x = 2$$

From $x = \sin^{-1} 0{\cdot}5$ we get $x = 30°$ and $150°$.
There is no real value of x which satisfies $\sin x = 2$ (yet $\sin x = 2$ is a solution to the equation). Hence the principal solution of the equation consists of the two principal roots $x = 30°, 150°$ and the complete solution

$$x = 30° \pm n(360°), 150° \pm n(360°) \qquad \text{Answer}$$

Exercise 13.5

Obtain complete solutions to the following equations:

1. $2 \cos^2 x - 5 \cos x + 2 = 0$
2. $\sin x + \cos x = 1$
3. $\sin x - \cos x = 1$
4. $\cos^2 x - \sin^2 x = 2$
5. $\tan x = \cot x$
6. $\sin^2 x = 9$
7. $2 \sin^2 x - \sin x = 0$
8. $\tan x - \cot x = 2$
9. $2 \sin^2 x + 3 \sin x - 2 = 0$

Graphs of Trigonometric Functions

Using typical values like those computed in preceding examples and exercises, it is a simple matter to plot the trigonometric functions of a variable angle x with respect to an x, y set of rectangular coordinate axes.

The graph of $\boldsymbol{y = \sin x}$ is the regular wave like curve of indefinite length drawn with an unbroken contour in the upper diagram of Fig. 47. Note how it fluctuates back and forth between the horizontal lines $y = \pm 1$, lying above the x axis in Quadrants I and II, lying below the x axis in Quadrants III and IV, and alternating in sign every 180° along the x axis on either side of these four quadrants.

Distinguished from the sine curve in the same diagram by broken-line drawing, the graph of $y = \operatorname{cosec} x$ is a sequence of open ended U-shaped curves; every other one inverted, but always on the same side of the x axis as the sine curve, and 'touching' it every 180° at the points where $\operatorname{cosec} x = \sin x = \pm 1$. Note how it changes in value from an indicated $+\infty$, down to $+1$, back up to an indicated $+\infty$, in the first two quadrants; next 'jumps' to an indicated $-\infty$, increases to -1, and returns to an indicated $-\infty$, in the next two quadrants. This is, of course, the pattern we should expect from the arithmetic of the formula, $\operatorname{cosec} x = 1/\sin x$.

The cos and sec columns in a table of natural trigonometric functions, you should recall, are merely the sin and cosec columns read in reverse order. This typographical saving of space is possible for the mathematical reason that $\cos x = \sin (90° - x)$ and $\sec x = \operatorname{cosec} (90° - x)$ (Fig. 42, page 152 above). Hence the graph of $\boldsymbol{y = \cos x}$ and the graph of $\boldsymbol{y = \sec x}$, as shown in the middle diagram of Fig. 47 are identical with the sin and cosec graphs in the upper diagram except that they are shifted one quadrant, or $90° = \pi/2$ to the left along the x axis.

The graph of $y = \tan x$, finally, is the sequence of upward sweeping curves shown intersecting the x axis every 180° in the lower diagram of Fig. 47, and the graph of $y = \cot x$ is the sequence of downward sweeping curves in the same diagram. Note how the tan curves pass in value every 180° from an indicated $-\infty$, to 0, on up to an indicated $+\infty$, typically in the second and

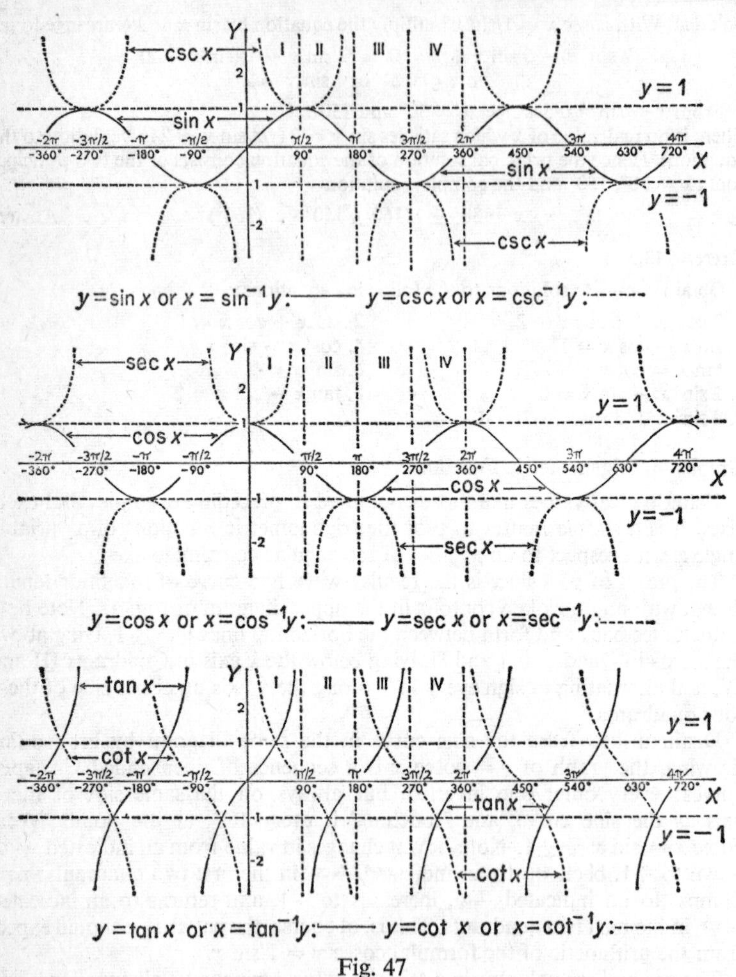

Fig. 47

third quadrants; and also how the cot curves pass in value every 180° from an indicated $+\infty$, to 0, on down to an indicated $-\infty$, typically in the first and second quadrants.

Check the patterns of values depicted by these graphs with Table 1 and Fig. 46 (page 159). Try reconstructing them for yourself until you have their contours firmly fixed in your mind's eye. Then you may be able to use them—perhaps even more conveniently than diagrams such as Fig. 46—as an aid in improvising formulae (page 159) to find the functions of angles outside the first quadrant (page 161), or to find the complete solutions of trigonometric equations (page 163).

Suppose, for example, that you need to find trigonometric functions of $x = -150°$, as in the example above (page 162). Glance down the vertical line, $x = -150°$, on the diagrams of Fig. 47. If necessary, use the right edge of a vertically placed ruler—preferably a transparent ruler—to guide your eye. From the intersection of this line with several graphs, and from the pattern of these graphs in each 180° interval of the x axis, you should be able to see at once that

$$\sin - 150° = -\sin 30°, \tan - 150° = \tan 30°, \text{etc.}$$

Or suppose that you need to find all the angles, x such that $x = \sin^{-1} \frac{1}{2}$, or $\sin^{-1} 2$. Glance across the horizontal lines, $y = \frac{1}{2}$ and $y = 2$, in the upper diagram of Fig. 47. From the intersections of the line $y = \frac{1}{2}$ with the sine curve you should be able to see at once that

$$x = \sin^{-1} \tfrac{1}{2} = 30° \pm n(360°), 150° \pm n(360°)$$

And from the failure of the line $y = 2$ to intersect the sine curve at any point, you can also see at once that there is no angle $x = \sin^{-1} 2$.

Exercise 13.6

1. Find graphically where $\tan x$ and $\cos x$ are equal.
2. Find graphically where $\cos x$ and x are equal, it being understood that x is measured in radians.

Variations of Trigonometric Functions

Each of the graphs in Fig. 47 repeats a pattern of values, called its **cycle**, at a regular interval, called its **period.** For this reason the trigonometric functions are classed as **periodic functions.**

Half the difference between the largest and smallest values which such a function attains in each cycle is called its **amplitude.**

Some idea of the physical meaning of these terms is suggested by the fact

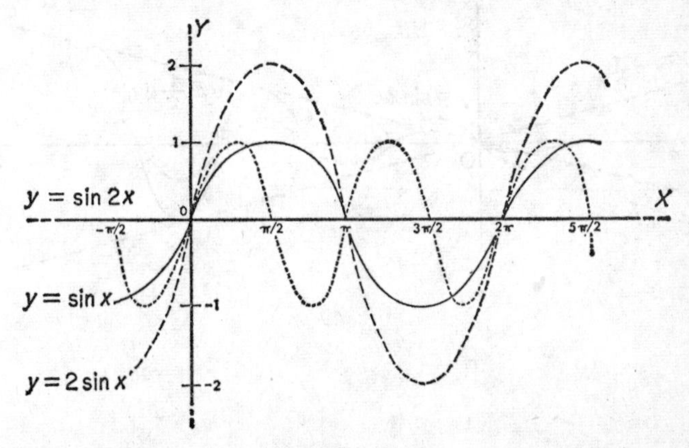

Fig. 48

that the movement of molecules of matter in sound waves can be depicted by modified sine or cosine curves, and have lower pitched sounds when the periods of these curves are longer, and louder sounds when the amplitudes of these curves are greater.

Here, however, we are primarily concerned with interpreting the mathematical devices by which these variations in periodic behaviour are expressed in periodic functions.

As in the case of other functions, the graph of any trigonometric function is simply raised or lowered c units by the addition of a constant $\pm c$ term. By tabulating and plotting values of $y = \sin x + c$, for example, we obtain a graph identical to that for the sine curve in Fig. 47 except that it is raised c units higher with respect to the x axis.

The addition of a constant term $\pm c\pi$ to the independent variable angle of a trigonometric function, however, displaces its phase that number of units to the left or right with respect to the vertical y axis. By tabulating and plotting values of $y = \sin (x + \pi/2)$, for example, we obtain the graph of the $\cos x$ curve in Fig. 47, which has already been observed to be identical to that of the $\sin x$ curve except that it is displaced $\pi/2$ units to the left (page 166).

Moreover, when the trigonometric functions are multiplied by a constant c their amplitude is multiplied by the same constant but, when their independent variable angle is multiplied by c, their period is divided by this constant. For instance, the graph of $y = 2 \sin x$ shown in Fig. 48 has twice the amplitude, but the same period, as that of $y = \sin x$. On the other hand, the graph of $y = \sin 2x$ has the same amplitude, but half the period, of that of $y = \sin x$.

When a trigonometric function is multiplied by a variable factor, however, the resulting product is no longer strictly a periodic function. But its pattern of variation may nevertheless reflect the periodicity of its trigonometric factor. For instance, the graph of $y = \frac{1}{2}x \sin x$, as shown in Fig. 49, intersects the x axis every 180° like the graph of $y = \sin x$, but its amplitude within each interval of 180° changes with the magnitude of its factor, $\frac{1}{2}x$.

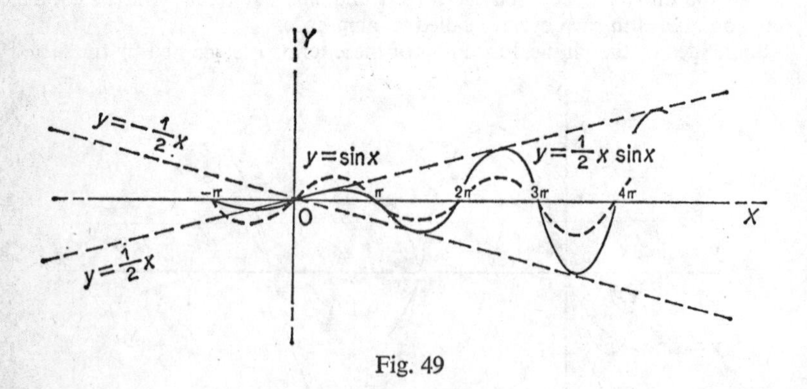

Fig. 49

Exercise 13.7

Sketch at least one cycle of the graphs of the following functions:

1. $y = -\sin x$, and $y = -\sin 2x$ $(0 \leqslant x \leqslant 2\pi)$
2. $y = \sin x$, and $y = -\sin (x - \pi/2)$ $(0 \leqslant x \leqslant 2\pi)$
3. $y = \frac{1}{2} \cos x$, and $y = \cos \frac{1}{2}x$ $(0 \leqslant x \leqslant 4\pi)$
4. $y = \cos 3x$, and $y = 3 \cos x$ $(0 \leqslant x \leqslant 2\pi)$
5. $y = \sin x + \cos x$, and $y = \sin x - \cos x$ $(0 \leqslant x \leqslant 5\pi/2)$
6. $y = \sin^2 x$, and $y = \sin^2 x + \cos^2 x$ $(0 \leqslant x \leqslant 2\pi)$
7. $y = \frac{1}{2} \tan \frac{1}{2}x$ $(0 \leqslant x \leqslant 4\pi)$
8. $y = \frac{1}{2} \sec x - 1$ $(0 \leqslant x \leqslant 2\pi)$

Regarding the functions in Examples 1–6 above as pairs of simultaneous trigonometric equations, what are their graphic solutions?

CHAPTER FOURTEEN

TRIGONOMETRIC FUNCTIONS OF COMPOUND ANGLES AND SOME SOLID GEOMETRY

The Cosine Rule

The cosine rule states that for any acute or obtuse angled triangle ABC

$$a^2 = b^2 + c^2 - 2bc \cos A$$
$$b^2 = c^2 + a^2 - 2ca \cos B$$
$$c^2 = a^2 + b^2 - 2ab \cos C$$

The result is proved in elementary mathematics: to recall briefly the application of the theorem consider the following example.

Example 1. In triangle ABC, $AC = 10$ cm, $BC = 12$ cm, angle $ACB = 130°$ Calculate the remaining sides and angles of the triangle.

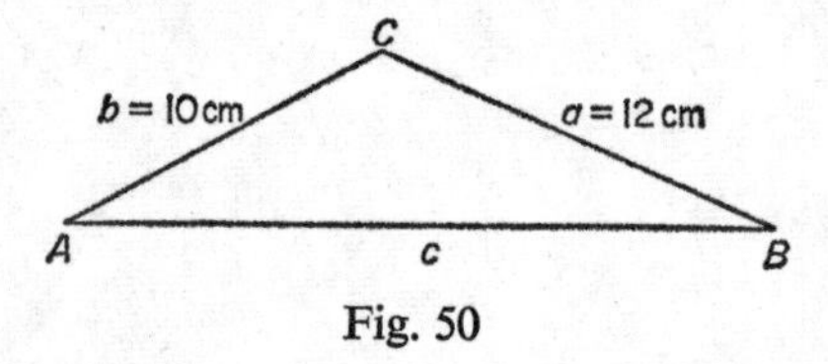

Fig. 50

Solution. Let Fig. 50 represent the problem. In the usual notation the sides opposite the angles A, B and C are given the lengths a, b and c respectively. The given information is $a = 12$ cm, $b = 10$ cm, $C = 130°$.

To find c: $c^2 = a^2 + b^2 - 2ab \cos 130°$ (cosine rule)

$$= 144 + 100 + 240 \cos 50° \quad (\cos 130° = -\cos 50° = -0{\cdot}6428)$$
$$= 244 + 154{\cdot}272$$
$$\therefore \quad c^2 = 398{\cdot}272$$
$$\therefore \quad c = 19{\cdot}96 \quad \text{(correct to 2 decimal places)}$$

To find A: we could use the sine rule $\dfrac{\sin A}{a} = \dfrac{\sin B}{b} = \dfrac{\sin C}{c}$ but we prefer to continue with the cosine rule in the form

$$a^2 = b^2 + c^2 - 2bc \cos A$$
$$\therefore \quad 144 = 100 + 398{\cdot}272 - 20 \times 19{\cdot}96 \cos A$$

Hence $$\cos A = \frac{498{\cdot}272 - 144}{20 \times 19{\cdot}96} = \frac{354{\cdot}272}{399{\cdot}2} = 0{\cdot}8874$$

$$\therefore \quad A = \cos^{-1} 0{\cdot}8874 = 27° \, 27'$$

With $A + C = 157° \, 27'$ we have $B = 180° - 157° \, 27' = 22° \, 33'$

$\therefore$ $c = 19{\cdot}96$ cm, angle $CAB = 27° \, 27'$, angle $ABC = 22° \, 33'$

Answer

Compound Angles

We shall now use the cosine rule to obtain the cosine of the compound angle $A - B$, by finding the distance between two points on a unit circle. For this purpose remember from page 53, that the distance between the two points $P(x_1, y_1)$ and $Q(x_2, y_2)$ is given by $PQ^2 = (x_1 - x_2)^2 + (y_1 - y_2)^2$. Consider any point on the unit circle in Fig. 51(*a*) which has the origin O as centre. The coordinates of a point on this circle are given by $x = \cos\theta$, $y = \sin\theta$, and θ is any angle between 0 and 360°. Let OP make an angle A with the positive x axis so that the coordinates of P are $x_1 = \cos A$ and $y_1 = \sin A$. With Q on the circle and OQ making an angle B with the positive x axis the coordinates of B are given by $x_2 = \cos B$, and $y_2 = \sin B$. In the triangle POQ we have angle $POQ = A - B$, $OP = OQ = 1$, so that on applying the cosine rule to find PQ we get

$$PQ^2 = 1 + 1 - 2\cos POQ = 2 - 2\cos(A - B) \qquad \text{(i)}$$

Using the expression $PQ^2 = (x_1 - x_2)^2 + (y_1 - y_2)^2$ for the distance PQ we get

$$PQ^2 = (\cos A - \cos B)^2 + (\sin A - \sin B)^2$$
$$= \cos^2 A - 2\cos A\cos B + \cos^2 B + \sin^2 A - 2\sin A\sin B + \sin^2 B$$
$$\therefore\quad PQ^2 = \cos^2 A + \sin^2 A + \cos^2 B + \sin^2 B - 2\{\cos A\cos B + \sin A\sin B\}$$

Using the result in (i) and $\sin^2 + \cos^2 = 1$ we have

$$2 - 2\cos(A - B) = 2 - 2\{\cos A\cos B + \sin A\sin B\}$$
$$\therefore \cos(A - B) = \cos A\cos B + \sin A\sin B \qquad \text{(ii)}$$

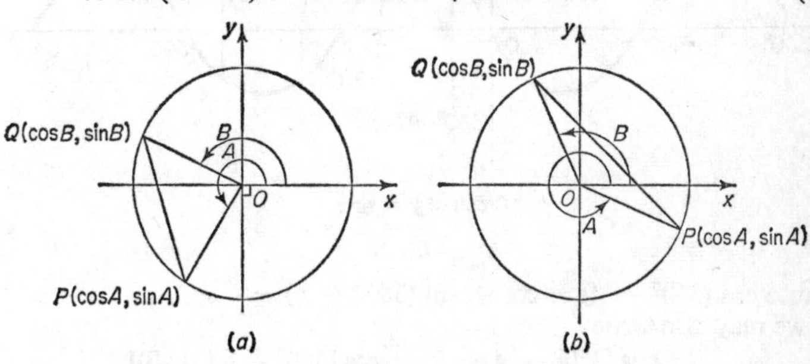

Fig. 51

a result which holds for any value of A and B, including cases in which $B > A$ because in this instance the angle POQ in Fig. 51(*a*) becomes $B - A$ and the result (ii) becomes

$$\cos(B - A) = \cos B\cos A + \sin B\sin A = \cos(A - B) \qquad \text{(Result 1)}$$

Fig. 51(*b*) illustrates the case of $A - B$ being a reflex angle, i.e. $180° < A - B < 360°$.

The angle POQ is now $360° - (A - B)$. An examination of the graph of $y = \cos x$ in Fig. 52 or 47 shows that if we take *any* value for θ and then find the point on the graph corresponding to $360° - \theta$ we discover that $\cos(360° - \theta) = \cos\theta$. In our case we are interested in $\theta = A - B$ giving

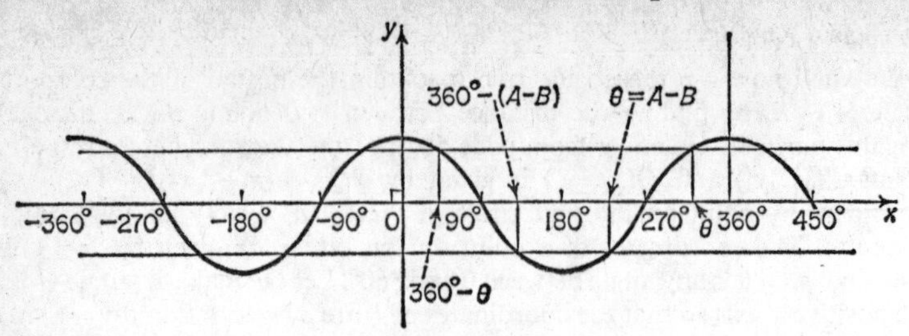

Graph of $y = \cos x$

Fig. 52

$\cos[360° - (A - B)] = \cos(A - B)$, i.e. $\cos POQ = \cos(A - B)$. Consequently, Result 1 follows as before.

By examining the graph of $y = \sin x$ in Fig. 53 or 47 we find that for *any* angle θ, $\sin(360° - \theta) = -\sin\theta$ in exactly the same manner as we found $\cos(360° - \theta) = \cos\theta$. With this result at hand we now consider $\cos(360° - A - B)$ in Result 1

$$\begin{aligned}\cos(360° - A) - B &= \cos(360 - A)\cos B + \sin(360 - A)\sin B \\ &= \cos A\cos B - \sin A\sin B\end{aligned}$$

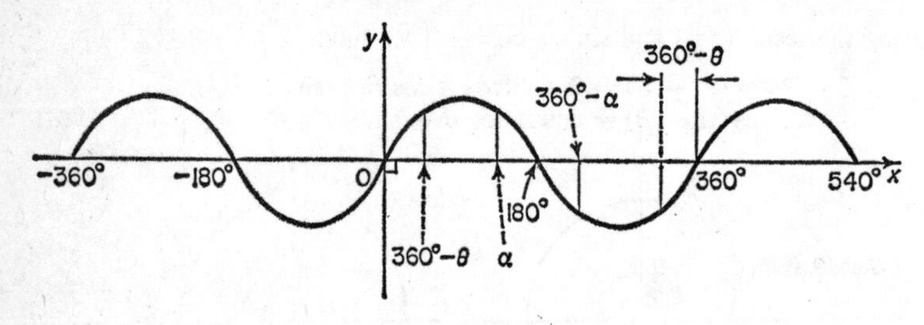

Graph of $y = \sin x$

Fig. 53

because $\cos(360° - A) = \cos A$, $\sin(360° - A) = -\sin A$.

But we may also write

$$\cos(360° - A - B) = \cos[360° - (A + B)]$$
$$\therefore \cos[360° - (A + B)] = \cos(A + B)$$

and we have a final result of

$$\cos A + B = \cos A\cos B - \sin A\sin B \qquad \text{(Result 2)}$$

Note that we could have obtained Result 2 from Result 1 by putting $-B$ instead of B, i.e.

$$\begin{aligned}\cos[A - (-B)] = \cos(A + B) &= \cos A\cos(-B) + \sin A\sin(-B) \\ &= \cos A\cos B - \sin A\sin B.\end{aligned}$$

The reason for not making this suggestion is that the methods of Fig. 51(*a*) and 51(*b*) are not quite so convincing if we involve the measurement of angle POQ with the mathematical device of negative angles. Furthermore, we can get much more general information from a study of the graphs of the functions.

Exercise 14.1

Find the remaining angles and sides in each of the following triangles ABC for Questions 1, 2, 3 and 4:

1. $A = 60°, b = 10, c = 12$
2. $A = 120°, b = 10, c = 12$
3. $B = 103°, a = 20, c = 30$
4. $a = 5, b = 6, c = 9$
5. Which of the following are equal to $\cos A$ when A is an acute angle?
   (i) $\cos(90° - A)$ (ii) $\sin(90° + A)$ (iii) $\sin(90° - A)$ (iv) $\cos(360° - A)$
   (v) $-\cos(180° - A)$
6. Which of the following are equal to $\cos A$ when A is an obtuse angle?
   (i) $\cos(A - 90°)$ (ii) $\cos(A + 90°)$ (iii) $\cos(180° - A)$
   (iv) $\cos(360° - A)$ (v) $-\sin(270° - A)$
7. Find $\cos(A + B) + \cos(A - B)$ and use the result to show that $\cos 100° + \cos 20° = \cos 40°$.
8. Find $\cos(A - B) - \cos(A + B)$ and use the result to show that $\cos 20° - \cos 100° = \sqrt{3} \sin 40°$
9. Prove that $\cos(135° + A) + \cos(A - 45°) = 0$.
10. Prove that $\cos 75° = \frac{1}{4}(\sqrt{6} - \sqrt{2})$ given that $\cos 45° = \frac{1}{2}\sqrt{2}$ and $\cos 30° = \frac{1}{2}\sqrt{3}$.

Having obtained Results 1 and 2 we can deduce a few more standard results which we may confirm by looking at the graphs on page 166. In the result for $\cos(A - B)$, which holds for all values of A and B, consider putting $A = 90°$ so that $\cos(90° - B) = \cos 90° \cos B + \sin 90° \sin B$, i.e.

$$\cos(90° - B) = \sin B \qquad \text{(i)}$$

If we put $B = 90° - C$, this last result becomes

$$\cos[90° - (90° - C)] = \cos C = \sin(90° - C) \qquad \text{(ii)}$$

We may take this even further by considering $\cos[90° - (A - B)] = \sin(A - B)$ from (i) so that

$$\begin{aligned} \sin(A - B) &= \cos(90° - A) + B \\ &= \cos(90° - A)\cos B - \sin(90° - A)\sin B \quad \text{(from Result 2)} \\ &= \sin A \cos B - \cos A \sin B \quad \text{(using (i) and (ii))} \end{aligned}$$

$$\therefore \sin(A - B) = \sin A \cos B - \cos A \sin B \qquad \text{(Result 3)}$$

Clearly we can obtain $\sin(A + B)$ by considering $\cos[90° - (A + B)]$ in the following manner.

$$\begin{aligned} \cos[90° - (A + B)] &= \sin(A + B) \quad \text{(from (i) above)} \\ \therefore \sin(A + B) &= \cos(90° - A) - B \\ &= \cos(90° - A)\cos B + \sin(90° - A)\sin B \quad \text{(from Result 1)} \\ &= \sin A \cos B + \cos A \sin B \quad \text{(from (i) and (ii))} \\ \therefore \sin(A + B) &= \sin A \cos B + \cos A \sin B \ldots \quad \text{(Result 4)} \end{aligned}$$

We could have obtained this result from Result 3 by putting $B = -B$. These four results are sufficiently important to justify writing them together and observing the comparisons in the statements

$$\cos(A - B) = \cos A \cos B + \sin A \sin B \quad (1)$$
$$\cos(A + B) = \cos A \cos B - \sin A \sin B \quad (2)$$
$$\sin(A - B) = \sin A \cos B - \cos A \sin B \quad (3)$$
$$\sin(A + B) = \sin A \cos B + \cos A \sin B \quad (4)$$

We need only remember the first result because each of the others may be derived from it by substituting $-B$, or $90° - B$, for B.

Multiple Angles

From the above results we deduce the trigonometric functions of multiple angles, meaning multiples of the same angle such as $2A$, $3A$ and so on. If we put $A = B$ in the results 2 and 4 we arrive at

$$\cos 2A = \cos^2 A - \sin^2 A \text{ and } \sin 2A = 2 \sin A \cos A$$

Since $1 = \cos^2 A + \sin^2 A$ we can modify the expression for $\cos 2A$ by the addition,

$$1 + \cos 2A = 2\cos^2 A,$$

or the subtraction,

$$1 - \cos 2A = 2\sin^2 A.$$

Taking this one stage further we can obtain $\cos 3A$ by considering

$$\begin{aligned} \cos 3A = \cos(2A + A) &= \cos 2A \cos A - \sin 2A \sin A \\ &= (\cos^2 A - \sin^2 A)\cos A - 2\sin A \cos A \sin A \\ &= \cos^3 A - 3\sin^2 A \cos A \end{aligned}$$

We may express $\cos 3A$ in terms of $\cos A$ by substituting $1 - \cos^2 A$ for $\sin^2 A$ in the last line.

$$\begin{aligned} \cos 3A &= \cos^3 A - 3(1 - \cos^2 A)\cos A \\ &= \cos^3 A - 3\cos A + 3\cos^3 A \\ \therefore \cos 3A &= 4\cos^3 A - 3\cos A \end{aligned}$$

In a similar way we can express $\sin 3A$ in terms of $\sin A$ by considering

$$\sin 3A = \sin(2A + A) = \sin 2A \cos A + \cos 2A \sin A$$

since $1 - 2\sin^2 A = \cos 2A$ we get

$$\begin{aligned} \sin 3A &= 2\sin A \cos A \cos A + (1 - 2\sin^2 A)\sin A \\ &= 2\sin A \cos^2 A + \sin A - 2\sin^3 A \\ &= 2\sin A(1 - \sin^2 A) + \sin A - 2\sin^3 A \\ \therefore \sin 3A &= -4\sin^3 A + 3\sin A \end{aligned}$$

Notice the close similarity between these results for $\cos 3A$ and $\sin 3A$. This similarity is only to be expected because

$$\begin{aligned} \cos(270° - 3A) &= \cos 270° \cos 3A + \sin 270° \sin 3A \\ &= 0 - 1 \sin 3A \\ \therefore -\sin 3A = \cos 3(90° - A) &= 4\cos^3(90° - A) - 3\cos(90° - A) \\ &= 4\sin^3 A - 3\sin A \\ &\quad \text{since } \cos(90° - A) = \sin A \\ \therefore \sin 3A &= -4\sin^3 A + 3\sin A \end{aligned}$$

Naturally we must expect comparable results for $\tan(A + B)$ and

$\tan(A - B)$, $\tan 2A$ and $\tan 3A$ from the results above. We shall work out one of these results

$$\tan(A - B) = \frac{\sin(A - B)}{\cos(A - B)} = \frac{\sin A \cos B - \cos A \sin B}{\cos A \cos B + \sin A \sin B}$$

Dividing the numerator and denominator by $\cos A \cos B$ we get

$$\tan(A - B) = \frac{\dfrac{\sin A \cos B}{\cos A \cos B} - \dfrac{\cos A \sin B}{\cos A \cos B}}{\dfrac{\cos A \cos B}{\cos A \cos B} + \dfrac{\sin A \sin B}{\cos A \cos B}} = \frac{\tan A - \tan B}{1 + \tan A \tan B}$$

Just as $\sin 3A$ can be expressed in terms of $\sin A$ so the sin, cos or tan of any angle can be expressed in terms of any basic known result without the aid of tables. This means that $\sin 45°$ can be expressed in terms of $\sin 15°$, and of course vice versa; similarly, for $\tan 75°$ in terms of $\tan 25°$ and so on. The merit of the work lies in its interest rather than the practical application for we have no intention of discarding the given trigonometric tables in favour of calculating our own.

Example 2. Simplify (i) $\cot B \cos A - \sin A$,
(ii) $\cos(B - A)\cos A - \sin(B - A)\sin A$,
(iii) $\cos(A - B)\cos(A + B) + \sin(A - B)\sin(A + B)$.

Solution.

(i) Writing $\cot B$ as $\dfrac{\cos B}{\sin B}$ we have

$$\frac{\cos B \cos A}{\sin B} - \sin A = \frac{\cos B \cos A - \sin B \sin A}{\sin B}$$

$$= \frac{\cos(A + B)}{\sin B} \quad \text{from Result 2} \qquad \text{Answer}$$

(ii) $\cos(B - A)\cos A - \sin(B - A)\sin A = \cos[(B - A) + A]$ from Result 2
$= \cos B$ Answer

(iii) $\cos(A - B)\cos(A + B) + \sin(A - B)\sin(A + B)$
$= \cos[(A - B) - (A + B)]$ from Result 1
$= \cos(-2B) = \cos 2B$ Answer

Example 3. Given that $\tan 45° = 1$ and $\tan 30° = 1/\sqrt{3}$ find $\tan 15°$.

Solution. In problems of this type the idea is to obtain the solutions in terms of $\tan 45°$ and $\tan 15°$. Thus $\tan(45° + 30°)$ would yield $\tan 75°$ and so on. Here we need $\tan(45° - 30°)$. From the result for $\tan(A - B)$ we have

$$\tan 15° = \frac{\tan 45° - \tan 30°}{1 + \tan 45° \tan 30°} = \frac{1 - \dfrac{1}{\sqrt{3}}}{1 + \dfrac{1}{\sqrt{3}}} = \frac{\sqrt{3} - 1}{\sqrt{3} + 1}$$

Rationalizing the denominator we get

$$\tan 15° = \frac{(\sqrt{3} - 1)(\sqrt{3} - 1)}{(\sqrt{3} + 1)(\sqrt{3} - 1)} = \frac{3 - 2\sqrt{3} + 1}{3 - 1} = 2 - \sqrt{3} \qquad \text{Answer}$$

Example 4. Prove that it is possible to find an acute angle p such that $y = a\cos x + b\sin x = \sqrt{(a^2 + b^2)}\sin(x + p)$, $a > 0$, $b > 0$, where $\tan p = a/b$. Also find a value of x which gives y its maximum value. (You may find it easier first to examine the numerical example below.)

Solution. Start by observing that if we could make a and b correspond to $\sin p$ and $\cos p$ respectively the result would be to make $y = \sin p\cos x + \cos p\sin x$.

Unfortunately this is impossible because a or b may be greater than 1.

Furthermore, we need $\sin^2 p + \cos^2 p = 1$ and $a^2 + b^2$ may not satisfy this requirement. However, these remarks do give a clue to the method which follows

$$y = \sqrt{(a^2 + b^2)}\left\{\frac{a}{\sqrt{(a^2 + b^2)}}\cos x + \frac{b}{\sqrt{(a^2 + b^2)}}\sin x\right\}$$

If we now put $\sin p = \dfrac{a}{\sqrt{(a^2 + b^2)}}$ and $\cos p = \dfrac{b}{\sqrt{(a^2 + b^2)}}$

we have $\sin^2 p + \cos^2 p = 1$, $-1 \leqslant \sin p \leqslant 1$, $-1 \leqslant \cos p \leqslant 1$, $\tan p = a/b$ as required. We now have

$$y = \sqrt{(a^2 + b^2)}\{\sin p\cos x + \cos p\sin x\} = \sqrt{(a^2 + b^2)}\sin(p + x).$$

Since the maximum value of $\sin(p + x)$ is 1, the maximum value of y is $\sqrt{(a^2 + b^2)}$. Finally, $\sin(p + x) = 1$ when $p + x = 90°$, $x = 90° - p$.

But, $\tan p = \dfrac{\sin p}{\cos p} = \left(\dfrac{a}{b}\right)$ so we have $p = \tan^{-1}\left(\dfrac{a}{b}\right)$ and $x = 90° - \tan^{-1}\left(\dfrac{a}{b}\right)$.

In conclusion, when $x = 90° - \tan^{-1}\left(\dfrac{a}{b}\right)$, y obtains its maximum value of $\sqrt{(a^2 + b^2)}$ Answer

A numerical example of the above in a slightly different form is to find the maximum value of $y = 5\sin x + 12\cos x$. Following the solution above we have

$$\sqrt{(5^2 + 12^2)} = 13 \text{ and } y = 13\{\tfrac{5}{13}\sin x + \tfrac{12}{13}\cos x\}.$$

With $\tan p = \frac{5}{12} = 0{\cdot}4167$ we have $p = 22°\,37'$ and our modification for y now reads

$$y = 13\{\sin 22°\,37'\sin x + \cos 22°\,37'\cos x\} = 13\cos(x - 22°\,37')$$

The maximum is 13 as expected and a value of x which gives the maximum is obtained from $\cos(x - 22°\,37') = 1$ when $x = 22°\,37'$ Answer

Exercise 14.2

1. Using the results for $\cos(A + B)$, $\cos(A - B)$, $\sin(A + B)$ and $\sin(A - B)$ calculate:
   (i) $\cos 105°$ and $\cos 75°$ with $\sin 60° = \frac{1}{2}\sqrt{3}$, $\sin 45° = \frac{1}{2}\sqrt{2}$
   (ii) $\sin 15°$ and $\sin 225°$
   (iii) $2\sin 15°\sin 45°$
2. Given that $\sin A = \frac{3}{5}$ and $\sin B = \frac{5}{13}$ find $\cos(A + B)$ and $\cos(A - B)$, without using tables. (Hint: draw a right-angled triangle and find $\cos A$, etc.)
3. Given that $\sin A = \frac{60}{61}$ and $\cos B = \frac{9}{41}$, find $\sin(A + B)$ and $\sin(A - B)$ without using tables.
4. Find the maximum and minimum value of each of the following and a value of x at which the maximum and minimum values occur:
   (i) $\cos x + \sqrt{3}\sin x$
   (ii) $\sin x + \cos x$
   (iii) $3\cos x + 4\sin x$
5. Find $\tan(A + B)$ using the results for $\cos(A + B)$ and $\sin(A + B)$.
6. Prove that $\tan(45° + B) = \dfrac{1 + \tan B}{1 - \tan B}$ and thereby find $\tan 105°$.
7. Prove that $\tan 2A = \dfrac{2\tan A}{1 - \tan^2 A}$ and use this result to find $\tan 22\frac{1}{2}°$.

8. Prove that $\dfrac{1}{\tan A} - \dfrac{1}{\tan 2A} = \dfrac{1}{\sin 2A}$

9. Use the result $1 + \cos 2A = 2\cos^2 A$ to find cos 15° given that cos 30° = 0·866.

10. Solve the equation $3 \sin x + 4 \cos x = 1$ for $0 \leqslant x \leqslant 360°$.

The Angles Between Two Straight Lines

In coordinate geometry the general equation of a straight line was given by $y = mx + c$, where $m = \tan \theta$ represents the gradient of the line with θ being the angle which the line makes with the direction of the positive x axis.

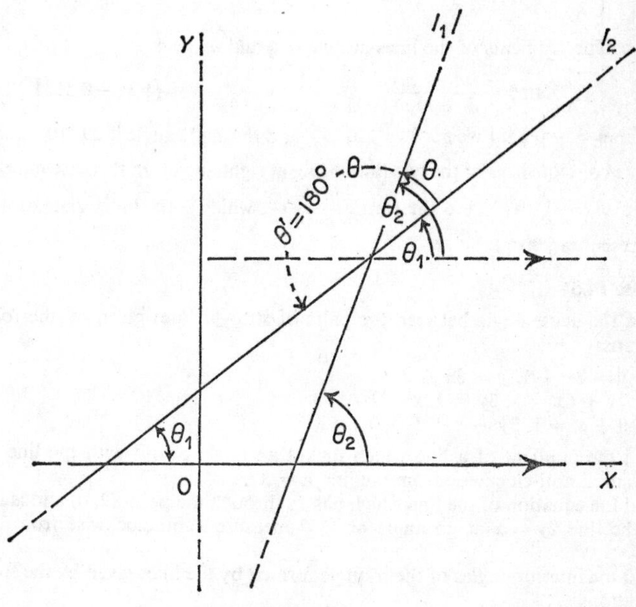

Fig. 54

The constant c is the intercept on the y axis; that is, the line $y = mx + c$ intersects the y axis in the point (O, c). With the formula

$$\tan (A - B) = \frac{\tan A - \tan B}{1 + \tan A \tan B}$$

now available we can express the tangent of the angle between two straight lines in terms of the gradients of the lines.

In Fig. 54 let l_1 and l_2 be any two non-parallel lines with gradients m_1 and m_2 and equations $y = m_1x + c_1$, $y = m_2x + c_2$ respectively. An auxiliary line parallel to the x axis has been drawn through their point of intersection to make clear that the angle θ, measured anti-clockwise from l_1 to l_2, is given by the relation

$$\theta = \theta_2 - \theta_1$$

$$\therefore \tan \theta = \tan (\theta_2 - \theta_1) = \frac{\tan \theta_2 - \tan \theta_1}{1 + \tan \theta_1 \tan \theta_2} = \frac{m_2 - m_1}{1 + m_1 m_2} \qquad \text{(i)}$$

Without the aid of a diagram we may well have interchanged the two lines and found $\tan\theta' = \frac{m_1 - m_2}{1 + m_1m_2}$, which is the result in (i) multiplied by -1, so that if the result (i) is positive this second result will be negative and thereby indicate that θ' is an obtuse angle, in which case we have found $\theta' = 180° - \theta$, as illustrated in Fig. 54. We are therefore able to take the lines in any order and leave the obtaining of the required acute angle to depend upon the sign of $\tan\theta$.

Example 5. Find the acute angle between the two lines $9y = 4x + 18$ and $3y = 4x + 6$.

Solution. The gradients of the lines are $m_1 = \frac{4}{9}$ and $m_2 = \frac{4}{3}$

$$\therefore \quad \tan\theta = \frac{m_1 - m_2}{1 + m_1m_2} = \frac{\frac{4}{9} - \frac{4}{3}}{1 + \frac{16}{27}} = -\frac{\frac{8}{9}}{\frac{43}{27}} = -\frac{24}{43} = -0{\cdot}5581$$

With $\tan\theta = -0{\cdot}5581$ we get $\theta = 150°\ 50'$ and the acute angle is $29°\ 10'$ Answer

Note the consequences of the two lines being at right angles. In this case $m_1m_2 = -1$ and $\tan(\theta_1 - \theta_2)$ will take the form $\frac{m_1 - m_2}{0}$, which is the indeterminate form ∞ to represent $\tan 90°$.

Exercise 14.3

1. Find the acute angle between the pairs of straight lines given by the following equations:

(i) $y = 2x + 1$, $y = 3x + 2$
(ii) $2y + 6x = 1$, $3y = 12x - 7$
(iii) $y + x = 1$, $3y - x + 1 = 0$

2. Find the gradient of a line which makes an angle of 45° with the line $y = 2x$ measured anti-clockwise from the line $y = 2x$.
3. Find the equation of the line which passes through the point (2, 0) and is inclined to the line $2y = x$ at an angle of 135° measured anti-clockwise from the line $2y = x$.
4. Find the interior angles of the triangle formed by the lines given by the following equations:

(i) $y = x$ (ii) $y = 3x$ (iii) $y + x = 2$

Sums and Products of Sines and Cosines

We use the four results on page 174 to produce relations between sums and products of sines and cosines as follows

$$\begin{aligned}
\cos(A - B) &= \cos A\cos B + \sin A\sin B && \text{(i)}\\
\cos(A + B) &= \cos A\cos B - \sin A\sin B && \text{(ii)}\\
\sin(A - B) &= \sin A\cos B - \cos A\sin B && \text{(iii)}\\
\sin(A + B) &= \sin A\cos B + \cos A\sin B && \text{(iv)}
\end{aligned}$$

$$\begin{aligned}
\cos(A - B) + \cos(A + B) &= 2\cos A\cos B && \text{from (i) + (ii)}\\
\cos(A - B) - \cos(A + B) &= 2\sin A\sin B && \text{from (i) − (ii)}\\
\sin(A + B) + \sin(A - B) &= 2\sin A\cos B && \text{from (iii) + (iv)}\\
\sin(A + B) - \sin(A - B) &= 2\cos A\sin B && \text{from (iv) − (iii)}
\end{aligned}$$

The second of the new results needs care in its application because it is contrary to the result we expect to get, as in the case $\sin(A + B) - \sin(A - B)$.

The reason is that the cosine of an angle decreases as the angle increases, so that to give a positive statement on the right-hand side we consider $\cos(A-B)-\cos(A+B)$ and not $\cos(A+B)-\cos(A-B)$. We can obtain a revised form of these new results by writing $T=A+B$ and $V=A-B$, whereupon $T+V=2A$ or $\frac{1}{2}(T+V)=A$, and $T-V=2B$ or $\frac{1}{2}(T-V)=B$

$$\cos(A-B)+\cos(A+B)=\cos V+\cos T=2\cos\tfrac{1}{2}(T+V)\cos\tfrac{1}{2}(T-V)$$
$$\cos(A-B)-\cos(A+B)=\cos V-\cos T=2\sin\tfrac{1}{2}(T+V)\sin\tfrac{1}{2}(T-V)$$
$$\sin(A+B)+\sin(A-B)=\sin T+\sin V=2\sin\tfrac{1}{2}(T+V)\cos\tfrac{1}{2}(T-V)$$
$$\sin(A+B)-\sin(A-B)=\sin T-\sin V=2\cos\tfrac{1}{2}(T+V)\sin\tfrac{1}{2}(T-V)$$

Some people seem able to commit this whole set of results to memory while the majority of us have to be content with only being able to remember the four basic results on page 174. Since the last four results are of such fundamental importance it is as well to attempt to absorb them with the following descriptive phrasing:

The sum of two cosines = *two, cos semi sum, cos semi difference*
The difference of two cosines = *two, sin semi sum, sin semi difference reversed*
The sum of two sines = *two, sin semi sum, cos semi difference*
The difference of two sines = *two, cos semi sum, sin semi difference*

Consider the following list of examples as illustrations of these results:

1. $\cos 60° + \cos 30° = 2\cos 45° \cos 15°$
2. $\cos 30° - \cos 60° = 2\sin 45° \sin 15°$
3. $\cos 0° + \cos A = 2\cos\frac{1}{2}A\cos\frac{1}{2}A$
   i.e. $1+\cos A = 2\cos^2\frac{1}{2}A$
4. $\sin 75° - \sin 45° = 2\cos 60° \sin 15°$
5. $2\cos 30° \sin 10° = \sin 40° - \sin 20°$
6. $\cos 160° + \cos 130° = 2\cos 145° \cos 15°$
7. $2\cos 60° \cos 120° = \cos 180° + \cos 60°$
   i.e. $-2\cos^2 60° = -1 + \cos 60°$
8. $\cos 130° - \cos 160° = 2\sin 145° \sin 15°$
9. $2\sin 50° \sin 20° = \cos 30° - \cos 70°$
10. $\sin 70° + \sin 10° = 2\sin 40° \cos 30°$
11. $2\sin 39° \cos 21° = \sin 60° + \sin 18°$
12. $\sin 110° - \sin 50° = 2\cos 80° \sin 30°$
13. $2\cos 70° \sin 40° = \sin 110° - \sin 30°$

Frequently we encounter an expression like $\sin A+\cos B$ or $\sin 40° + \cos 20°$ which at first inspection does not appear to fit any of the above results. A subsequent rewriting of either sin or cos in terms of the complementary angle will reveal the relevant simplification. Thus $\sin 40° + \cos 20°$ becomes $\cos 50° + \cos 20° = 2\cos 35° \cos 15°$, or alternatively $\sin 40° + \cos 20°$ becomes $\sin 40° + \sin 70° = 2\sin 55° \cos 15°$. These results are the same since $\cos 35° = \sin 55°$.

Example 6. In the triangle ABC the perpendicular from A to BC meets BC at a point N as shown in Fig. 55. $AB = 61$ cm, $AN = 60$ cm, $AC = 65$ cm. Calculate the lengths of BN and NC and the values of

(i) $\sin A$ (ii) $\sin 2C$
(iii) $2 \sin \frac{1}{2}(B - C) \cos \frac{1}{2}(B + C)$ (iv) $\tan \frac{1}{2}C$

Solution. Applying Pythagoras's Theorem in triangle ABN we have $61^2 = 60^2 + BN^2$

$$\therefore \quad BN^2 = 61^2 - 60^2 = (61 - 60)(61 + 60) = 121 \quad \therefore \; BN = 11 \text{ cm}$$

Answer

In triangle ANC the same theorem yields $65^2 = 60^2 + NC^2$

$$\therefore \quad NC^2 = 65^2 - 60^2 = (65 - 60)(65 + 60) = 5 \times 125 \quad \therefore \; NC = 25 \text{ cm}$$

Answer

With these results we now have

$$\cos B = \frac{11}{61}, \quad \sin B = \frac{60}{61}, \quad \cos C = \frac{25}{65} = \frac{5}{13}, \quad \sin C = \frac{60}{65} = \frac{12}{13}$$

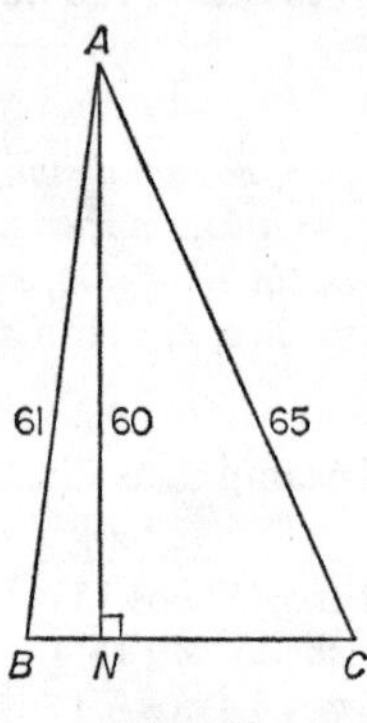

Fig. 55

Unfortunately we are unable to obtain $\sin A$ directly from the triangle so we use the result $\sin A = \sin (180° - A) = \sin (B + C)$.

$$\therefore \quad \sin (B + C) = \sin B \cos C + \cos B \sin C = \frac{60}{61} \cdot \frac{5}{13} + \frac{11}{61} \cdot \frac{12}{13}$$

$$= \frac{300 + 132}{793} = \frac{432}{793} \qquad \text{Answer}$$

$$\sin 2C = 2 \sin C \cos C = 2 \times \frac{12}{13} \times \frac{5}{13} = \frac{120}{169} \qquad \text{Answer}$$

$$2 \sin \tfrac{1}{2}(B - C) \cos \tfrac{1}{2}(B + C) = \sin B - \sin C = \frac{60}{61} - \frac{12}{13} = \frac{780 - 732}{793} = \frac{48}{793}$$

Answer

$$\tan \tfrac{1}{2}C = \frac{\sin \frac{1}{2}C}{\cos \frac{1}{2}C} = \frac{\sin \frac{1}{2}C \cos \frac{1}{2}C}{\cos^2 \frac{1}{2}C} = \frac{\sin C}{2 \cos^2 \frac{1}{2}C}$$

Not only do we use the result $2 \sin \frac{1}{2}C \cos \frac{1}{2}C = \sin C$ but we also recall the result (page 174) $\cos 2A = 2 \cos^2 A - 1$, which is modified to give $\cos C = 2 \cos^2 \frac{1}{2}C - 1$

$$\therefore \quad 2 \cos^2 \tfrac{1}{2}C = 1 + \cos C = 1 + \frac{5}{13} = \frac{18}{13}$$

Substitution in the expression for $\tan \frac{1}{2}C$ gives

$$\tan \tfrac{1}{2}C = \frac{12}{13} \div \frac{18}{13} = \frac{12}{18} = \frac{2}{3} \qquad \text{Answer}$$

Example 7. Simplify the expression $x = \dfrac{\sin 5A - \sin 3A}{\cos 5A - \cos 3A}$.

Solution. $\sin 5A - \sin 3A = 2 \sin A \cos 4A$

$$\cos 5A - \cos 3A = 2 \sin (-A) \sin 4A = -2 \sin A \sin 4A$$

We must append the negative sign because on the left-hand side of this expression we have $5A > 3A$. For example if $A = 10°$ then $\cos 50° < \cos 30°$ and $\cos 50° - \cos 30°$ is negative

$$\therefore \quad x = -\frac{\cos 4A}{\sin 4A} = -\cot 4A \qquad \text{Answer}$$

Example 8. Prove that $\cos 20° + \cos 100° + \cos 140° = 0$.
Solution. $\cos 100° + \cos 140° = 2 \cos 120° \cos 20° = -\cos 20°$ because $\cos 120° = -0{\cdot}5$

$$\cos 20° + \cos 100° + \cos 140° = 0 \qquad \text{Q.E.D.}$$

Example 9. Prove that if $A + B + C = 180°$ then

$$\tan \tfrac{1}{2}A \tan \tfrac{1}{2}B + \tan \tfrac{1}{2}B \tan \tfrac{1}{2}C + \tan \tfrac{1}{2}C \tan \tfrac{1}{2}A = 1.$$

Solution. Because $A + B + C = 180°$ we have $\sin A = \sin (180° - A) = \sin (B + C)$, $\cos A = \cos [180° - (B + C)] = -\cos (B + C)$. Hence $\tan A = -\tan (B + C)$.

If we now consider the similar half-angle relationship we obtain

$$\tfrac{1}{2}A + \tfrac{1}{2}B + \tfrac{1}{2}C = 90°$$

and thereafter

$$\sin \tfrac{1}{2}A = \sin \tfrac{1}{2}[180° - (B + C)] = \sin [90° - \tfrac{1}{2}(B + C)] = \cos \tfrac{1}{2}(B + C)$$
$$\therefore \quad \cos \tfrac{1}{2}A = \sin \tfrac{1}{2}(B + C)$$

Hence

$$\tan \tfrac{1}{2}A = \cot \tfrac{1}{2}(B + C) = \frac{1}{\tan \frac{1}{2}(B + C)}$$

This last result may be written as $\tan \frac{1}{2}A \tan \frac{1}{2}(B + C) = 1$ and then $\tan \frac{1}{2}(B + C)$ expanded to give

$$\tan \tfrac{1}{2}A \,.\, \frac{(\tan \frac{1}{2}B + \tan \frac{1}{2}C)}{1 - \tan \frac{1}{2}B \tan \frac{1}{2}C} = 1$$

Multiply both sides of the equation by $1 - \tan \frac{1}{2}B \tan \frac{1}{2}C$

$$\therefore \quad \tan \tfrac{1}{2}A \tan \tfrac{1}{2}B + \tan \tfrac{1}{2}A \tan \tfrac{1}{2}C = 1 - \tan \tfrac{1}{2}B \tan \tfrac{1}{2}C$$

Hence $$\tan \tfrac{1}{2}A \tan \tfrac{1}{2}B + \tan \tfrac{1}{2}B \tan \tfrac{1}{2}C + \tan \tfrac{1}{2}C \tan \tfrac{1}{2}A = 1 \qquad \text{Q.E.D.}$$

It is worth noting the symmetry of this last result. To say the result is symmetrical in the three variables A, B, C means that any two of the variables may be interchanged without altering the value of the expression. For example, an interchange of A and B or B and C or C and A leaves the left-hand side unchanged. Similarly, the value of the expression

$$\sin (A + B) = \sin A \cos B + \cos B \sin A$$

is unaltered when A and B are interchanged. We say therefore that the expression is symmetrical in the two variables A and B.

Exercise 14.4

1. Prove that (i) $\cos 5° - \sin 25° = \cos 55°$
   (ii) $\sin 40° - \sin 80° + \sin 20° = 0$
2. Prove that $\sin A + \sin (A + 120°) = \sin (A + 60°)$.
3. In the triangle ABC, the perpendicular from A to BC meets BC at a point N. $AB = 13$ cm, $AN = 12$, $CN = 9$ cm. Calculate the lengths of AC and BN and find the values of: (i) $\cos A$ (ii) $2 \sin \frac{1}{2}A \cos \frac{1}{2}(B - C)$ (iii) $\sin 2B$ (iv) $\tan \frac{1}{2}B$
4. Prove that $\dfrac{\sin 3A - \sin A}{\cos A - \cos 3A} = \cot 2A$
5. Prove that $\dfrac{\sin (3x + 2y) + \sin (2x + 3y)}{\cos (2x + 3y) + \cos (3x + 2y)} = \tan \frac{1}{2}(5x + 5y)$
6. Find the simplest form of $\dfrac{\sin 3A + \sin A + \sin 5A + \sin 7A}{\cos A + \cos 3A + \cos 5A + \cos 7A}$
7. Given that $A + B + C = 180°$ prove that

$$\tan A + \tan B + \tan C = \tan A \tan B \tan C.$$

   (Hint: $\tan A = -\tan (180° - A)$.)
8. Solve the equation $\sin 2x + \sin 4x = \cos x$ for $0 \leqslant x \leqslant 360°$.
9. Solve the equation $\sin 2x = \cos 3x$ for $0 < x < 90°$.
10. Solve the equation $20 \cos (20° + x) \sin (x - 5°) = 1$ for $0 < x < 90°$.

The Half-Angle Formulae for a Triangle

This chapter began with the cosine rule and it is to this rule that we turn again to deduce some more trigonometric relations in a triangle. We have deduced on page 174 that we may write $2 \sin^2 \frac{1}{2}A = 1 - \cos A$ and $2 \cos^2 \frac{1}{2}A = 1 + \cos A$ for any angle A.

Consider the triangle ABC in the usual notation of A, B, C and a, b, c

$$\cos A = \frac{b^2 + c^2 - a^2}{2bc}$$

$$\therefore 1 + \cos A = 1 + \frac{b^2 + c^2 - a^2}{2bc} = \frac{b^2 + 2bc + c^2 - a^2}{2bc}$$

$$= \frac{(b + c)^2 - a^2}{2bc} = \frac{(b + c + a)(b + c - a)}{2bc}$$

From this result it follows that

$$\tfrac{1}{2}(1 + \cos A) = \cos^2 \tfrac{1}{2}A = \frac{(b + c + a)(b + c - a)}{4bc}$$

We now introduce the standard notation of letting $2s = a + b + c$, i.e. s is the semi-perimeter of the triangle. With $2s = a + b + c$ it follows that

$$2s - 2a = a + b + c - 2a = b + c - a = 2(s - a)$$
$$2s - 2b = a + b + c - 2b = a - b + c = 2(s - b)$$
$$2s - 2c = a + b + c - 2c = 2(s - c)$$
$$\therefore \cos^2 \tfrac{1}{2}A = \frac{2s . 2(s - a)}{4bc}$$

and thus $\quad \cos \frac{1}{2}A = \sqrt{\dfrac{s(s - a)}{bc}} \qquad$ (i) taking the positive square root because $\frac{1}{2}A$ is an acute angle. Similarly

$$1 - \cos A = 1 - \frac{(b^2 + c^2 - a^2)}{2bc} = \frac{2bc - b^2 - c^2 + a^2}{2bc} = \frac{a^2 - (b - c)^2}{2bc}$$

because

$$2bc - b^2 - c^2 = -\{b^2 - 2bc + c^2\} = -(b - c)^2$$

$$\therefore 1 - \cos A = 2\sin^2 \tfrac{1}{2}A = \frac{[a - (b - c)][a + (b - c)]}{2bc}$$

$$= \frac{(a - b + c)(a + b - c)}{2bc}$$

$$\therefore \sin^2 \tfrac{1}{2}A = \frac{2(s - b)2(s - c)}{4bc}$$

$$\therefore \sin \tfrac{1}{2}A = \sqrt{\frac{(s - b)(s - c)}{bc}} \qquad \text{(ii)}$$

again taking the positive root because $\frac{1}{2}A$ is an acute angle. Examination of the results for (i) and (ii) shows immediately the corresponding results for $\cos \frac{1}{2}B$, $\sin \frac{1}{2}B$ and so on. Furthermore, we can combine (i) and (ii) to get a similar expression for

$$\tan \tfrac{1}{2}A = \frac{\sin \frac{1}{2}A}{\cos \frac{1}{2}A} = \sqrt{\frac{(s - b)(s - c)}{s(s - a)}}$$

If we look back to Example 6 on page 180 we see that part (iv) required $\tan \frac{1}{2}C$. From the work above we note that $\tan \frac{1}{2}C = \sqrt{\frac{(s - a)(s - b)}{s(s - c)}}$ and that in this case $a = 36$, $b = 65$, $c = 61$, $s = \frac{1}{2}(a + b + c) = 81$. Again $\tan \frac{1}{2}C$ is positive because $\frac{1}{2}C$ is an acute angle.

$$\therefore \tan \tfrac{1}{2}C = \sqrt{\frac{(81 - 36)(81 - 65)}{81(81 - 61)}} = \sqrt{\frac{45 \times 16}{81 \times 20}} = \sqrt{\frac{4}{9}} = \frac{2}{3}$$

and the result $\tan \frac{1}{2}C = \frac{2}{3}$ is the same as before.

The Area of a Triangle

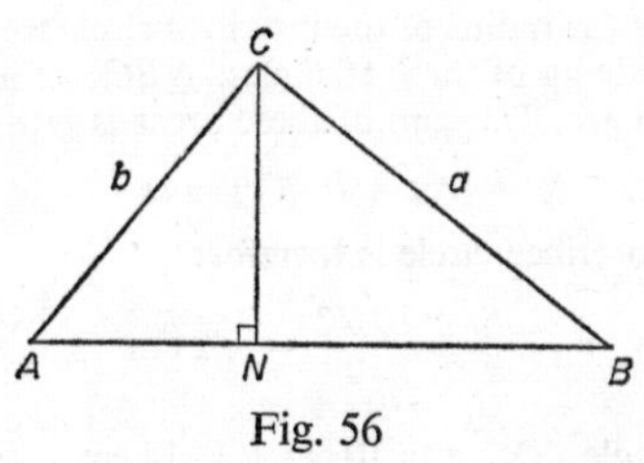

Fig. 56

The area of a triangle ABC drawn as in Fig. 56 is $\frac{1}{2}$ base $\times$ perpendicular height given by $\frac{1}{2}\,AB.NC$ but since NC is $b \sin A$ and $AB = c$ we have

$$\Delta = \text{Area of triangle } ABC = \tfrac{1}{2}bc \sin A$$

Again the arrangement of the letters suggests the two other alternative expressions for the area. However, on noting that $\Delta = bc \sin \frac{1}{2}A \cos \frac{1}{2}A$ we may now use the half-angle formulae to obtain the result

$$\Delta = bc\sqrt{\frac{(s-b)(s-c)}{bc}}\sqrt{\frac{s(s-a)}{bc}}$$

$$\therefore \Delta = \sqrt{[s(s-a)(s-b)(s-c)]}$$

usually referred to as Hero's formula.

Example 10. In triangle ABC, $a = 6$ cm, $b = 9$ cm, $c = 5$ cm. Find (i) the area of the triangle ABC (ii) $\sin B$ (iii) the radius of the inscribed circle.
Solution. Let Fig. 57 represent the problem. Since $2s = 6 + 9 + 5 = 20$ we have $s = 10$.

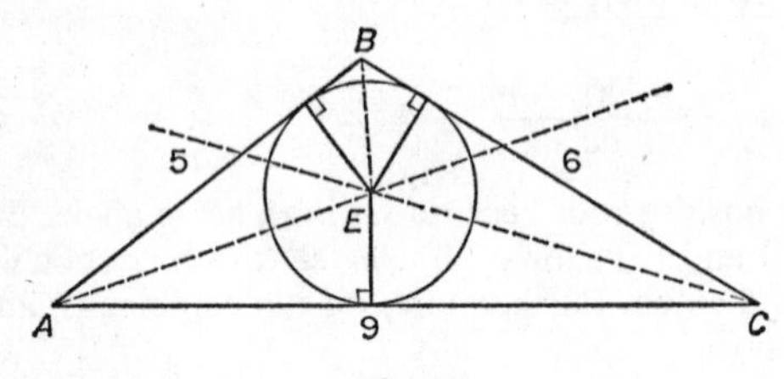

Fig. 57

(i) The area of the triangle ABC is

$$\Delta = \sqrt{[10.(10-6)(10-9)(10-5)]} = \sqrt{200} = 10\sqrt{2}\ \text{cm}^2$$

Answer

(ii) We may reason that

$$\tfrac{1}{2}ca \sin B = \Delta$$

$$\therefore \quad \sin B = \frac{10\sqrt{2}}{\frac{1}{2} \times 5 \times 6} = \frac{2\sqrt{2}}{3}$$

Answer

Alternatively we may use

$$\sin B = 2 \sin \tfrac{1}{2}B \cos \tfrac{1}{2}B = 2\sqrt{\frac{s(s-b)(s-a)(s-c)}{ac\,.\,ac}}$$

(iii) The centre of the inscribed circle is at E, the point of intersection of the bisectors of the angles. If r is the radius of the inscribed circle we see that the area of the triangle ABC is made up of three triangles. $\triangle BCE$ of area $\frac{1}{2}ra$, $\triangle CEA$ of area $\frac{1}{2}rb$, $\triangle AEB$ of area $\frac{1}{2}rc$. The sum of these areas is $\frac{1}{2}r(a + b + c)$

$$\therefore \quad \Delta = \tfrac{1}{2}r(a + b + c) = rs$$

The radius of the inscribed circle is therefore

$$r = \frac{\Delta}{s} = \frac{10\sqrt{2}}{10} = \sqrt{2}\ \text{cm}$$

Answer

Example 11. In the triangle ABC, $a = 10$ cm, $b = 14$ cm, $c = 16$ cm. Find the radius of the circumscribed circle to the triangle.

Solution. The centre of the circumscribed circle is F the point of intersection of the bisectors of the sides of the triangle (not shown). Thus CFD is a diameter of the circle in Fig. 58, so that $CD = 2R$ where R is the radius of the circle. Furthermore, $\angle CDA = \angle CBA = B$ (angles at the circumference standing on the same arc AC)

and $\angle DAC = 90°$ (angle in a semicircle). In the right-angled triangle ADC, $AC = CD \sin CDA$.

$$\therefore \quad b = 2R \sin B, \text{ and } R = \frac{b}{2 \sin B}$$

We can obtain another form of this answer by using the fact that $\Delta = \frac{1}{2}ca \sin B$ (page 183) in which case

$$\sin B = \frac{2\Delta}{ac} \quad \text{and} \quad R = \frac{b}{2} \div \frac{2\Delta}{ac} = \frac{abc}{4\Delta}$$

For this example

$$s = 20 \quad \text{and} \quad \Delta = \sqrt{[20(20-10)(20-14)(20-16)]}$$

$$\therefore \quad \Delta = \sqrt{(20 \times 10 \times 6 \times 4)} = \sqrt{(100 \times 16 \times 3)} = 40\sqrt{3} \text{ cm}^2$$

Hence $\quad R = \frac{10 \times 14 \times 16}{4 \times 40\sqrt{3}} = \frac{14}{\sqrt{3}} \frac{\sqrt{3}}{\sqrt{3}} = \frac{14\sqrt{3}}{3}$ cm $\quad$ Answer

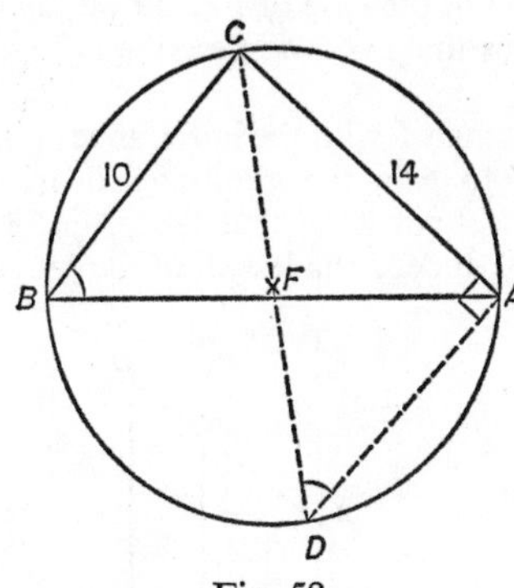

Fig. 58

Exercise 14.5

Calculate (i) the area (ii) $\sin \frac{1}{2}A$ (iii) $\cos \frac{1}{2}A$ (iv) r (v) R, for each of the following triangles in Questions 1 to 4:

1. $a = b = 10$ m, $c = 16$ m
2. $a = 3$ m, $b = 4$ m, $c = 3$ m
3. $a = 10$ cm, $b = 8$ cm, $c = 12$ cm
4. $a = 6$ cm, $b = 16$ cm, $c = 18$ cm
5. Write down the half angle formula for $\sin \frac{1}{2}B$, $\cos \frac{1}{2}B$, and $\tan \frac{1}{2}C$.
6. Find the radius of the circumscribed circle and the area of the triangle ABC with $a = 10$ m, $c = 20$ m, $\angle ABC = 60°$.
7. In the triangle ABC, angle $BAC = 60°$. Prove that $4s(s - a) = 3bc$.
8. Prove that the area of a triangle ABC is $\frac{1}{2}b^2 \frac{\sin A \sin C}{\sin (A + C)}$
   (Hint: $A + B + C = 180°$.)
9. Two circles of radii 8 cm and 10 cm intersect in the points A and B. Find the length of AB if the centres are 12 cm apart.
10. Find the areas of the equilateral triangles which can be (i) inscribed and (ii) circumscribed to a circle of radius 10 cm.

CHAPTER FIFTEEN

GEOMETRY IN THREE DIMENSIONS

Orthogonal Projection

Examining the plane diagram in Fig. 59 we see that from A and B lines have been drawn perpendicular to the line l at P and Q. If any point C were chosen between A and B on the line then the perpendicular from C to l would meet l at a point R in between P and Q. We describe such a construction as an **orthogonal projection** and in the case of Fig. 59 we say we have projected the line segment AB on to the line l and we describe PQ as being the orthogonal projection of AB on l.

If AB makes an acute angle θ with the line l, as seen in Fig. 60, then we may draw AN parallel to l to gain the right-angled triangle BAN, in which $AN = PQ$. But $AN = AB \cos \theta$, so that the projection AB on to the line l has a length of $AB \cos \theta$. Indeed the length of the projection of AB onto any

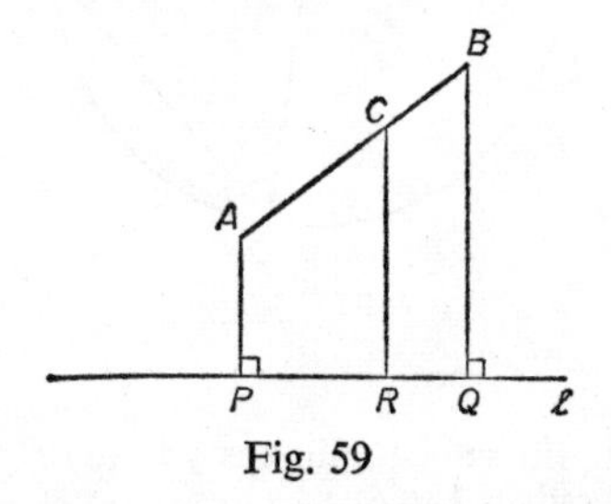

Fig. 59

other line inclined at angle α to AB is $AB \cos \alpha$. It would not be possible to project AB on to a line which is not in the same plane, for in this case AB and l would be described as **skew** lines, e.g. a diagonal of the front cover of this book and any one of the edges of the back cover are skew lines. Of course, any right-angled triangle contains the spirit of orthogonal projection since each side is related to the hypotenuse by the general relations $b = c \cos \theta$, $a = c \cos (90° - \theta)$.

The Angle Between a Line and a Plane

We define the angle between a line and a plane as the angle between the line and its projection on the plane. (From hereon in this text 'projection' will be taken to mean orthogonal projection.) This means that we now move the arrangement of Fig. 60 on to the plane of Fig. 61(a). Thus perpendiculars are drawn from A and B to the plane at P and Q, and PQ is now the projection of AB on to the plane. If BA and QP are produced to intersect at T then $\angle ATP$ is the angle between the line AB and the plane.

By a perpendicular to a plane we mean a line which is **perpendicular to every**

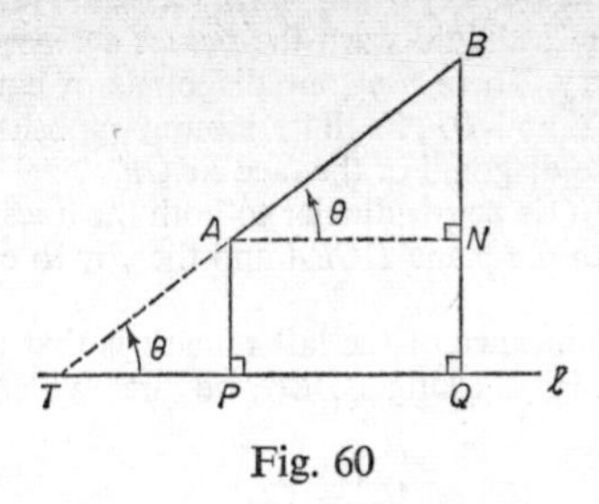

Fig. 60

line in the plane. Fortunately we do not have to check the perpendicularity for every line in the plane—just two different lines through the foot of the perpendicular will be sufficient. Thus, Fig. 61(*b*) indicates that *BQ* is perpendicular to the plane *k* because it is perpendicular to the two distinct lines *RQV* and *LQM* in the plane.

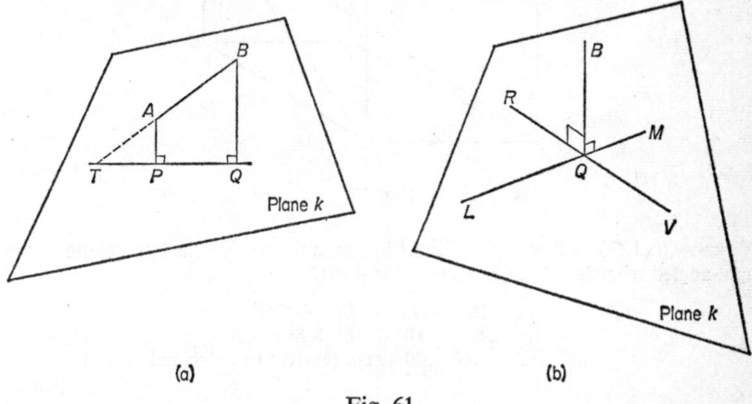

Fig. 61

Returning to Fig. 61(*a*), we therefore note that the angle between the line *AB* and the plane *k* will be given by

$$\tan\theta = \frac{PA}{TP} \text{ or } \sin\theta = \frac{QB}{TB} \text{ and so on}$$

Example 1. *ABCDEFGH* is a box with rectangular faces: $AB = 10$ cm, $BC = 16$ cm and $EA = 8$ cm. Find the length of a diagonal of the box, and the angle made by each of the lines *AG* and *BG* with the plane *ABCD*.

Solution. Let Fig. 62 represent the box. Before proceeding with the solution let us identify some of the geometrical ideas so far. The two lines *EG* and *AC* are parallel and therefore in the same plane, but *EG* and *BD* are skew lines because they are not in the same plane. The two lines *FB* and *AD* are also skew lines and so are *FB* and *AG*. The interesting feature here is that *FB*, being skew to *AD* and *AG*, does not imply that *AD* and *AG* are skew lines. The line *FB* is parallel to *EA* and *EA* is parallel to *HD* so *FB* is also parallel to *HD*. The relation of being 'perpendicular to' is not quite the same as that of being 'parallel to'. Thus *FB* is perpendicular to *BC* and *BC* is perpendicular to *CG* and *CH*, but *FB* is not perpendicular to *CG* or *CH*.

These last three examples should warn the reader against jumping to false conclusions in solid geometry. There are four diagonals of equal length to this box and they are AG, BH, CE and DF, i.e. lines joining opposite vertices. The line BG would be described as the diagonal of the face $BCGF$.

Finally, we note that GC is perpendicular to both the lines CB and CD, in which case it is perpendicular to the plane $DCBA$ and thereby to every line in $DCBA$ including CA.

To find AG.—The significance of the last remark is that we now see AG as the hypotenuse of a right-angled triangle ACG, so that we can apply Pythagoras's Theorem to gain

$$AG^2 = AC^2 + CG^2$$

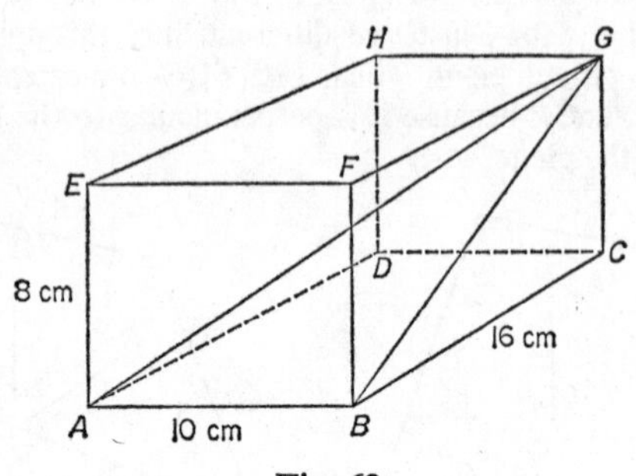

Fig. 62

We know that $CG = 8$ cm but AC must be found from being the hypotenuse of the right-angled triangle ABC, i.e. $AC^2 = AB^2 + BC^2$

$$\therefore \quad AG^2 = AB^2 + BC^2 + CG^2$$
$$\text{i.e.} \quad AG^2 = 10^2 + 16^2 + 8^2 = 420$$
$$\therefore \quad AG = 20{\cdot}49 \text{ cm (correct to 2 decimal places)}$$

Answer

Notice that the diagonals bisect each other at the centre of the box and that the angle between the diagonals AG and EC is twice angle GAC.

Find the angle between AG and plane ABCD.—We must first project AG on to the plane. A is already in the plane and the perpendicular from G is GC. The line AC is therefore the projection of AG on the plane $ABCD$, so that $\angle GAC$ is the required angle.

In the right-angled triangle ACG

$$\sin GAC = \frac{CG}{AG} = \frac{8}{20{\cdot}49}$$

$$\sin GAC = \frac{1}{2{\cdot}561} = 0{\cdot}3904 \text{ (from reciprocal tables)}$$

$$\therefore \quad \angle GAC = 23° \, 59'$$

$\therefore$ AG is inclined at 22° 59′ to plane $ABCD$ — Answer

To find the angle between BG and plane ABCD.—The projection of BG on the plane $ABCD$ is BC, so we require the angle GBC in the right-angled triangle GBC.

$$\tan GBC = \frac{CG}{BC} = \frac{8}{16} = 0{\cdot}5 \quad \therefore \quad \angle GBC = 26° \, 34'$$

$\therefore$ BG is inclined at 26° 34′ to the plane $ABCD$ — Answer

Exercise 15.1

1. Points A and B have coordinates (5, 9), (8, 16) respectively. Find the projection of AB on (i) the x axis, (ii) the y axis.
2. $ABCDEFGH$ is a rectangular box (right prism) with $AB = 20$ cm, $BC = 30$ cm, $BF = 12$ cm. Find: (i) the length of the diagonal BH; (ii) the angle between the diagonals BH and FD, and BH and AG; (iii) the angles made by each of the lines AG and BG with the plane $ABCD$.
3. A right-angled triangle ABC has $B = 90°$, $AB = 5$ m and $CB = 12$ m. A pole BD of length 10 m is erected at B perpendicular to the plane of ABC. Find: (i) angle BAD; (ii) angle DCB; (iii) the position of the point M on AC for which the angle between MD and the plane ABC is a maximum.
4. A ladder rests against a vertical wall. The ladder is 5 m long and is placed with its foot 2 m horizontally from the wall and the top of the ladder 4 m above the ground. Find the angle of inclination of the ladder to the horizontal.
5. A straight stick 1 m long is placed inside a tank in the form of a cube with edges of 75 cm. Find the position of the stick which gives a minimum angle of inclination with the base of the tank.

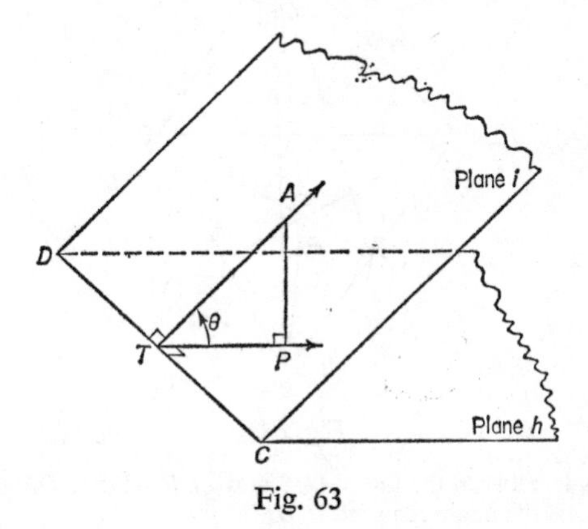

Fig. 63

The Angle Between Two Planes

Two non-parallel planes intersect in a common line. We define the angle between the two planes to be the angle between two lines, one in each plane, drawn perpendicular to the common line from a point on the line. (This angle is sometimes referred to as the dihedral angle.) In Fig. 63 we have CD as the common line of the two planes designated h for horizontal and i for inclined. The point T on CD has been chosen from which to draw the two perpendiculars to CD. TA is in the i plane and TP is the h plane. If TP is the projection of TA then $TP = TA \cos \theta$ gives the angle between planes. If we look back to Fig. 62 on page 188 we can see that the planes $ABCD$ and ABG have AB as their common line and since $\angle ABC = 90°$ (rectangular base) and $\angle ABG = 90°$ (AB perpendicular to face $CBFG$) the angle between BC and BG will be the angle between the planes $ABCD$ and ABG.

Example 2. A regular tetrahedron $ABCD$ is a four-faced pyramid with each face being an equilateral triangle. If the length of an edge is 10 cm find: (i) the angle between the edge AD and the base ABC; (ii) the angle between two faces. (The regular tetrahedron is sometimes called a right pyramid.)

Solution. Let Fig. 64 represent the tetrahedron with DE the perpendicular from D to the base ABC. Since DE is perpendicular to the base the angles DEA, DEB and DEC are 90° each. Since $DA = DB = DC$ the triangles DEA, DEB, DEC are congruent (r.h.s.) so $AE = BE = CE$. Hence E is the circumcentre of the triangle ABC so that CE produced is perpendicular to AB and F is the midpoint of AB. With F the midpoint of AB it follows that DF is perpendicular to AB, because triangle DAB is isosceles or equilateral.

To find the angle between AD and ABC.—From Fig. 64 this angle is given by $\angle DAE$, and $\cos DAE = AE/AD = AE/10$. Since AE is a radius of the circumcircle of $\triangle ABC$ then $AE = \dfrac{b}{2 \sin B} = \dfrac{10}{2 \sin 60^\circ}$ (page 185)

$$\therefore \quad \cos DAE = \frac{AE}{AD} = \frac{10}{20 \sin 60^\circ} = \frac{1}{\sqrt{3}} = \frac{\sqrt{3}}{3} = 0{\cdot}5774$$

$$\therefore \quad \angle DAE = 54^\circ\, 44, \qquad \text{Answer}$$

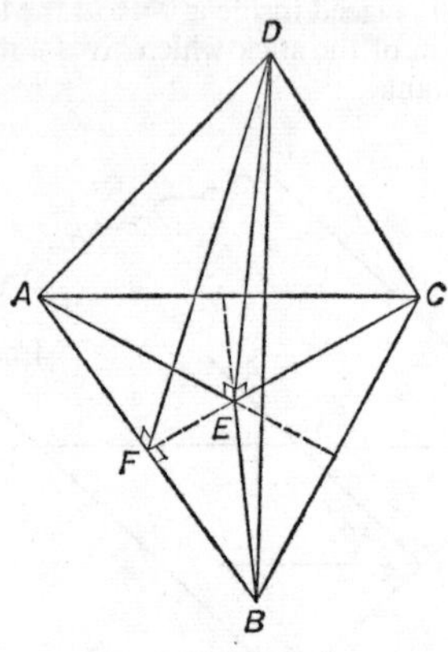

Fig. 64

To find the angle between the faces ADB and ACB.—Since DF and CF are each perpendicular to AB the angle required is DFE.

In $\triangle DAF$, $DF = AD \sin 60^\circ = 10 \sin 60^\circ = 5\sqrt{3}$ cm. Similarly $FD = FC = 5\sqrt{3}$ cm. The cosine rule in $\triangle DFC$ gives $DC^2 = DF^2 + FC^2 - 2DF.FC \cos DFC$.

$$\therefore \quad 100 = 75 + 75 - 2.75 \cos DFC \; ; (FD^2 = 75)$$

$$\therefore \quad \cos DFC = \tfrac{1}{3}$$

$$\therefore \quad \angle DFC = 70^\circ\, 32' \qquad \text{Answer}$$

Volumes

A **prism** is a solid figure with two congruent end faces whose side faces are **parallelograms.** When all the side faces are rectangles the solid is called a **right prism**. The end faces of a prism may be triangles, quadrilaterals or polygons. If the end faces are circles then we call the figure a cylinder.

The volume of a prism is equal to the product of the area of one of its end faces and its perpendicular distance from the opposite end face. In Fig. 65 the volume of the prisms is $V = Ah$, or A_1h.

The Volume of a Right Prism

Consider the triangular prism $ABCDEF$ in Fig. 66 and let CN be the perpendicular from C to AB. The figure, being a right prism, has faces that are

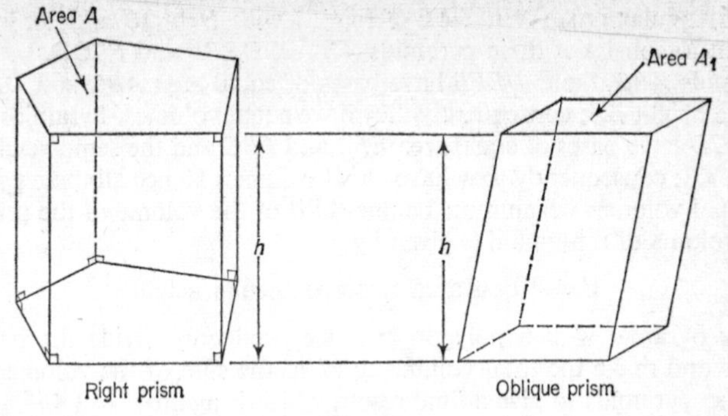

Fig. 65

rectangles so that finding the surface area of the prism is an elementary matter.

In order to find the volume we consider the prism as part of the rectangular block *ABLM DEOP* by making *AM* and *BL* equal and parallel to *NC*.

Clearly the volume of the prism $= \frac{1}{2}$ (volume of the rectangular block)
$= \frac{1}{2}$ (area *ABLM*) $\times$ height
$\therefore V =$ area of *ABC* $\times$ height
or $V = Ah$ (in the usual notation)

Any prism on a polygonal base may be considered as a sum of triangular prisms. A cylinder in particular may be considered as a prism with a polygon base which has an infinite number of equal sides to give an area of πr^2 and the final volume of $\pi r^2 h$.

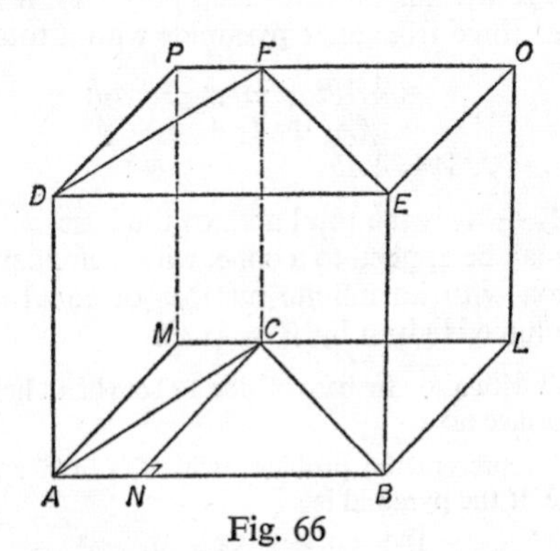

Fig. 66

The Volume of a Pyramid

A pyramid is formed by joining the vertices of any polygon to a point outside the plane of the polygon. For the moment we shall consider a pyramid with a triangular base and to save some tedious work we shall *assume* that triangular pyramids with equal heights and bases of equal area are equal in volume.

The triangular prism $ABCDEF$ of Fig. 67 with DB, DC and BF joined is now seen to consist of three pyramids $ABCD$, $DEFB$ and $FBCD$.

Pyramids $ABCD$ and $DEFB$ have bases of equal area ABC and DEF and the same height BE; consequently they have equal volumes. Pyramids $DEFB$ and $FBCD$ have bases of equal area EFB and FBC and the same height from D to $BCFE$; consequently they have equal volumes. Hence all three pyramids have equal volumes which must be one-third of the volume of the prism.

The volume of a pyramid is given by

$$V = \tfrac{1}{3}\text{ base area} \times \text{perpendicular height}$$

For any pyramid with a polygon base we need only divide the base into triangles and make the total volume equal to the sum of the volumes of the triangular pyramids to give a final result, which is again $V = \frac{1}{3}Ah$.

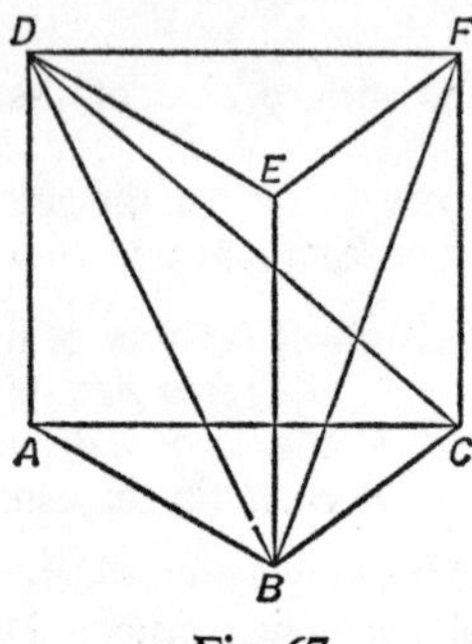

Fig. 67

Suppose the base is a pentagon (five-sided polygon): it is then possible to consider the solid as three triangular pyramids with a total volume of

$$V = \tfrac{1}{3}A_1h + \tfrac{1}{3}A_2h + \tfrac{1}{3}A_3h$$
$$= \tfrac{1}{3}(A_1 + A_2 + A_3)h$$
$$\therefore\ V = \tfrac{1}{3}Ah$$

where $A = A_1 + A_2 + A_3$ is the total area of the base.

The same result may be applied to a cone, whose circular base may be considered as a polygon with an infinite number of equal sides. In this case $A = \pi r^2$ and the volume is given by $V = \frac{1}{3}\pi r^2h$.

Example 3. A pyramid with a square base of side 100 cm has a height of 30 cm. Find: (i) the volume; (ii) surface area.

Solution. Let Fig. 68 represent the problem, with $EN = 30$ cm the height of the pyramid. The volume of the pyramid is

$$\tfrac{1}{3} \times 100 \times 100 \times 30 = 10^5 \text{ cm}^3 \qquad \text{Answer}$$

To find the surface area of the face EBC we shall need the height ME. Since triangle EBC is an isosceles triangle (i.e. $EB = EC$) the perpendicular height is ME, where M is the midpoint of BC.

With the right-angled triangle ENM we have $NE = 30$ cm, $NM = 50$ cm

$$\therefore\quad ME^2 = 30^2 + 50^2 \quad \text{(Pythagoras's Theorem)}$$
$$= 3400$$
$$\therefore\quad ME = 10\sqrt{34} = 58{\cdot}31 \text{ cm (correct to 2 decimal places)}$$

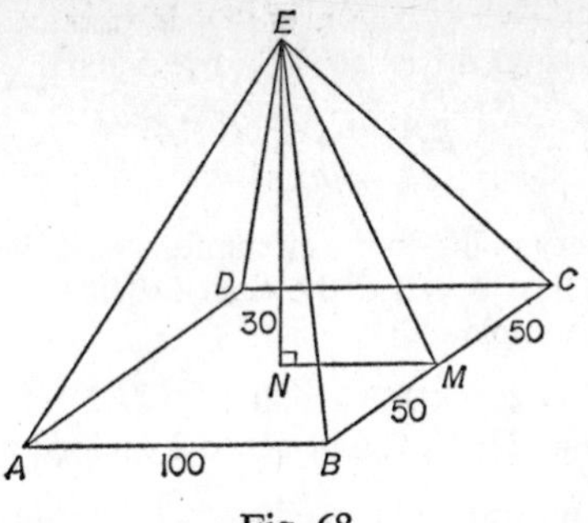

Fig. 68

The area of the four congruent side faces is

$$4 \times \frac{1}{2} \times 100 \times 58{\cdot}31 = 11\,662 \text{ cm.}$$

The total external surface area of the pyramid, which includes the area of the base, is therefore given by

$$S = 11\,662 + 10\,000 = 21\,662 \text{ cm}^2 = 2{\cdot}1662 \text{ m}^2$$

$\therefore$ Total surface area of the pyramid is $2{\cdot}17$ m² (correct to 2 decimal places)

Answer

The Volume and Curved Surface Area of a Cone

Example 4. Find the volume and curved surface area of a right circular cone of height 40 cm and base radius of 10 cm. (Recall that a right circular cone is such that a perpendicular from the vertex will meet the base at its centre—otherwise it would be called an oblique cone.) Use 3·14 as an approximation for π.

Solution. The volume is $V = \frac{1}{3} Ah = \frac{1}{3}\pi r^2 h$, where r is the radius of the base and h the perpendicular height

$$\therefore \quad V = \frac{3{\cdot}14 \times 10 \times 10 \times 40}{3} = 4187 \text{ cm}^3 \text{ [nearest cm}^3\text{]} \quad \text{Answer}$$

To find the area of curved surface we examine Fig. 69(*a*), which shows the cone with vertex T and circular base centre C. If A is any point on the circum-

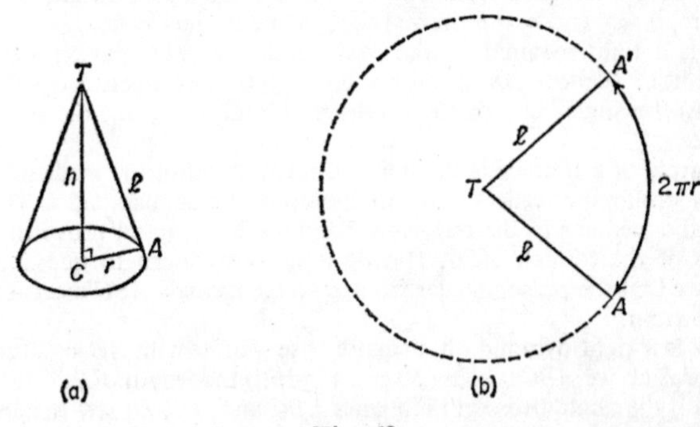

Fig. 69

ference of the base then the length of TA is obtained from Pythagoras's Theorem in the right-angled triangle TCA as

$$\begin{aligned} TA^2 &= TC^2 + CA^2 \\ &= h^2 + r^2 \end{aligned}$$

thus showing that every point on the circumference is the same distance from the vertex. This distance is called the **slant height** and given the symbol l, i.e. $TA = l$. In this example

$$l^2 = 40^2 + 10^2 = 500$$
$$\therefore l = 22{\cdot}36 \text{ cm (correct to 2 decimal places)}$$

It follows that if we unwrap the curved surface from the cone the shape will be that of a sector of a circle. This is because the locus of points which are the same distance l from the fixed point T is the arc of a circle radius l and centre T (page 62). The circumference of the base of the cone, which was $2\pi r$ in length, has now become the arc of the sector AA' in Fig. 69(b). We require the area of this sector since it is the area of the curved surface of the cone.

The circumference of the circle centre T and radius l is $2\pi l$ and its area is πl^2. It follows therefore that since the arc AA' is of length $2\pi r$ the sector TAA' is a fraction $2\pi r/2\pi l$ of the area of the complete circle which has an area of πl^2

$$\therefore \text{Area of sector } TAA' = \frac{2\pi r}{2\pi l} \times \pi l^2 = \pi rl$$

$\therefore$ The area S of the curved surface of a right circular cone is πrl.

In this example

$$S = 3{\cdot}14 \times 10 \times 22{\cdot}36 = 702 \text{ cm}^2 \text{ (to the nearest cm}^2\text{)} \qquad \text{Answer}$$

It follows that $S + \pi r^2$ gives the total surface area of the cone, a result which may be factorised as $\pi r(l + r)$. Answer

Exercise 15.2

1. A right prism has a height of 20 cm and faces which are equilateral triangles of side 10 cm. Find the volume and total surface area of the prism.
2. A pyramid stands on a triangular base ABC and has a height of 10 cm. If $a = 6$ cm, $b = 4$ cm, $c = 8$ cm find the volume of the pyramid.
3. $ABCD$ is a right pyramid on the equilateral base ABC having $AB = 20$ cm. The height of D above ABC is 30 cm. Find: (i) the angle between AD and plane ABC; (ii) the angle between the planes ADB and ABC; (iii) the volume of the pyramid.
4. The **centroid** of a triangle is the point of intersection of the medians, which are the lines joining the vertices to the midpoints of the opposite sides. The centroid divides the medians in the ratio 1:2. Examine Fig. 64 and prove that E is the centroid of the triangle ABC. Having done this, return to question 3 above and prove that the perpendicular from D to the triangle ABC meets the triangle at its centroid.
5. $ABCDE$ is a right pyramid on a square base $ABCD$ with $AB = 20$ cm and the height of E above $ABCD$ being 50 cm. Find: (i) the length of EB; (ii) the angle EBD; (iii) the angle between the planes EBC and $ABCD$; (iv) the angle AEC; (v) the volume of the pyramid.

6. A tetrahedron $ABCD$ has an equilateral base ABC of side 12 cm and a height of 23 cm. Find the volume of the tetrahedron.
   If the figure is now placed with BCD as base how high is the vertex A above BCD?
7. Find the volume and surface area of a right circular cone of base radius 5 cm and height 12 cm (take $\pi = 3{\cdot}14$).
8. In the tetrahedron $ABCD$, AB, AC and AD are each perpendicular to the others (i.e. mutually perpendicular). $AB = AC = 4$ m and $AD = 3$ m. Calculate the angle between the planes BCD and ABC, and the length of the perpendicular from A to BCD.
9. A sector of a circle of radius 20 cm includes an angle of 216°. If this sector is taken to be the curved surface of a cone what is the volume of the cone? (Take $\pi = 3{\cdot}14$.)
10. A right circular cone of height 18 cm and base radius 3 cm is cut by a plane parallel to the base at a point 12 cm above the base. If the top is now removed find the volume of the remainder. (The remaining part is called a **frustum** of the cone.)

The Sphere

The intersection of a plane with a sphere is a circle. When the plane also passes through the centre of the sphere the circle is referred to as a **great circle**, otherwise it is called a **small circle**. We can prove that the intersection

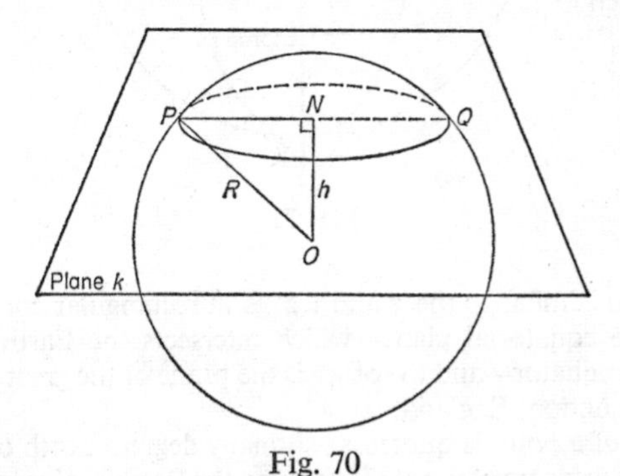

Fig. 70

is a circle as follows. In Fig. 70 let the plane k intersect the sphere of radius R in a curve and let P be any point on that curve. Now let the perpendicular from the centre of the sphere to the plane meet the plane at N. With this construction we have obtained a right-angled triangle OPN, with $OP = R$ (a radius of the sphere) and $ON = h$ (which is the only perpendicular from O to the plane)

$\therefore NP^2 = R^2 - h^2$ (Pythagoras's Theorem)
$\qquad = $ a constant for all positions of P on the dotted curve, because R and h are constant.

We therefore have a point, which moves on a plane k, such that its distance from the fixed point N is constant. The locus of P is therefore a circle centre N and radius $\sqrt{(R^2 - h^2)}$.

If $h = 0$, N coincides with O and the plane passes through the centre of the sphere and the locus of P is a circle centre O and radius R, i.e. a great circle. Thus on the Earth, which is assumed to be a sphere, circles of latitude are imaginary small circles parallel to the equator, while circles of longitude are all great circles through the north and south poles.

Latitude and Longitude

The position of any point on the Earth's surface is given by the two coordinates of latitude and longitude, which are given with respect to two

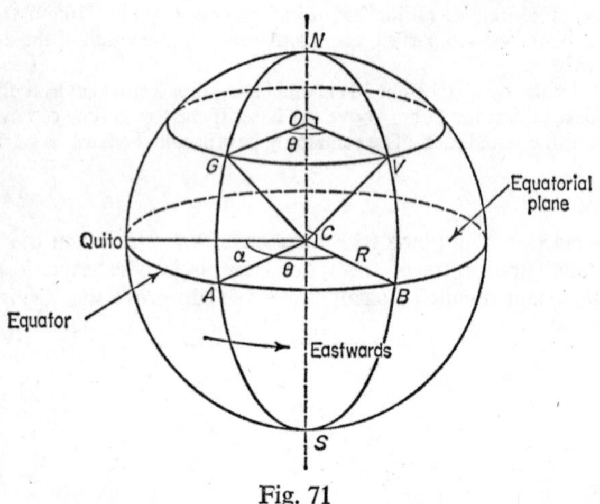

Fig. 71

reference planes (similar to the x and y axes in rectangular coordinates); one of which is the equatorial plane—which intersects the Earth sphere in the circle called the equator—and the other is the plane of the great circle through Greenwich in London, England.

The latitude of a point is quoted as so many degrees north or south of the equator, being a measure of the angle between the Earth radius to the point and the equatorial plane. The latitude of Greenwich is 51° 29′ N, which means that with the position of Greenwich represented by G in Fig. 71 and C taken as the centre of the Earth, the angle $ACG = 51° 29'$. $NGAS$ is the circle of longitude passing through Greenwich, which means that the projection of CG on the equatorial plane lies along CA. Of course any point on the circle of latitude passing through G will have the same latitude; for example, the point V will also have a latitude of 51° 29′ N, i.e. angle $VCB = 51° 29'$.

To find the longitude of a point we need to calculate the angle θ between the circles of longitude through Greenwich and the point. Thus Tokyo has a latitude of 35° 48′ N and a longitude of 139° 45′ E, and both readings tell us that Tokyo is nearer the Equator than Greenwich (by $51° 29' - 35° 48'$) and that the angle θ in Fig. 71 is 139° 45′ in an easterly sense.

The town of Quito in Ecuador has a position given by 0° 20′ S, 78° 15′ W, which means that it is just below the Equator and on a circle of longitude given by $\alpha = 78° 15'$ in a westerly sense from Greenwich. We see therefore that the latitude of a point is given by the angle between a line and a plane, whereas the longitude of a point is given by the angle between two planes.

Distance Between Two Points on a Sphere

An angle of 1 degree of longitude at the equator gives a greater distance over the Earth's surface than at any other latitude. Indeed this distance decreases as we get closer to the poles.

If we take the Earth as a sphere of radius 6368 km, then the circumference at the equator is of length $2\pi \times 6368$ km. A difference of 1 degree in longitude at the equator will therefore represent a surface distance of

$$\frac{2\pi \times 6368 \text{ km}}{360} = 111 \text{ km}$$

to the nearest kilometre (with $\pi = 3{\cdot}14$). To move through 1 degree of longitude on a circle of latitude through Greenwich is clearly a shorter distance. To calculate this distance we need the radius OG shown in Fig. 71. Since G has a latitude of $51° 29'$ it follows that angle CGO is also $51° 29'$ (alternate angles between AC parallel to GO) so that from the right-angled triangle GOC we have $OG = R \cos$ (latitude) $= 6368 \cos 51° 29' = r$ (say).

The circumference of this circle of latitude is $2\pi r$, so that one degree change in longitude on this circle represents a surface distance of

$$\frac{2\pi r}{360} = \frac{2\pi \times 6368 \cos 51° 29'}{360} = 69 \text{ km to the nearest kilometre}$$

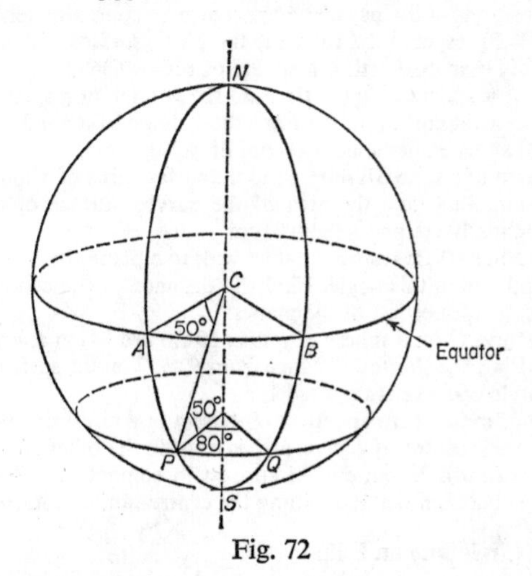

Fig. 72

Example 5. Find the shorter distance measured on the small circle of latitude between the points P (50° S, 20° W) and Q (50° S, 60° E), assuming the Earth to be a sphere of radius 6368 km.

Solution. Let the problem be represented by Fig. 72. Since the latitude of both P and Q is quoted as 50° S this means that both points are south of the equator and lie on a circle of latitude such that the join of P or Q to C, the Earth's centre, makes an angle of 50° with the equatorial plane. With the circle of longitude through P intersecting the equator at A we have angle $PCA = 50°$. The radius of the circle of latitude through P and Q is $CP \cos 50° = 6368 \cos 50°$.

Since P is west of Greenwich and Q is east of Grenwich the angle between their respective circles of longitude is $60 + 20 = 80°$. Therefore the shorter distance between P and Q is 80/360 of the complete circumference of the circle of latitude through P and Q.

This distance is given by

$d = \frac{80}{360} \times 2\pi\, 6368 \cos 50° = 5708$ km to the nearest kilometre (taking $\pi = 3{\cdot}14$).

Exercise 15.3 (Use the approximation of 3·14 for π.)

1. A point A which is 10 cm above a horizontal plane is the centre of a sphere of radius 26 cm. Find the radius of a circle of intersection. If P is a point on the circle of intersection find the angle between OP and the horizontal.
2. Using the diagram in Fig. 70, let Q be the end of the diameter obtained by producing PN. Calculate the distance on the surface of the sphere in travelling from P to Q along (i) the small circle centre N, (ii) the great circle centre O, when $R = 41$ cm, $ON = 9$ cm.
3. Find the distances (on the surface of the Earth, assumed to be a sphere of radius 6368 km) between the following pairs of points:

   (i) P(60° N, 10° W), Q(60° N, 80° E)
   (ii) A(60° N, 10° W), B(60° S, 10° W)
   (iii) C(30° N, 11° E), D(30° N, 71° E)
4. Taking the positions of Sydney and Cape Town as approximately (34° S, 151° E) and (34° S, 18° E) respectively, calculate the Earth surface distance between the towns assuming that the Earth is a sphere of radius 6368 km.
5. The volume of a sphere is $\frac{4}{3}\pi r^3$: (i) Find the volume of a sphere with a radius of 10 cm (to nearest cm³), and (ii) Find the volume of a solid sphere of radius 12 cm which has an empty spherical core of radius 2 cm.
6. The surface area of a sphere is $4\pi r^2$. Find the surface area of a sphere which has a radius of 10 cm. Calculate the area of the curved surface of a right circular cylinder of radius 10 cm and height 20 cm.
7. A sphere of radius 10 cm rests in contact with two planes inclined at an angle of 60 degrees and within this angle. Find the distance of the centre of the sphere from the line of intersection of the planes.
8. Two points P and Q on a sphere of radius 10 cm are 10 cm apart as joined by a straight line. Find the shortest distance from P to Q on the surface of the sphere. (The great circle distance is required.)
9. Two spheres of radii 14 cm and 10 cm intersect in a circle of radius 8 cm. How far apart are their centres if each centre is outside the other sphere?
10. Two spheres of radii 15 cm and 25 cm rest in contact on a horizontal plane. Find the angle between the line joining the centres and the horizontal plane.

Projection of a Circle into an Ellipse

Returning to the first idea in this chapter, we now consider the orthogonal projection of a circle into an ellipse. We shall restrict our attention to producing the ellipse

$$\frac{x^2}{a^2} + \frac{y^2}{b^2} = 1 \text{ from the circle } \frac{x^2}{a^2} + \frac{y^2}{a^2} = 1$$

We consider two planes i and h, one of which is chosen to be horizontal for visual convenience. The circle is drawn in the i plane together with its x and y axes respectively parallel and perpendicular to the line of intersection of the two planes.

If α is the angle between the two planes then any line AB parallel to the y axis will project into a length $AB \cos \alpha$, but any line CD parallel to the x axis will remain unchanged in length because it is also parallel to the h plane. Any point P on the circle of radius a centre O has coordinates given by

$$x = ON = a \cos \theta, \quad y = NP = a \sin \theta \qquad \text{(page 158)}$$

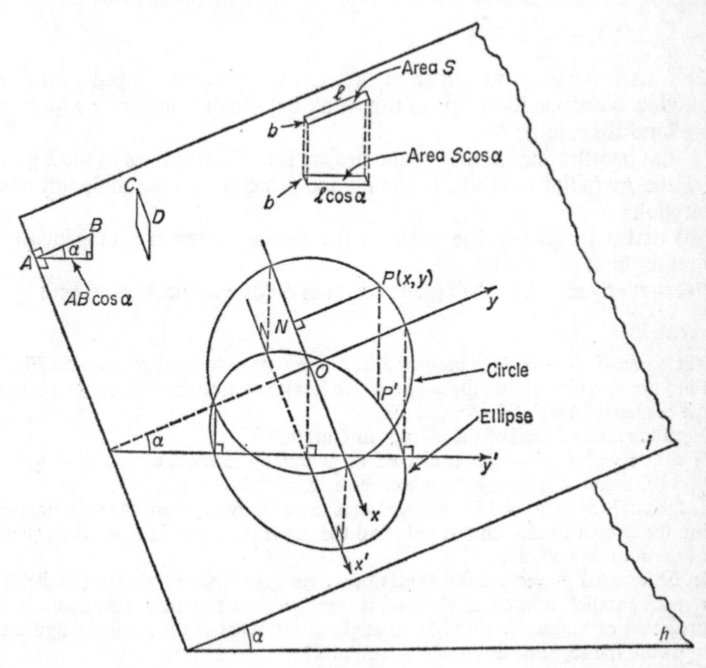

Fig. 73

The orthogonal projection of the x and y axes in the i plane gives rise to x' and y' axes in the h plane as shown so any x coordinate remains unchanged $x = x'$, but the y coordinate becomes $y' = y \cos \alpha$ or $y'/\cos \alpha = y$. Moreover

$$\frac{x^2}{a^2} + \frac{y^2}{a^2} = 1 \text{ on the } i \text{ plane becomes } \frac{x'^2}{a^2} + \frac{y'^2}{a^2 \cos^2 \alpha} = 1$$

Since we now go on to suggest that we choose the angle α so that $\cos \alpha = b/a$, $0 < b < a$, then the circle $\frac{x^2}{a^2} + \frac{y^2}{a^2} = 1$ has been projected into the figure $\frac{x'^2}{a^2} + \frac{y'^2}{b^2} = 1$, a figure which we call an ellipse. Numerically this means that we may project a circle $\frac{x^2}{4} + \frac{y^2}{4} = 1$ (say), into an ellipse $\frac{x^2}{4} + \frac{y^2}{1} = 1$ by choosing the angle α such that $\cos \alpha = 1/2$ ($\alpha = 60°$).

The Area of an Ellipse

If we consider a rectangle whose length l is parallel to the y axis in Fig. 73 and whose breadth b is parallel to the x axis as shown in the corner of Fig. 73, this rectangle will become another rectangle whose breadth is unchanged but whose length becomes $l \cos \alpha$. The area of the rectangle on the h plane is therefore $lb \cos \alpha$. Thus any area S in the i plane is projected orthogonally into an area $S \cos \alpha$ in the h plane.

With the circle of area πa^2 becoming an ellipse of area $\pi a^2 \cos \alpha$, on substituting the suggestion that $\cos \alpha = b/a$, we see that the area of any ellipse $\frac{x^2}{a^2} + \frac{y^2}{b^2} = 1$ is $\pi a^2 \frac{b}{a} = \pi ab$.

The more detailed investigation of what remains unchanged during the projection is outside the scope of this book but for the moment we quote the more intuitive results:

(i) that parallel lines in the i plane project into parallel lines in the h plane;

(ii) the midpoint of a line in the i plane projects into the midpoint of the projection;

(iii) that a tangent to the circle in the i plane becomes a tangent to the ellipse in the h plane; and

(iv) that any area S in the i plane becomes $S \cos \alpha$ in the h plane, and so on.

Exercise 15.4

1. The circle $x^2 + y^2 = 25$ is projected from the i plane to the h plane as in Fig. 73. Find the equation of the ellipse into which it is projected when the angle α between the planes is (i) 60°, (ii) 45°, (iii) 30°.
2. Find the area of each of the ellipses in Question 1.
3. If a circle $x^2 + y^2 = 4$ is projected orthogonally into an ellipse $x^2 + 9y^2 = 4$ find the angle of the projection and the area of the ellipse.
4. If the circle $x^2 + y^2 = 16$ is reduced in area to 4π by orthogonal projection leaving the x coordinates unchanged, find the angle of projection and the equation of the resulting ellipse.
5. In orthogonal projection the angles of a figure are changed but the parallel lines remain parallel. A circle of radius 3 is inscribed in a square. If this figure is now projected orthogonally through an angle of 60° what is the resulting figure and what are the separate areas of its two parts?

CHAPTER SIXTEEN

RATES AND LIMITS

Rates of Change

The calculus is sometimes referred to as the mathematics of change because it involves measuring rates at which one quantity may change with respect to another. Everyday examples of such measurements are speed, in which the distance travelled changes with respect to time; petrol consumption measured by kilometres per litre; birth or death rates measured by so many people per 1000 head of population; money in the bank increasing by a rate of interest described in terms of a percentage per annum, and so on.

Frequently we may consider a quantity to change with respect to more than one other quantity, a situation which is illustrated by the example of a stone which falls from rest under gravity only. The distance it has fallen in time t is given by $s = \frac{1}{2}gt^2$, and its speed at this time is given by $v = gt$ or $v = \sqrt{(2gs)}$. The last two equations show that v may be considered to change with respect to time or with respect to the square root of the distance fallen. Alternatively the area of a rectangle $A = lb$ may vary with respect to its length, or its breadth, or both.

Thinking in general terms, when we write $y = f(x)$ we are saying that variations in y will depend upon changes in the independent variable x (page 12). Finding the rate at which y changes with respect to x is a typical problem of the differential calculus.

Constant Rates

We are familiar with fixed (i.e. constant) rates of interest over a period of time, e.g. interest on savings being at a fixed rate of 6 per cent for three years (say), meaning that equal increases or increments will be added at the end of equal intervals or increments of time. We know that the speed of a body is defined as the distance travelled per unit time, i.e. the rate of change of distance with respect to time. When we speak of a constant speed of 10 m s^{-1} we mean that during any one second of moving, the object concerned will move through a distance of 10 m, or in general terms the object will move through equal distances in equal intervals of time as long as the distance is increasing with respect to time at the constant rate of 10 m s^{-1}. Therefore in any 0·5 s of movement at this speed the body will travel a distance of 5 m. Similarly in any period of 2 s the body will travel a distance of 20 m. (We are not arguing as to whether it is at all practically possible to travel at a constant speed—we are assuming that it is.)

In algebraic terms the rate at which y changes with respect to x is found by dividing the change in x into the corresponding change which it produces in y. Consider the equation $y = 5x$ from the point of view of discovering whether y will increase at a constant rate with respect to x. Thus we are asking, 'Does y change by equal increments when x changes by equal increments?' We can start by suggesting that x changes by equal increments of 1; that is, when x changes to $x + 1$ we consider what will be the corresponding change in y.

Commencing with $y = 5x$ it follows that the new value of y will be $5(x + 1)$. The change in the value of y is therefore $5(x + 1) - 5x = 5$. This means that whatever the value of x we start with (-8, -7, 11, 91, etc.) an increase in x of 1 will lead to an increase in y of 5. Of course this does not prove that y increases at a constant rate with respect to x, only that it increases at a constant rate of 5 units for every 1 unit of x, i.e. we need to obtain the same rate for any sized increment of x. The only way we can confirm that the rate of change is constant is by starting x anywhere in the domain of discussion and considering the results of increments of x of any size and not just an increment of size 1.

Since this method of enquiry is continually employed in the calculus we use special symbols to represent the increments. We refer to a change in x as 'an increment of x' and we write this as δx. This is the Greek letter 'delta' placed before x and read 'delta x', meaning an increment of x. For instance, the example we have just discussed suggested $\delta x = 1$. There is no restriction on the sign of δx, e.g. $\delta x = -1$ would have been equally acceptable. Similarly, we represent an increment of y by δy and again in the example just discussed we obtained $\delta y = 5$ for it is clear that just as y depends upon x so δy depends upon δx. As for δx, so δy is read 'delta y' meaning an increment of y. Just as we refer to a man as Mr . . ., in which M and r are inseparable and Mr is taken as one symbol, so we refer to the increment of x as the one symbol δx. As in the case of Mr we are using an abbreviation δx or δy or δu and so on. Thus, δx is one symbol and does not indicate a multiplication of δ by x, and the same holds for δy.

Finding the Rate

If a body moves through a distance of 12 m in 3 s then we say that it was moving at an average rate of $\frac{12}{3} = 4 \text{ m s}^{-1}$ during the three seconds. If the body had taken 5 s to move through a distance of 12 m then the average rate would have been $\frac{12}{5} = 2{\cdot}4 \text{ m s}^{-1}$ during the 5 s. In the first case we are not saying that the body moved 4 m in each second or even 2 m in each half second. In the second case we are not saying that the body moved 2·4 m in each second. The results which we have just obtained are the **average rates** over the interval of time concerned in each case.

Similarly, if the area of a square increases from 36 m² to 100 m² in 12 s the average rate at which the area increases is $\dfrac{100 - 36}{12} = 5\frac{1}{3}$ square metres per second during or over the interval of 12 s. In the same time the side of the square changes from 6 m to 10 m so we can also calculate the average rate at which the area changes with respect to the length of the side of the square as $\dfrac{100 - 36}{10 - 6} = \dfrac{64}{4} = 16$ square metres per metre. Again this is an average rate of change over this interval of 4 m change in the length of the side.

We now return to $y = 5x$ to consider the rate of change of y with respect to x in more general terms.

Let us suppose that x is changed to $x + \delta x$. Starting with $y = 5x$ we argue that changing x to $x + \delta x$ will change y to $y + \delta y$ so that

$$y + \delta y = 5(x + \delta x)$$

Substituting $y = 5x$ in this result we get

$$5x + \delta y = 5x + 5\delta x$$
$$\therefore \delta y = 5\delta x$$

Again we must be aware that $5\delta x$ is $5 \times (\delta x)$ and not $5 \times \delta \times x$.

The rate at which y changes with respect to x over the interval δx is therefore given by

$$\frac{\delta y}{\delta x} = 5$$

and since 5 is constant we can say that y changes with respect to x at a constant rate of 5. Note that this is the average rate of change over the interval or increment δx.

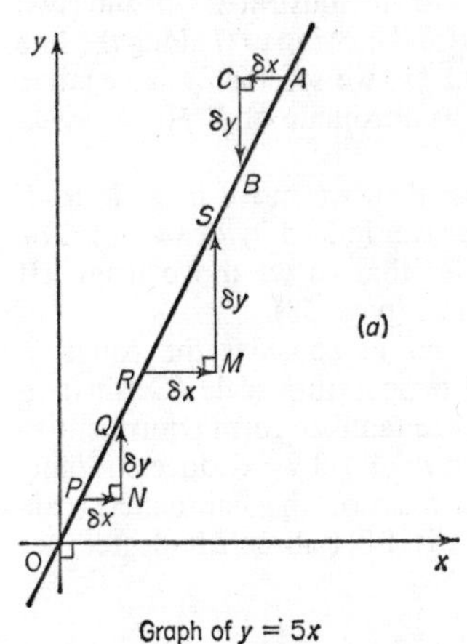

Graph of $y = 5x$

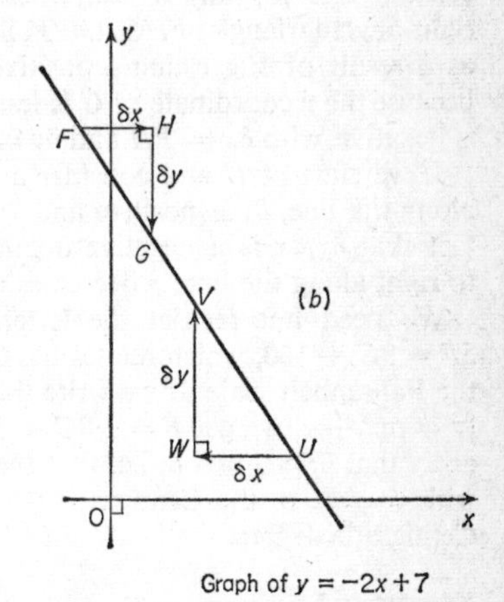

Graph of $y = -2x + 7$

Fig. 74

In Fig. 74(*a*) we have the sketch graph of $y = 5x$. The result $\delta y/\delta x = 5$ for any δx anywhere on the straight line is illustrated by the three right-angled triangles PNQ, RMS, ACB. Starting at P on the line, the increment δx takes us to Q and in so doing increases the y coordinate by δy. To put it another way, if P is the point (x, y) then Q is the point $(x + \delta x, y + \delta y)$ and in the diagram $PN = \delta x$, and $NQ = \delta y$. With R as starting point and a different δx we move from R to S along the line and $RM = \delta x$, $MS = \delta y$. In both cases $\delta y/\delta x = 5$. If we start at A and consider δx negative then this increment will take us from A to B along the line where B is again the point $(x + \delta x, y + \delta y)$, but this time δx is negative, $AC = \delta x$, and $CB = \delta y$ both being negative, so that $\delta y/\delta x = 5$ once again.

The fact that $\delta y/\delta x$ is a positive quantity indicates that as we move from left to right along the line, y increases as x increases (page 58). If the equation had been $y = mx + c$ instead of $y = 5x$ the same procedure of suggesting that x changes to $x + \delta x$ making y change to $y + \delta y$ would have yielded the following results

$$y + \delta y = m(x + \delta x) + c = mx + m\delta x + c$$

Since we started with $y = mx + c$ this last result may be written

$$mx + c + \delta y = mx + m\delta x + c$$

whence

$$\delta y = m\delta x$$

and

$$\frac{\delta y}{\delta x} = m$$

Again we see that the rate of change of y with respect to x over the increment δx is the constant m in the equation $y = mx + c$. Since we already know that m is the gradient of the line $y = mx + c$ we realise that it is appropriate to relate the rate given by $\delta y/\delta x$, to the gradient of a straight line.

For another example, if we have $y = -2x + 7$ then anywhere on this line the rate at which y is changing with respect to x is -2, i.e. $\delta y/\delta x = -2$.

In Fig. 74(*b*) we have the sketch graph of $y = -2x + 7$. The result $\delta y/\delta x = -2$ for any δx anywhere on the line is illustrated in the two right-angled triangles *FHG*, *UWV*. Starting at *F* and moving to *G* along the line as a result of suggesting a positive increment δx, we see the δy is negative because the y coordinate of *G* is less than the y coordinate of *F*. Hence $\delta y/\delta x$ is negative, with $\delta x = FH$ and $\delta y = HG$.

If we start at *U* and consider a negative δx then we move from *U* to *V* along the line, δy is positive and $\delta y/\delta x$ is negative, indeed $\delta y/\delta x = -2$. The fact that $\delta y/\delta x$ is a negative quantity indicates that as we move from left to right along the line, y decreases as x increases (page 58).

We need not restrict the lettering to x and y: consider the equation $5F = 9C + 160$, which relates the Centigrade temperature scale (Celsius) to the Fahrenheit scale. If we write this in the more familiar form equivalent to '$y = mx + c$' we get $F = 1{\cdot}8C + 32$ and with $m = 1{\cdot}8$ we deduce straightaway that $\delta F/\delta C = 1{\cdot}8$, i.e. that the rate of increase on the Fahrenheit scale with respect to the Centigrade (Celsius) scale is 1·8 Fahrenheit degree per Centigrade degree.

Exercise 16.1

1. Find $\delta y/\delta x$ in each of the following examples.
   (i) $y = 10x + 3$ (ii) $y = -3x + 10$
   (iii) $3y = x + 10$ (iv) $3y = -x - 10$
2. If $\delta y/\delta x = 2$ for all values of x find an equation relating y and x, given that $y = 3$ when $x = 0$.
3. If $\delta y/\delta x = -3$ for all values of x find the change in y when x increases by 5.
4. The constant rate of change of y with respect to time t is 6. What is the value of $\delta y/\delta t$?
5. As a point (x, y) moves along a straight line, y increases as x increases at the rate of 3 units of y for every 1 unit of x. What is the value of $\delta y/\delta x$? If the line passes through the point (1, 1) what is its equation?
6. The y coordinate of a point $P(x, y)$ decreases as the point moves from left to right along the line. What can you say about the value of $\delta y/\delta x$? If y decreases by 4 units each time x is increased by 3 units, find the equation of the line if it also passes through (0, 0).

The Gradient of a Curve

We have just seen that in the case of y being a linear function of x, i.e. $y = mx + c$, the rate of change of y with respect to x is measured by the gradient m of the line. Our next problem is to try to find if there is an equally recognisable measure of this rate of change when y is not a linear function of x, i.e. when the x, y graph of the equation is a curve.

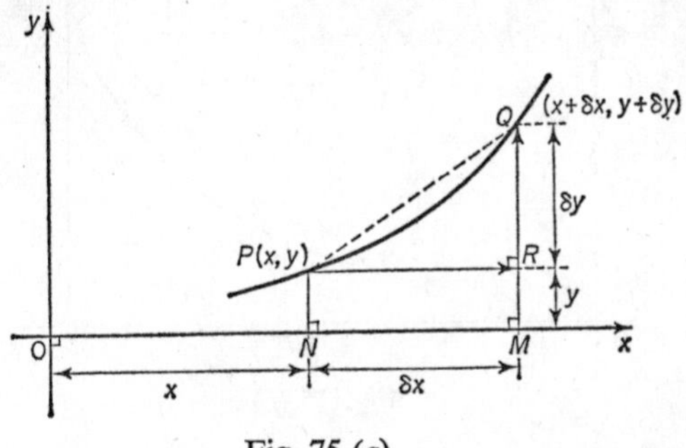

Fig. 75 (*a*)

Consider the point $P(x, y)$ to lie on the given curve in Fig. 75(*a*), and in the same manner as before let us increase x to $x + \delta x$ and y to $y + \delta y$ by moving along the curve from P to Q. In the usual notation of coordinate geometry with $P(x, y)$ we have $ON = x$ and $NP = y$; similarly with $Q(x + \delta x, y + \delta y)$ we have $OM = x + \delta x$ and $MQ = y + \delta y$. With PR parallel to the x axis and thereby perpendicular to MQ we have $PR = NM = \delta x$, and finally with triangle PRQ right-angled at R we have the result $\delta y/\delta x = RQ/PR$.

At first sight it appears that we have found the rate at which y changes with respect to x, until we realise that had we chosen a different δx the result would not have been the same as RQ/PR. Consider taking δx a little smaller than before so that we only move from P to Q_1 along the curve in Fig. 75(*b*).

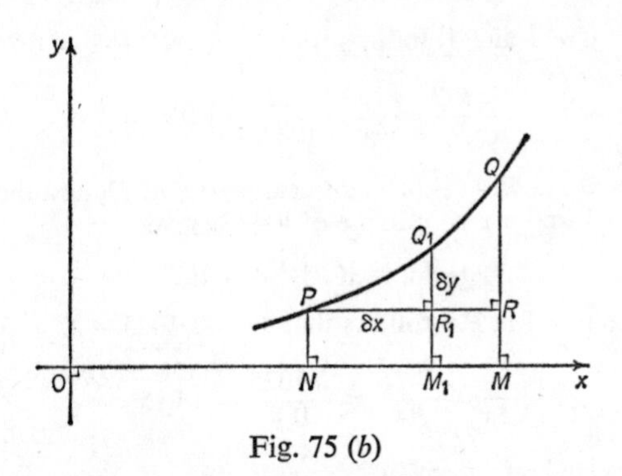

Fig. 75 (*b*)

By the method already explained we shall obtain

$$\frac{\delta y}{\delta x} = \frac{R_1Q_1}{PR_1}$$

and again we see that the result depends upon the size of δx. If in Fig. 75(*c*

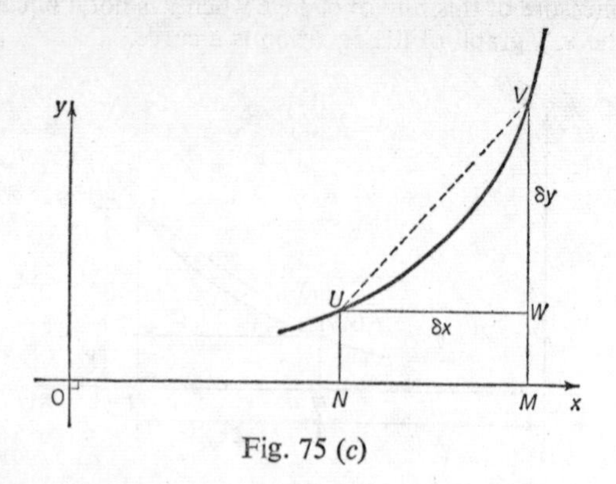

Fig. 75 (*c*)

we start from another point $U(x, y)$ on the curve and move along the curve to $V(x + \delta x, y + \delta y)$ with the same δx as in the first case of Fig. 75(*a*), then we get $\delta y/\delta x = WV/UW$ which is again a different result from the first one but for a different reason. In this case the value of $\delta y/\delta x$ on the curve is different because we have started with a different x, i.e. from a different point on the curve. All three cases show that $\delta y/\delta x$ depends not only on δx but also on x, i.e. its whereabouts on the curve.

We can illustrate this numerically by considering the curve $y = \frac{1}{4}x^2$ at the point $P(2, 1)$ and choosing $\delta x = 0{\cdot}2$. Referring to Fig. 75(*a*) with $PR = \delta x = 0{\cdot}2$, the point Q on the curve $y = \frac{1}{4}x^2$ has an x coordinate of $2 + 0{\cdot}2$ and a y coordinate of $y + \delta y$ given by

$$y + \delta y = \tfrac{1}{4}(2 + 0{\cdot}2)^2 = \tfrac{1}{4}(2{\cdot}2)^2 = 1{\cdot}21.$$

Therefore since $y = 1$ at P it follows that $\delta y = 1{\cdot}21 - 1 = 0{\cdot}21$ and so

$$\frac{RQ}{PR} = \frac{\delta y}{\delta x} = \frac{0{\cdot}21}{0{\cdot}20} = 1{\cdot}05 \qquad \text{(i)}$$

If we take $\delta x = 0{\cdot}1$ then Q is brought nearer to P at Q_1 and the coordinates of Q_1 are $x = 2{\cdot}1$ and a y coordinate of $y + \delta y$ given by

$$y + \delta y = \tfrac{1}{4}(2{\cdot}1)^2 = 1{\cdot}1025.$$

Therefore, since $y = 1$ at P it follows that $\delta y = 1{\cdot}1025 - 1 = 0{\cdot}1025$ and so

$$\frac{\delta y}{\delta x} = \frac{R_1Q_1}{PR_1} = \frac{0{\cdot}1025}{0{\cdot}1} = 1{\cdot}025 \qquad \text{(ii)}$$

a result which is different from (i).

Moving along the curve to U let us suppose that this is the point (6, 9) and that we take $\delta x = 0{\cdot}2$ as for (i) in Fig. 75(*a*). Then

$$y + \delta y = \tfrac{1}{4}(6 + 0{\cdot}2)^2 = 9{\cdot}61$$

with $y = 9$ and $y + \delta y = 9{\cdot}61$ we get $\delta y = 0{\cdot}61$ and consequently

$$\frac{\delta y}{\delta x} = \frac{WV}{UW} = \frac{0{\cdot}61}{0{\cdot}2} = 3{\cdot}05 \qquad \text{(iii)}$$

a result which is different from both (i) and (ii) and therefore illustrates numerically how the rate of change of y with respect to x depends not only upon δx but also upon x.

In each case, $\delta y/\delta x$ gave the gradient of the line respectively through the two points P, Q; P, Q_1; U, V on the curve. We therefore call $\delta y/\delta x$ the **average gradient** of the curve over the interval δx.

While this gives a satisfactory answer with reference to an interval δx it does not tell us what the rate of change is for any particular value of x and y. For example, we are still unable to answer the question 'At what rate is y increasing with respect to x at the point $P(2, 1)$?' At the moment we can only answer 'that over the interval $\delta x = 0{\cdot}1$ starting at $x = 1$ the average rate of change is . . .'. However, we can see that if we make δx smaller and smaller then we shall be closer to being able to say what is happening at $P(2, 1)$. But, of course as δx is made smaller so Q gets closer to P and the line PQ approaches the position of the tangent to the curve at the point P. It is this reasoning which leads us to the following definition.

Definition. The gradient of a curve at any point on the curve is the gradient of the tangent at that point.

We have already discussed the idea of a tangent line intersecting a curve in two coincident points (page 71) but here we shall look again at both the diagrammatic and algebraic approach to limits, in which a line becomes a tangent.

Limits

The tangent to a curve at a point P is thought of as the limiting position of the line PQ as Q approaches P. In Fig. 76 we have a sequence of positions for the point Q on the curve such that the point P is the limit point of these positions. The closer that Q gets to P the nearer does the line PQ become the tangent PT at P inasmuch as it becomes a line intersecting the curve in two points at P. We summarise this procedure by saying that as the sequence of

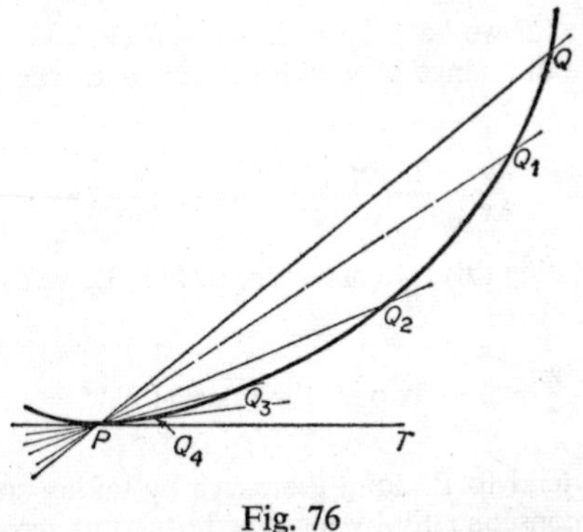

Fig. 76

points $\{Q_n\}$ approaches the limit point P so the sequence of lines $\{PQ_n\}$ approaches the limit line PT which is the tangent at P. At the same time the gradient of the line PQ tends to the gradient of the tangent at P, or in other words the gradient of the tangent to the curve at P is the limit of the gradient of the line PQ as Q tends to P.

For other limits associated with diagrams consider the following exercise.

Exercise 16.2

1. AB is a straight line of length 10 cm. A sequence of points $\{M_n\}$ is chosen such that M_1 is the midpoint of AB, M_2 is the midpoint of M_1B, M_3 is the midpoint of M_2B and so on. The sequence $\{M_n\}$ has a limit point, where is it?
2. What is the limit point of the sequence $\{M_n\}$ in Question 1, if M_2 is the midpoint of AM_1, M_3 is the midpoint of AM_2 and so on.
3. A n sided regular polygon is inscribed in a circle of radius 10 cm. To what figure does the polygon tend as n tends to infinity?
4. $ABCD$ is a square. A sequence of points $\{M_n\}$ is chosen on BC such that M_1 is the midpoint of BC, M_2 is the midpoint of BM_1, M_3 is the midpoint of BM_2 and so on. Another sequence of points $\{N_n\}$ is chosen on DC such that N_1 is the midpoint of DC, N_2 is the midpoint of CN_1, N_3 is the midpoint of CN_2 and so on. What is the limiting position of the triangle AM_nN_n as n tends to infinity?
5. Find the result in Question 4 when M_2 is the midpoint of CM_1, M_3 is the midpoint of CM_2 and so on.
6. P and Q are two points on the circumference of a circle such that the line PQ is always parallel to a fixed diameter AB. What are the limiting positions of the line PQ?

The Gradient at the Point (1, 3) on the curve $y = x^2 + 2$

We have discussed what we mean by the gradient at a point on a curve and also how the tangent is obtained as a limit position of the lines drawn through the point. What we need now is some way of making these lines approach the tangent position by controlling the algebra behind the equations of the lines. We illustrate the approach by considering the particular curve given by the equation $y = x^2 + 2$ and trying to find the gradient of the tangent (curve) at the point (1, 3)—Fig. 77.

We start by choosing Q on the curve in exactly the same way as for Fig. 75(*a*), i.e. $Q(x + \delta x, y + \delta y)$, and similarly we obtain

$$y + \delta y = (x + \delta x)^2 + 2 = x^2 + 2x\,\delta x + (\delta x)^2 + 2.$$

With $y = x^2 + 2$ at P we have $\delta y = 2x\,\delta x + (\delta x)^2$.

The average rate of change of y with respect to x over the interval NM is therefore

$$\frac{\delta y}{\delta x} = \frac{2x\,\delta x + (\delta x)^2}{\delta x} = 2x + \delta x \qquad \text{(i)}$$

Since we are calculating this rate of change at $P(1, 3)$, we can substitute $x = 1$ so that

$$\frac{\delta y}{\delta x} = 2 + \delta x \text{ over the interval } NM = \delta x \qquad \text{(ii)}$$

We now make Q tend to P along the curve by taking smaller and smaller values of δx, and consequently by making δx tend to zero we make the line

PQ tend to the tangent at P and, by so doing, the value of $\delta y/\delta x$ will give us the gradient of the tangent at P and, thereby, the rate at which y is changing with respect to x at P. By letting $\delta x \to 0$ we see that $\delta y/\delta x \to 2$, but there is a slight difficulty for, as $\delta x \to 0$ so does $\delta y \to 0$ and we apparently have an expression which reads $0/0 = 2$, but this is not so for we are speaking of the limit of $\delta y/\delta x$ as $\delta x \to 0$ and **not** the value gained by substituting $\delta x = 0$, $\delta y = 0$, a substitution which is meaningless. The result which follows from (ii) above is therefore written as $\lim_{\delta x \to 0} \delta y/\delta x = 2$ at the point $P(1, 3)$. (In words this reads 'the limit of $\delta y/\delta x$ as $\delta x \to 0$, is 2 at the point $P(1, 3)$'.)

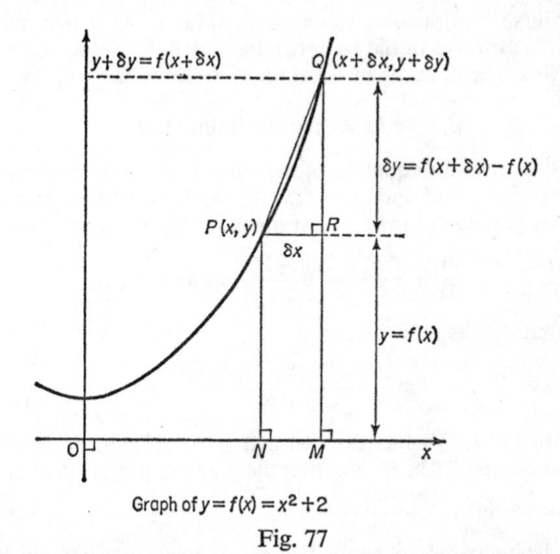

Fig. 77

The result which follows from (i) above is similarly written as

$$\lim_{\delta x \to 0} \delta y/\delta x = 2x \text{ at the point } (x, y).$$

Again, approaching the result (ii) numerically, consider the following table of values based on $\delta y = 2x\,\delta x + (\delta x)^2 = 2\delta x + (\delta x)^2$ at the point $P(1, 3)$.

$\delta x = 0{\cdot}1$	0·01	0·001	0·0001	0·000 01
$\delta y = 0{\cdot}21$	0·0201	0·002 001	0·000 200 01	0·000 020 000 1
$\dfrac{\delta y}{\delta x} = 2{\cdot}1$	2·01	2·001	2·0001	2·000 01

We see numerically that the closer we take δx to 0 the closer $\delta y/\delta x$ gets to 2, indeed we can get $\delta y/\delta x$ as close to 2 as we please by taking δx small enough. We must remind ourselves that δx may be positive or negative and that the same limit must be obtained when we approach P from either the right (δx positive) or the left (δx negative). The approach to P from the left by taking negative values of δx is indicated in the following table, again with $\delta y = 2\delta x + (\delta x)^2$ at $P(1, 3)$ and we see numerically that the closer we take

δx to 0 (this time through negative values, i.e. from the left of 0) the closer $\delta y/\delta x$ gets to 2.

$\delta x = -0{\cdot}1$	$-0{\cdot}01$	$-0{\cdot}001$	$-0{\cdot}0001$
$\delta y = -0{\cdot}19$	$-0{\cdot}0199$	$-0{\cdot}001\,999$	$-0{\cdot}000\,199\,99$
$\frac{\delta y}{\delta x} = 1{\cdot}9$	$1{\cdot}99$	$1{\cdot}999$	$1{\cdot}9999$

It is, of course, not possible to choose a value of δx which makes $\delta y/\delta x$ equal to two, because we would be in trouble with $\delta y/\delta x$ being of the form 0/0, so we have to speak in terms of the limit of $\delta y/\delta x$ as $\delta x \to 0$, and write this as

$$\lim_{\delta x \to 0} \delta y/\delta x = 2 \text{ at the point } (1, 3).$$

If we consider that the equation of the curve is given in the form $y = f(x)$ then $y + \delta y = f(x + \delta x)$ and $\delta y = f(x + \delta x) - f(x)$. Written in this form it means that the gradient of the curve at the point (x, y) is given by the limit of

$$\frac{\delta y}{\delta x} = \frac{f(x + \delta x) - f(x)}{\delta x} \text{ as } \delta x \to 0,$$

written alternatively as

$$\lim_{\delta x \to 0} \frac{f(x + \delta x) - f(x)}{\delta x}$$

Since the writing of δx is sometimes distracting to the beginner we could adopt another notation and let $\delta x = h$ so that the gradient of the curve at the point (x, y) is given by $\lim_{h \to 0} \frac{f(x + h) - f(x)}{h}$

We must be prepared to apply this result to a considerable variety of functions and if possible to handle the procedure of finding the limit without the aid of a diagram of the graph, which may in itself be very difficult to obtain. The following examples should clarify the general method of application.

Example 1. Find the gradient of the curve $y = f(x) = 3x^2 + 8x$ at the point (2, 28).

Solution. The gradient of the curve at the point (2, 28) is equal to

$$\lim_{h \to 0} \frac{f(2 + h) - f(2)}{h} = \lim_{h \to 0} \frac{NQ}{PN} \quad \text{(Fig. 78)}$$

$$f(2 + h) = 3(2 + h)^2 + 8(2 + h) = 3(4 + 4h + h^2) + 16 + 8h = 28 + 20h + 3h^2$$

$$f(2) = 28$$

$$\therefore \; f(2 + h) - f(2) = 20h + 3h^2$$

$$\frac{f(2 + h) - f(2)}{h} = \frac{20h + 3h^2}{h} = 20 + 3h$$

$$\therefore \; \lim_{h \to 0} \frac{f(2 + h) - f(2)}{h} = \lim_{h \to 0} (20 + 3h) = 20$$

Thus, the gradient of the curve at the point (2, 28) is 20. Another way of stating this result is to say that the gradient of the tangent to the curve at the point (2, 28) is 20.

Alternatively, y is increasing with respect to x at the rate of 20 units of y for 1 unit of x.

An examination of the graph of $f(x) = 3x^2 + 8x$ in Fig. 78 shows how once we have calculated the gradient we are then able to draw the tangent at the point (2, 28) by constructing the right-angled triangle PRT with angle $RPT = \theta$ and $\tan \theta = 20$ given by $RT = 20$ and $PR = 1$. Answer

We can get very close to the tangent at (2, 28) by taking even a crude δx. With $\delta x = 1$, we move from P to S on the curve and so $\delta y = 51 - 28 = 23$.

$$\therefore \quad \frac{\delta y}{\delta x} = \frac{RS}{PR} = \frac{23}{1}$$

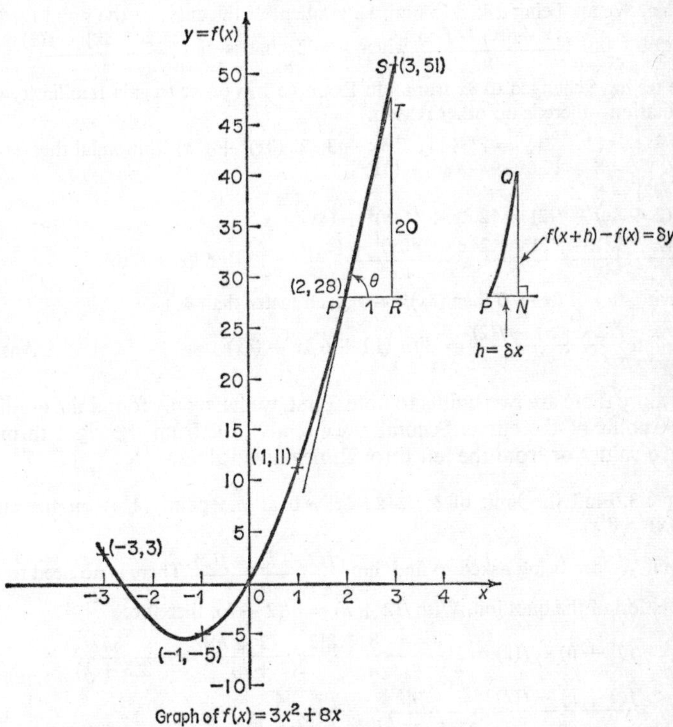

Fig. 78

There is a loss of generality in the above example because we could gain much more information by finding the gradient of the curve at any point (x, y) on the curve. To this end let us consider $\lim\limits_{h \to 0} \dfrac{f(x+h) - f(x)}{h}$

$$\begin{aligned}
f(x+h) &= 3(x+h)^2 + 8(x+h) \\
&= 3(x^2 + 2xh + h^2) + 8x + 8h \\
&= 3x^2 + 6xh + 3h^2 + 8x + 8h \\
f(x) &= 3x^2 + 8x \\
\therefore f(x+h) - f(x) &= 6xh + 8h + 3h^2 \\
\frac{f(x+h) - f(x)}{h} &= 6x + 8 + 3h
\end{aligned}$$

$$\therefore \lim_{h \to 0} \frac{f(x+h) - f(x)}{h} = \lim_{h \to 0} (6x + 8 + 3h) = 6x + 8$$

Now, this result tells us that the gradient of the curve at any point (x, y) is $6x + 8$. In particular at the point $(2, 28)$ this gradient is $12 + 8 = 20$, as found above. At the point $(-3, 3)$ the gradient is $6(-3) + 8 = -10$, the meaning of negative gradient being understood from the discussion on page 58.

Example 2. Find the rate at which y is increasing with respect to x on the curve $y = f(x) = x^3$ at the point $(2, 8)$.

Solution. We are being asked to find the gradient of the curve at the point $(2, 8)$ so we require $\lim\limits_{\delta x \to 0} \dfrac{f(x + \delta x) - f(x)}{\delta x}$ when $x = 2$, that is $\lim\limits_{\delta x \to 0} \dfrac{f(2 + \delta x) - f(2)}{\delta x}$

(Here we have changed to δx from h in Example 1 in order to gain familiarity with the notation—there is no other reason.)

$$f(2 + \delta x) = (2 + \delta x)^3 = 2^3 + 3 \,.\, 2^2\, \delta x + 3 \,.\, 2 \,.\, (\delta x)^2 + (\delta x)^3 \text{ (binomial theorem)}$$
$$= 8 + 12\, \delta x + 6(\delta x)^2 + (\delta x)^3$$
$$f(2) = 8$$
$$\therefore\ f(2 + \delta x) - f(2) = 12\, \delta x + 6(\delta x)^2 + (\delta x)^3$$
$$\frac{f(2 + \delta x) - f(2)}{\delta x} = \frac{12\delta x + 6(\delta x)^2 + (\delta x)^3}{\delta x} = 12 + 6\, \delta x + (\delta x)^2$$

Observing that if $\delta x \to 0$ then $(\delta x)^2 \to 0$ (even faster than δx)

$$\therefore\ \lim_{\delta x \to 0} \frac{f(2 + \delta x) - f(2)}{\delta x} = \lim_{\delta x \to 0} (12 + 6\, \delta x + (\delta x)^2) = 12 \qquad \text{Answer}$$

Once more there are two points to note. First, we have only found the gradient at one point of the curve. Secondly, δx tends to 0 from the right through positive values or from the left through negative values.

Example 3. Find the limit of $\delta y/\delta x$ as $\delta x \to 0$ at the point $(2, 4)$ on the curve $y = f(x) = 8/x$.

Solution. We are being asked to find $\lim\limits_{h \to 0} \dfrac{f(2 + h) - f(2)}{h}$. There is no need to use the notation of the question. With $f(2 + h) = 8/(2 + h)$, therefore

$$f(2 + h) - f(2) = \frac{8}{2 + h} - \frac{8}{2} = \frac{8[2 - (2 + h)]}{2(2 + h)} = -\frac{8h}{2(2 + h)}$$
$$\frac{f(2 + h) - f(2)}{h} = -\frac{4h}{(2 + h)h} = -\frac{4}{(2 + h)}$$
$$\therefore\ \lim_{h \to 0} \frac{f(2 + h) - f(2)}{h} = \lim_{h \to 0} -\frac{4}{2 + h} = -\frac{4}{2} = -2$$

This means that y is decreasing with respect to x at the point $(2, 4)$ at the rate of two units of y for every one unit of x. Answer

Example 4. Given the graph of $y = x^3$ in Fig. 79 find the difference between $\delta y/\delta x$ when $\delta x = \frac{1}{2}$ and $\lim\limits_{\delta x \to 0} \delta y/\delta x$, at $x = 1$.

Solution. When $x = 1$ we have $y = 1$ so that the point at which we start with $\delta y/\delta x$ is $P(1, 1)$. With $\delta x = \frac{1}{2}$ we get $y + \delta y = (1 + \frac{1}{2})^3 = \frac{27}{8} = 3\frac{3}{8}$, i.e. the y coordinate of the point Q is 3·375.

$$\therefore\ \text{If } y = 1, \text{ and } y + \delta y = 3\tfrac{3}{8} \text{ then } \delta y = 2\tfrac{3}{8}$$
$$\therefore\ \frac{\delta y}{\delta x} = \frac{19}{8} \div \frac{1}{2} = \frac{19}{4}$$
$$\therefore\ \delta y/\delta x = 4{\cdot}75 \text{ at } (1, 1) \text{ over the interval } \delta x = \tfrac{1}{2}$$

In order to find $\lim_{\delta x \to 0} \delta y/\delta x$ we must generalise and find $y + \delta y = (x + \delta x)^3$ for any δx

$\therefore \quad y + \delta y = x^3 + 3x^2\,\delta x + 3x\,(\delta x)^2 + (\delta x)^3$ (binomial expansion), and since $y = x^3$

we have

$$\delta y = 3x^2\,\delta x + 3x(\delta x)^2 + (\delta x)^3$$

$$\therefore \quad \frac{\delta y}{\delta x} = 3x^2 + 3x\,\delta x + (\delta x)^2$$

Now we let $\delta x \to 0$ (hence $(\delta x)^2 \to 0$) so that $\lim_{\delta x \to 0} \delta y/\delta x = 3x^2$ (remember that this gives us the gradient of the curve at any point (x, y) of the curve). Hence, when $x = 1$, $\lim_{\delta x \to 0} \delta y/\delta x = 3$, i.e. the tangent at the point (1, 1) has a gradient of 3.

Referring to the diagram in Fig. 79, we observe that $\delta y/\delta x$ is associated with the

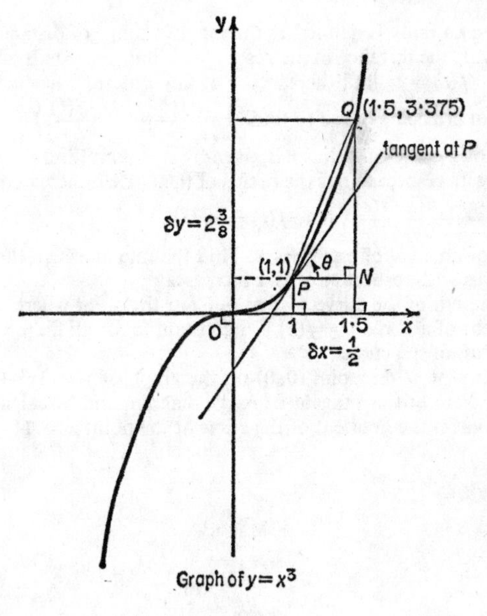

Fig. 79

gradient of the line PQ (and was 4·75) whereas the $\lim_{\delta x \to 0} \delta y/\delta x$ is associated with the gradient of the tangent at P.

Thus, $\tan NPQ = 4{\cdot}75$, $\tan \theta = 3$ in Fig. 79.

Exercise 16.3

1. Find the values of $f(x) = \dfrac{x^2 - 1}{x - 1}$ when $x = 1{\cdot}1, 1{\cdot}01, 1{\cdot}001, 1{\cdot}0001$, and say what happens to $f(x)$ as x gets closer to 1. Does $f(1)$ exist? What is $\lim_{x \to 1} \dfrac{x^2 - 1}{x - 1}$?

2. Examine the values of $f(h) = \dfrac{(1+h)^2 - 1}{h}$ for $h = 0{\cdot}1, 0{\cdot}01, 0{\cdot}001$, and say what happens to $f(h)$ as h gets closer to 0. Does $f(0)$ exist? What is $\lim\limits_{h\to 0} \dfrac{(1+h)^2 - 1}{h}$?
3. The points $P(1, 1)$ and $Q(1 + h, (1 + h)^2)$ lie on the curve $y = x^2$. Find the gradient of the line PQ. If $h \to 0$ to what limit does the line PQ tend?
4. Two points P and Q on the curve $y = x^2$ have x coordinates of $1 - h$ and $1 + h$ respectively where h is positive. To what point on the curve do P and Q tend as $h \to 0$? Draw a sketch for P and Q on the graph and find the gradient of the line PQ. What is the limiting position of the line PQ as $h \to 0$?
5. Find $\lim\limits_{h\to 0} \dfrac{f(a+h) - f(a)}{h}$ in each of the following cases:
   (i) $y = f(x) = x^2$ with $a = 1$
   (ii) $y = f(x) = kx^2$ with $a = 1$, k being a constant
   (iii) $y = f(x) = x^2 + x$ with $a = 3$
   (iv) $y = f(x) = \dfrac{1}{x}$ with $a = 3$
6. The speed v of a body is defined as the rate of change of distance s with respect to the time t. If the distance in metres and the time in seconds are related by the equation $s = f(t) = t^2$ find the speed v at any time t. What is the speed when $t = 3$? $\left(\text{You are being asked to find } \lim\limits_{h\to 0} \dfrac{f(t+h) - f(t)}{h}.\right)$
7. The area A of a circle of radius r is given by $A = \pi r^2$. Find the rate at which A will change with respect to r as the radius of the circle is increased. $\left(\text{Find } \lim\limits_{h\to 0} \dfrac{f(r+h) - f(r)}{h} \text{ where } f(r) = \pi r^2.\right)$
8. The length of an edge of a cube is x. Find the rate at which the volume of the cube increases with respect to x as x increases.
9. Find the gradient of the curve $y = x^2 - 6x$ at the point where $x = 3$.
10. If the gradient of a curve $y = f(x)$ is a constant m for all finite values of x what is the equation of the curve?
11. Draw the tangent at the point $(0, 0)$ on the graph of $y = 3x^2 + 8x$ in Fig. 78. Find the gradient of this tangent from the diagram and by calculation.
12. In Fig. 78 what is the gradient of the curve at the point $x = 1\frac{1}{3}$, $y = 16$?

CHAPTER SEVENTEEN

DIFFERENTIATION

In Chapter Sixteen we saw that we could think of the rate of change of y with respect to x in two different ways. Either we draw a graph and find the limiting position of lines joining points P and Q on the curve as on page 207, or we ignore the graph and consider the algebraic exercise of finding the limit of $\delta y/\delta x$ as $\delta x \to 0$, or, what amounts to the same thing, finding

$$\lim_{h\to 0} \frac{f(x+h)-f(x)}{h}.$$

Learning to think in algebraic terms will be more useful in the long run and certainly more economical in time, since it avoids the need for curve plotting. Unfortunately, although the basic ideas are elementary at this stage, the notation—especially its variety—is difficult to absorb. One cannot overemphasise the need for patience in order to let the notation grow into acceptance. Of course, each notation is seeking to say the same thing, so let us summarise what we have so far and try to develop it a little further.

Everything depends upon finding $\delta y/\delta x$, whether by graph or by algebraic substitution, and having found $\delta y/\delta x$ we must then let $\delta x \to 0$. We can write this up in three different ways as follows:

Notation

1. Express δy as $y + \delta y - y$. If we are given that $y = x^4$ or any similar function of x then we can put $y + \delta y = (x + \delta x)^4$ since we are suggesting that the increment δx, when added to x and substituted in the equation, produces the result $y + \delta y$.

We then arrive at the form

$$\delta y = (y + \delta y) - y = (x + \delta x)^4 - x^4$$

Dividing through by δx we get

$$\frac{\delta y}{\delta x} = \frac{(y+\delta y)-y}{\delta x} = \frac{(x+\delta x)^4 - x^4}{\delta x}$$

and we are now ready to let $\delta x \to 0$. This will then give us the limit of $\delta y/\delta x$ at the point (x, y). If we require the limit at the point $x = a$ we must replace x by a and consider the limit of $\dfrac{(a+\delta x)^4 - a^4}{\delta x}$ as $\delta x \to 0$.

2. In this case we concentrate on using the notation $y = f(x)$ with the result that instead of $y + \delta y$ we write $f(x + \delta x)$. In the case of $y = x^4$ we have $y = f(x) = x^4$ and $y + \delta y = f(x + \delta x) = (x + \delta x)^4$. (If we had suggested $y = ax^4 + bx^2$ then we would have had

$$y + \delta y = f(x + \delta x) = a(x + \delta x)^4 + b(x + \delta x)^2.)$$

In the first case $\delta y = (y + \delta y) - y = f(x + \delta x) - f(x) = (x + \delta x)^4 - x^4$.

Dividing through by δx we get

$$\frac{\delta y}{\delta x} = \frac{(y + \delta y) - y}{\delta x} = \frac{f(x + \delta x) - f(x)}{\delta x} = \frac{(x + \delta x)^4 - x^4}{\delta x}$$

and we are again ready to let $\delta x \to 0$. At first sight it looks as though we are writing down even more than before. Remember that if we were to concentrate on $f(x) = x^4$ the first two fractions would not be written and we would go straight to

$$\frac{f(x + \delta x) - f(x)}{\delta x} = \frac{(x + \delta x)^4 - x^4}{\delta x}$$

and then let $\delta x \to 0$ in order to obtain the limit of $\delta y/\delta x$ at the point (x, y). If we require the limit at the point $x = a$ we must replace x by a and consider the limit of $\dfrac{f(a + \delta x) - f(a)}{\delta x}$ as $\delta x \to 0$.

3. In this third presentation we try to avoid the awkwardness of writing δx by substituting h for δx. Thus h is the increment of x and with $y = x^4$ instead of $y + \delta y = (x + \delta x)^4$ we write $y + \delta y = f(x + h) = (x + h)^4$ and

$$\frac{\delta y}{\delta x} = \frac{(y + \delta y) - y}{\delta x} = \frac{f(x + h) - f(x)}{h} = \frac{(x + h)^4 - x^4}{h}$$

and we are again ready to let $h \to 0$ and find the limit of $\delta y/\delta x$ at the point (x, y). If we require the limit at the point $x = a$ we must replace x by a and consider the limit of $\dfrac{f(a + h) - f(a)}{h}$ as $h \to 0$.

Whenever the limit of $\delta y/\delta x$ exists as $\delta x \to 0$ we shall represent that limit by the one symbol $\dfrac{dy}{dx}$. We refer to this as one symbol when it represents the limit because it must *not* be considered as $dy \div dx$. The limit $\dfrac{dy}{dx}$ is called the **derivative** of y with respect to x at the point (x, y) and we say that the function $y = f(x)$ is **differentiable** with respect to x at the point (x, y). Glancing back to case 2 above we see that we dropped the use of y in the final form for $\delta y/\delta x$; consequently whenever the limit of $\dfrac{f(x + \delta x) - f(x)}{\delta x}$ exists as $\delta x \to 0$ we shall write this as $\dfrac{df}{dx}$ or $\dfrac{df(x)}{dx}$, i.e. a straightforward substitution of f or $f(x)$ for y. Similarly for case 3, as $h \to 0$ if the limit of $\delta y/\delta x$ exists we shall write it as $\dfrac{df}{dx}$ or $\dfrac{df(x)}{dx}$.

When $\dfrac{df(x)}{dx}$ is left as a function of x we refer to it as the derived function, and in keeping with this we use the notation $f'(x)$ (read as 'f dash of x') to represent the derived function.

Taking the increment of x to be positive in each case, Fig. 80 illustrates the relation between the algebra and the geometry.

In Fig. 80(*a*) $P(x, y)$, $Q(x + \delta x, y + \delta y)$, $PN = \delta x$, $NQ = \delta y$. PT is the tangent at P and consequently the limiting position of the line PQ as Q approaches P. Thus $\frac{\delta y}{\delta x}$ is the gradient of PQ, $\frac{dy}{dx} = \tan\theta$ is the gradient of the

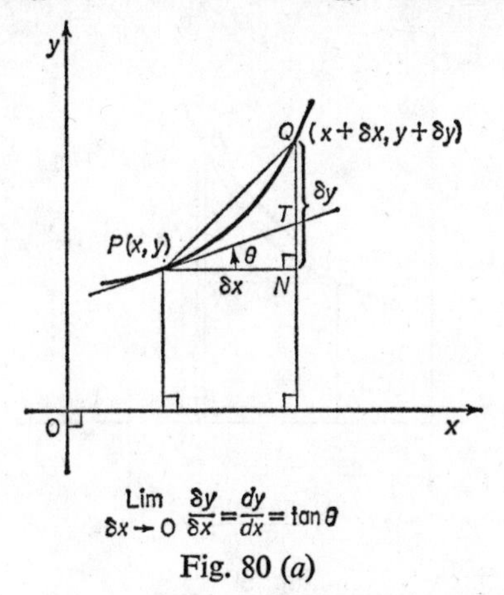

Fig. 80 (*a*)

tangent at P; that is, as $\delta x \to 0$ algebraically so Q approaches P along the curve geometrically.

In Fig. 80(*b*)

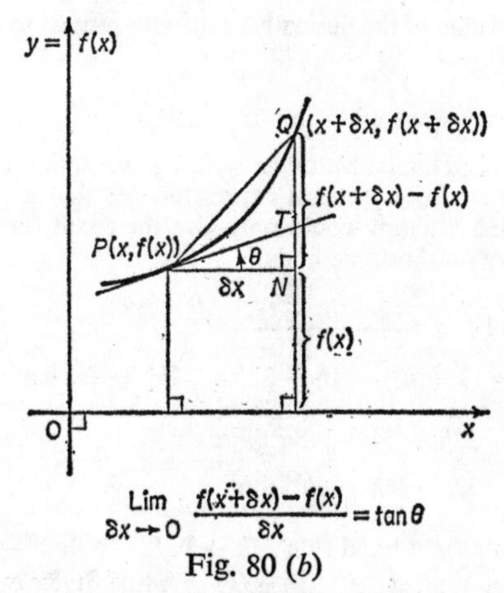

Fig. 80 (*b*)

$$P(x, f(x)),\ Q(x + \delta x), f(x + \delta x)),\ PN = \delta x,\ NQ = f(x + \delta x) - f(x)$$

PT is the tangent at P and consequently the limiting position of the line PQ as Q approaches P. Thus $\frac{f(x + \delta x) - f(x)}{\delta x}$ is the gradient of PQ, $\frac{df(x)}{dx} = \tan\theta$ (or $f'(x) = \tan\theta$) is the gradient of the tangent at P; that is, as $\delta x \to 0$ algebraically so Q approaches P along the curve geometrically.

In Fig. 80(*c*), $P(x, y)$, $Q(x + h, f(x + h))$, $PN = h$, $NQ = f(x + h) - f(x)$. PT is the tangent at P and consequently the limiting position of the line PQ

as Q approaches P. Thus $\dfrac{f(x+h)-f(x)}{h}$ is the gradient of PQ, $\dfrac{\mathrm{d}f(x)}{\mathrm{d}x} = \tan\theta$ (or $f'(x) = \tan\theta$) is the gradient of the tangent at P; that is, as $h \to 0$ algebraically so Q approaches P along the curve geometrically.

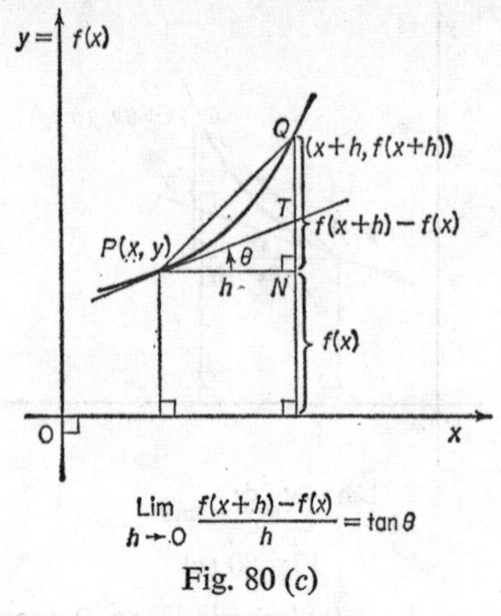

Fig. 80 (*c*)

Again, do not bother to commit this jargon to memory at this stage—just grow used to it by working examples.

Example 1. Find the value of the derivative of y with respect to x at the point $x = 2$ on the curve $y = f(x) = x^4$.

Solution. There are two ways of attempting such a problem. Either we find $\dfrac{dy}{dx}$ at $x = 2$ or we find $f'(x)$ and then substitute $x = 2$. The advantage of the second attempt would be that of obtaining a general expression for the derivative at any point (x, y), whereas the first attempt would only give the result for the particular point $x = 2$. We shall carry out both methods.

Method 1. We require $\lim\limits_{h\to 0} \dfrac{f(2+h)-f(2)}{h}$

$$f(2+h) = (2+h)^4 = 16 + 32h = 24h^2 + 8h^3 + h^4 \text{ (binomial expansion)}$$

$$f(2) = 16$$

$$f(2+h) - f(2) = 32h + 24h^2 + 8h^3 + h^4$$

$$\frac{f(2+h)-f(2)}{h} = 32 + 24h + 8h^2 + h^3$$

As h tends to 0 on the right hand side $24h \to 0$, $8h^2 \to 0$, $h^3 \to 0$

$$\therefore \quad \lim_{h\to 0} \frac{f(2+h)-f(2)}{h} = 32$$

$\therefore$ The derivative of $f(x) = x^4$ at the point (2, 16) is 32.

Or to put it another way, the gradient of the tangent at the point (2, 16) on the curve $f(x) = x^4$ is 32. Answer

Method 2. We require $\lim\limits_{h\to 0} \dfrac{f(x+h)-f(x)}{h}$. Having found the limit we will substitute $x = 2$

$$f(x+h) = (x+h)^4 = x^4 + 4x^3h + 6x^2h^2 + 4xh^3 + h^4, \; f(x) = x^4$$
$$\therefore \; f(x+h) - f(x) = 4x^3h + 6x^2h^2 + 4xh^3 + h^4$$
$$\frac{f(x+h) - f(x)}{h} = 4x^3 + 6x^2h + 4xh^2 + h^3$$

As h tends to 0 on the right hand side $6x^2h \to 0$, $4xh^2 \to 0$ and $h^3 \to 0$. $4x^3$ remains unchanged

$$\therefore \; \lim_{h \to 0} \frac{f(x+h) - f(x)}{h} = 4x^3$$

Here we have obtained the derived function $f'(x) = 4x^3$.

At $x = 2$, $f'(2) = 4 \times 2^3 = 32$ as above. Thus $f'(x) = 4x^3$ is the gradient of the tangent at any point (x, y) on the curve, or $f'(a) = 4a^3$ is the gradient of the tangent to the curve at the point $x = a$. Answer

The reader is probably beginning to realise that we are trying to build up a store of results for finding the derivatives of powers of x, i.e. x^n for $n =$ an integer. So far we have found the results related to x, x^2, x^3 and x^4. As the power of x increases so we rely more and more on the binomial theorem, but once the pattern of the proof is seen the general result follows very easily. We shall now obtain the derivative of x^5 and x^n.

Example 2. Find the derivative of the function $f(x) = x^5$ at any point of its graph. In particular find the gradient of the graph of $f(x) = x^5$ at the point (2, 32).

Solution. We require $\lim_{h \to 0} \frac{f(x+h) - f(x)}{h}$

$f(x+h) = (x+h)^5 = x^5 + 5x^4h + 10x^3h^2 + 10x^2h^3 + 5xh^4 + h^5, f(x) = x^5$

It is at this stage that we pause for thought. We know we are about to divide by h and then let $h \to 0$, so why bother to expand $(x+h)^5$ past the term in h? Hence we adopt the following argument

$$f(x+h) = (x+h)^5 = x^5 + 5x^4h + (\text{terms in } h^2 \text{ and higher powers of } h),$$
$$f(x) = x^5$$
$$\therefore \; f(x+h) - f(x) = 5x^4h + (\text{terms in } h^2 \text{ and higher powers of } h)$$
$$\therefore \; \frac{f(x+h) - f(x)}{h} = 5x^4 + (\text{terms in } h \text{ and higher powers of } h)$$

As we let $h \to 0$ on the right-hand side so all the terms which occur in the bracket must tend to zero

$$\therefore \; \lim_{h \to 0} \frac{f(x+h) - f(x)}{h} = f'(x) = 5x^4$$ Answer

At the point (2, 32), $f'(2) = 5 \times 16 = 80$. This means that the value of the derived function at $x = 2$ is 80, or that the gradient of the tangent at the point (2, 32) of the curve $y = x^5$ is 80. Answer

This list of results which we have deduced so far reads as follows:

(i) $f(x) = x$; $f'(x) = 1$
(ii) $f(x) = x^2$; $f'(x) = 2x$
(iii) $f(x) = x^3$; $f'(x) = 3x^2$
(iv) $f(x) = x^4$; $f'(x) = 4x^3$
(v) $f(x) = x^5$; $f'(x) = 5x^4$

If the pattern of these results is to continue we would expect that when $f(x) = x^n$ then $f'(x) = nx^{n-1}$. For example, the next entry in the table would be $f(x) = x^6$, $f'(x) = 6x^5$ and so on; that is, we observe that we decrease the power of x by one to $(n - 1)$ and multiply the result by the original power n.

A further extension of these results is worth noting, and that is if k is a non-zero constant then the following results arise:

(i) $f(x) = kx$, $f'(x) = k$
(ii) $f(x) = kx^2$, $f'(x) = 2kx$
(iii) $f(x) = kx^3$, $f'(x) = 3kx^2$
(iv) $f(x) = kx^4$, $f'(x) = 4kx^3$ and so on

The Derivative of x^n for n a Positive Integer

We require $\lim\limits_{h\to 0} \dfrac{f(x + h) - f(x)}{h}$

$$f(x + h) = (x + h)^n = x^n + nx^{n-1}h + (\text{terms in } h^2 \text{ and higher powers of } h);$$

$$f(x) = x^n$$

$$\therefore f(x + h) - f(x) = nx^{n-1}h + (\text{terms in } h^2 \text{ and higher powers of } h)$$

$$\therefore \frac{f(x + h) - f(x)}{h} = nx^{n-1} + (\text{terms in } h \text{ and higher powers of } h)$$

As $h \to 0$ so each term in the bracket will also tend to 0, and we shall be left with nx^{n-1}, which will remain unchanged as $h \to 0$

$$\therefore \lim_{h\to 0} \frac{f(x + h) - f(x)}{h} = f'(x) = nx^{n-1}$$

Similarly when $f(x) = kx^n$ then $f'(x) = nkx^{n-1}$.

In fact **this result holds true for any non-zero rational value of n.** For example, $n = \frac{1}{2}, \frac{2}{3}, -\frac{5}{6}$ and so on, an extension of the result which we shall have to assume in the interests of making more general progress in other directions (see page 257).

The Derivative of the Sum and Difference of Two Functions

The result for two functions $f(x)$ and $g(x)$ may be stated symbolically as follows

$$\frac{d[f(x) \pm g(x)]}{dx} = \frac{df(x)}{dx} \pm \frac{dg(x)}{dx}$$

We have already deduced one or two simple instances of this theorem. On page 208 we found $\dfrac{d(x^2 + 2)}{dx} = 2x$, which is the same as

$$\frac{d(x^2)}{dx} + \frac{d(2)}{dx} = 2x + 0.$$

On page 210 we found that $\dfrac{d(3x^2 + 8x)}{dx} = 6x^2 + 8$, which is the same as

$$\frac{d(3x^2)}{dx} + \frac{d(8x)}{dx} = 6x + 8.$$

This theorem enables us to write down the derivative of a sum or difference of any number of functions. In particular we are able to find the derivative of a polynomial as in the following example.

Example 3. Differentiate (i.e. find the derivative of) the function

$$f(x) = 4x^{10} + 7x^3 - 10x - 19.$$

Solution. We know that

$$\frac{d(4x^{10})}{dx} = 40x^9, \frac{d(7x^3)}{dx} = 21x^2,$$

$$\frac{d(10x)}{dx} = 10, \frac{d(19)}{dx} = 0$$

By the theorem stated above we may add and subtract these results as appropriate to obtain

$$\frac{df(x)}{dx} = 40x^9 + 21x^2 - 10$$

This result means that if we had been able to draw the graph of this function (given the time) then the gradient of the tangent at any point (x, y) on the curve would be given by the derived function

$$f'(x) = 40x^9 + 21x^2 - 10$$

For example,

when $x = 1$, we get $f'(1) = 40 + 21 - 10 = 51$
when $x = -1$ we get $f'(-1) = -40 + 21 - 10 = -29$

and so on for any real value of x. Answer

Example 4. Obtain the derived function of $f(x) = 5x + \frac{3}{x}$ and find the values of x for which the gradient of the tangent to the curve $y = f(x)$ is zero.

Solution. On page 212 we obtained the derivative of $f(x) = \frac{8}{x}$ at the point (2, 4) and not the derived function, consequently we do not have a general result with which to obtain the derivative of $f(x) = \frac{3}{x}$. On page 220 we remarked that for $f(x) = x^n$ we have $f'(x) = nx^{n-1}$ for all rational values of n. Here we have $f(x) = \frac{3}{x} = 3x^{-1}$ (see page 116), which is equivalent to $n = -1$. Application of the result $f'(x) = nx^{n-1}$ when $f(x) = x^n$ gives $f'(x) = 3 \times (-1) \times x^{-1-1} = -3x^{-2}$

$$\therefore f'(x) = -\frac{3}{x^2}$$

Alternatively we could work from first principles with

$$f(x + h) - f(x) = \frac{3}{x + h} - \frac{3}{x} = 3\left\{\frac{(x) - (x + h)}{x(x + h)}\right\}$$

$$\therefore \quad f(x + h) - f(x) = -\frac{3h}{x(x + h)}$$

$$\therefore \quad \frac{f(x + h) - f(x)}{h} = -\frac{3}{x^2 + xh}$$

As $h \to 0$ on the right-hand side $xh \to 0$ so $f'(x) = -\frac{3}{x^2}$ as obtained above.

With $\frac{d(5x)}{dx} = 5$ and $\frac{d\frac{3}{x}}{dx} = -\frac{3}{x^2}$ we use the theorem about the derivative of a sum to yield

$$\frac{df(x)}{dx} = f'(x) = 5 - \frac{3}{x^2}$$ Answer

Furthermore, $f'(x) = 0$ when $5 = \frac{3}{x^2}$ or $5x^2 = 3$ or $x^2 = 0{\cdot}6$. Taking a square root we get $f'(x) = 0$ when $x = \pm\sqrt{0{\cdot}6} = \pm 0{\cdot}7746$ Answer

Note: in the above example $f(x)$ does not exist at $x = 0$, so neither does $f'(x)$ at $x = 0$. Alternatively we may describe $f(0)$ as not defined and in consequence $f'(0)$ cannot be obtained.

Exercise 17.1

1. Find from first principles the derivative of the function $f(x)$ at the point $x = 1$ when $f(x)$ is equal to (i) x^6, (ii) $\frac{1}{x^2}$
2. Using the result that if $f(x) = x^n$ then $f'(x) = nx^{n-1}$ for all non-zero n write down $f'(x)$ in each of the following cases.

   (i) $f(x) = x^{21}$ (ii) $f(x) = x^{-2}$ (iii) $f(x) = x^{-3}$
   (iv) $f(x) = x^{\frac{1}{3}}$ (v) $f(x) = x^{\frac{1}{2}}$
3. Write down $f'(x)$ when $f(x)$ is equal to (i) kx^{-2} (ii) kx^{-3} (iii) $kx^{\frac{1}{2}}$ where k is a constant in each case.
4. Find the gradient of the tangent at the point $x = 4$ on the graph of each of the following.

   (i) $y = x^2$ (ii) $y = \frac{1}{x^2}$ (iii) $f(x) = \frac{1}{x^2}$ (iv) $f(x) = x^{\frac{1}{2}}$
5. Look back to Fig. 78 and draw the tangent to the curve $y = 3x^2 + 8x$ at the point (1, 11) and use a right-angled triangle like *PRT* to find the gradient.
6. Find the derivative of $3x^2 + 8x$ at the point (1, 11) and check your answer to the previous question.
7. Repeat questions 5 and 6 for the tangent at the point (0, 0).
8. Given that $f(x) = 16x + \frac{1}{x^2}$ find the values of x for which $f'(x) = 0$.
9. Find the points on the graph of $y = x^2$ at which the tangent has a gradient of (i) 2, (ii) 6, (iii) -4.
10. Given that $y = x(x - 7)$ find the points at which $\frac{dy}{dx}$ is equal to (i) 0, (ii) 1, (iii) -1.
11. For a certain function $f(x)$ the value of $f'(x)$ is 2 for all values of x. Write down an expression for $f(x)$.
12. A curve is given by the equation $y = x^2 - x - 2$. Find the equation of the tangents at the points where the curve intersects the x axis.

Coordinate Geometry and the Use of the Derivative

It is of interest at this stage to look back on some of the coordinate geometry results in Chapter Seven, concerning the equation of the tangents to one or two curves. For the moment we shall re-examine the tangents to the rectangular hyperbola $xy = k^2$ and the parabola $y = kx^2$ or $y^2 = kx$.

Whenever we can express y as a function of x by writing $y = f(x)$ we say that we are expressing y explicitly in terms of x. This is not always as easy as it might appear as the equation

$$ax^2 + 2hxy + by^2 + 2gx + 2fy + c = 0$$

is quick to show. In the case of the circle $x^2 + y^2 = a^2$ we may proceed along the lines

$$y^2 = a^2 - x^2$$
$$\therefore y = \sqrt{(a^2 - x^2)}$$

but unfortunately our work so far is not sufficiently advanced to enable us to find $\frac{dy}{dx}$ (see page 250). It is the elementary nature of the equation $xy = k^2$ of the rectangular hyperbola which makes it a first choice because we obtain $y = f(x) = k^2/x$ straight away.

On page 89 we saw that the gradient of the tangent at the point (x_1, y_1) was $-y_1/x_1$. In this chapter with $y = k^2/x$ we have

$$\frac{dy}{dx} = k^2\left(-\frac{1}{x^2}\right) = -\frac{k^2}{x^2}$$

By substituting $k^2 = xy$ we get

$$\frac{dy}{dx} = -\frac{xy}{x^2} = -\frac{y}{x}$$

It follows therefore that at the point (x_1, y_1) the gradient of the tangent is $-y_1/x_1$, which is the same result as on page 89.

In a subsequent worked example we required the tangent to $xy = 36$ at the point (12, 3), an enquiry which may now proceed as follows:
from $xy = 36$, we get

$$y = \frac{36}{x} \text{ and } \frac{dy}{dx} = -\frac{36}{x^2}$$

at the point (12, 3) we have

$$\frac{dy}{dx} = -\frac{36}{144} = -\frac{1}{4}$$

the equation of the tangent at (12, 3) is therefore

$$\frac{y-3}{x-12} = -\frac{1}{4}$$

which becomes

$$4y - 12 = -x + 12$$
$$4y + x = 24 \text{ (as on page 89)}$$

We had two simple forms for the equation of a parabola and they were either $4ay = x^2$ (page 85) or $y^2 = 4ax$ (page 82). The first form presents no difficulty for we see immediately that from $4ay = x^2$ we can obtain $y = \frac{x^2}{4a}$ and equally quickly we have $\frac{dy}{dx} = \frac{2x}{4a} = \frac{x}{2a}$. This result means that at the point (x_1, y_1) we have $\frac{dy}{dx} = \frac{x_1}{2a}$ and the equation of the tangent at this point is

$$\frac{y - y_1}{x - x_1} = \frac{x_1}{2a}$$
$$\therefore 2ay - 2ay_1 = xx_1 - x_1^2$$

Substituting $x_1^2 = 4ay_1$ this becomes

$$2ay - 2ay_1 = xx_1 - 4ay_1$$

and finally we have

$$2a(y + y_1) = xx_1 \text{ (as on page 85)}$$

The second form of the equation $y^2 = 4ax$ is less straightforward because in order to obtain y as an explicit function of x we must take a square root to gain $y = (2\sqrt{a})\sqrt{x}$, which we write in index form in order to use the differentiation rule on page 220

$$\therefore y = (2\sqrt{a})x^{\frac{1}{2}} \qquad \text{(i)}$$

$$\therefore \frac{dy}{dx} = 2\sqrt{a}\left(\frac{1}{2}x^{-\frac{1}{2}}\right) = \frac{\sqrt{a}}{x^{\frac{1}{2}}} = \frac{\sqrt{a}}{\sqrt{x}} \qquad \text{(ii)}$$

We may tidy up this last result by either substituting for $\sqrt{a} = \frac{y}{2x^{\frac{1}{2}}}$ from (i) so that (ii) now reads

$$\frac{dy}{dx} = \frac{y}{2x}$$

or substituting for $x^{\frac{1}{2}} = \frac{y}{2\sqrt{a}}$ from (i) so that (ii) now reads

$$\frac{dy}{dx} = \sqrt{a}.\frac{2\sqrt{a}}{y} = \frac{2a}{y}$$

This result now means that the tangent at the point (x_1, y_1) on the parabola $y^2 = 4ax$ has a gradient given by $\frac{dy}{dx} = \frac{2a}{y_1}$.

The equation of the tangent at the point (x_1, y_1) is therefore

$$\frac{y - y_1}{x - x_1} = \frac{2a}{y_1}$$

$$\therefore yy_1 - y_1^2 = 2ax - 2ax_1$$

Since $y_1^2 = 4ax_1$ this may be rewritten as

$$yy_1 = 2ax + y_1^2 - 2ax_1$$

i.e. $$yy_1 = 2ax + 2ax_1 \text{ (as on page 84)}$$

With $\frac{dy}{dx}$ so easy to obtain as the gradient of the tangent at a point it follows that we are able to obtain the equation of the normal equally easily by using the result $m_1m_2 = -1$ on page 59.

Referring to Fig. 81, we have drawn the tangent to the curve $y = f(x)$ at $P(x, y)$ to intersect the x axis at L. The normal to the curve at P, being perpendicular to the tangent, has been drawn to intersect the x axis at M. On the diagram the gradient of LP is $\tan\theta$ and the gradient of MP is $\tan\alpha$. We know that the product of the gradients of two perpendicular lines is -1, so $\tan\theta.\tan\alpha = -1$, and since $\frac{dy}{dx} = \tan\theta$ it follows that

$$\tan\alpha = -\frac{1}{\tan\theta} = \frac{-1}{\frac{dy}{dx}}$$

Example 5. Find the equation of the tangent and normal to the curve $y = x^3 + 5x^2 + 3$ at the point (1, 9).

Solution. With $y = f(x) = x^3 + 5x^2 + 3$ we find

$$f'(x) = \frac{d(x^3)}{dx} + \frac{d(5x^2)}{dx} + \frac{d(3)}{dx} = 3x^2 + 10x + 0$$

Hence, at the point (1, 9) the gradient of the tangent is $f'(1) = 13$.

From this result and '$m_1 m_2 = -1$' it follows that the gradient of the normal must be $-\frac{1}{13}$.

The equation of the tangent at the point (1, 9) is

$$\frac{y-9}{x-1} = 13$$

i.e. $y = 13x - 4$ Answer

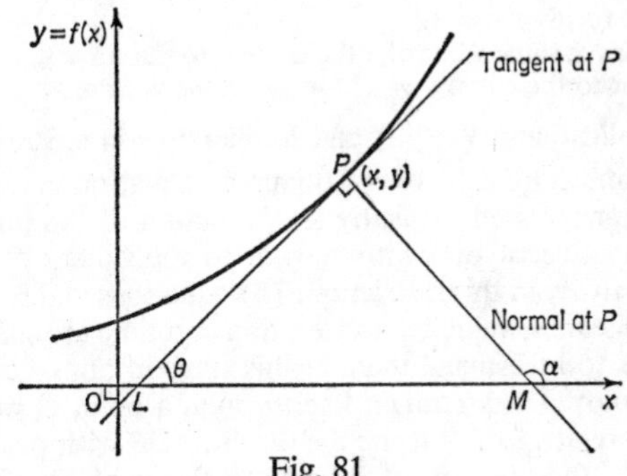

Fig. 81

The equation of the normal at the point (1, 9) is

$$\frac{y-9}{x-1} = -\frac{1}{13}$$

i.e. $$13y = -x + 118$$ Answer

Example 6. Find the angle between the curves $y = 3x^3 - x$ and $y = 5x^2 + 6x - 9$ at their point of intersection (1, 2).

Solution. The angle at which two curves intersect is the angle between their tangents at the point of intersection.

For $y = 3x^3 - x$, $\frac{dy}{dx} = 9x^2 - 1$. The gradient of the tangent at the point (1, 2) is 8.

For $y = 5x^2 + 6x - 9$, $\frac{dy}{dx} = 10x + 6$. The gradient of the tangent at the point (1, 2) is 16. We now recall the formula $\tan\alpha = \frac{m_1 - m_2}{1 + m_1 m_2}$ (page 177 with α the angle between the tangents). Taking $m_1 = 16$, $m_2 = 8$ we arrive at

$$\tan\alpha = \frac{16-8}{1+128} = \frac{8}{129} = 0{\cdot}0620 \text{ (correct to 4 decimal places)}$$

$$\therefore \quad \alpha = 3° 33' \quad \text{or} \quad \tan^{-1}\frac{8}{129}$$ Answer

Exercise 17.2

1. Find the gradient of the tangent to the rectangular hyperbola $xy = 4$ at the points (i) (1, 4) (ii) (−1, −4) (iii) (4, 1) (iv) (−4, −1). Which of these tangents are parallel?

2. Show that on the curve $xy = k^2$ no tangent has a positive gradient.
3. Find the points of the curve $xy = 36$ at which the tangents have a gradient of -9.
4. Find the gradient of the tangent to the parabola $y^2 = 4x$ at the points: (i) $(1, 2)$ (ii) $(1, -2)$ (iii) $(16, 8)$ (iv) $(4, 4)$
5. At which points on the curve $y^2 = 16x$ does the tangent have a gradient of (i) 2, (ii) $\frac{2}{3}$?
6. Find the angle between the tangents at the point of intersection $(2, 16)$ of the curves $y = 4x^2$ and $xy = 32$.
7. At what rate is y increasing with respect to x at the point $(2, 5)$ on the curve $xy = 10$?
8. Find the equations of the normals to the parabola $y = x^2/4$ at the points $(-2, 1)$, $(-4, 4)$. Prove that the point of intersection of these normals lies on the parabola.
9. The curve $y = f(x) = 12x - x^2$ intersects the x axis at $P(0, 0)$ and $Q(12, 0)$. Prove that: (i) $f'(x) > 0$ for $0 \leqslant x < 6$ (ii) $f'(x) = 0$ for $x = 6$ (iii) $f'(x) < 0$ for $6 < x \leqslant 12$
10. Prove that for the same value of x the tangent to the curve $y = x^2 + k$ is parallel to the tangent to the curve $y = x^2$ for any value of k.

Elementary Applications: Velocity and Acceleration in a Straight Line

At the moment, as long as we can obtain an equation in two variables such that one can be expressed explicitly as a function of the other then we can obtain the derivative of one with respect to the other. We shall see later (page 255) how to avoid this restriction. The equations and relations involving displacement, distance, velocity, acceleration and time are sufficiently elementary to allow us to investigate them at this stage in our calculus work.

If a body P moves in a **straight line** through a point O we define the displacement of P relative to O to be the distance OP with positive or negative value to indicate the position of P being to the right or left of O. In other words, the position of P is registered in exactly the same way as a point on the x axis relative to the origin O. We usually represent the displacement by the letter s.

The **velocity** of P at any time is defined as the rate of change of the displacement of P with respect to time. This means that just as $\frac{dy}{dx}$ gives the rate of change of y with respect to x at the point (x, y), so $\frac{ds}{dt}$ gives the rate of change of displacement s with respect to time at the position s on the straight line. We therefore write the velocity v as

$$v = \frac{ds}{dt}$$

The **acceleration** of a body P is defined as the rate of change of its velocity with respect to time. Using the letter a to represent the acceleration and $\frac{dv}{dt}$ to represent the rate of change of velocity with respect to time we therefore write

$$a = \frac{dv}{dt}$$

One small point should be noted before moving on: a is the derivative of v with respect to t and v is the derivative of s with respect to t. This is interesting to us because it indicates that we have a meaning here for a 'derivative of a derivative', i.e. a second derivative. In this language we can describe the acceleration a as the second derivative of the displacement s with respect to time. We shall later introduce a new notation (page 241) to represent such a result. In the meantime consider the information to be gained from plotting the graphs of displacement and velocity against the time.

Displacement–time Graphs

In Fig. 82(*a*) the curve is given by $s = f(t)$, i.e. s as a function of t. Imagine that the body moves in the **straight line** through O and that it starts its motion at L when $t = 0$. Thus all the displacements are measured from O but the time is measured from the starting position at L. If the velocity is positive then the direction of motion is from left to right.

At the point $P(t, s)$ on the curve the gradient of the tangent is $\frac{ds}{dt} = \tan\theta$ as before in all the x, y curve situations. Thus, the velocity at a point P' on

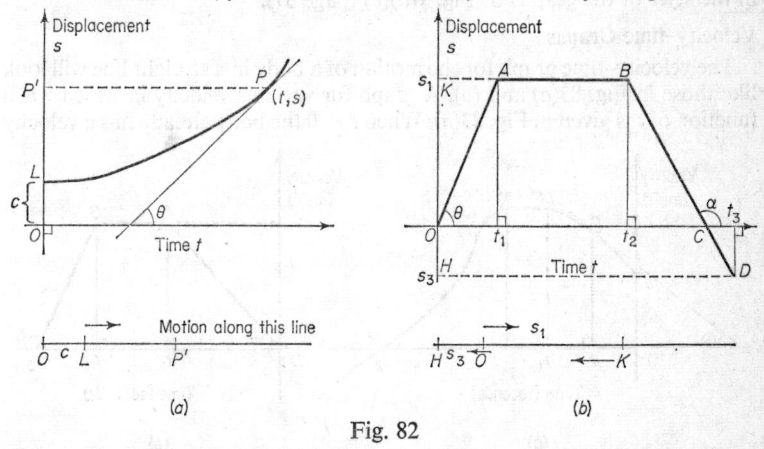

Fig. 82

the line of motion at a given time t, is given by the gradient of the tangent at that point. Since the value of $\tan\theta$ is clearly varying as t increases, then the form of the graph in Fig. 82(*a*) represents a motion in which the velocity is variable, (i.e. not constant); furthermore, we see that when $t = 0$ the distance from the origin O is c.

In Fig. 82(*b*) we have a displacement–time graph for a motion which falls into three parts. On the section represented by O to A the gradient of the line is constant, this means that $\frac{ds}{dt} = v = \tan\theta = \text{constant}$.

The body starts from O when $t = 0$ and continues with a constant velocity represented by the gradient of OA, for a time t_1. The section AB in which the displacement from O remains the same (being parallel to the t axis) means that $\frac{ds}{dt} = v = 0$, i.e. the body stops for a time $t_2 - t_1$. The section BC, being a straight line, represents the body returning to the point O from which it started (i.e. $s = 0$). On this section the gradient of the line is a constant and $\frac{ds}{dt} = v = \tan\alpha$ a negative quantity, indicating that the body is returning to the point from which it started.

The point D on the graph shows that at time t_3 the body has gone past the starting point a distance of s_3. The actual motion which takes place on the straight line through O must have been as described below.

Start at O with a constant velocity given by $v = \tan\theta$ and move left to right through a distance $OK = s_1$. The point K is reached in time t_1. Stop at K for a time of $t_2 - t_1$ and then return to O with a velocity of $v = \tan\alpha$. (Note that $\tan\alpha$ will be negative indicating that the body is now moving from right to left.) At time t_3 the body is at H where $OH = s_3$. The negative value of s_3 indicating that the body is on the left of O.

A body which moved with a constant velocity of say 2 m s^{-1} would give a straight line displacement–time graph like the section OA, in which case $\tan\theta = 2$. Of course if the body starts at a positive distance c from the origin O then the line section OA would be raised through a distance representing c in the style of the graphs of Fig. 1(*a*)(*c*) (page 37).

Velocity–time Graphs

The velocity–time graph for the motion of a body in a straight line will look like those in Fig. 83(*a*) and (*b*). A graph for variable velocity in which v is a function of t is given in Fig. 83(*a*). When $t = 0$ the body already has a velocity

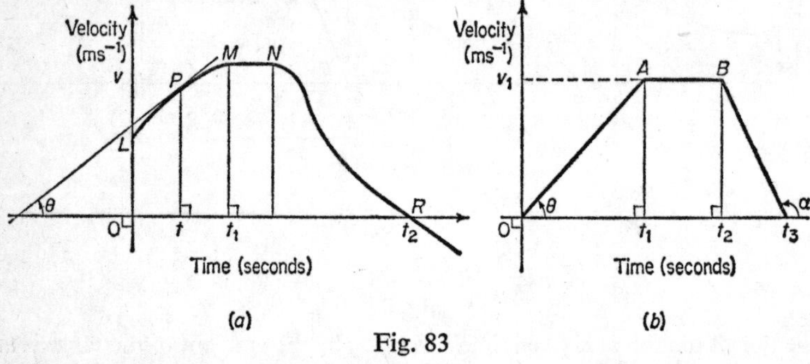

Fig. 83

in the positive direction (i.e. a positive velocity) represented by the length of OL on the graph. For the first t_1 seconds the velocity is increasing so that at any time t the gradient of the tangent to the curve at P is given by $\frac{dv}{dt} = \tan\theta$, so that $\tan\theta$ represents the acceleration at the time t. From M to N on the graph the velocity remains constant and the gradient of the curve is zero, i.e. $\frac{dv}{dt} = 0$, or in other words the acceleration is zero when the velocity is constant. From N to R on the graph the velocity is decreasing so that $\frac{dv}{dt} < 0$ (in which case the acceleration is negative). At R the velocity is zero and the body is at rest at time t_2. After t_2 the velocity is negative and the body is on its way back to the starting point.

The velocity–time graph of Fig. 83(*b*) looks like Fig. 82(*b*) but is *not* related. When $t = 0$ we have $v = 0$, i.e. the body starts from rest at $t = 0$. For the first t_1 seconds the body moves with a constant acceleration given by $\frac{dv}{dt} = \tan\theta$ (i.e. constant because the gradient of the straight line OA is a constant). From t_1 to t_2 the body moves with constant velocity v_1, then from t_2

to t_3 the velocity decreases uniformly to zero, i.e. $\frac{dv}{dt} = \tan \alpha$ is a constant negative acceleration ('uniformly' means 'at a constant rate').

Area Under the Velocity–time Graph

As we shall see later in Chapter Twenty, the area under the velocity–time graph for motion in a straight line gives the displacement. As an elementary illustration of this point consider the area under the part AB of the graph in Fig. 83(*b*). Since the velocity is v_1 (and this means that a distance v_1 metres is travelled each second), it follows that the distance moved in $(t_2 - t_1)$ seconds is $v_1(t_2 - t_1)$ metres, which is the area of the graph 'under AB'.

In summary therefore we have the following facts for motion in a straight line:

The gradient of the displacement–time graph, or curve, for time t gives the velocity at time t.

The gradient of the velocity–time graph, or curve, for time t gives the acceleration at time t.

The area under the velocity–time graph gives the displacement.

Example 7. A body moves in a straight line with an initial velocity of 5 m s^{-1} when $t = 0$. If the distance s from the starting point at time t is given by the equation $s = 5t - t^2$ find the velocity v at any time t and show that $v^2 = 25 - 4s$.

Solution. The velocity is given by $v = \frac{ds}{dt} = \frac{d(5t - t^2)}{dt} = 5 - 2t$ Answer

With $v = 5 - 2t$ we get

$$v^2 = (5 - 2t)^2 = 25 - 20t + 4t^2 = 25 - 4s$$ Answer

We can get much more out of this problem as follows:
The acceleration is

$$a = \frac{dv}{dt} = \frac{d(5 - 2t)}{dt} = -2$$

-2 is a constant and the negative sign tells us that the motion is being retarded, i.e. that the velocity decreases as time increases.

From the result $v = 5 - 2t$ we see that the body comes to rest when $v = 0$ and $t = 2\frac{1}{2}$ seconds. When t exceeds $2\frac{1}{2}$ seconds the velocity is negative. In $2\frac{1}{2}$ seconds the body has travelled a distance given by $s = 5 \times 2\frac{1}{2} - 2\frac{1}{2} \times 2\frac{1}{2} = 6\frac{1}{4}$ m.

In 5 seconds the body is at a distance s from the starting point given by

$$s = 5 \times 5 - 5^2 = 0$$

All this information enables us to describe the motion as follows in Fig. 84. The body

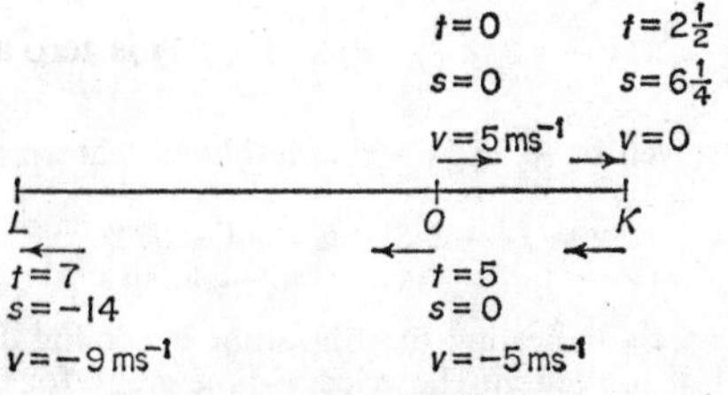

Fig. 84

starts at O when $t = 0$ with a velocity of 5 m s^{-1}, i.e. it moves from O to the right. The body reaches the point K in 2·5 seconds and is instantaneously at rest at K. After 2·5 seconds the body starts coming back to O.

When $t = 5$ seconds, $s = 0$ and the body is back at O and moving right to left with a velocity of 5 m s^{-1}.

When $t = 7$ seconds, $s = 35 - 49 = -14$ m and $v = 5 - 14 = -9 \text{ m s}^{-1}$, i.e. the body is gaining speed and getting further away from O as t increases.

When $t = 7$, the body is at L.

Example 8. A stone is thrown vertically upwards with an initial velocity of 14 m s^{-1}. The distance s metres from the point of projection is given by the equation $s = 14t - 4{\cdot}9t^2$. Calculate: (i) the time taken to reach a point which is 5 m above the point of projection; (ii) the greatest height to which the stone will rise; (iii) the velocity of the stone when $t = 6$. Draw a velocity–time graph of the motion.

Solution. When $s = 5$ we have $5 = 14t - 4{\cdot}9t^2$

i.e.
$$4{\cdot}9t^2 - 14t + 5 = 0$$

This is a quadratic equation in t with two solutions, as discussed on page 21. The solutions are

$$t = \frac{14 \pm \sqrt{(14^2 - 4 \times 5 \times 4{\cdot}9)}}{2 \times 4{\cdot}9}$$

$$= \frac{14 \pm \sqrt{98}}{9{\cdot}8} = \frac{14 \pm 7\sqrt{2}}{9{\cdot}8} = \frac{2 \pm \sqrt{2}}{1{\cdot}4}$$

With $\sqrt{2} = 1{\cdot}414$ we have

$$t = \frac{3{\cdot}414}{1{\cdot}4} \quad \text{or} \quad \frac{0{\cdot}586}{1{\cdot}4} = 2{\cdot}44 \quad \text{or} \quad 0{\cdot}42$$

(correct to 2 decimal places)

The two different times mean that at 0·42 seconds the stone is 5 m above the point of projection and travelling upwards. At $t = 2{\cdot}44$ seconds the stone is 5 m above the point of projection but is now travelling downwards. Answer

The stone will have reached its greatest height when it stops rising any further and starts coming down. At this moment the velocity is zero, i.e. $v = 0$. Now, with $s = 14t - 4{\cdot}9t^2$ it follows that

$$v = \frac{ds}{dt} = 14 - 2 \times 4{\cdot}9t$$

$v = 0$ when $14 - 9{\cdot}8t = 0$, i.e.

$$t = \frac{14}{9{\cdot}8} = \frac{1}{0{\cdot}7} = 1\tfrac{3}{7}$$

The stone stops rising when $t = \frac{10}{7}$ seconds. (Incidentally, we notice that the acceleration is $a = \frac{dv}{dt} = 9{\cdot}8$.) The distance above the point of projection at this time is given by $s = t(14 - 4{\cdot}9t)$ when $t = \frac{10}{7}$. Therefore the greatest height reached is

$$\frac{10}{7}\left(14 - 4{\cdot}9 \times \frac{10}{7}\right) = \frac{10}{7}(14 - 7) = 10 \text{ m}$$ Answer

With the velocity v given by $\frac{ds}{dt} = 14 - 9{\cdot}8t$ it follows that when $t = 6$ the velocity

$$v = 14 - 9{\cdot}8 \times 6 = 14 - 58{\cdot}8$$
$$\therefore \quad v = -44{\cdot}8 \text{ m s}^{-1}$$

The negative sign merely indicating that the stone is coming down. Indeed it has gone past the point of projection. The velocity–time graph for this motion is given in Fig. 85.

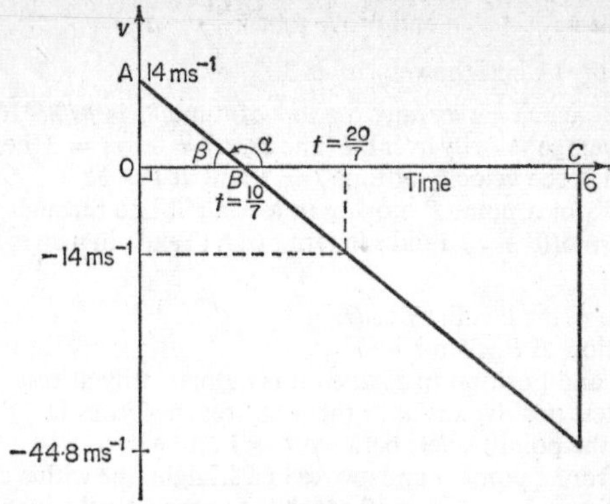

Fig. 85

Of course the knowledge that $v = 14 - 9{\cdot}8t$, tells us that the v–t graph is a straight line—think of $y = mx + c$ or $y = c + mx$ with a gradient of $-9{\cdot}8$ and an intercept on the v axis of 14 (given by OA) (page 65). This means that on the diagram $\tan \alpha = -9{\cdot}8$ or $\tan \beta = 9{\cdot}8$. Furthermore, from the results obtained in (ii) $v = 0$ when $t = 1\frac{3}{7}$ so that B on the t axis gives $OB = 1\frac{3}{7}$. The graph for the first 6 seconds of the motion is completed at C. We shall have more to say about the area under the line graph on page 263. Answer

Exercise 17.3

1. Differentiate with respect to t, each of the following expressions for a displacement s in terms of the time t and find the values of t for which the velocity is zero:
   (i) $s = 2t^3 + t^2 - 1$ (ii) $s = 5t - t^2$ (iii) $s = 3t + 1$
2. Which of the following equations in velocity v, displacement s and time t give motion with a uniform velocity?
   (i) $s = 16 + 3t$ (ii) $s = t^2 - 2t$
   (iii) $v = 4t$ (iv) $v = 6$
3. A body moves with velocity v in a straight line and its distance from a point O in the line is s.
   (i) If $\frac{dv}{dt} > 0$ what can you say about the motion?
   (ii) If $\frac{ds}{dt} > 0$ what can you say about the motion?
   (iii) If $\frac{ds}{dt} = 1\text{ m s}^{-1}$ for all t what is $\frac{dv}{dt}$?
4. A particle moves in a straight line with a velocity given by $v = 4 - 4t^2$. Find the acceleration and the times when the particle is momentarily at rest. Explain the meaning of the negative velocity when $t = 3$.
5. Starting from the equation $s = \frac{1}{2}at^2$ with s the distance fallen by a body in time t and a a constant
   (i) find v in terms of t and prove that the acceleration $\frac{dv}{dt} = a$

(ii) show that $v = s^{\frac{1}{2}}\sqrt{2a}$ and prove that $v\frac{dv}{ds} = a$

(iii) find the time when the velocity is 3.

6. The average velocity over any period of time δt is $\delta s/\delta t$. If $s = 16t^2$ metres what is the average velocity over the time from $t = 1$ to $t = 3$ (i.e. find $(s + \delta s - s) \div \delta t$)? What is the velocity at time $t = 1$ and at $t = 3$?

7. The distance s of a point P moving in a straight line through O where $OP = s$ is given by $s = 5t(8 - t)$. Find v in terms of t. (Take s in metres and t in seconds.) Calculate:

   (i) when the point P is back at O.
   (ii) the position of P when $t = 6$
   (iii) the time and position of P when it is momentarily at rest
   (iv) the greatest positive distance the point reaches from O
   (v) how far the point travels between $t = 1$ and $t = 6$

8. A car starts from a point O and moves in a straight line with a constant acceleration of magnitude 1 m s^{-2} for 10 seconds to reach a velocity v. For the next 20 seconds it moves at this constant velocity and in the following 30 seconds slows to rest with a constant retardation. Draw a velocity–time graph for the complete motion and find (i) v, (ii) the distance travelled at constant velocity, (iii) $\frac{dv}{dt}$ over the last 30 seconds.

9. A stone is projected vertically upwards with an initial speed of 24·5 m s^{-1}. The distance s metres from the point of projection is given by $s = 24{\cdot}5t - 4{\cdot}9t^2$ at time t seconds. Calculate:

   (i) the velocity v and the acceleration at any time t
   (ii) the greatest height reached by the stone

10. A body moves along a straight line so that its distance s from a point O is given by the equation $s = 4t^3 - 24t^2 + 45t$. Find:

    (i) the velocity when $t = 1, 2, 3$
    (ii) when and where the body is at an instantaneous rest

    Describe the motion.

11. A stone is thrown downwards on a paper parachute from a roof top 100 metres above ground and after t seconds it is s metres below, where $s = 6t + 1{\cdot}6t^2$. Find:

    (i) the height and velocity when $t = 3$
    (ii) when the stone hits the ground
    (iii) the acceleration

CHAPTER EIGHTEEN

MAXIMA AND MINIMA

Assumptions about $y = f(x)$

In this chapter we shall increase the scope of differentiation to gain the most interesting and powerful results in any elementary course in calculus, namely, the maximum and minimum values of a function. Although we are not at this stage in a position to appreciate the significance of some of our assumptions it is as well to bear in mind that all the functions we shall be dealing with are quite 'well behaved' for the places or values of x we are investigating.

For example, one value of x gives rise to one and only one value of y when we plot the graph of the function. There are no loops in graphs such as Fig. 86(*a*) where we would be in difficulty in obtaining the gradient of the tangent (which one?) at the point P. There are no gaps or missing points in graphs like Fig. 86(*b*) or Fig. 86(*c*). In all our work we shall assume that $y = f(x)$ has a derivative at the points of the intervals we are dealing with. Consequently we shall be excluding a graph like Fig. 86(*d*), which is $y =$ positive root of $\sqrt{x^2}$ or $y = |x|$, and restricting our attention to functions, which have a graph like Fig. 86(*e*) and have a derivative for all the points on the graph. We must, of course, still avoid those occasions when the function takes on infinite value, i.e. tends to infinity. For example, $y = f(x) = 1/x$ (the rectangular hyperbola in Fig. 86(*f*)) tends to plus infinity as $x \to 0$ from the right and minus infinity as $x \to 0$ from the left, so we shall have to exclude the point $x = 0$ from any differentiation of $f(x) = 1/x$.

Investigations of the difficulties involved in the cases mentioned above have to be reserved for more advanced courses in calculus. It is only necessary at this stage to be aware of the difficulties which can arise once we step outside the set of 'well behaved' functions we are dealing with. In the language of the calculus, functions such as the one shown in Fig. 86(*e*) would be described as continuous and differentiable (page 216) for all values of x in between A and B for which it has been drawn. Again, do not worry about the jargon, just capture the idea.

The Greatest or Least Value of the Quadratic $f(x) = ax^2 + bx + c$

The graph of any quadratic function takes the form of a parabola (page 36). We can get most of the information we need by looking at $y = x^2$ or $y = -x^2$ in Fig. 87. With $y = f(x) = x^2$ we get $\frac{dy}{dx} = f'(x) = 2x$ in Fig. 87(*a*).

The significance of $\frac{dy}{dx} = 2x$ is that $\frac{dy}{dx}$ is positive whenever x is positive, or to put it another way the gradient of the tangent is positive whenever x is positive. Whenever x is negative the gradient of the tangent is negative.

Consider moving along the curve from left to right. Starting at A we have $f'(x) < 0$, and $f(x)$ is decreasing until we get O. After O and on to B we

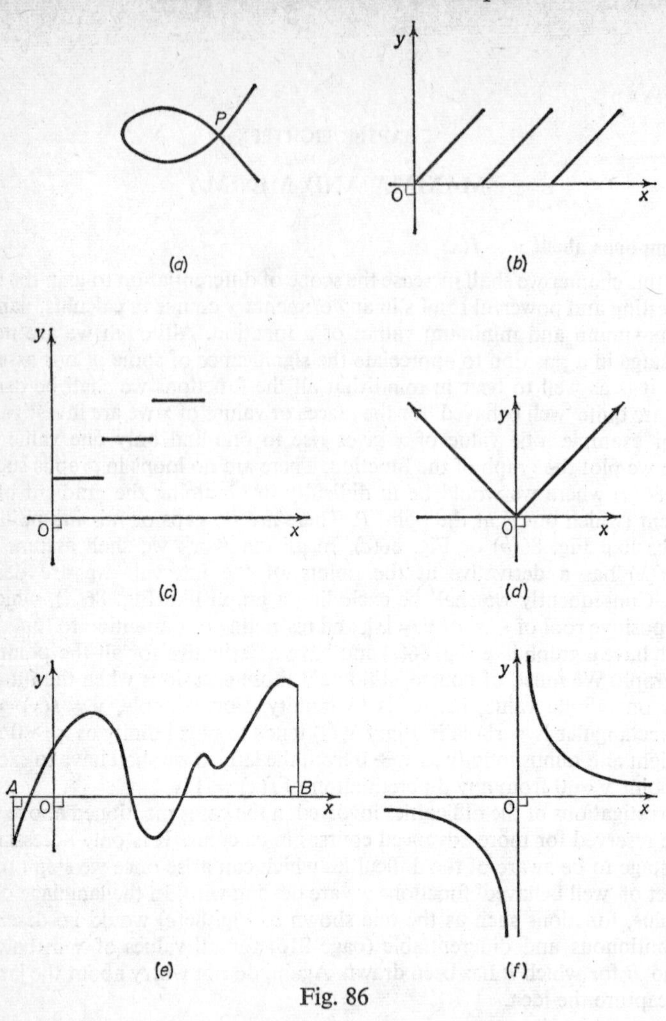

Fig. 86

have $f'(x) > 0$ and $f(x)$ is increasing. At $x = 0$ we have $f'(x) = 0$ and $f(x)$ takes its least value.

With $y = f(x) = -x^2$ in Fig. 87(*b*) we get $\frac{dy}{dx} = f'(x) = -2x$. Now we have $f'(x)$ positive for negative values of x and negative for positive values of x, or expressed in abbreviated form: $f'(x) > 0$ whenever $x < 0$; $f'(x) < 0$ whenever $x > 0$; and at $x = 0$ (the point at which $f'(x)$ is about to change sign) $f'(x) = 0$, as before. As the graph clearly shows $f(x)$ takes its greatest value at $x = 0$.

Putting all these observations together we see that what both functions or graphs had in common was $f'(x) = 0$ at their greatest or least values.

The difference between having a greatest or least value at a point is associated with the change in $f'(x)$ as the values of x proceed through the point;

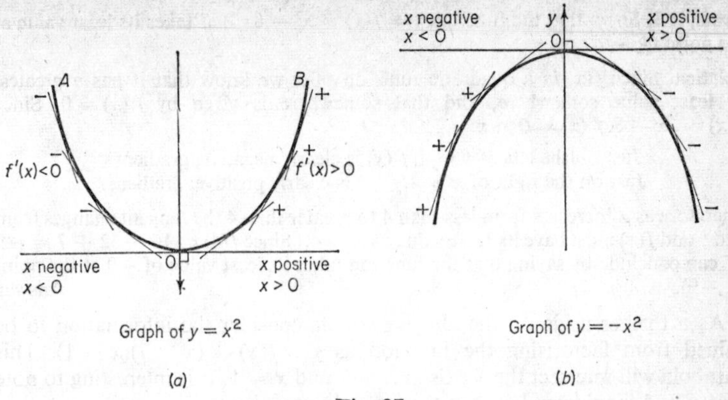

Fig. 87

that is, (i) if $f'(x)$ changes from positive to negative as x increases through the point then $f(x)$ has a greatest value at the point; (ii) if $f'(x)$ changes from negative to positive as x increases through the point then $f(x)$ has a least value at the point.

The graph of general quadratic $y = f(x) = ax^2 + bx + c$ is a similar shape to those in Fig. 87. If a is positive we have the form of Fig. 87(a) and if a is negative we have the form of Fig. 87(b). Fig. 88 shows two general positions for $y = ax^2 + bx + c$. As before, at both places of greatest or least value, we have $f'(x) = 0$. Moreover, as before at Q, where a least value is taken, we see that $f'(x)$ is negative on the left and positive on the right of Q. At P, where a greatest value is taken, we see that $f'(x)$ is positive on the left and negative on the right of P.

We now apply these findings to some examples.

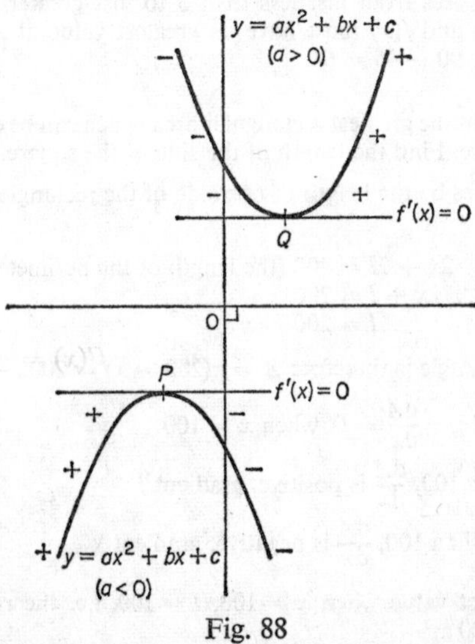

Fig. 88

Example 1. Show that the function $y = f(x) = x^2 - 8x + 7$ takes its least value at the point $(4, -9)$.

Solution. Since $f(x)$ is a quadratic function of x we know that it has a greatest or least value somewhere, and that somewhere is given by $f'(x) = 0$. Since $f'(x) = 2x - 8, f'(x) = 0$ at $x = 4$.

Just on the left of $x = 4, f'(x)$ is clearly negative, gradient \
Just on the right of $x = 4, f'(x)$ is clearly positive, gradient /

Therefore as x increases from less than 4 to greater than 4 the tangent changes from \ to / and $f(x)$ must have its least value at $x = 4$. Since $f(4) = 16 - 32 + 7 = -9$ we can conclude by saying that the function takes its least value of -9 at the point $(4, -9)$.

Answer

As a further aside on the above example consider the information to be gained from factorising the function as $y = f(x) = (x - 7)(x - 1)$. This parabola will intersect the x axis at $x = 7$ and $x = 1$. It is interesting to note that $x = 4$ is midway between these two intersections.

An alternative line of thought arises from completing the square (page 16), which means writing

$$\begin{aligned} y = f(x) &= x^2 - 8x + 16 - 9 \\ &= (x - 4)^2 - 9 \end{aligned}$$

Since $(x - 4)^2$ is a perfect square and its least value is 0 it follows that the least value of $f(x)$ is -9 when $x = 4$.

Example 2. Find the greatest value of $y = f(x) = 19 + 30x - 5x^2$.

Solution. $f'(x) = 30 - 10x$

$$\therefore \quad f'(x) = 0 \text{ when } x = 3.$$

Just on the left of $x = 3, f'(x)$ is positive, gradient /
Just on the right of $x = 3, f'(x)$ is negative, gradient \

Therefore as x increases from just less than 3 to just greater than 3 the tangent changes from / to \ and $f(x)$ must have its greatest value at $x = 3$. The greatest value is $f(3) = 19 + 90 - 45 = 64$.

Answer

Example 3. Prove that the greatest rectangular area which can be enclosed with 400 m of fencing is a square. Find the length of the side of the square.

Solution. Let x metres be the length of one side of the rectangle and l the length of the other side.

$$\begin{aligned} \therefore \quad 2x + 2l &= 400 \text{ (the length of the perimeter)} \\ \therefore \quad x + l &= 200 \\ l &= 200 - x \end{aligned}$$

The area of the rectangle is therefore $A = x(200 - x) = 200x - x^2$

With $\frac{dA}{dx} = 200 - 2x, \quad \frac{dA}{dx} = 0$, when $x = 100$

With x just less than 100, $\frac{dA}{dx}$ is positive, gradient /

With x just greater than 100, $\frac{dA}{dx}$ is negative, gradient \

$\therefore$ A has its greatest value when $x = 100$, $l = 100$, i.e. the rectangle becomes a square of area 10 000 m^2

Answer

Notice that in the example just completed we had A instead of y or $f(x)$, so that we write $\frac{dA}{dx}$ instead of $\frac{dy}{dx}$ or $f'(x)$.

Exercise 18.1

Find the greatest or least values of each of the following quadratic functions:

1. $x^2 + x + 1$
2. $x^2 - x + 1$
3. $-2x^2 + 8x - 1$
4. $16x^2 + 64x + 10$
5. $(x - 1)(x + 2)$
6. $(x + 2)(4 - x)$
7. A stone is projected vertically upwards so that its distance s metres from the point of projection at any time is given by $s = 30t - 5t^2$. Find the greatest height to which the stone will rise.
8. Find the greatest rectangular area which can be enclosed by 1 200 metres of fencing: (i) by using the fencing on all four sides; (ii) by using the fencing for three sides and an existing brick wall for the fourth side.
9. Show that all the parabolas given by $y = 3x^2 + 4x + c$ for different values of c have their least values at the same x.
10. If $A = 42u - 4 - 7u^2$ find the greatest value of A.

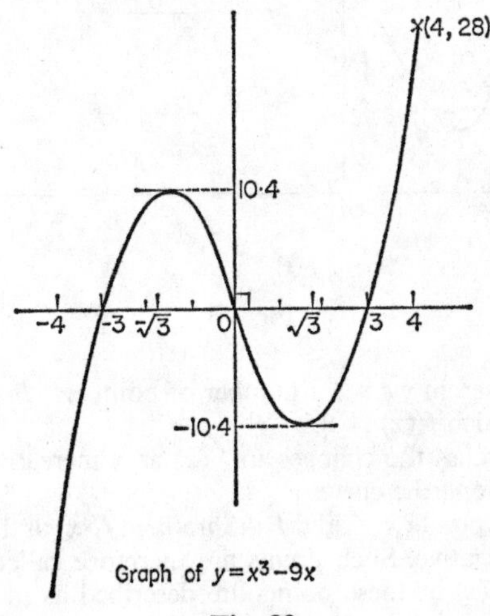

Fig. 89

Maxima and Minima

We chose quadratic functions in the last section because such functions of x have either one greatest or one least value at which $f'(x) = 0$. A function like $f(x) = x^3 - 9x$ has values of x for which $f'(x) = 0$ and yet such values of x will not necessarily give the greatest or least values of $f(x)$.

Fig. 89 shows a sketch of the graph of $y = f(x) = x^3 - 9x$.

$$f'(x) = 3x^2 - 9 = 3(x^2 - 3) = 0 \text{ when } x^2 = 3, \text{ i.e. } x = \pm\sqrt{3}.$$

When x is just less than $-\sqrt{3}$ then x^2 is just less than 3 and $f'(x)$ is negative. When x is just greater than $\sqrt{3}$ then x^2 is just greater than 3 and $f'(x)$ is positive.

It follows that $x = -\sqrt{3}$ gives? (a greatest value?),

$$f(-\sqrt{3}) = 6\sqrt{3} = 10{\cdot}4 \text{ (1 decimal place)}.$$

Similarly, $x = +\sqrt{3}$ gives? (a least value?), $f(\sqrt{3}) = -6\sqrt{3} = -10{\cdot}4$ (1decimal place).

The uncertainty is underlined by considering:
(i) $x = 4$ and $f(4) = 4^3 - 36 = 28$ (which is greater than $f(-\sqrt{3}) = 10{\cdot}4$); and (ii) $x = -4$ and $f(-4) = -4^3 + 36 = -28$ (which is less than $f(\sqrt{3}) = -10{\cdot}4$).

Clearly we can no longer conclude that $f'(x) = 0$ will give a greatest or least value. Faced with this difficulty, we invent the notion of maxima and minima. Thus we refer to $f(-\sqrt{3})$ as a maximum and $f(\sqrt{3})$ as a minimum of the function $f(x)$. Before defining these terms let us consider the more general

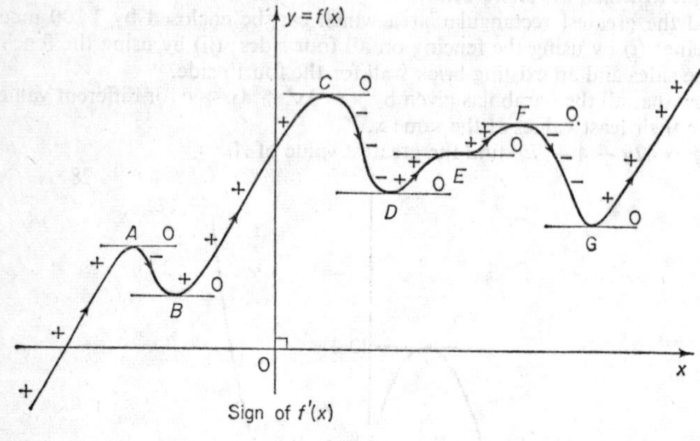

Fig. 90

curve of Fig. 90 wherein we see a number of points A, B, C, D, E, F and G on the curve for which $f'(x) = 0$.

As always we discuss the changes in $f'(x)$ as x increases, i.e. as we move from left to right along the curve.

(a) At each of the points A, C and F the gradient $f'(x)$ of the tangent changes from positive to negative. Such points are therefore called **maximum points** and the values of $f(x)$ at these points are described as maximum values or maxima of the function $f(x)$.

(b) At each of the points B, D and G the gradient $f'(x)$ changes from negative to positive. Such points are therefore called the **minimums point** and the values of $f(x)$ at these points are described as minimum values or minima of the function $f(x)$. At this moment do note that we can have a maximum (at A) which is less than a minimum (at D).

(c) At the point E the gradient is positive, then zero, then again positive. This change in $f'(x)$ does not satisfy the requirements for either a maximum or a minimum of $f(x)$. Such a point is called a **point of inflexion.** Any S-bend part of a curve indicates a point of inflexion and it is not necessary to have $f'(x) = 0$.

At any point of the curve $y = f(x)$ where $f'(x) = 0$ the function $f(x)$ is said to have a **stationary value.** Thus at all the points A, B, C, D, E, F, G, $f(x)$ has a stationary value.

Maximum or minimum points are called **turning points** and the values of $f(x)$ at these points are called **turning values.**

Again, do not bother to commit these descriptive phrases to memory at this stage. Be content to let the terms become familiar through working examples.

Example 4. Find the stationary values of $f(x) = x^3 + 3x^2 - 9x + 1$ and determine whether they are maxima or minima.

Solution. $f'(x) = 3x^2 + 6x - 9 = 3(x^2 + 2x - 3) = 3(x + 3)(x - 1)$

$\therefore \quad f'(x) = 0$ when $x = 1$ or -3

(i) $x = 1$. $f'(x)$ is negative on the left and positive on the right.
Try $x = 0{\cdot}9$ on the left. $(x + 3)(x - 1)$ is $(+)(-) = (-)$.
Try $x = 1{\cdot}1$ on the right $(x + 3)(x - 1)$ is $(+)(+) = (+)$

$\therefore\ f(1)$ is a minimum $\quad f(1) = 1 + 3 - 9 + 1 = -4$

(ii) $x = -3$. $f'(x)$ is positive on the left and negative on the right of $x = -3$.
Try $x = -3{\cdot}1$ on the left of $x = -3$, $(x + 3)(x - 1)$ is $(-)(-) = (+)$.
Try $x = -2{\cdot}9$ on the right of $x = -3$, $(x + 3)(x - 1)$ is $(+)(-) = (-)$

$\therefore\ f(-3)$ is a maximum $\qquad f(-3) = -27 + 27 + 27 + 1 = 28$

$\therefore$ Stationary values are a maximum of 28 at $x = -3$ and a minimum of -4 at $x = 1$

Answer

Example 5. Find the turning points on the graph of $y = 24x - 9x^2 - 2x^3$ and determine their type. (We use y instead of $f(x)$ merely as a change in notation, no other reason.)

Solution. $$\frac{dy}{dx} = 24 - 18x - 6x^2 = 6(4 - 3x - x^2) = 6(4 + x)(1 - x)$$

Thus $$\frac{dy}{dx} = 0 \text{ when } x = -4 \text{ or } 1$$

(i) To the left of $x = -4$, $\frac{dy}{dx}$ is negative. To the right of $x = -4$, $\frac{dy}{dx}$ is positive.

Try $x = -4{\cdot}1$ on the left of $x = -4$, $(4 + x)(1 - x)$ is $(-)(+) = (-)$
Try $x = -3{\cdot}9$ on the right of $x = -4$, $(4 + x)(1 - x)$ is $(+)(+) = (+)$.
Hence y has a minimum value at $x = -4$. This value is

$$y = -96 - 144 + 128 = -112$$

(ii) To the left of $x = 1$, try $x = 0{\cdot}9$, $(4 + x)(1 - x)$ is $(+)(+) = (+)$

$$\therefore \quad \frac{dy}{dx} \text{ is positive.}$$

To the right of $x = 1$, try $x = 1{\cdot}1$, $(4 + x)(1 - x)$ is $(+)(-) = (-)$

$$\frac{dy}{dx} \text{ is negative.}$$

Hence y has a maximum value at $x = 1$. This value is $y = 24 - 9 - 2 = 13$.
The turning points are a maximum at $(1, 13)$ and a minimum at $(-4, -112)$.

Answer

Example 6. Find the turning values of $f(x) = x^4 - 8x^3 + 22x^2 - 24x$.

Solution. We are being asked to find the maximum and minimum values of $f(x)$ and also the values of x from which they arise. This means that we must start by finding the solution to $f'(x) = 0$

$$f'(x) = 4x^3 - 24x^2 + 44x - 24$$
$$= 4(x^3 - 6x^2 + 11x - 6)$$

(We have deliberately chosen this expression.) The remainder theorem (page 28) tells us that $x - 1$ is a factor of $x^3 - 6x^2 + 11x - 6$

$$\therefore \quad x^3 - 6x^2 + 11x - 6 = (x - 1)(x^2 - 5x + 6)$$
$$= (x - 1)(x - 2)(x - 3)$$
$$\therefore \quad f'(x) = 0 \text{ when } x = 1, 2 \text{ and } 3.$$

With $f'(x) = 4(x - 1)(x - 2)(x - 3)$

(i) We see that $f'(x) = (-)(-)(-)$ is negative just on the left of $x = 1$, and positive $(+)(-)(-)$ just on the right

$$\therefore \quad f(1) \text{ is a minimum} \qquad f(1) = 1 - 8 + 22 - 24 = -9$$

(ii) $f'(x) = (+)(-)(-)$ is positive just to the left of $x = 2$ and negative $(+)(+)(-)$ just to the right of $x = 2$

$$\therefore \quad f(2) \text{ is a maximum} \qquad f(2) = 16 - 64 + 88 - 48 = -8$$

(iii) $f'(x) = (+)(+)(-)$ is negative just to the left of $x = 3$ and positive $(+)(+)(+)$ just to the right of $x - 3$

$$\therefore \quad f(3) \text{ is a minimum} \qquad f(3) = 81 - 216 + 198 - 72 = -9$$

The turning values are minimum -9, maximum -8, minimum -9 at $x = 1, 2, 3$ respectively.

The turning points are $(1, -9)$, $(2, -8)$, $(3, -9)$ Answer

Exercise 18.2

Find the stationary values of each of the following functions 1 to 7:

1. x^3
2. $x^3 - 1$
3. $2x^3 - 3x^2$
4. $x^3 - 3x + 7$
5. $12x - x^3 - 24$
6. $x + \frac{1}{x}$, $(x \neq 0)$
7. $\frac{1}{2}x^4 - x^2$

Find the coordinates of the turning points on the following curves 8 to 11.

8. $y = 3x^3 - 9x + 7$
9. $y = x^3 - 3x^2 + 3x + 4$
10. $y = x^2 - x - \frac{1}{3}(x^3 + 5)$
11. $y = \frac{x^4}{4} + x^3 + \frac{3x^2}{2} + x$
12. Fig. 91 represents a rectangular card from which squares of side x have been cut in each corner. The card is then folded along the dotted lines to form an open topped box of volume V. If the size of the original rectangle had been 16 cm by 10 cm show that $V = (16 - 2x)(10 - 2x)x$. Find the maximum value of V.

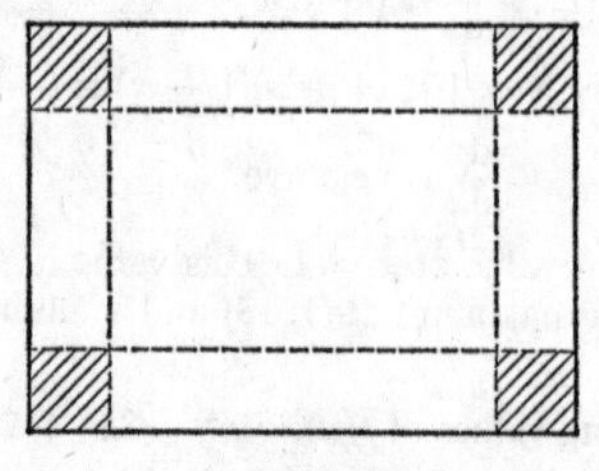

Fig. 91

13. If the size of the original rectangle in Question 12 had been 24 cm by 9 cm what would have been the maximum possible volume V?
14. An open-topped rectangular water tank of height 1 m is to have a volume of

2 m^3. What are the dimensions of the tank if the total surface area is a minimum. (Hint: length and breadth of the base are x and $\frac{2}{x}$.)

15. The volume of a right circular cylinder tin is $V = \pi r^2 h$ and the surface area of this cylinder is $A = 2\pi r^2 + 2\pi rh$, where r is the radius of the end faces and h the height
    (i) Show that $A = 2\pi r^2 + \frac{2V}{r}$
    (ii) With V kept constant find $\frac{dA}{dr}$
    (iii) Show that $h = 2r$ gives a minimum for A for a constant V.
    (iv) Find the radius and height of the tin which has a minimum area of metal for a volume of 269·5 cm^3.

The Second Derivative

Given a function of x like $5x^3 + 7x^2 - x + 2$ there is no reason why we should not differentiate the function more than once. For example, if

$$f(x) = 5x^3 + 7x^2 - x + 2$$

then

$$f'(x) = 15x^2 + 14x - 1$$

If we had started with $g(x) = 15x^2 + 14x - 1$ then clearly we could have obtained $g'(x) = 30x + 14$.
If we had started with $h(x) = 30x + 14$ then clearly we could have obtained $h'(x) = 30$.

The process of renaming each function as $g(x)$ and then $h(x)$ becomes tedious so we relate the results to the original function $f(x)$ which we started with.

We call $g'(x)$ the second derivative of $f(x)$ with respect to x and we write this as $f''(x)$ and read it as 'f, two dash of x' or 'f, double dash of x'.

We may go on to call $h'(x)$ the third derivative of $f(x)$ with respect to x and write this as $f'''(x)$, but we shall not be concerned with this order of differentiation.

Just as $f'(t)$ means the first derivative of $f(t)$ with respect to t so $f''(t)$ will mean the second derivative of $f(t)$ with respect to t.

Example 7. With $f(t) = t^4 + t^3 + t^2 + t + 1$ find $f''(t)$.

Solution. $f'(t) = 4t^3 + 3t^2 + 2t + 1$

$f''(t) = 12t^2 + 6t + 2$ Answer

Example 8. Find $f''(x)$ when $f(x) = 3x^3 + \frac{4}{x}$

Solution.

$$f'(x) = 9x^2 - \frac{4}{x^2}$$

$$f''(x) = 18x + \frac{8}{x^3}$$ Answer

Because we also have the notation $\frac{dy}{dx}$ to represent the first derivative of y with respect to x it follows that we need a similar form for the second derivative. This is written $\frac{d^2y}{dx^2}$ and read as 'd two, y, d, x squared', and as in the case for $\frac{dy}{dx}$ is used as one symbol so that d^2y is not to be separated from dx^2 and dx^2 does not mean $d \times x \times x$ and d^2y does not meand $\times$ d $\times$ y.

Example 9. Find $\frac{d^2y}{dx^2}$ when $y = 6x^3 - 7x$.

Solution.
$$\frac{dy}{dx} = 18x^2 - 7$$
$$\frac{d^2y}{dx^2} = 36x \quad \text{Answer}$$

Rates of Change and the Second Derivative

It will be remembered from page 58 that when the curve $y = f(x)$ has a positive gradient it indicates that y is increasing as x increases. Similarly, when the curve has a negative gradient y is decreasing as x increases. We subsequently removed our thoughts from the graphs or curves and referred only to the function $f(x)$ increasing when $f'(x)$ is positive and decreasing when $f'(x)$ is negative.

The same argument applies to any function $f(x)$ so consider

$$f(x) = \frac{x^3}{3} - \frac{3x^2}{2} + 2x \text{ and } g(x) = x^2 - 3x + 2 = (x - 2)(x - 1)$$

We shall consider $g(x)$ with special reference to $x = 1$ and $x = 2$. From $g'(x) = 2x - 3$ we have $g'(1) = -1$ and $g'(2) = 1$.

These last two results tell us that $g(x)$ decreases as x passes through $x = 1$ (because $g'(x)$ is negative) and increases as x passes through $x = 2$ (because $g'(x)$ is positive). (Recall that when we speak of 'x passing through $x = 1$' we mean x starts with values just to the left of $x = 1$, then proceeds left to right through $x = 1$.)

Now $f'(x) = g(x)$ and $f''(x) = g'(x)$, so what we have just said about $g(x)$ may be said about $f'(x)$ and we may use $f''(x)$ to say it. Thus $f'(x)$ is increasing when $f''(x) > 0$ and decreasing when $f''(x) < 0$. Looking back to Fig. 90 we notice that near each of the maximum points A, C, F the gradient of the curve changes from positive to negative as x passes through the point; in other words $f'(x)$ is decreasing. And if $f'(x)$ is decreasing then $f''(x) < 0$.

Again in Fig. 90, near each of the minimum points B, D, G the gradient of the curve changes from negative to positive; in other words $f'(x)$ is increasing. And if $f'(x)$ is increasing then $f''(x) > 0$.

We can summarise these findings as follows:

(i) If $f'(x) = 0$ for $x = a$, and $f''(a) < 0$ then $f(a)$ is a maximum.
(ii) If $f'(x) = 0$ for $x = b$, and $f''(b) > 0$ then $f(b)$ is a minimum.
(iii) If $f''(x) = 0$ for $x = c$, and $f''(c) = 0$ then we are unable to conclude that $f(c)$ is either a maximum or a minimum. It **may** be a point of inflexion, e.g. $f(x) = x^3$ has a point of inflexion at $x = 0$ but $f(x) = x^4$ has a minimum of $x = 0$ yet $f''(0) = 0$ in both cases.

Example 10. Find the maximum and minimum values of $f(x) = \frac{x^3}{3} - \frac{3x^2}{2} + 2x$.

Solution. $f'(x) = x^2 - 3x + 2 = (x - 2)(x-1)$
$f''(x) = 2x - 3$

The stationary values are given by $f'(x) = 0$, i.e. $x = 1$ or 2

Since $f''(1) = -1$ is negative, $f(1) = \frac{5}{6}$ is a maximum — Answer
Since $f''(2) = 1$ is positive, $f(2) = \frac{2}{3}$ is a minimum — Answer

Example 11. The same example as on page 239: find the turning points on the graph of $y = 24x - 9x^2 - 2x^3$

Solution.
$$\frac{dy}{dx} = 24 - 18x - 6x^2 = 6(4 + x)(1 - x)$$
$$\frac{d^2y}{dx^2} = -18 - 12x$$

The turning points are given by $\frac{dy}{dx} = 0$, $x = 1$ or -4.

When $x = -4$; $\frac{d^2y}{dx^2} = -18 + 48 = 30$, positive, so that y is a minimum for $x = -4$.

When $x = 1$; $\frac{d^2y}{dx^2} = -18 - 12 = -30$, negative, so that y is a maximum for $x = 1$.

The turning points are $(1, 13)$ and $(-4, -112)$ as before. Answer

Example 11. Find the stationary values of $f(x) = 3x^4 + 4x^3 - 30x^2 + 36x$.

Solution.
$$\begin{aligned} f'(x) &= 12x^3 + 12x^2 - 60x + 36 \\ &= 12(x^3 + x^2 - 5x + 3) \\ &= 12(x - 1)(x - 1)(x + 3) \quad \text{(by remainder theorem)} \\ f''(x) &= 36x^2 + 24x - 60 \\ &= 12(3x^2 + 2x - 5) \\ &= 12(3x + 5)(x - 1) \end{aligned}$$

The stationary values arise from $f'(x) = 0$, i.e. $x = 1$, $x = -3$. We decide whether the values are a maximum or a minimum by examining the sign of $f''(1)$ and $f''(-3)$.

$f''(-3) = 12(-4)(-4)$, is positive. Hence $f(-3)$ is a minimum

$f''(1) = 0$, and we cannot conclude whether $f(1)$ is a maximum or a minimum.

In this case we examine $f'(x) = 12(x - 1)^2(x + 3)$ for values of x on either side of $x = 1$. On the left of $x = 1$, $f'(x)$ is positive since $(x - 1)^2$ is a perfect square and $x + 3$ is positive. On the right of $x = 1$, $f'(x)$ is positive since $(x - 1)^2$ is a perfect square and $x + 3$ is positive. Hence $f(x)$ has a point of inflexion at $x = 1$, just like the point E on Fig. 90.

Exercise 18.3

Find the stationary values of the following functions:

1. $3x^4 - 4x^3 + 1$
2. x^3
3. $(x - 1)^3$
4. x^4
5. $\frac{x^3}{3} + 3x^2 - 40x + 160$
6. The sum of the two shorter sides of a right-angled triangle is 20 cm. Find the maximum area of the triangle.
7. Prove that when $a + b = 10$, the minimum value of $a^2 + b^2$ is 50.
8. A right circular cylinder has a radius r and a height h. Given that $h + r = 30$ cm find the maximum volume of the cylinder.
9. Two numbers have a sum of $2n$. Find the two numbers such that the sum of their cubes is a minimum.
10. A body moves in a straight line so that its distance from a point O on the line is given by $s = t^3 - 12t^2 + 45t - 450$. Find the times for the maximum and minimum values of s. Is the maximum for s the greatest distance that the body gets from O?

CHAPTER NINETEEN

FURTHER METHODS OF DIFFERENTIATION

In this chapter we shall discuss the differentiation of (i) $\sin x$ and $\cos x$, (ii) a function of a function and (iii) products of functions. The last two sections will increase the application of our work quite considerably as we shall see.

In the differentiation of $\sin x$ from first principles we shall need to use a particular limit result of considerable importance so we start by considering this limit separately.

The Limit of $\frac{\sin \theta}{\theta}$ as θ Tends to 0

Consider the circle of unit radius and centre O in Fig. 92 with angle $AOP = \theta$ radians, angle $PNO = 90°$, angle $OPB = 90°$. Relying upon the eye for the length comparisons, we assert that the following inequality of lengths will be true

$$NP < \text{arc } AP < BP \qquad \text{(i)}$$

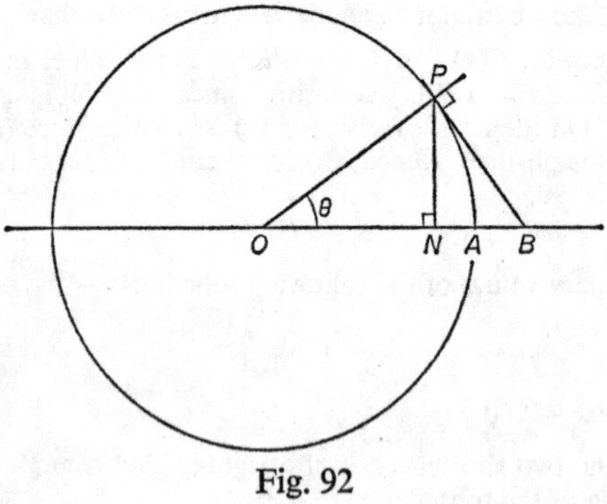

Fig. 92

In the right-angled triangle ONP we have $NP = OP \sin \theta = \sin \theta$ since $OP = 1$. Taking the angle θ to be measured in radians means that the length of the arc $AP = \theta$ (see page 157). From the right-angled triangle OPB we have $BP = OP \tan \theta = \tan \theta$ since $OP = 1$.

Collecting these results together in the inequality (i) above we have

$$\sin \theta < \theta < \tan \theta$$

Dividing by $\sin \theta$ which is positive

$$1 < \frac{\theta}{\sin \theta} < \frac{\tan \theta}{\sin \theta}$$

Since all the quantities are positive we may invert each of them to obtain

$$1 > \frac{\sin\theta}{\theta} > \frac{\sin\theta}{\tan\theta}$$

$$\therefore 1 > \frac{\sin\theta}{\theta} > \cos\theta, \quad \left(\frac{\sin\theta}{\tan\theta} = \cos\theta\right)$$

This shows that for all acute angles θ, $\frac{\sin\theta}{\theta}$ lies between 1 and $\cos\theta$.

We now let θ tend to 0 so as to make $\cos\theta$ get as close to $\cos\theta = 1$ as we please. It follows therefore that as $\cos\theta$ gets closer to 1 so $\frac{\sin\theta}{\theta}$ which is in between 1 and $\cos\theta$—must also get closer to 1, indeed as close to 1 as we please.

To put it more formally we have

(a) $\cos\theta \to 1$ as $\theta \to 0$

(b) Since $1 > \frac{\sin\theta}{\theta} > \cos\theta$ we deduce that

$$\frac{\sin\theta}{\theta} \to 1 \text{ as } \theta \to 0$$

(c) If θ is small and negative then (i) $\sin\theta$ is negative, so $\frac{\sin\theta}{\theta}$ remains positive; (ii) $\cos\theta$ is positive.

Hence the inequality in (b) will still be true for a negative acute angle and $\frac{\sin\theta}{\theta} \to 1$ as θ tends to 0 from the left or the right, i.e. through positive or negative values.

(d) The results just obtained may now be written

$$\lim_{\theta\to 0} \frac{\sin\theta}{\theta} = 1$$

(e) Other forms in which this last result may appear are

$$\lim_{\delta x\to 0} \frac{\sin(\delta x)}{\delta x} = 1, \quad \lim_{h\to 0} \frac{\sin(h)}{h} = 1, \quad \lim_{h\to 0} \frac{\sin(\frac{1}{2}h)}{\frac{1}{2}h} = 1$$

In each case it must be remembered that θ, δx, h are measured in radians because $\lim_{\theta\to 0} \frac{\sin\theta}{\theta}$, where θ is measured in degrees, is equal to $\frac{\pi}{180}$ and not 1.

The Derivative of $f(x) = \sin x$

We recall that the definition of the derivative is $\lim_{h\to 0} \left(\frac{f(x+h) - f(x)}{h}\right)$.

With $f(x) = \sin x$ we have $f(x+h) = \sin(x+h)$

$$\therefore f(x+h) - f(x) = \sin(x+h) - \sin x$$
$$= 2\cos(x + \tfrac{1}{2}h)\sin\tfrac{1}{2}h$$

(page 179, $\frac{1}{2}(x + h + x) = x + \frac{1}{2}h$)

$$\frac{f(x+h) - f(x)}{h} = \cos(x + \tfrac{1}{2}h)\frac{\sin(\frac{1}{2}h)}{\frac{1}{2}h}$$

$$\therefore \lim_{h\to 0}\frac{f(x+h) - f(x)}{h} = \lim_{h\to 0}\cos(x + \tfrac{1}{2}h)\frac{\sin(\frac{1}{2}h)}{\frac{1}{2}h}$$

$$= \lim_{h\to 0}\cos(x + \tfrac{1}{2}h)\lim_{h\to 0}\frac{\sin(\frac{1}{2}h)}{\frac{1}{2}h}$$

Notice that we have assumed that the limit of the product is equal to the product of the limits in this last statement. It appears to be intuitively obvious but nevertheless needs to be proved.

Unfortunately (or fortunately according to your point of view), an elementary treatment does not contain such work, consequently the 'proofs' offered at this level must sometimes make one or two assumptions like this last statement.

As $h \to 0$, $x + \frac{1}{2}h \to x$ and $\cos(x + \frac{1}{2}h) \to \cos x$

As $h \to 0$, $\lim_{h\to 0}\frac{\sin(\frac{1}{2}h)}{\frac{1}{2}h} = 1$

$$\therefore \lim_{h\to 0}\frac{f(x+h) - f(x)}{h} = \cos x$$

and since this limit exists we agree to write it as the derived function

$$f'(x) = \cos x, \text{ when } f(x) = \sin x$$

Alternatively we may write

$$\frac{d(\sin x)}{dx} = \cos x, \ (x \text{ measured in radians})$$

or, with $y = \sin x$ $\qquad \frac{dy}{dx} = \cos x$

The Derivative of $f(x) = \cos x$

Similarly, we may obtain the derivative of $\cos x$ as $\frac{d(\cos x)}{dx} = -\sin x$. Briefly

$$f(x+h) - f(x) = \cos(x+h) - \cos x$$
$$= -2\sin(x + \tfrac{1}{2}h)\sin(\tfrac{1}{2}h) \qquad \text{(page 179)}$$

$$\therefore \frac{f(x+h) - f(x)}{h} = -\sin(x + \tfrac{1}{2}h)\frac{\sin(\frac{1}{2}h)}{\frac{1}{2}h}$$

$$\therefore \lim_{h\to 0}\frac{f(x+h) - f(x)}{h} = \lim_{h\to 0} -\sin(x + \tfrac{1}{2}h)\frac{(\sin\frac{1}{2}h)}{\frac{1}{2}h}$$

$$= \lim_{h\to 0} -\sin(x + \tfrac{1}{2}h)\lim_{h\to 0}\frac{\sin(\frac{1}{2}h)}{\frac{1}{2}h}$$

We now leave the reader to fill in the missing parts of the final conclusions.

As $h \to 0$, $x + \frac{1}{2}h \to$? and $-\sin(x + \frac{1}{2}h) \to$?

As $h \to 0$, $\lim_{h \to 0} \dfrac{\sin(\frac{1}{2}h)}{\frac{1}{2}h} = ?$

$$\therefore \lim_{h \to 0} \frac{f(x+h) - f(x)}{h} = ?$$

$$\therefore f'(x) = ? \text{ when } f(x) = \cos x$$

Both the above proofs involve ideas which are not easy to absorb at a first reading. Again you are advised to concentrate on the final results and reread the proofs at a later stage.

Example 1. Find the second derivative of (i) $f(x) = \sin x$, and (ii) $g(x) = \cos x$.

Solution. As we have just proved (with certain assumptions), if $f(x) = \sin x$ then

$$f'(x) = \cos x$$

$$\therefore \quad f''(x) = \frac{d(\cos x)}{dx} = -\sin x \quad \text{(from above)}$$

$\therefore$ The second derivative of $f(x) = \sin x$ is $f''(x) = -\sin x$ **Answer**

With $g(x) = \cos x$ we get $g'(x) = -\sin x$

$$\therefore \quad g''(x) = \frac{d(-\sin x)}{dx} = -\frac{d(\sin x)}{dx} = -\cos x$$

$\therefore$ The second derivative of $g(x) = \cos x$ is $g''(x) = -\cos x$ **Answer**

Alternatively with $y = \sin x$, we write

$$\frac{dy}{dx} = \cos x \text{ and } \frac{d^2y}{dx^2} = -\sin x$$

Example 2. Find the maximum value of $f(x) = 12 \sin x + 5 \cos x$ for $0 \leqslant x \leqslant \pi/2$.

Solution.

$$f'(x) = 12 \cos x - 5 \sin x$$
$$f''(x) = -12 \sin x - 5 \cos x$$

The maximum value of $f(x)$ will be given by $f'(x) = 12 \cos x - 5 \sin x = 0$.

$$\therefore \quad 12 \cos x = 5 \sin x$$

i.e.

$$\tan x = \tfrac{12}{5}$$

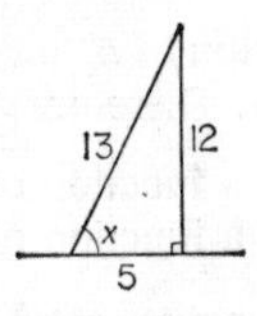

As the adjoining right-angled triangle shows, when $\tan x = \frac{12}{5}$ we have $\sin x = \frac{12}{13}$ and $\cos x = \frac{5}{13}$; $0 \leqslant x \leqslant \pi/2$. We now consider the sign of $f''(x)$ when $\tan x = \frac{12}{5}$.

$f''(x) = -12 \sin x - 5 \cos x$ is clearly negative for acute angles x and therefore indicates that $f(x)$ has a maximum value for x given by $\tan x = \frac{12}{5}$; $0 \leqslant x \leqslant \pi/2$. This maximum is

$$12 \times \frac{12}{13} + 5 \times \frac{5}{13} = \frac{144 + 25}{13} = \frac{169}{13} = 13$$ **Answer**

Compare the treatment above with that on page 176 for similar problems.

Exercise 19.1

1. With $y = x + \sin x$ show that $\frac{d^2y}{dx^2} + y = x$.

2. With $y = 3 \sin x + \cos x$ show that $\frac{d^2y}{dx^2} + y = 0$.

3. With $y = x - \sin x + \cos x$ show that $\frac{d^2y}{dx^2} + y = x$.

Find the maximum or minimum values of each of the functions in questions 4–9 with the restriction that $0 \leqslant x \leqslant \pi$:

4. $\sin x + \cos x$, $\left(\tan \frac{\pi}{4} = 1, \sin \frac{\pi}{4} = \cos \frac{\pi}{4} = \frac{1}{\sqrt{2}}\right)$
5. $3 \sin x + 4 \cos x$, (use a 3, 4, 5 triangle)
6. $4 \sin x + 3 \cos x$
7. $\sin x - \cos x$, $\left(\tan \frac{3\pi}{4} = -1\right)$
8. $x - 2 \sin x$, $\left(\cos \frac{\pi}{3} = 0{\cdot}5\right)$
9. $x + 2 \cos x$, $\left(\sin \frac{\pi}{6} = 0{\cdot}5, \sin \frac{5\pi}{6} = 0{\cdot}5\right)$
10. Prove from first principles that the derivative of $f(x) = \sin 2x$ is $2 \cos 2x$.
11. Find the maximum and minimum values of $f(x) = x - \sin 2x$.

Function of a Function

Examples of $y = f(u)$, i.e. y is a function of u, are $y = u^2$, $y = u^3$, $y = u^2 + u^3$, $y = \sin u$, $y = u - 4 \sin u + 1/u$, and so on for an endless list of similar examples. Suppose we now consider that u is a function of x, i.e. $u = g(x)$. Examples of this suggestion are $u = 1 + x^2$, $u = x^5$, $u = \sin x + \cos x$ and so on. Again the list of possibilities is endless.

Consider the following suggestion

$$y = f(u) = u^2 \text{ and } u = g(x) = 1 + 3x \text{ (say)}$$

Then we may relate y to x by the substitution

$$y = u^2 = (1 + 3x)^2$$

so that we start with y as a function of u, and because u is a function of x we may express y as a function of x. Therefore

(i) y is a function of u
(ii) u is a function of x

Hence y is a function of (a function of x), and this is how the expression function of a function arises. Here we are interested in differentiating y with respect to x, i.e. differentiating a function of a function.

We already know how to find $\frac{dy}{du}$ and $\frac{du}{dx}$, what we want to find is $\frac{dy}{dx}$. The illustration $y = u^2$, $u = 1 + 3x$ shows that it is possible to find $\frac{dy}{dx}$ because we obtained

$$y = (1 + 3x)^2 = 1 + 6x + 9x^2$$

So we may find

$$\frac{dy}{dx} = 6 + 18x = 6(1 + 3x)$$

We are more interested in learning something from the fact that $\frac{dy}{du} = 2u = 2(1 + 3x)$; $\frac{du}{dx} = 3$ so that with $\frac{dy}{dx} = 6(1 + 3x)$ it would appear that $\frac{dy}{dx} = \frac{dy}{du}\cdot\frac{du}{dx}$. Of course, one example does not constitute a proof but it does indicate a possibility worth investigating.

Example 3. With $y = \sin u$ and $u = 2x$ show that $\frac{dy}{dx} = \frac{dy}{du}\cdot\frac{du}{dx}$.

Solution. In the last exercise we deduced that $\frac{d(\sin 2x)}{dx} = 2\cos 2x$ from first principles, so we already know that $\frac{dy}{dx} = 2\cos 2x$.

With $y = \sin u$, $\frac{dy}{du} = \cos u = \cos 2x$.

With $u = 2x$, $\frac{du}{dx} = 2$

Hence $$\frac{dy}{du}\cdot\frac{du}{dx} = 2\cos 2x = \frac{dy}{dx} \qquad \text{Q.E.D.}$$

Example 4. With $y = u^3$ and $u = x^2 + 1$ confirm that $\frac{dy}{dx} = \frac{dy}{du}\cdot\frac{du}{dx}$.

Solution. We have $y = u^3 = (x^2 + 1)^3$ which we can expand binomially as

$$y = x^6 + 3x^4 + 3x^2 + 1$$

$$\therefore \quad \frac{dy}{dx} = 6x^5 + 12x^3 + 6x = 6x(x^4 + 2x^2 + 1) = 6x(x^2 + 1)^2$$

With $y = u^3$ we have $\frac{dy}{du} = 3u^2 = 3(x^2 + 1)^2$

With $u = x^2 + 1$ we have $\frac{du}{dx} = 2x$

Hence $$\frac{dy}{du}\cdot\frac{du}{dx} = 6x(x^2 + 1)^2 = \frac{dy}{dx} \qquad \text{Q.E.D.}$$

Exercise 19.2

Confirm that $\frac{dy}{dx} = \frac{dy}{du}\cdot\frac{du}{dx}$ in each of the following examples:

1. $y = u^2$, $u = x^3 + 1$
2. $y = u^3$, $u = 2x + 1$
3. $y = 6 + u^2$, $u = x^3 + 1$
4. $y = 6 + 5u^3$, $u = x - 1$
5. $y = \frac{1}{u}$, $u = x^2$
6. $y = u^2$, $u = 1 - 5x^3$
7. $y = u^5$, $u = x^{\frac{1}{2}}$

We can explain the meaning of the results in differentiating a function of a function above in the following terms. Suppose we have three cars A, B and C in a race so that at a particular time A is travelling three times as fast as B and B is travelling four times as fast as C. It follows that A is travelling 12 times as fast as C. A numerical identification may make this clearer. If C is travelling at 1 m s^{-1} then B is travelling at 4 m s^{-1}. Since A is travelling three times as fast as B then A must be travelling at 12 m s^{-1}. Therefore A is travelling 12 times as fast as C. We now generalise this to the following:

if y increases p times as fast as u and
u increases q times as fast as x, then
y increases pq times as fast as x

Returning to the representation of y as a function of u and u as a function of x the above may now be thought of as

(i) $$\frac{dy}{du} = p$$

(ii) $$\frac{du}{dx} = q$$

(iii) $$\frac{dy}{dx} = pq$$

Thus $$\frac{dy}{dx} = \frac{dy}{du} \cdot \frac{du}{dx}$$

Thinking in the language of increments we may write

$$\frac{\delta y}{\delta x} = \frac{\delta y}{\delta u} \frac{\delta u}{\delta x}$$

assuming that the increment δx does not make $\delta u = 0$.

Assuming again the result about the limit of a product being equal to the product of the limits as $\delta x \to 0$ (as we did on page 246), this last result gives

$$\frac{dy}{dx} = \frac{dy}{du} \cdot \frac{du}{dx}$$

We have widened the range of application of differentiation enormously by this theorem.

Example 5. Find $\frac{dy}{dx}$ for $y = (3x + 5)^{10}$.

Solution. Before we obtained the above result this problem would have been done by expanding $(3x + 5)^{10}$ binomially and then differentiating each term separately. Here we put $u = 3x + 5$ and thereby $y = u^{10}$.

$$\therefore \quad \frac{dy}{dx} = \frac{dy}{du} \frac{du}{dx} = 10u^9 \,.\, 3 = 30\,(3x + 5)^9 \qquad \text{Answer}$$

Example 6. Find $\frac{dy}{dx}$ for $y = \sqrt{(5x^2 + 3)}$.

Solution. Put $u = 5x^2 + 3$ so that $y = u^{\frac{1}{2}}$

$$\therefore \quad \frac{dy}{du} = \tfrac{1}{2}u^{-\frac{1}{2}} = \frac{1}{2u^{\frac{1}{2}}} = \frac{1}{2\sqrt{(5x^2 + 3)}}, \quad \frac{du}{dx} = 10x$$

$$\therefore \quad \frac{dy}{dx} = \frac{dy}{du} \cdot \frac{du}{dx} = \frac{1}{2\sqrt{(5x^2 + 3)}} \times 10x = \frac{5x}{\sqrt{(5x^2 + 3)}} \qquad \text{Answer}$$

Example 7. Find $\frac{dy}{dx}$ when $y = \sin(3x^2 + 1)$.

Solution. Put $u = 3x^2 + 1$, and thereby $y = \sin u$

$$\frac{dy}{du} = \cos u = \cos(3x^2 + 1), \frac{du}{dx} = 6x$$

$$\therefore \quad \frac{dy}{dx} = \frac{dy}{du} \cdot \frac{du}{dx} = 6x \cos(3x^2 + 1). \qquad \text{Answer}$$

Example 8. Find $\frac{dy}{dx}$ when (i) $y = \sin^2 x$

(ii) $y = \cos^3 x$

Solution.

(i) Put $u = \sin x$ so that $y = u^2$

$$\therefore \quad \frac{dy}{du} = 2u = 2 \sin x \text{ and } \frac{du}{dx} = \cos x$$

$$\therefore \quad \frac{dy}{dx} = \frac{dy}{du} \cdot \frac{du}{dx} = 2 \sin x \cos x \qquad \text{Answer}$$

(ii) Put $v = \cos x$ so that $y = v^3$

$$\therefore \quad \frac{dy}{dv} = 3v^2 = 3 \cos^2 x \text{ and } \frac{dv}{dx} = -\sin x$$

$$\therefore \quad \frac{dy}{dx} = -3 \cos^2 x \sin x \qquad \text{Answer}$$

Exercise 19.3

Find $\frac{dy}{dx}$ in each of the examples 1 to 13:

1. $(x^2 + 1)^{10}$
2. $(3x^2 + 4)^5$
3. $(4 - x^2)^3$
4. $\frac{1}{(x + 1)^4}$; $(x \neq -1)$
5. $\sin (x^2)$
6. $(x^2 + x + 1)^5$
7. $\frac{1}{x^2 + 1}$
8. $\frac{1}{x^3 + 1}$; $(x \neq -1)$
9. $\sqrt{(x + 1)}$
10. $\sqrt{(x^2 + 1)}$
11. $\cos^2 x$
12. $\sin^3 x$
13. $\sin^2 x + \cos^2 x$
14. Find the maximum and minimum values of:

(i) $x + \frac{1}{x - 1}$; $(x \neq 1)$

(ii) $(x + 1)^3 + \frac{3}{(x + 1)}$; $(x \neq -1)$

Draw a sketch of both these curves to show why the maximum is less than the minimum.

Differentiation of a Product

Suppose that $y = f(x) . g(x)$, the product of two functions of x. A change from x to $x + \delta x$ will yield the usual change of y to $y + \delta y$.

$$\therefore y + \delta y = f(x + \delta x) . g(x + \delta x)$$

For convenience we now represent $f(x + \delta x)$ as $f + \delta f$ and $g(x + \delta x)$ as $g + \delta g$. Similarly we write $y = fg$

$$\therefore y + \delta y = (f + \delta f)(g + \delta g) = fg + f\delta g + g\,\delta f + \delta f \times \delta g$$

Hence $\qquad \delta y = f\,\delta g + g\,\delta f + \delta f . \delta g$

Dividing by δx we get

$$\frac{\delta y}{\delta x} = f\frac{\delta g}{\delta x} + g\frac{\delta f}{\delta x} + \frac{\delta f}{\delta x}\delta g \qquad \text{(i)}$$

We now let δx tend to zero, so that

$$\frac{\delta y}{\delta x} \to \frac{dy}{dx}, \quad \frac{\delta g}{\delta x} \to \frac{dg}{dx}, \quad \frac{\delta f}{\delta x} \to \frac{df}{dx}$$

For $\frac{\delta f}{\delta x} \times \delta g$ as $\frac{\delta f}{\delta x} \to \frac{df}{dx}$ it is multiplied by δg which tends to zero,

$\therefore \frac{\delta f}{\delta x} \times \delta g \to 0.$

As $\delta x \to 0$ (i) becomes

$$\frac{dy}{dx} = \frac{d(fg)}{dx} = f\frac{dg}{dx} + g\frac{df}{dx}$$

which shows that we can differentiate a product by keeping each factor constant in turn and multiplying it by the derivative of the other factor and adding the two results.

Example 9. With $y = (3x + 1)(2x + 11)$ find $\frac{dy}{dx}$.

Solution. Comparing with the result just found consider $y = fg$ where $f(x) = (3x + 1)$ and $g(x) = (2x + 11)$.

$$\therefore \quad f'(x) = 3, \ g'(x) = 2$$

Hence

$$\frac{dy}{dx} = 3(2x + 11) + 2(3x + 1) = 12x + 35$$

Alternatively $y = 6x^2 + 35x + 11$ whence $\frac{dy}{dx} = 12x + 35$ Answer

Example 10. Find the derivative of $y = (3x^2 + 1)(2x + 11)$ with respect to x.

Solution.
$$\begin{aligned} \frac{dy}{dx} &= (2x + 11)\frac{d(3x^2 + 1)}{dx} + (3x^2 + 1)\frac{d(2x + 11)}{dx} \\ &= (2x + 11)\,6x + (3x^2 + 1)2 \\ &= 12x^2 + 66x + 6x^2 + 2 \end{aligned}$$

$\therefore \quad \frac{dy}{dx} = 18x^2 + 66x + 2$ Answer

Example 11. Differentiate the product $y = (3x^2 - 5)(x + 1)^2$.

Solution. Put $f(x) = 3x^2 - 5$ and $g(x) = (x + 1)^2$

$\therefore \quad f'(x) = 6x$ and $g'(x) = 2(x + 1)$

$\therefore \quad \frac{dy}{dx} = g\frac{df}{dx} + f\frac{dg}{dx} = (x + 1)^2 6x + (3x^2 - 5)2(x + 1)$

$\therefore \quad \frac{dy}{dx} = 12x^3 + 18x^2 - 4x - 10$ Answer

Example 12. Find the derivative of $y = x \sin x$.

Solution. $\frac{dy}{dx} = x\frac{d(\sin x)}{dx} + (\sin x)\frac{dx}{dx}$ (i.e. $f(x) = x$, $g(x) = \sin x$)

$= x \cos x + \sin x$ Answer

Example 13. Differentiate the product $y = (3x^2 - 5)(x^3 + 1)^2$.

Solution. Put $f(x) = 3x^2 - 5$ and $g(x) = (x^3 + 1)^2$

$$\therefore \quad f'(x) = 6x$$

To find $g'(x)$ we shall suggest $u = x^3 + 1$ and use the relation

$$\frac{dg}{dx} = \frac{dg}{du} \cdot \frac{du}{dx}$$

With $g(u) = u^2$, $\dfrac{dg}{du} = 2u = 2(x^3 + 1)$, and $\dfrac{du}{dx} = 3x^2$

$$\therefore \quad g'(x) = \frac{dg}{dx} = 2(x^3 + 1)3x^2$$

$$\therefore \quad \frac{dy}{dx} = g\frac{df}{dx} + f\frac{dg}{dx} = 6x(x^3 + 1)^2 + (3x^2 - 5)(6x^2)(x^3 + 1) \qquad \text{Answer}$$

Example 14. Differentiate the product of three functions with respect to x.

Solution. Let $f(x)$, $g(x)$, $h(x)$ be the three functions, which we shall refer to as f, g, h. With $y = fgh$ we lock the two functions g and h together as one function $u = gh$

$$\therefore \quad y = uf \text{ and } \frac{dy}{dx} = u\frac{df}{dx} + f\frac{du}{dx} \qquad \text{(i)}$$

But since u is a product of two functions gh then

$$\frac{du}{dx} = g\frac{dh}{dx} + h\frac{dg}{dx} \qquad \text{(ii)}$$

Substituting (ii) in (i) we get

$$\frac{dy}{dx} = u\frac{df}{dx} + f\left\{g\frac{dh}{dx} + h\frac{dg}{dx}\right\}$$

$$= gh\frac{df}{dx} + fg\frac{dh}{dx} + fh\frac{dg}{dx} \qquad \text{Answer}$$

Example 15. Find $\dfrac{dy}{dx}$ when $y = (2x + 1)(3x + 1)(4x + 1)$.

Solution. We differentiate each factor in turn, leaving the others alone, and add the results

$$\therefore \quad \frac{dy}{dx} = 2(3x + 1)(4x + 1) + (2x + 1)(3)(4x + 1) + (2x + 1)(3x + 1)4 \quad \text{Answer}$$

Exercise 19.4

Find the derivative with respect to x in each of the following examples 1 to 8:

1. $(x + 1)(x + 2)$
2. $x(2 - x)$
3. $(3x + 1)(x + 4)$
4. $x^2 \sin x$
5. $x \sin (x^2)$
6. $\sin x \cos x$
7. $(4x + 1)(5x + 2)(6x + 3)$
8. $(x^2 + 1)^2(x^3 + 1)$

Differentiation of a Quotient

A quotient of two functions is $y = \dfrac{f(x)}{g(x)}$ but we may treat this as a product $y = f(x) \times \dfrac{1}{g(x)}$ for which we already have a rule of differentiation.

Some readers may prefer to start by reading Example 18 and applying the simpler form of the differentiation of a quotient to confirm the results in Examples 16 and 17.

Example 16. Find the derivative of $y = \frac{x^2}{x^3 + 1}$ with respect to x.

Solution. Consider y as the product $x^2\left(\frac{1}{x^3 + 1}\right)$.

$$\therefore \quad \frac{dy}{dx} = x^2 \frac{d[1/(x^3 + 1)]}{dx} + \frac{1}{(x^3 + 1)} \cdot \frac{d(x^2)}{dx}$$

Now $\quad \frac{d[1/(x^3 + 1)]}{dx} = \frac{-3x^2}{(x^3 + 1)^2} \quad \left(\text{put } u = x^3 + 1 \text{ and find } \frac{d(1/u)}{du} \cdot \frac{du}{dx}\right)$

$$\therefore \quad \frac{dy}{dx} = \frac{-3x^4}{(x^3 + 1)^2} + \frac{2x}{x^3 + 1}$$

We may simplify this result by making $(x^3 + 1)^2$ a common denominator

$$\therefore \quad \frac{dy}{dx} = \frac{-3x^4 + 2x(x^3 + 1)}{(x^3 + 1)^2} = \frac{-x^4 + 2x}{(x^3 + 1)} \qquad \text{Answer}$$

Example 17. Find the derivative of y with respect to x when

$$y = \frac{3x^2 + 1}{x^2 - 1}, \quad (x^2 \neq 1).$$

Solution. Arranging this in the form of a product we have $y = (3x^2 + 1)\left(\frac{1}{x^2 - 1}\right)$.

Now $\frac{d(3x^2 + 1)}{dx} = 6x$ is elementary but for $\frac{d[1/(x^2 - 1)]}{dx}$ we need to use the method of differentiating a function of a function.

Put $\; f(x) = \frac{1}{x^2 - 1}$ and $u = x^2 - 1$

$$\therefore \quad \frac{df(x)}{dx} = \frac{df}{du} \cdot \frac{du}{dx} = \frac{d\left(\frac{1}{u}\right)}{du} \cdot \frac{du}{dx}$$

$$= -\frac{1}{u^2} \cdot 2x = -\frac{2x}{(x^2 - 1)^2}$$

$$\therefore \quad \frac{dy}{dx} = 6x\left(\frac{1}{x^2 - 1}\right) + (3x^2 + 1)\left(-\frac{2x}{(x^2 - 1)^2}\right)$$

$$= \frac{6x(x^2 - 1) - 2x(3x^2 + 1)}{(x^2 - 1)^2} = -\frac{8x}{(x^2 - 1)^2} \qquad \text{Answer}$$

Example 18. If u and v are functions of x find $\frac{dy}{dx}$ when $y = \frac{u}{v}$.

Solution. Expressing y as $u \times \frac{1}{v}$ we differentiate y as a product.

$$\therefore \quad \frac{dy}{dx} = \frac{du}{dx} \times \frac{1}{v} + u \times \frac{d(1/v)}{dx}$$

But $$\frac{d(1/v)}{dx} = \frac{d(1/v)}{dv} \cdot \frac{dv}{dx} = -\frac{1}{v^2} \cdot \frac{dv}{dx}$$

$$\therefore \quad \frac{dy}{dx} = \frac{1}{v}\frac{du}{dx} - \frac{u}{v^2}\frac{dv}{dx}$$

$$\therefore \quad \frac{d\left(\frac{u}{v}\right)}{dx} = \frac{1}{v^2}\left\{v\frac{du}{dx} - u\frac{dv}{dx}\right\} \qquad \text{Answer}$$

This is an extremely useful result.

Example 19. Find $\frac{d \tan\theta}{d\theta}$ by writing $\tan\theta = \frac{\sin\theta}{\cos\theta}$, $(\cos\theta \neq 0)$.

Solution. Comparison with the previous example suggests that $u = \sin\theta$, $v = \cos\theta$

With $\dfrac{du}{d\theta} = \cos\theta$, $\dfrac{dv}{d\theta} = -\sin\theta$ we may immediately write

$$\frac{d(\tan\theta)}{d\theta} = \left\{\cos\theta\,\frac{d(\sin\theta)}{d\theta} - \sin\theta\,\frac{d(\cos\theta)}{d\theta}\right\}\frac{1}{\cos^2\theta}$$

$$= \frac{\cos^2\theta + \sin^2\theta}{\cos^2\theta} \text{ (now use } \cos^2\theta + \sin^2\theta = 1\text{, page 152)}$$

$$\therefore \quad \frac{d(\tan\theta)}{d\theta} = \frac{1}{\cos^2\theta} = \sec^2\theta$$ Answer

Exercise 19.5

1. Differentiate each of the following expressions with respect to x.

(i) $\dfrac{x}{x^2+1}$ (ii) $\dfrac{x^2}{x^3+1}$

(iii) $\dfrac{3x+4}{5x-7}$, $(x \neq 1{\cdot}4)$

(iv) $\dfrac{x^2+1}{x^2-1}$, $(x^2 \neq 1)$ (v) $\cot\theta$, $(0 < \theta < \pi)$

2. Find the maximum and minimum values of each of the following

(i) $\dfrac{3x+4}{x^2+1}$ (ii) $\dfrac{4x+5}{x^2-1}$, $(x \neq \pm 1)$

Implicit Functions

Expressions such as $y = x + 1$, $y = x^2 + 7x$, $y = 1/x^2 + x \sin x$, are said to express y **explicitly** as a function of x. That is, we are able to isolate y by itself on one side of the equation leaving a function of x on the other side. An expression such as $y^2 + xy + \sin xy = x$ or $x^2 + y^2 = 49$ expresses y **implicitly** in terms of x, in the sense that we must manipulate the equation before being able to express y explicitly in terms of x, assuming that this is possible.

For example, from the implicit function $x^2 + y^2 = 49$ we may obtain $y^2 = 49 - x^2$ and then the two explicit functions $y = \sqrt{(49 - x^2)}$ and $y = -\sqrt{(49 - x^2)}$. The attempt to obtain y as an explicit function of x from the example $y^2 + xy + \sin xy = x$ will fail, yet we may still require to find $\dfrac{dy}{dx}$. That we may do so can be seen from the following example.

Example 20. Find the derivative with respect to x of (i) y^2 (ii) yx (iii) $\dfrac{x}{y}$.

Solution. For (i) we require $\dfrac{d(y^2)}{dx}$

Recall that $\dfrac{d(u^2)}{dx} = \dfrac{d(u^2)}{du}\cdot\dfrac{du}{dx} = 2u\dfrac{du}{dx}$ (page 249)

There is no difference between u and y in the use of this formula.

$$\therefore \quad \frac{d(y^2)}{dx} = \frac{d(y^2)}{dy}\cdot\frac{dy}{dx} = 2y\frac{dy}{dx}$$ Answer

For (ii) we require the derivative of a product.

$$\therefore \quad \frac{d(yx)}{dx} = y\frac{dx}{dx} + x\frac{dy}{dx} = y + x\frac{dy}{dx}$$ Answer

(For iii) we require the derivative of a quotient.

$$\therefore \quad \frac{d(x/y)}{dx} = \left(y\frac{dx}{dx} - x\frac{dy}{dx}\right)\frac{1}{y^2} = \left(y - x\frac{dy}{dx}\right)\frac{1}{y^2}$$

or
$$\frac{d(x/y)}{dx} = \frac{1}{y} - \frac{x}{y^2}\frac{dy}{dx} \qquad \text{Answer}$$

Example 21. Find $\frac{dy}{dx}$ when $xy + y^2 = x$.

Solution. We merely differentiate both sides of the equation as it stands, i.e. we make no attempt to obtain an explicit expression for y. Thus

$$\frac{d(xy + y^2)}{dx} = \frac{dx}{dx}$$

i.e.
$$y\frac{dx}{dx} + x\frac{dy}{dx} + 2y\frac{dy}{dx} = 1$$

$$\therefore \quad \frac{dy}{dx}(x + 2y) = 1 - y$$

Hence,
$$\frac{dy}{dx} = \frac{1 - y}{x + 2y} \text{ (dividing by } x + 2y) \qquad \text{Answer}$$

Example 22. Find the equation of the tangent to the circle $x^2 + y^2 = 169$ at the point (5, 12).

Solution. The gradient of the tangent to the curve at the point (x, y) is given by $\frac{dy}{dx}$.
From $x^2 + y^2 = 169$ we have

$$\frac{d(x^2 + y^2)}{dx} = \frac{d(169)}{dx}$$

$$\therefore \quad 2x + 2y\frac{dy}{dx} = 0 \quad \text{(i.e. the derivative of a constant is zero)}$$

$$\therefore \quad \frac{dy}{dx} = -\frac{x}{y} \quad \text{(dividing by } 2y)$$

At the point (5, 12) this becomes $\frac{dy}{dx} = -\frac{5}{12}$ and the equation of the tangent is

$$\frac{y - 12}{x - 5} = -\frac{5}{12}$$

i.e.
$$12y - 144 = 25 - 5x \qquad \text{Answer}$$
$$12y + 5x = 169$$

(Examine the work on page 224 in the light of this example.)

Example 23. Find dy/dx from the relation $y^p = x^m$, where m and p are positive integers.

Solution. We have already proved that if m is a positive integer then

$$\frac{d(x^m)}{dx} = mx^{m-1} \qquad \text{(page 220)}$$

Also

$$\frac{d(y^p)}{dx} = \frac{d(y^p)}{dy} \cdot \frac{dy}{dx} \qquad \text{(page 250)}$$

$$= py^{p-1} \cdot \frac{dy}{dx}$$

Therefore, differentiation of the equation $y^p = x^m$ yields

$$p\,y^{p-1}\frac{dy}{dx} = mx^{m-1}$$

Dividing by py^{p-1}

$$\frac{dy}{dx} = \frac{m}{p}\frac{x^{m-1}}{y^{p-1}}$$

Writing $\frac{x^m}{x}$ for x^{m-1} and $\frac{y}{y^p}$ for $\frac{1}{y^{p-1}}$ we get

$$\frac{dy}{dx} = \frac{m}{p}\cdot\frac{x^m}{x},\frac{y}{y^p} = \frac{m}{p}\cdot\frac{1}{x}\cdot y \text{ (because } y^p = x^m)$$

With $y^p = x^m$ we get $y = x^{m/p}$ by taking the pth root of each side. Combined with $\frac{1}{x}$ as x^{-1} we obtain

$$\frac{dy}{dx} = \frac{m}{p}x^{\frac{m}{p}-1}$$

We have just proved that the formula (of page 220) $\frac{dx^n}{dx} = nx^{n-1}$ holds for $n =$ any positive rational number, i.e.

$$\frac{dx^n}{dx} = nx^{n-1}$$

Answer

for any positive integer or rational number.
This result has already been applied in cases such as

$$\frac{d(x^{\frac{1}{3}})}{dx} = \frac{1}{3}x^{-\frac{2}{3}} \text{ or } \frac{d(x^{\frac{1}{2}})}{dx} = \tfrac{1}{2}x^{-\frac{1}{2}} = \frac{1}{2x^{\frac{1}{2}}}$$

and so on.

Exercise 19.6

Find $\frac{dy}{dx}$ in each of the expressions 1 to 6.

1. $x + y^2 = 9$
2. $x^2 - y^2 = x$
3. $x + \sin y = x^2$
4. $xy + \sin x = 3$
5. $xy^2 + \cos x = 0$
6. $y^2 \sin x = x$
7. Find the equation of the tangent at the point $(-2, 0)$ on the circle $x^2 + y^2 + 2x + 2y = 0$.
8. Find the equation of the tangent at the point $(4, 4)$ on the parabola $y^2 = 4x$ (see page 83).
9. Show that the gradient of the tangent at any point (x, y) on the parabola

$$y^2 = 4ax \text{ is (i) } \frac{2a}{y}, \text{ (ii) } \frac{y}{2x} \quad \text{(see page 84).}$$

10. By differentiating $yx^m = 1$, where m is a positive integer, prove that the formula $\frac{dx^n}{dx} = nx^{n-1}$ holds when n is a negative integer (see page 256).

11. By considering the speed v as a function of s and s as a function of t show that the acceleration $\frac{dv}{dt} = v\frac{dv}{ds}$ (see page 231, Question 5).
12. Find the derivative with respect to x of each of the following: (i) $y = (x^2 + 1)^{\frac{4}{3}}$; (ii) $y = (2x + 1)^{-5}$ $(x \neq -\frac{1}{2})$; (iii) $y = \frac{1}{(x^2 + 1)^2}$.
13. Find the number which exceeds its square by the greatest number possible.
14. The sum of the height and circumference of a right circular cylinder is 3 metres. Find its maximum volume.
15. The volume of water in a container is given by $V = 2s^3 + 2s^2 + 9s$ cubic centimetres when the depth of the water is s cm. Water is added to the container at a constant rate of 50 cm^3 s^{-1}. Find the rate at which the level is rising when $s = 3$ cm.
16. If the volume of a sphere is increasing at the rate of 10 cm^3 s^{-1} find the rate at which the radius is increasing with respect to time when the radius is 5 cm $(V = \frac{4}{3}\pi r^3)$.

CHAPTER TWENTY

INTEGRATION

Differentiation Reversed

The process of finding $f(x)$ by starting from $f'(x)$ is called **integration.** Having found $f(x)$ we are said to have **integrated** $f'(x)$ with respect to x. It is obvious why such a process is called the reverse of differentiation.

For example, suppose we are given that $f(x) = 5x + 3$; the derived function then is $f'(x) = 5$. If, therefore, we reverse this process and start with $f'(x) = 5$ shall we then get $f(x) = 5x + ?$ At this stage we encounter the

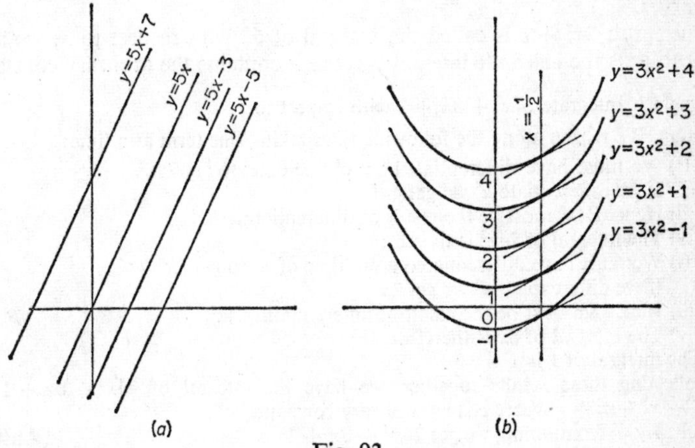

Fig. 93

difficulty of discovering how to get back to $f(x) = 5x + 3$ without any further information, since $f(x) = 5x + c$ for all values of c will lead to the derived function in $f'(x) = 5$.

We are thus left with the fact that the integration of $f'(x) = 5$ does not lead to a unique result unless we have more information. In lieu of that information we agree to leave the answer as $f(x) = 5x + c$ and refer to c as an arbitrary constant. The result $5x + c$ is called the **integral** of 5 with respect to x. As Fig. 93(a) shows, $y = f(x) = 5x + c$ gives a set of parallel lines of gradient 5. If we had been asked to integrate $f'(x) = 5$ such that $f(0) = 7$ we would obtain the result in two steps:

(i) Integration of $f'(x) = 5$ leads to the general solution $f(x) = 5x + c$, where c is an arbitrary constant;

(ii) $f(0) = 0 + c = 7$, so the final result of the integration is $f(x) = 5x + 7$.

Using the notation of $\frac{dy}{dx}$, we restate the solution as follows.

Given that $\frac{dy}{dx} = 5$ we integrate $\frac{dy}{dx}$ with respect to x to obtain

$$y = 5x + c, \text{ where } c \text{ is an arbitrary constant.}$$

If the condition that $y = 7$ when $x = 2$ is applied then we obtain the result

$$y = 5x + 7 \text{ as stated above.}$$

We know from the results of differentiation that if $y = f(x) = 3x^2 + c$ where c is any constant number, then $\frac{dy}{dx} = f'(x) = 6x$.

The set of curves $y = 3x^2 + c$ is a set of parallel parabolas some of which can be seen in Fig. 93(*b*). The result $\frac{dy}{dx} = 6x$ means that at $x = \frac{1}{2}$ (for example) on *each* of the parabolas the gradient of the tangent is 3. Integration of the equation $\frac{dy}{dx} = 6x$ therefore leads to the result $y = 3x^2 + c$, and again we need more information to obtain a particular member of this set of curves.

The result $3x^2 + c$ is called the integral of $6x$ with respect to x, we also refer to it as the **indefinite integral** because it contains the arbitrary constant.

Example 1. Integrate $24x^3 + 8x + 1$ with respect to x.

Solution. We reason along the following lines taking one term at a time:

(a) (i) We must have differentiated a multiple of x^4 to get $24x^3$
  (ii) If we differentiate x^4 we get $4x^3$
  (iii) Hence $24x^3$ must be the result of differentiating $6x^4$
  (iv) The integral of $24x^3$ is therefore $6x^4$

(b) (i) We must have differentiated a multiple of x^2 to get $8x$
  (ii) If we differentiate x^2 we get $2x$
  (iii) Hence $8x$ must be the result of differentiating $4x^2$
  (iv) The integral of $8x$ is therefore $4x^2$

(c) The integral of 1 is x

Collecting these results together, we have the integral of $24x^3 + 8x + 1$ as $6x^4 + 4x^2 + x + c$ where c is an arbitrary constant.

Check: By differentiating, we get $24x^3 + 8x + 1$ Answer

Example 2. Integrate $x + \cos x$ with respect to x.

Solution. The integral of x is $x^2/2$. We have to differentiate $\sin x$ to get $\cos x$, hence the integral of $\cos x$ is $\sin x$.

The integral of $x + \cos x$ with respect to x is $x^2/2 + \sin x$. Answer

Exercise 20.1

1. Integrate the following with respect to x:
   (i) $2x$ (ii) $3x^2$ (iii) $4x^3$ (iv) $5x^4$ (v) $6x^5$ (vi) nx^{n-1} (n a positive integer)
2. Find the integrals of the following with respect to t:
   (i) $2t + 3t^2$ (ii) $24t^2 + \cos t$ (iii) $24t^3 - \sin t$ (iv) t^{-2} (v) $2t^{-3}$ (vi) $1/t^2$
3. Given that $\frac{dy}{dx} = 16x$ find y such that $y = 3$ when $x = 0$.
4. Integrate the equation $\frac{dy}{dx} = 6x^2 + x$ given that $y = 19$ when $x = 2$.
5. Find y when $\frac{dy}{dx} = ax^2 + b$ where a, b are constants.

Differential Equations

Any equation involving $\frac{dy}{dx}$ is called a differential equation and the removal of $\frac{dy}{dx}$ by integration is called solving the differential equation. For example, the solution of the differential equation $\frac{dy}{dx} = 3x^2 + 5$ is $y = x^3 + 5x + c$ where c is an arbitrary constant.

Until extra information is given such as $y = 1$ when $x = 0$, we must use the arbitrary constant to represent the solution in general, i.e. we say that the solution can be any one of the functions given by $x^3 + 5x + c$ for different c. $y = x^3 + 5x + c$ is therefore called the **general solution** of the differential equation $\frac{dy}{dx} = 3x^2 + 5$.

Just as we may differentiate a function twice to obtain $\frac{d^2y}{dx^2}$ so we may start with a differential equation like $\frac{d^2y}{dx^2} = 6x$ and integrate twice to solve the equation, i.e. obtain y as a function of x.

The integration of $\frac{d^2y}{dx^2} = 6x$ with respect to x gives

$$\frac{dy}{dx} = 3x^2 + c_1, \text{ where } c_1 \text{ is an arbitrary constant.}$$

The integration of $\frac{dy}{dx} = 3x^2 + c_1$ with respect to x gives

$$y = x^3 + xc_1 + c_2 \text{ where } c_2 \text{ is another arbitrary constant.}$$

We notice that integrating twice introduces two arbitrary constants into the general solution of the differential equation. If we add two conditions like $y = 0$ when $x = 0$ and $y = 5$ when $x = 1$ then we must have $c_2 = 0$ and $c_1 = 4$ and the solution becomes $y = x^3 + 4x$. It follows that the solution of differential equations like $\frac{d^2y}{dx^2} = f(x)$ does not present too many difficulties when of the above form.

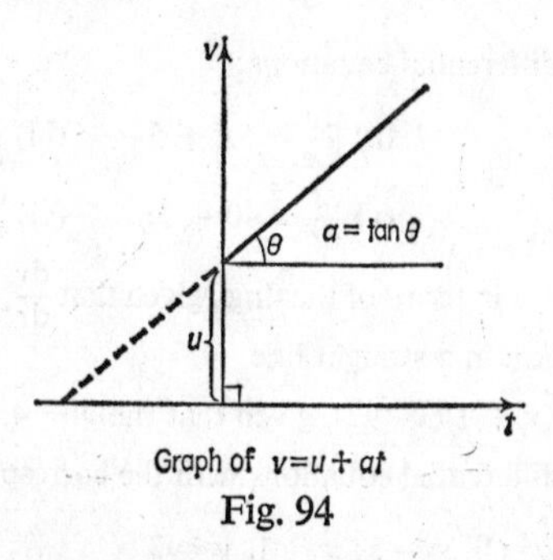

Graph of $v = u + at$

Fig. 94

Acceleration, Velocity, Displacement

The basic equation for motion in a **straight line** under constant acceleration is given by the differential equation $\frac{dv}{dt} = a$ (a constant). Integration of this equation gives $v = at + c_1$, c_1 an arbitrary constant.

We notice immediately that when $t = 0$, $v = c_1$ so the significance of c_1 is that it represents the velocity at the beginning of the timing of the motion, that is the initial velocity, which is usually called u. Notice also that we have a form of '$y = mx + c$' so that the graph of v against t is a straight line when the acceleration is constant. Our velocity time equation has now become $v = u + at$, and its graph is shown in Fig. 94, a straight line with gradient $\tan\theta = a$. Substituting $v = \frac{ds}{dt} = u + at$ we may now integrate once more to gain

$$s = ut + \tfrac{1}{2}at^2 + c \text{ (} c \text{ an arbitrary constant).}$$

If we now suggest that $s = 0$ when $t = 0$, then $c = 0$. The original differential equation $\frac{dv}{dt} = a$ thus produces the equations of motion

$$v = u + at \qquad \text{(i)}$$

and

$$s = ut + \tfrac{1}{2}at^2 \qquad \text{(ii)}$$

We may obtain a third equation by substitution

From (i) $\qquad v^2 = (u + at)^2 = u^2 + 2uat + a^2t^2$

$$\therefore v^2 - u^2 = 2uat + a^2t^2$$

From (ii) $\qquad 2as = 2aut + a^2t^2$

Comparison of these two results gives

$$v^2 - u^2 = 2as$$

or

$$v^2 = u^2 + 2as \qquad \text{(iii)}$$

These are the three equations for the motion in a straight line with constant acceleration, a.

Exercise 20.2

1. Solve the following differential equations:

(i) $\frac{dy}{dx} = 8x$ (ii) $\frac{dy}{dx} = 8x + 5$ (iii) $\frac{dy}{dx} = 8x^2 + 5x$

(iv) $\frac{dy}{dx} = x(x + 1)$ (v) $\frac{ds}{dt} = 40 + 3t$ (vi) $\frac{ds}{dt} = 8 - 4{\cdot}9t$

2. Find the displacement s in terms of the time, given that $\frac{dv}{dt} = 4$, $u = 10$ and $s = 0$ when $t = 0$, for motion in a straight line.

3. Find y from $\frac{dy}{dx} = 6(x + 1)(x + 2)$, given that the $y = 4$ when $x = 0$.

4. Solve the following differential equations with the corresponding conditions:

(i) $\frac{d^2y}{dx^2} = 6x - 2$; $x = 0, y = 3$; $x = 1, y = 3$

(ii) $\frac{d^2y}{dx^2} = -\sin x$; $x = 0, y = 0$; $x = 1, y = \frac{\pi}{2}$

(iii) $\frac{d^2y}{dx^2} = 6x - \sin x$; $x = 0$, $\frac{dy}{dx} = 5$, $x = 0$, $y = -2$

(iv) $\frac{d^2A}{dx^2} = -\sin x + \cos x$; $x = 0$, $\frac{dA}{dx} = 1$; $x = 0$, $A = 3$.

5. A stone is projected vertically upwards from O with an initial velocity of $u = 49\ \text{m s}^{-1}$. With the distance s and velocity v measured positive in the upwards direction and negative in the downwards direction, deduce from the acceleration due to gravity $\frac{dv}{dt} = -4{\cdot}9\ \text{m s}^{-2}$ the equations for v and s in terms of t. Find where the stone is at $t = 1$, $t = 20$ and find the greatest height reached taking the initial condition that $s = 0$ when $t = 0$.
6. A car starts from rest and moves in a straight line with an acceleration of $0{\cdot}1(5 - t)$ metres per second for the first five seconds. Find the velocity of the car when $t = 2$ and the distance travelled in 5 seconds.
7. Given that $s = 3t + 4t^2$ metres find the distance travelled between $t = 3$ and $t = 5$.
8. The velocity of a body is given by $v = 3t^2 - 16t$ metres per second. If $s = 9$ when $t = 0$ find the distance travelled between $t = 6$ and $t = 10$.

Area Under a Straight Line

Consider the graph of $y = 2x + 6$ shown in Fig. 95. By the area under the straight line we mean the area between the graph and the x axis. Selecting two points $A(1, 8)$ and $B(4, 14)$ on the graph we refer to the area under the

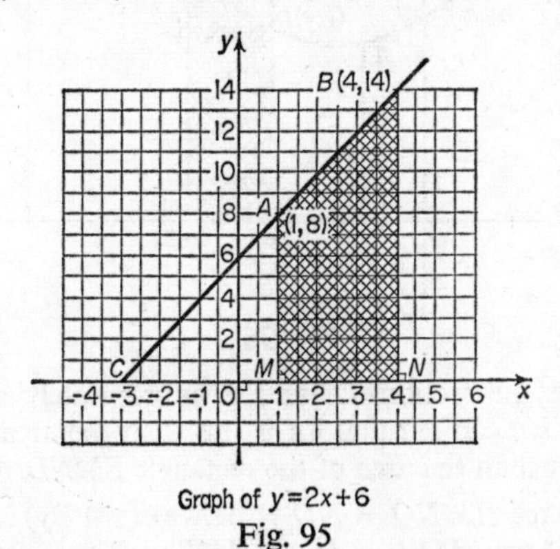

Fig. 95

line between A and B as the area of the figure $AMNB$, which is a trapezium in this example with AM and BN perpendicular to the x axis.

Since $AM = 8$ and $BN = 14$ and $MN = 3$,

$$\text{the area } AMNB = \tfrac{1}{2}(3)(8 + 14) = 33 \text{ square units.}$$

Integrating the function $2x + 6$ with respect to x gives $x^2 + 6x + c$, where c is an arbitrary constant.

$$\text{When } x = 4,\ x^2 + 6x + c = 16 + 24 + c = 40 + c$$
$$\text{When } x = 1,\ x^2 + 6x + c = 1 + 6 + c = 7 + c$$

Subtracting these results we have $(40 + c) - (7 + c) = 33 =$ the area $AMNB$, i.e. the change in the integral as x increases from 1 to 4 gives the area $AMNB$.

Using the same Fig. 95 suppose we find the area of triangle MAC.

With $AM = 8$ and $CM = 4$ the area of triangle $MAC = 16$. The integral of $2x + 6$ with respect to x being $x^2 + 6x + c$, the change in the integral as x increases from -3 to 1 is given by

$$(1 + 6 + c) - (9 - 18 + c) = 16 = \text{the area of the triangle } MAC.$$

Area Under a Curve

Consider the curve of Fig. 96 given by the general equation $y = f(x)$, e.g. $y = 6x^3 + 7x$ or $y = 5x^2 + 9$ and so on. We shall find the area A under the curve from T to V. On this curve we shall take two points $P(x, y)$ and $Q(x + \delta x, y + \delta y)$. MP and NQ are perpendicular to the x axis and because P is the point (x, y) we have $MP = y$. Also, because Q is the point $(x + \delta x, y + \delta y)$ we have $NQ = y + \delta y$. MP is produced to H and PL is parallel to MN to create the two rectangles $HMNQ$, $PMNL$.

The shaded area under the section PQ of the curve is an increment of the

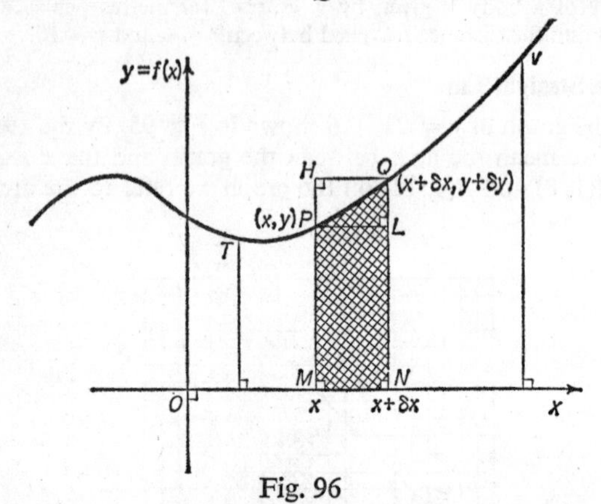

Fig. 96

total area A under the section TV and is appropriately measured by δA. From the diagram we can see that δA is less than the area of the rectangle $HMNQ$ but greater than the area of the rectangle $PMNL$.

$$\text{Area } HMNQ = NQ \times MN = (y + \delta y)\,\delta x$$
$$\text{Area } PMNL = MP \times MN = y\delta x$$

Therefore $y\delta x < \delta A < (y + \delta y)\delta x$

With δx positive we may divide through by δx and obtain

$$y < \frac{\delta A}{\delta x} < y + \delta y \qquad \text{(i)}$$

As $\delta x \rightarrow 0$, $\delta y \rightarrow 0$ we have

$$(y + \delta y) \rightarrow y \text{ and } \frac{\delta A}{\delta x} \rightarrow \frac{dA}{dx}$$

In (i) we now let $\delta x \rightarrow 0$ so that $\delta A/\delta x$ has a value which is always between y and $y + \delta y$, and since $y + \delta y$ becomes as close as we please to y it follows that

$$\frac{dA}{dx} = y$$

With y as a function of x we have $\frac{dA}{dx} = f(x)$ so that the area A will be the integral of $f(x)$ with respect to x.

In terms of the last example in Fig. 95 this means that $\frac{dA}{dx} = 2x + 6$.

Example 3. Suppose the curve in Fig. 96 has the equation $y = 12(x^3 + 7x^2 - x)$ and that we require the area under the part of the curve given by going from $x = 1$ to $x = 6$, i.e. the x coordinates of T and V are 1 and 6 respectively.

Solution. We now know that $\frac{dA}{dx} = 12(x^3 + 7x^2 - x)$ so that by integration $A = 3x^4 + 28x^3 - 6x^2 + c$, ($c$ an arbitrary constant).
The area A starts at zero when $x = 1$

$$\therefore \quad A = 3 + 28 - 6 + c = 0$$
$$\therefore \quad c = -25$$

Hence $$A = 3x^4 + 28x^3 - 6x^2 - 25$$

The area we require is given by A when $x = 6$

$$A = 3888 + 6048 - 216 - 25 = 9695$$ Answer

We may generalise the above ideas for finding the area under the curve $y = f(x)$ between two coordinates $x = a$ and $x = b$ in the following way:

(i) Integrate $\frac{dA}{dx} = y$ to obtain A
(ii) Put $A = 0$ when $x = a$ and find the arbitrary constant
(iii) The required area is the value of A when $x = b$.

Example 4. Find the area under the curve given by $y = 6x^2 + 4x^3$ from $x = 1$ to $x = 3$.

Solution. $\frac{dA}{dx} = 6x^2 + 4x^3$
(i) Integrating $\frac{dA}{dx}$ with respect to x we get

$$A = 2x^3 + x^4 + c$$

(ii) $A = 0$ when $x = 1$ $\quad \therefore \quad 0 = 2 + 1 + c$, hence $c = -3$

$$\therefore \quad A = 2x^3 + x^4 - 3$$

(iii) When $x = 3$ $\quad A = 54 + 81 - 3 = 132$

The area under the curve from $x = 1$ to $x = 3$ was 132. Answer

Symbols for Integration

The reader is warned that we are about to introduce some more new terms and symbols with which he needs to become familiar. Familiarity comes more easily from working the examples than by committing the notation to memory before doing the examples.

The symbol $\int y \, dx$ represents the integral of y with respect to x. The instruction to integrate with respect to x is represented by $\int \, dx$ and must be treated as one symbol. The symbol $\int$ is referred to as the 'integral sign'.

Examples we have already met are:

$$\int 6x \, dx = 3x^2 + c$$

$$\int (x + \cos x) \, dx = \frac{x^2}{2} + \sin x + c$$

$$\int (6x^2 + 4x^3) \, dx = 2x^3 + x^4 + c$$

$$\int (2t + 3t^2) \, dt = t^2 + t^3 + c$$

$$\int \frac{1}{t^2} \, dt = -\frac{1}{t} + c$$

with c an arbitrary constant in each case.

When the integral contains the arbitrary constant we refer to it as an **indefinite integral.** Each of the five results above is an indefinite integral.

Exercise 20.3

Find the following indefinite integrals

1. $\int x \, dx$ 2. $\int 4 \, dx$ 3. $\int (x + 4) \, dx$ 4. $\int (x - 4) \, dx$
5. $\int 2x \, dx$ 6. $\int \cos x \, dx$ 7. $\int (2x + \cos x) \, dx$ 8. $\int (2x - \cos x) \, dx$
9. $\int 1/x^2 \, dx$ 10. $\int (\sin x + x \cos x) \, dx$ (Hint: Try differentiating $x \cos x$ or $x \sin x$.)

The Definite Integral

In the last worked example we evaluated $\int (6x^2 + 4x^3) \, dx$ between $x = 1$ and $x = 4$, by finding the value of the arbitrary constant. This is not necessary, for suppose that the value of $A = g(x) + c$ and we follow the summary of page 265.

(ii) $A = 0$ when $x = 1$ $\therefore 0 = g(1) + c$ i.e. $c = -g(1)$
$\therefore A = g(x) - g(1)$

(iii) The value of A when $x = 4$ is
$A = g(4) - g(1)$

We may generalise this result again. To find the area under the curve $y = f(x)$ between $x = a$ and $x = b$ we need only

(i) Integrate $\frac{dA}{dx} = y$ to obtain A

(ii) Find A with $x = a$ and subtract the result from A with $x = b$.

We represent this difference by the symbol $\Big[A\Big]_a^b$ or $\int_a^b y \, dx$, and we call this the **definite integral.**

Thus $\int_a^b y \, dx$ is the definite integral and means that the integral of y with respect to x has taken place between the **limiting values** of $x = a$ and $x = b$. b is called the **upper limit** and a is the **lower limit** of the integral.

Example 5. Find $\int_1^3 (2x + 8x^3)\,dx$.

Solution. We first obtain the integral of $2x + 8x^3$ without the arbitrary constant. This is $x^2 + 2x^4$.

$$\therefore \quad \int_1^3 (2x + 8x^3)\,dx = \Big[x^2 + 2x^4\Big]_1^3$$

As already introduced, this square bracket symbol represents the subtraction of $x^2 + 2x^4$ at $x = 1$ from $x^2 + 2x^4$ at $x = 3$

i.e. $$\Big[x^2 + 2x^4\Big]_1^3 = (9 + 162) - (1 + 2) = 168$$

$$\therefore \quad \int_1^3 (2x + 8x^3)\,dx = 168 \qquad \text{Answer}$$

Note that this also means that the area under the curve $y = 2x + 8x^3$ from $x = 1$ to $x = 3$ is 168 square units.

Example 6. Calculate the area under the graph of $y = 6x(x + 1)$ from $x = -2$ to $x = 3$.

Solution. We require $\int_{-2}^3 6(x^2 + x)\,dx = A$ say

$$\int_{-2}^3 6\,(x^2 + x)\,dx = \Big[2x^3 + 3x^2\Big]_{-2}^3 = (54 + 27) - (-16 + 12)$$

$$\therefore \quad A = 81 + 4 = 85 \text{ square units} \qquad \text{Answer}$$

The next example is extra important because it introduces the idea of a negative area.

Example 7. Find the area under the graph of $y = x - 2$ from

(i) $x = -4$ to $x = 2$
(ii) $x = 1$ to $x = 3$

Solution. (i) the required area is $A = \int_{-4}^2 (x - 2)\,dx = \Big[\frac{x^2}{2} - 2x\Big]_{-4}^2$

$$= (2 - 4) - \left(\frac{16}{2} + 8\right)$$

$$\therefore \quad A = -18 \qquad \text{Answer}$$

(ii) The required area is $A = \int_1^3 (x - 2)\,dx = \Big[\frac{x^2}{2} - 2x\Big]_1^3 = \left(\frac{9}{2} - 6\right) - \left(\frac{1}{2} - 2\right)$

$$\therefore \quad A = -1\tfrac{1}{2} - (-1\tfrac{1}{2}) = 0 \qquad \text{Answer}$$

To give a meaning to these two results consider the graph of $y = x - 2$ in

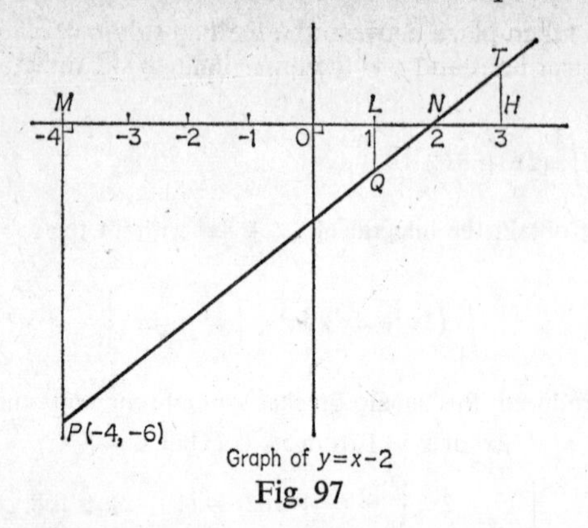

Graph of $y=x-2$

Fig. 97

Fig. 97. The first area asked for is the area of triangle *MNP*, which is below the x axis. The second area asked for is the sum of the areas of triangles *LNQ* and *NHT*. We see therefore that the area under the graph is given as negative by the integral when it is below the x axis. The triangles *LNQ* and *NHT* are congruent and have the same area numerically but under the sign convention where the integral represents the area under the curve we give area *LNQ* a negative sign and area *NHT* a positive sign and the process of integration gives the sum of zero.

Example 8. Find the area under the curve $y = 4(x - 1)^2(x - 4)$ from

$$\text{(i) } x = 1 \text{ to } x = 4$$
$$\text{(ii) } x = 3 \text{ to } x = 5$$

Solution. The graph of $y = 4(x - 1)^2(x - 4)$ is given in Fig. 98. From the graph we shall expect to obtain a negative result for (i) and the sum of a negative and positive area for (ii) if we integrate without thought.

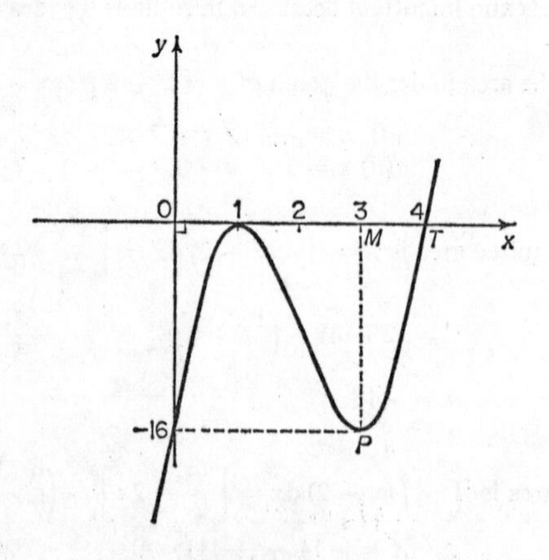

Graph of $y=4(x-1)^2(x-4)$

Fig. 98

(i) With $y = 4x^3 - 24x^2 + 36x - 16$ we need

$$A = \int_1^4 (4x^3 - 24x^2 + 36x - 16)\,dx$$

$$\therefore\ A = \Big[x^4 - 8x^3 + 18x^2 - 16x\Big]_1^4$$

$$= (256 - 512 + 288 - 64) - (1 - 8 + 18 - 16)$$

$$A = -32 - (-5) = -27 \text{ square units}$$

Thus the area concerned measures 27 square units. The integration gives the result of -27 because the area is below the x axis and so we must employ a sign convention in order to get the true measure of the area.

(ii) $A = \int_3^5 y\,dx$ will not give us the measure of the areas we require, because part of the area being below the x-axis will be given a negative sign by the integral.

We need $-\int_3^4 y\,dx + \int_4^5 y\,dx$ and not $\int_3^5 y\,dx$

$$\int_3^4 y\,dx = \Big[x^4 - 8x^3 + 18x^2 - 16x\Big]_3^4 = (-32) - (81 - 216 + 162 - 48)$$

$$= -32 - (-21) = -11$$

The area between the section of the curve PT and the x axis measures 11 square units

$$\int_4^5 y\,dx = \Big[x^4 - 8x^3 + 18x^2 - 16x\Big]_4^5 = (625 - 1000 + 450 - 80) - (-32)$$

$$= -5 + 32 = 27$$

As expected we have obtained a positive result. Thus the sum of the areas is 38 square units.

The value of $\int_3^5 y\,dx = 27 - 11 = 16$ Answer

The reader needs to be on his guard for examples like this.

Example 9. Show that the area under a velocity–time graph represents the distance travelled. Hence find the distance travelled from $t = 1$ to $t = 3$ when $v = t^2 + 6t$.

Solution. The area under the curve is given by $\int v\,dt$ with the appropriate sign convention.

With

$$\frac{ds}{dt} = v \left(\text{compare } \frac{dA}{dx} = y\right)$$

we get

$$s = \int v\,dt \text{ (compare } A = \int y\,dx)$$

When $v = t^2 + 6t$,

$$s = \int_1^3 (t^2 + 6t)\,dt = \left[\frac{t^3}{3} + 3t^2\right]_1^3 = (9 + 27) - \left(\frac{1}{3} + 3\right)$$

$$\therefore\ s = 36 - 3\tfrac{1}{3} = 32\tfrac{2}{3} \text{ in the appropriate units}$$ Answer

Exercise 20.4

1. Evaluate the following definite integrals.

(i) $\int_1^2 x^3\,dx$ (ii) $\int_{-1}^1 x^3\,dx$ (iii) $\int_1^5 1/x^2\,dx$

(iv) $\int_0^3 (u^2 + 2u)\,du$

2. Find the area under the following graphs between the given limits.

(i) $y = x^4$; $x = -1$, $x = 1$ (ii) $y = x^3$; $x = -1$, $x = 1$

(iii) $y = 3x + \frac{1}{x^2}$; $x = 1$, $x = 4$ (iv) $y = \cos x$; $x = 0$, $x = \frac{\pi}{2}$

3. Find the area under the curve $y = x^2(x - 3)$ from

(i) $x = -1$ to $x = 0$ (ii) $x = 0$ to $x = 3$

(iii) $x = 2$ to $x = 4$

after drawing a sketch like Fig. 98.

4. A rocket moves in a straight line with a velocity given by $v = 10t - \frac{4}{t^3}$ for the period $t = 2$ to $t = 10$. Find the distance travelled if the velocity is measured in metres per second.

5. Given that $A = \frac{x^2 + 1}{x^2 - 1}$ find $\frac{dA}{dx}$ and hence find $\int_2^3 -\frac{4x\,dx}{(x^2 - 1)^2}$.

6. Given that $A = \frac{x^2 - 1}{x^2 + 1}$ find $\frac{dA}{dx}$ and hence find $\int_0^1 \frac{4x\,dx}{(x^2 + 1)^2}$.

7. Find the area beneath each of the following graphs from $x = 0$ to $x = 8$.

(i) $y = x$ (ii) $y = 9x - x^2$ (parabola)

Draw a sketch of both graphs and hence find the closed area between the straight line and the parabola from $x = 0$ to $x = 8$.

8. Find the closed area between the straight line $y = 4x$ and $y = x^2$ (Find the points of intersection of the line with the parabola.)

9. The velocity–time graphs for two different motions arise from A: $v = 4t$, B: $v = t^2$. Find how much further A travels than B in the first 4 seconds of the motion if v is measured in metres per second.

10. Find the time t at which A and B have travelled the same distance in question 9 and calculate the distance.

SOLUTIONS

Exercise 1.1

1. $x = 3, y = 2$ 2. $x = 4, y = 1$ 3. $x = 6, y = -5$
4. $x = 2, y = 0$ 5. $x = 1, y = 2$ 6. $x = 2, y = 2$

Exercise 1.2

1. $x = 6, y = 2$ 2. $x = 4, y = 4$ Consider $3y = 16 - x = 20 - 2x$.
3. $x = 1, y = 3$ Consider $4x = 1 + y = 13 - 3y$.
4. $x = 2, y = 1$ Consider $2x = 13 - 9y = 9y - 5$.
5. $x = 1{\cdot}5, y = 2{\cdot}5$ Consider $y = 4 - x = 7 - 3x$.
6. $x = 2, y = 1{\cdot}5$ Consider $\frac{1}{x} = 2 - y = 11 - 7y$.

Exercise 1.3

1. $4x = 16$ 2. $2x = 10$ 3. $2x = 6$ 4. $-4y = 4$
5. $6x - 10y = 46$ 6. $16x = 56$ 7. $-16y = 24$

Exercise 1.4

1. $x = 1, y = 6$ Either multiply (i) by 2 and subtract (ii) to get $9x = 9$, or multiply (ii) by 2 and subtract (i) to get $3y = 18$
2. $x = 2, y = 2$ Either multiply (i) by 2 and (ii) by 3 and subtract the results to get $5y = 10$. Or multiply (i) by 3 and (ii) by 7 and subtract the results to get $-5x = -10$.
3. $x = 2, y = -3$ Multiply (ii) by 2 and subtract from (i) to get $-7x = -14$.
4. $x + y = 19, x + 8 = 2y$ $x = 10, y = 9$
5. Age of the husband is $x + 5$. The sum of their ages is $x + x + 5 = 89$. $x = 42$.
6. $x = 10, y = 1$ 7. $3x + 8y = 38$, $8x + 3y = 83$, $x = 10, y = 1$ 8. $x = 100, y = 10, z = 1$
9. $2x + 5y + 7z = 257$; $2x + 7y + 5z = 275$; $5x + 7y + 2z = 572$; $5x + 2y + 7z = 527$; $7x + 5y + 2z = 752$; $7x + 2y + 5z = 725$
Any three of the six possible equations have the solution $x = 100, y = 10, x = 1$.

Exercise 1.5

1. $10 = a + b, 40 = 16a + b$ Hence $30 = 15a$ and $a = 2, b = 8$. When $M = 10$, $C = 28$
2. $g - b = 5, g + b = 18$, so that $2g = 23, g = 11\frac{1}{2}, b = 6\frac{1}{2}$
3. Let the side of the square be of length w. $7w = 350$. The rectangle has a length of 100 m, and a breadth of 50 m, and an area of 5000 m^2.
4. (a) The equations are inconsistent. (b) As in (a).
   (b) The equations are dependent, i.e. the same relation as the first equation. There are as many solutions as we please, e.g. $x = 1, y = 8$; $x = 2, y = 7$; $x = 3, y = 6$; etc.
   (d) $x = 2, y = \frac{1}{2}$ The equations do not have whole number solutions for both x and y. (e) $x = 10, y = 3$.
5. $2 = 4m + c, -4 = 2m + c, m = 3, c = -10$
   When $x = 100$, $y = 3x - 10$ gives $y = 290$.
6. $x + y = 100,\ x - 8 = 2(y - 8)$. The second equation becomes $2y - x = 8$, $y = 36,\ x = 64$. Alternatively we may think along the following lines. Eight

years ago the sum of their ages was 84 years and this represented $3(y - 8)$. Therefore $3(y - 8) = 84$, etc.

Exercise 1.6

1. (i) As x takes all possible values greater than 2 so y may take all possible negative and positive values. Thus $x = 2{\cdot}1$ allows any value of y greater than $3{\cdot}9$ and $x = 106$ allows any value of y greater than -100 and so on. $y > 6 - x$.
   (ii) $y > 4$, $y > 6 - x$.
   (iii) Similar to (i). As x takes all possible values greater than 6 so y may take all possible negative and positive values. Thus $x = 7$ allows any value of y greater than -1, $x = -5$ allows any value of y greater than 11, and so on.
   (iv) $y > 0$, $y > 6 - x$.
2. (i) $x > 2$, $x > 1 + y$.
   (ii) As y takes all possible values less than 1 so x may take all possible negative and positive values such that $x > 1 + y$.
   (iii) $x > 2$
   (iv) $x > 3$, $x > 1 + y$.
3. We may add the two inequalities to get $3x > 6$. We may then divide by the positive number 3 to get $x > 2$.
4. (i) $1 < x + y < 10$
   (ii) $0 < 3x < 9$ and $2 < 2y < 14$. The least value of $3x - 2y$ is $0 - 14 = -14$. The greatest value of $3x - 2y$ is $9 - 2 = 7$.
   (iii) $3 < 3y < 21$ and $0 < 4x < 12$ The least value of $3y - 4x$ is $3 - 12 = -9$. The greatest value of $3y - 4x$ is $21 - 0 = 21$.
   (iv) $1 < x + 1 < 4$ and $1 < y < 7$. Therefore $\frac{1}{4} < \frac{1}{x+1} < 1$ and $\frac{1}{7} < \frac{1}{y} < 1$.

$$\text{Hence} \quad \frac{1}{4} + \frac{1}{7} < \frac{1}{x+1} + \frac{1}{y} < 2$$

$$\text{i.e.} \quad \frac{11}{28} < \frac{1}{x+1} + \frac{1}{y} < 2$$

   (v) The least value is $\frac{1}{4} - 1 = -\frac{3}{4}$ and the greatest value is $1 - \frac{1}{7} = \frac{6}{7}$.

5. (i) $-\frac{1}{2} < \frac{1}{x} < -\frac{1}{5}$ and $-1 < \frac{1}{y} < -\frac{1}{3}$
   Least value is $-\frac{1}{2} - 1 = -1\frac{1}{2}$. Greatest value is $-\frac{1}{5} - \frac{1}{3} = -\frac{8}{15}$.
   (ii) Least value is $-1 - (-\frac{1}{5}) = -\frac{4}{5}$. Greatest value is $-\frac{1}{3} - (-\frac{1}{2}) = \frac{1}{6}$.
   (iii) Least value is $(-2)(-1) = 2$. Greatest value is $(-5)(-3) = 15$.
   (iv) $\frac{x}{y} = x\left(\frac{1}{y}\right)$ Least value is $(-2)(-\frac{1}{3}) = \frac{2}{3}$.
   Greatest value is $(-5)(-1) = 5$.
   (v) $-8 < x + y < -3$ so that $-\frac{1}{3} < \frac{1}{x+y} < -\frac{1}{8}$.

Exercise 2.1

1. $x = \pm 2{\cdot}5$
2. $x = \pm 0{\cdot}4$
3. $x = \pm 2{\cdot}5$ This is the same equation as in Question 1.
4. $x = \pm 0{\cdot}4$ This is the same equation as in Question 2.
5. $x = \pm 2$
6. $x = \pm 7$
7. $x + 1 = \pm 4$, $x = 3$ or -5
8. $x + 2 = \pm 4$, $x = 2$ or -6
9. $x - 2 = \pm 3$, $x = 5$ or -1
10. $(x - 1)^2 = 4$, $x - 1 = \pm 2$, $x = 3$ or -1

Exercise 2.2

1. $x^2 + 6x + 9 - 9 = (x + 3)^2 - 9$
2. $x^2 + 10x + 25 - 24 = (x + 5)^2 - 24$
3. $x^2 + 18x + 81 - 1 = (x + 9)^2 - 1$
4. $(x + \frac{1}{2})^2$
5. $(x + \frac{3}{2})^2$
6. $x^2 + x + \frac{1}{4} + \frac{3}{4} = (x + \frac{1}{2})^2 + \frac{3}{4}$
7. $x^2 + 2x + 1 + 1 = (x + 1)^2 + 1$
8. $x^2 + 12x + 36 + 4 = (x + 6)^2 + 4$
9. $x^2 + 3x + \frac{9}{4} - \frac{5}{4} = (x + \frac{3}{2})^2 - \frac{5}{4}$
10. $x^2 + 5x + \frac{25}{4} - \frac{25}{4} = (x + \frac{5}{2})^2 - \frac{25}{4}$
11. $2(x^2 + 6x + 9) = 2(x + 3)^2$
12. $2(x^2 + 20x + 100) = 2(x + 10)^2$
13. $x^2 + bx + \dfrac{b^2}{4} - \dfrac{b^2}{4} = \left(x + \dfrac{b}{2}\right)^2 - \dfrac{b^2}{4}$
14. $\left(x + \dfrac{b}{2}\right)^2 - \dfrac{b^2}{4} + 1$
15. $\left(x + \dfrac{b}{2}\right)^2 - \dfrac{b^2}{4} + c$
16. $x^2 - 2x + 1$
17. $x^2 - 4x + 4$
18. $9 - 6x + x^2$
19. $x^2 + bx + \dfrac{b^2}{4}$
20. $x^2 - bx + \dfrac{b^2}{4}$
21. $x^2 - \dfrac{bx}{a} + \dfrac{b^2}{4a^2}$

Exercise 2.3

1. $x^2 - 6x + 8 = 0$; $x^2 - 6x + 9 = -8 + 9$; $(x - 3)^2 = 1$; $x = 4$ or 2
   Alternatively $x^2 - 6x + 8 = (x - 4)(x - 2)$. Hence $x = 4$ or 2.
2. $x^2 - 10x = -3$, $x^2 - 10x + 25 = -3 + 25$, $(x - 5)^2 = 22$; $x = 5 \pm \sqrt{22}$
   $x = 9{\cdot}69$ or $0{\cdot}31$ (correct to 2 decimal places)
3. $x^2 - 2x - 9 = 0$; $x^2 - 2x + 1 = 9 + 1$; $(x - 1)^2 = 10$; $x = 1 \pm \sqrt{10}$
   $x = 4{\cdot}16$ or $-2{\cdot}16$ (correct to 2 decimal places)
4. $x = \dfrac{+10 \pm \sqrt{(100 + 108)}}{6} = \dfrac{10 \pm \sqrt{208}}{6} = \dfrac{10 \pm 14{\cdot}42}{6} = 4{\cdot}07$ or $-0{\cdot}74$
   (correct to 2 decimal places)
5. $x = \dfrac{-11 \pm \sqrt{(121 + 84)}}{14} = \dfrac{-11 \pm \sqrt{205}}{14} = \dfrac{-11 \pm 14{\cdot}32}{14} = 0{\cdot}24$ or $-1{\cdot}81$
   (correct to 2 decimal places)
6. $x = \dfrac{-2 \pm \sqrt{(4 + 88)}}{4} = \dfrac{-2 \pm \sqrt{92}}{4} = \dfrac{-2 \pm 9{\cdot}59}{4} = -2{\cdot}90$ or $1{\cdot}90$
   (correct to 2 decimal places)
7. $x(x + 2) = 13$; $x^2 + 2x + 1 = 14$; $(x + 1)^2 = 14$; $x + 1 = \pm 3{\cdot}74$, $x = -4{\cdot}74$ or $2{\cdot}74$ (correct to 2 decimal places). The sides of the rectangle are of lengths 2·74 cm and 4·74 cm.
8. $12 = 32t - 16t^2$. Rewrite and divide by 4. $4t^2 - 8t + 3 = 0$. This factorises to give $(2t - 1)(2t - 3) = 0$. $t = \frac{1}{2}$ or $1\frac{1}{2}$ s. i.e. $\frac{1}{2}$ s to 12 m from 0 on the way out and $1\frac{1}{2}$ s to 12 m from 0 on the way back.
9. (a) $(x - 5)^2 - 4$, least value -4 (b) $(x + 1)^2 + 2$, least value 2
   (c) $(x + 4)^2 + 3$, least value 3

Exercise 2.4

1. (a) $x = 4$ or $2{\cdot}5$ (b) $x = 3{\cdot}5$ or -1 (c) $x = 1$ or $-1{\cdot}5$
   (d) $x = -0{\cdot}5$ or $-1\frac{2}{3}$ (e) $x = -0{\cdot}8$ or $0{\cdot}5$ (f) $x = 1\frac{1}{3}$ or $0{\cdot}25$
2. (a) $(3x + 2)(x - 1) = 0$ $x = -\frac{2}{3}$ or 1
   (b) $(5x - 1)(x - 7) = 0$ $x = 0{\cdot}2$ or 7
   (c) $(11x + 2)(x + 1) = 0$ $x = -\frac{2}{11}$ or -1
   (d) $(7x - 1)(x - 11) = 0$ $x = \frac{1}{7}$ or 11
   (e) $(7x - 3)(x + 1) = 0$ $x = \frac{3}{7}$ or -1
   (f) $(2x - 1)(x + 3) = 0$ $x = 0{\cdot}5$ or -3
3. (a) $(2x - 5)(x - 4) = 0$ $x = 2{\cdot}5$ or 4
   (b) $(3x + 5)(2x + 1) = 0$ $x = -1\frac{2}{3}$ or $-0{\cdot}5$
   (c) $(5x + 4)(2x - 1) = 0$ $x = -0{\cdot}8$ or $0{\cdot}5$
   (d) $(3x + 2)(2x + 3) = 0$ $x = -\frac{2}{3}$ or $-1{\cdot}5$

(e) $(7x + 3)(x + 5) = 0 \qquad x = -\frac{3}{7}$ or -5
(f) $(4x + 3)(3x - 2) = 0 \qquad x = -0.75$ or $\frac{2}{3}$

Exercise 2.5

1. $x = \dfrac{-1 \pm \sqrt{(1-4)}}{2} = -0.5 \pm i\sqrt{3} = -0.5 \pm 1.73i$ (correct to 2 decimal places)
2. $x = \dfrac{-2 \pm \sqrt{(4-8)}}{2} = -1 \pm 1i$
3. $x = \dfrac{-1 \pm \sqrt{(1-60)}}{2} = -0.5 \pm 7.68i$ (correct to 2 decimal places)
4. $\alpha + \beta = -1.5,\ \alpha\beta = 2.5$
5. $\alpha + \beta = -\frac{2}{3},\ \alpha\beta = \frac{11}{3}$
6. $\alpha + \beta = -1\frac{1}{3},\ \alpha\beta = \frac{1}{3}$
7. $x^2 - 7x + 10 = 0;\ (x - 5)(x - 2);\ \alpha = 5,\ \beta = 2$
8. $x^2 + 11x + 18 = 0;\ (x + 9)(x + 2);\ \alpha = -9,\ \beta = -2$
9. $x^2 - 7x + 12 = 0;\ (x - 4)(x - 3);\ \alpha = 4,\ \beta = 3$
10. $x^2 + \dfrac{x}{2} - \dfrac{21}{2} = 0;\ 2x^2 + x - 21 = 0 = (2x + 7)(x - 3);\ \alpha = -3.5,\ \beta = +3$
11. $p = -7,\ q = 12$
12. $x = 9$ or 4. Put $w^2 = x$ so that the solution is $w^2 = 4$ or 9 and therefore $w = \pm 2$ or ± 3.

Exercise 2.6

1. $a = 1,\ b = 2m,\ c = m^2.\ \therefore b^2 - 4ac = 4m^2 - 4m^2 = 0$
   $\therefore x = -m$, and the roots are real if m is real.
2. $a = 1,\ b = m - 2,\ c = 2 - m.\ \therefore b^2 - 4ac = (m - 2)^2 - 4(2 - m) = m^2 - 4.$
   $b^2 - 4ac \geqslant 0$ if $m \geqslant 2$ or $m \leqslant -2$. If $-2 < m < 2$ then $m^2 - 4$ is negative and the roots are not real.
3. $\alpha + \beta = -\dfrac{2}{10},\ \alpha\beta = \dfrac{3}{10}.\ \therefore 2\alpha + 2\beta = -\dfrac{4}{10},\ 2\alpha 2\beta = \dfrac{12}{10}$
   The required equation is $10x^2 + 4x + 12 = 0$.
4. $\dfrac{12}{y^2} + \dfrac{10}{y} + 4 = 0$. Therefore $4y^2 + 10y + 12 = 0$ is the required equation.
   When $x = \alpha$ we get $y = \dfrac{2}{\alpha}$ when $x = \beta$ we get $y = \dfrac{2}{\beta}$. Alternatively, $\alpha + \beta = -\dfrac{5}{3}$;
   $\alpha\beta = \dfrac{4}{3};\ \dfrac{1}{\beta} + \dfrac{1}{\alpha} = -\dfrac{5}{4};\ \dfrac{2}{\alpha} + \dfrac{2}{\beta} = -\dfrac{5}{2};\ \dfrac{1}{\alpha\beta} = \dfrac{3}{4};\ \dfrac{4}{\alpha\beta} = 3.$
5. $2r + 3 = -2m;\ r^2 + 3r = n,\ 4m^2 = 4r^2 + 12r + 9 = 4n + 9$
6. $x = \dfrac{3}{y}$
7. If α and β are the roots of $x^2 - 8x + 15 = 0$ then $x = \alpha$ gives $y^2 = \alpha$ and $y = \pm\sqrt{\alpha}$. The roots are $\pm\sqrt{5}$ or $\pm\sqrt{3}$.
8. Pythagoras's Theorem enables us to write $(x + 2)^2 = (x + 1)^2 + x^2$, which reduces to $x^2 - 2x - 3 = (x - 3)(x + 1) = 0$. There is only one acceptable solution $x = 3$. The sides of the triangle are 3, 4, 5.

Exercise 3.1

1. $3x^2 + 4x + 1 = (x - 1)(3x + 7) + 8$. Substituting $x = 10$ this result becomes $3(10^2) + 4(10) + 1 = 341 = (9)(37) + 8$.
2. $(3x^2 + 4x + m) = (x - 1)(3x + 7) + m + 7.\ m + 7 = 13$ implies $m = 6$
3. (i) $n = 7 \times 3 + 6 = 27$, (ii) 3
4. $3x^3 + x + 1 = (x + 2)(3x^2 - 6x + 13) - 25$ (25 should be added)
5. $4x^3 + 3x + 9 = (4x^2 - 8x + 19)(x + 2) - 29$
6. -3. 0. This second result means that $x - 1$ and $x^2 + 3x + 2 = (x + 1)(x + 2)$ are factors of $f(x) + 3$.

Exercise 3.2

1. $f\left(\frac{3}{2}\right) = 2.\frac{27}{8} - \frac{9}{4} - \frac{21}{2} + 6 = 0$ so that $x - \frac{3}{2}$ or $2x - 3$ is a factor
$f(x) = (2x - 3)(x^2 + x - 2) = (2x - 3)(x - 1)(x + 2)$
2. $f(1) = 0, f(-1) = 0, f(2) = 0.\ \therefore f(x) = (x - 1)(x + 1)(x - 2)$
3. $f(x) = n(x - 1)(x - 3)(x + 7)$, where n is any number, is a function which has the factors $x - 1$, $x - 3$ and $x + 7$. For $f(0) = 63$ we must have $n = 3$. Final result $f(x) = 3(x - 1)(x - 3)(x + 7)$.
4. For $x^2 + x + 1 = 0$ $a = 1$, $b = 1$, $c = 1$ and $b^2 - 4ac = -3$, so the factors of $x^2 + x + 1$ cannot be real. Therefore $f(n)$ cannot be zero because this would imply that $x^2 + x + 1$ has a real factor $x - n$.
5. $f(1) = 1 + p + q + r = 0$ (i)
$f(-1) = -1 + p - q + r = 0$ (ii)
$f(3) = 27 + 9p + 3q + r = 8$ (iii)
From (i) + (ii) we get $2p + 2r = 0.\ \therefore p = -r$, and $q = -1$. Substituting these results in (iii), $27 - 9r - 3 + r = 8$ gives $r = 2.\ \therefore p = -2, q = -1, r = 2$
6. $x^2 - 1 = (x - 1)(x + 1)\ \therefore f(1) = 0$ and $f(-1) = 0$, $m = n = -1$
7. The quadratic is $f(x) = n(x - 1)(x + 2)$. $f(3) = 10n = 20$, $n = 2$.
8. $f(1) = m(0) + n(0) = 0$ for all values of m and n
9. (i) $x^3(a + d) + x^2(b + c) + x(c + b) + d + a = f(x)$. $f(-1) = 0$ therefore $x + 1$ is a factor. Now put $x = 10$ and we have shown that the number was divisible by 11.
(ii) $px^2 + qx + r + rx^2 + qx + p = x^2(p + r) + x(2q) + p + r = f(x)$
$f(-1) = 2p + 2r - 2q$, which is zero only if $p + r = q$, thus $462 + 264 = 66 \times 11$ but $461 + 164 = 625$ is not divisible by 11.
10. $g(4) = f(4) - f(4) = 0\ \therefore\ x - 4$ is a factor of $g(x)$

Exercise 3.3

1. (i) $(x - y + z)^2$, (ii) $(-x + y + z)^2$, (iii) $(x + y - z)^2$, (iv) $(x + y + z + w)^2$
2. $a^2 + 2a + 1 - b^2 = (a + 1 - b)(a + 1 + b)$
3. $2a + b - 3$
4. Sum is $(y - 1)(x^2 + 4x + y - 1)$. Quotient is $x^2 + 4x + y - 1$
5. $x^3 - y^3 = (x - y)(x^2 + xy + y^2)$
6. $x^3 + y^3 = (x + y)(x^2 - xy + y^2)$
7. (i) $(4a - 3b)(16a^2 + 12ab + 9b^2)$ (ii) $(2p + y)(4p^2 - 2py + y^2)$
8. $[(2x + y + 1) - (x - y + 1)]\,[(2x + y + 1) + (x - y + 1)]$
$= (x + 2y)(3x + 2)$
9. $4x^4 - 9x^2 - 1 + 6x = (2x^2 + 3x - 1)(2x^2 - 3x + 1)$
$= (2x^2 + 3x - 1)(2x - 1)(x - 1)$
10. $f(b + c) = 2b^2 + 4bc + 2c^2 + b^2 + cb - bc - c^2 - 3b^2 - c^2 - 4bc = 0$
$\therefore\ x - b - c$ is a factor. The other factor is $2x + 3b + c$.

Exercise 4.1

1. $x = 61$, $y = 60$
2. $x = 25$, $y = 24$
3. $x = 0{\cdot}5$, $y = 4$; $x = 2$, $y = 1$. Substitute $2x = 5 - y$.
4. $x = 2$, $y = 1$; $x = -\frac{5}{9}$, $y = -\frac{14}{9}$. Substitution of $x = y + 1$ produces $9y^2 + 5y - 14 = 0 = (9y + 14)(y - 1)$.
5. $x = 2$, $y = -2$. $4x^2 - 4yx - 3y^2 = (2x - 3y)(2x + y)$
6. $x = -1$, $y = -2$; $x = 2\frac{12}{13}$, $y = -\frac{9}{13}$. Substitution of $x = 5 + 3y$ produces $13y^2 + 35y + 18 = 0 = (13y + 9)(y + 2)$.
7. $x = 4$, $y = 5$; $x = -4$, $y = 5$; $x = +\sqrt{5}$, $y = -6$; $x = -\sqrt{5}$, $y = -6$. Substitution of $x^2 = 11 + y$ produces $y^2 + y - 30 = 0 = (y + 6)(y - 5)$.
8. $x = 8$, $y = 6$; $x = 8$, $y = -6$; $x = -8$, $y = 6$; $x = -8$, $y = -6$. Addition of the two equations produces $2x^2 = 128$, etc.
9. $x - y = \pm 4$, $x + y = \pm 2$; $x = 1$, $y = -3$; $x = -1$, $y = 3$; $x = 3$, $y = -1$; $x = -3$, $y = 1$

10. Subtraction of the two equations gives $y + \frac{1}{y} = 2$.

$\therefore\ y^2 - 2y + 1 = 0 = (y - 1)^2$. Solution $x = \frac{\pm 1}{\sqrt{2}}, y = 1$.

Exercise 4.2

1. $x = 4, y = 2; x = 6, y = 1\frac{1}{3}; x = -6, y = -1\frac{1}{3}; x = -4, y = -2$. Substitute $y = \frac{8}{x}$ to get $x^4 - 52x^2 + 576 = 0 = (x^2 - 16)(x^2 - 36)$. $x = \pm 4, \pm 6$.
2. Subtract the equations to reveal this question to be the same as question 1. This is an ellipse $x^2 + 9y^2 = 25$ intersected in four real points by a rectangular hyperbola $xy = 8$.
3. Multiply the first equation by 3 and subtract the second from the result to get $2x^2 - 5xy + 2y^2 = (2x - y)(x - 2y) = 0$. $\therefore\ y = 2x$ or $x = 2y$. Substitute these results in (i) to get $x^2 - 6x^2 + 4x^2 = -1$; $x^2 = 1$, $x = \pm 1$, $y = \pm 2$ and $4y^2 - 6y^2 + y^2 = -1$; $y^2 = 1$, $y = \pm 1$, $x = \pm 2$.
4. Multiply (i) by 2 and add the results to (ii). Hence $7x^2 = 28$, $x = \pm 2$, $y = \pm 2$, ∓ 2. Solutions are $x = 2, y = 2; x = 2, y = -2; x = -2, y = 2; x = -2, y = -2$.
5. Multiply (i) by 11 and subtract (ii) from the result to get $14x^2 - 13xy + 3y^2 = 0$ after dividing by 24. This equation becomes $(2x - y)(7x - 3y) = 0$. Solutions are $x = 1, y = 2; x = -1, y = -2; x = 3, y = 7; x = -3, y = -7$.
6. Multiply (i) by 11 and (ii) by 10 and subtract the results to get $12x^2 + 23yx + 10y^2 = (4x + 5y)(3x + 2y) = 0$. Substitute $x = -\frac{5y}{4}$ or $-\frac{2y}{3}$ in (i). Solution $x = 5, y = -4; x = -5, y = 4; x = 2, y = -3; x = -2, y = 3$.
7. Multiply (i) by 13 and subtract (ii) from the result to get $3x^2 + 10xy + 3y^2 = 0 = (3x + y)(x + 3y)$. Solutions are $x = 3, y = -1; x = -3, y = 1; x = 1, y = -3; x = -1, y = 3$.
8. From equation (ii) $x = 8$ or $y = 2$. Substitute in (i) $y = 0$; or $x = 0$. Solution $x = 8, y = 0; x = 0, y = 2$. The two lines $x = 8$, $y = 2$ are tangents to the ellipse $x^2 + 16y^2 = 64$.
9. Substitute $y = 2x + 11$ in (i) to get $x^2 - 2x + 2 = 0$. Solutions $x = 1 + i$, $y = 13 + 2i$; $x = 1 - i$, $y = 13 - 2i$.
10. Substitute $x = 2y + 5$ in (i) to get $y^2 - 2y + 5 = 0$. $y = 1 \pm 2i$. Solutions $x = 7 + 4i; y = 1 + 2i; x = 7 - 4i, y = 1 - 2i$.

Exercise 4.3

1. (a) 7, (b) 17, (c) 143, (d) 19

We can economise on the layout as follows (d)

```
2261) 2337 (1
      2261
        76) 2261 (29
            152
            741
            684
             57) 76 (1
                 57
                 19) 57 (3
                     57
                     00
```

Alternatively after the first stage when we obtain the remainder 76 we may reason that since $76 = 19 \times 4$ and since 2337, being an odd number, cannot have 4 as a factor it follows that if a common factor exists it must be 19. We confirm the result by dividing 2337 by 19.

2. (a) $2x + 3$ (b) $3x + 5$ (c) $x + 3$
   (d) $6x^2 + 13x - 28 - 2(3x^2 - 11x - 4) = 35x - 20 = 5(7x - 4)$. If there is a common factor it must be $7x - 4$. But, $7x - 4$ is not a factor of $3x^2 - 11x - 4$, hence there is no common factor.
   (e) $3x + 7$. Find the common factor of the first two expressions then see if it is also a factor of the third expression.
3. (a) $x^2 + x + 1$ (b) $x + 4$ (c) $2x + 3$ (d) $3x - 7$ (e) $3x + 8$

Exercise 4.4

1. $x^2 + 3x + 2$. After the second stage of the division, simplify the remainder by dividing by 2 to give $x^2 + 3x + 2$.
2. On subtracting the two expressions we get $x - 1$ so this is the only possible common factor. Applying the remainder theorem we find that $f(1)$ is not zero so $x - 1$ is not a factor.
3. $5x^2 - 1$. $20x^4 + x^2 - 1 = (5x^2 - 1)(4x^2 + 1)$;
   $25x^4 + 5x^3 - x - 1 = (5x^2 - 1)(5x^2 + x + 1)$
4. $f(x) = (x^2 + x - 6)(x^2 - 1) + x - 2 = x^4 + x^3 - 7x^2 + 4$
   $f(x) = (x - 2)(x^3 + 3x^2 - x - 2)$
5. $(x^2 - x + 1)(ax^2 + bx + x)$. Choose any value for a, b, c such that $a \neq 0$.
6. $(x^2 + x + 1)(x + 1)$, $(x^2 + x + 1)(x - 1)$, $(x^2 - x + 1)(x - 1)$
7. $x^2 + 7x + 12 = (x + 3)(x + 4)$. $k = 3$ or 4
8. $x^2 + 4x + 3$
9. $x^3 + 3x^2 + 3x + 1$. These expressions are $(x + 1)^5$, $(x + 1)^4$ and $(x + 1)^3$ respectively.
10. $x + 1$. $x^6 - 1 = (x^3 - 1)(x^3 + 1) = (x - 1)(x + 1)(x^2 + x + 1)(x^2 - x + 1)$
    $x^3 + 3x^2 + 3x + 1 = (x + 1)^3$

Exercise 5.1

1. (a) The point (2, 5) lies on the line KL and does not belong to the set (a).
   (b) Yes (c) No (d) $(-1, -1)$, $(-\frac{1}{2}, -\frac{1}{2})$, etc.
   (e) $(-3, -1)$, $(-6, -4)$, etc. shading for (d) extended outside the square.

See Fig. S.1

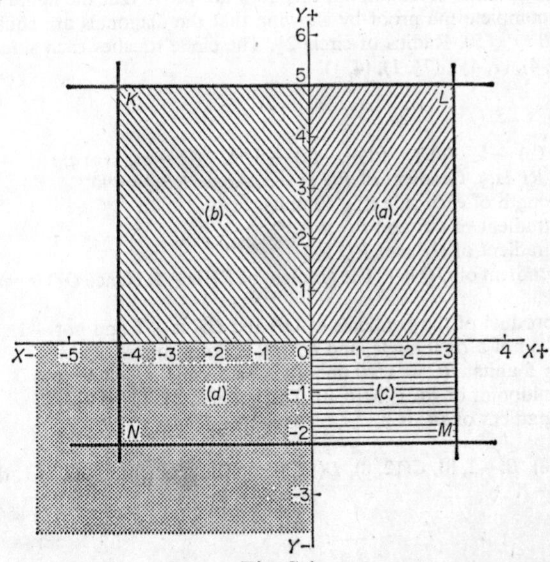

Fig. S.1

2. $(-1, -1)$, $(-2, -1)$, $(-1, -1{\cdot}9)$, $(-\frac{1}{2}, -1\frac{1}{4})$, etc.
3. $x = -4$ 4. $y = 5$
5. (a) Midpoint of OK has the coordinates $(-2, 2\frac{1}{2})$.
   Midpoint of OL has the coordinates $(1\frac{1}{2}, 2\frac{1}{2})$.
   Midpoint of OM has the coordinates $(1\frac{1}{2}, -1)$.
   Midpoint of ON has the coordinates $(-2, -1)$.
   (b) On OK. $(-1, 1\frac{1}{4})$, $(-3, 3\frac{3}{4})$, etc. Any point (x, y) such that $4y = -5x$.
   On OL. $(1, 1\frac{2}{3})$, $(2, 3\frac{1}{3})$, etc. Any point (x, y) such that $3y = 5x$.
   On OM. $(1, -\frac{2}{3})$, $(2, -1\frac{1}{3})$, etc. Any point (x, y) such that $3y = -2x$.
   On ON. $(-1, -\frac{1}{2})$ $(-3, -1\frac{1}{2})$, etc. Any point (x, y) such that $2y = x$.
6. (a) Midpoint of KL is $(-\frac{1}{2}, 5)$; midpoint of LM is $(3, 1\frac{1}{2})$; midpoint of MN is $(-\frac{1}{2}, -2)$; midpoint of NK is $(-4, 1\frac{1}{2})$.
   (b) 7 units
   (c) $OL^2 = 3^2 + 5^2 = 34$. $\therefore OL = \sqrt{34} = 5{\cdot}83$ (correct to 2 decimal places)
   $OM^2 = 3^2 + 2^2 = 13$. $\therefore OM = \sqrt{13} = 3{\cdot}61$ (correct to 2 decimal places)

Exercise 5.2

1. $D(6, 2)$. The centre of the square is the midpoint of AC or BD, $(4, 4)$.
2. $AB = \sqrt{32} = 4\sqrt{2}$, $AC = \sqrt{200} = 10\sqrt{2}$, $AD = 20$, $AE = \sqrt{464} = 4\sqrt{29}$, $BC = \sqrt{136} = 2\sqrt{34}$, $BD = \sqrt{464} = 4\sqrt{29}$, $BE = \sqrt{592} = 4\sqrt{37}$, $CE = 6\sqrt{10}$, $CD = 10\sqrt{2}$, $DE = \sqrt{32} = 4\sqrt{2}$
3. $D(4, 0)$ The centre of the square is $(4, 3)$. Since the x coordinates of B and D are the same then BD is parallel to the y axis.
4. x axis $(\pm 5, 0)$, y axis $(0, \pm 5)$
5. $AB^2 = 8^2 + 4^2 = 80$ $\therefore AB = \sqrt{80}$ The centre of the circle is $(2, 5)$.
6. $OA^2 = 2^2 + 11^2 = 125$, $OA = 11{\cdot}18$
   $OB^2 = 6^2 + 9^2 = 117$, $OB = 10{\cdot}82$, $OB - OA = 0{\cdot}36$ (correct to 2 decimal places). The midpoint of AB is $M(4, 5)$. $OM^2 = 4^2 + 5^2 = 41$.
   M is closer to the origin than either A or B.
7. Centre $(6, 6)$. Touches at $(0, 6)(6, 0)$. Other ends of the diameters $(12, 6)$ $(6, 12)$.
8. Each side is 5 units in length, but this does not prove that the figure is a square. We may complete the proof by showing that the diagonals are equal in length $AC = DB = \sqrt{50}$. Radius of circle $2\frac{1}{2}$. The circle touches each side at its midpoint $(3\frac{1}{2}, 4)$, $(7, 4\frac{1}{2})$, $(7\frac{1}{2}, 1)$, $(4, \frac{1}{2})$.

Exercise 5.3

1. (a) -2, (b) -4, (c) $\frac{1}{3}$, (d) $\frac{7}{3}$, (e) $-1/m$, (f) $-m$, (g) q/p
2. Gradient RP is $\frac{1}{3}$. Gradient of PQ is -3. Angle $RPQ = 90°$
3. (i) The length of each side is $\sqrt{40}$ units.
   (ii) The gradient of OR and PQ is 3.
   (iii) The gradient of OP and RQ is $\frac{1}{3}$.
   (iv) The gradient of OQ is 1. The gradient of RP is -1. Hence OQ is perpendicular to RP.
   (v) The product of the gradients of OP and OR is $+1$ and not -1 so the angle ROP is not a right angle, and $OPQR$ is not a square.
4. (i) $AB = 5$ units; $BC = \sqrt{20}$ units; $CA = 5$ units $\therefore AB = AC$
   (ii) The midpoint of BC is $M(6, 0)$. The gradient of AM is $-\frac{1}{2}$.
   The gradient of BC is 2. $\therefore AM$ is perpendicular to BC.
5. $y = 4$ 6. $x = 1$
   $A(-2, -6)$, $B(-2, 8)$, $C(12, 8)$, $D(12, -6)$. The gradient of AC is 1, the gradient f BD is -1.

8. The gradient of the line AB must be 5. If we put $A(x_1, y_1)$, $B(x_2, y_2)$ then
$\frac{y_2 - y_1}{x_2 - x_1} = 5.$
Suggest $B(3, ?)$, $A(2, ?)$ then $B(3, 8)$, $A(2, 3)$
or $B(3, 9)$, $A(2, 4)$ and so on.

Exercise 6.1

1. A line segment like AB in Fig. 15(a) but through the centre of the page.
2. The circumference of a circle centre A and radius 6 cm.
3. One straight line parallel to PQ and 30 cm away.
4. A straight line at right angles to AB and passing through the midpoint of AB, which is the point (3, 4). Unlike the locus in Fig. 15(c) this locus is unbounded, i.e. it is anywhere on the line and not confined to a line segment. The equation of the line is $y = 4$.
5. The locus consists of two straight lines parallel to the line $x = 7$ and 3 units away on either side. The equations of the two lines are $x = 4$ and $x = 10$.

Exercise 6.2

1. AB: $y = 4x$; BC: $7y = 5x$; CA: $6y = x + 23$
2. $\frac{y - 7}{x - 2} = 3$. Final form $y = 3x + 1$.
3. The gradient of the line is 13. Therefore $\frac{y - 5}{x - 1} = 13$ or $y = 13x - 8$.
4. The line intersects the x axis at (2, 0) and the y axis at (0, 5).
   (i) $\frac{x}{6} + \frac{y}{5} = 1$ (ii) $\frac{x}{7} + \frac{y}{8} = 1$
   (iii) $\frac{x}{-2} + \frac{y}{3} = 1$ or $\frac{y}{3} - \frac{x}{2} = 1$
   (iv) $\frac{x}{-4} + \frac{y}{-5} = 1$ or $\frac{x}{4} + \frac{y}{5} = -1$ (v) $\frac{x}{a} + \frac{y}{b} = 1$
5. The gradient of the perpendicular is $\frac{1}{2}$. $2y = x + 1$
6. Rewrite the equation as $by = -ax - c$; $y = -\frac{ax}{b} - \frac{c}{b}$. The gradient is $-\frac{a}{b}$
   (v), (ii), (i), (iii), (iv).
7. If c is positive the line intersects the y axis above the origin and if c is negative the point of intersection is below the origin. If $c = 0$ then the line passes through the origin. The line intersects the y axis at the point $(0, c)$.
8. The gradient of AB is 2, so the gradient of the perpendicular bisector is $-\frac{1}{2}$, and this must pass through the midpoint of AB, (3, 5). $2y = -x + 13$ is the final form.
9. (i), (iv) parallel (ii), (viii) perpendicular
   (iii), (vi) perpendicular (i), (v) perpendicular, also (iv), (v)
10. The perpendicular bisector of AB is required, i.e. $y = -x + 4$. To be equidistant from B and C the perpendicular bisector of BC is required, i.e. $y = 7$. The required point is (−3, 7).

Exercise 6.3

1. (i) $(x - 2)^2 + (y - 6)^2 = 16$ or $x^2 + y^2 - 4x - 12y + 24 = 0$
   (ii) $(x + 1)^2 + (y + 1)^2 = 9$ or $x^2 + y^2 + 2x + 2y - 7 = 0$
   (iii) $(x + 5)^2 + (y - 5)^2 = 25$ or $x^2 + y^2 + 10x - 10y + 25 = 0$
   (iv) $x^2 + y^2 + 12x + 16y = 0$ or $(x + 6)^2 + (y + 8)^2 = 100$
   Centre (−6, −8) radius 10.
   (v) $x^2 + (y - 4)^2 = 16$ or $x^2 + y^2 - 8y = 0$
   (vi) $(x - 4)^2 + y^2 = 16$ or $x^2 + y^2 - 8x = 0$

2. Centre (5, 6). Radius 5. $(x - 5)^2 + (y - 6)^2 = 25$
   or $x^2 + y^2 - 10x - 12y + 36 = 0$
3. Centre (9, 1). $(x - 9)^2 + (y - 1)^2 = 50$ or $x^2 + y^2 - 18x - 2y + 32 = 0$
4. (i) Centre (1, 3). Radius 10 (ii) Centre $(-1, -3)$. Radius 10
   (iii) Centre $(-4, 5)$. Radius 8
5. (a) Put $x = 1$ $\therefore y^2 - 6y - 91 = 0 = (y - 13)(y + 7)$. $\therefore y = 13$ or -7. Points are (1, 13), (1, -7).
   (b) As a different method from (a) use the form $(x + 1)^2 + (y + 3)^2 = 100$ and now substitute $y = 5$ to get $(x + 1)^2 + 64 = 100$ $\therefore$ $x + 1 = \pm 6$. $(-7, 5)$, $(+5, 5)$
   (c) Substitute $y = x + 1$ in the equation of the circle to get
   $x^2 + (x + 1)^2 + 8x - 10(x + 1) - 23 = 0$ which reduces to $x^2 = 16$, $x = \pm 4$; with $y = x + 1$ the points of intersection are (4, 5), $(-4, -3)$.
6. $(-2, 8)$; (5, 1)
7. $PO^2 = x^2 + y^2$, $PB^2 = (x - 8)^2 + y^2 = x^2 - 16x + y^2 + 64$. Equation of the locus is $2x^2 + 2y^2 - 16x + 64 = 64$, i.e. $x^2 + y^2 - 8x = 0 = (x - 4)^2 + y^2 - 16$, which is the equation of a circle centre (4, 0) radius 4.
8. $PO^2 = x^2 + y^2, PB^2 = x^2 + (y - 6)^2$. $\therefore 2x^2 + 2y^2 - 12y + 36 = 50$. Equation of the locus is $x^2 + y^2 - 6y - 7 = 0$. $x^2 + (y - 3)^2 = 16$, i.e. a circle as described. Neither (0, 0) nor (0, 6) satisfy the equation.

Exercise 6.4

1. (i) The line $x = 10$ intersects the circle where $y^2 = 0$, i.e. two coincident points (10, 0).
   (ii) As in (i). Tangent at $(-10, 0)$.
   (iii) Tangent at (0, 10). (iv) Tangent at $(0, -10)$.
   (v) $9x^2 = (50 - 4y)^2 = 2500 - 400y + 16y^2$; $9x^2 = 900 - 9y^2$
   $\therefore$ $25y^2 - 400y + 1600 = 0 = (5y - 40)(5y - 40)$, $y = 8$
   Two equal roots therefore the line is a tangent. Point of contact is (6, 8).
   (vi) Tangent at $(6, -8)$.
2. The centre of the circle is the point $C(3, 9)$. Radius 4. (i) $y = 13$, (ii) $x = 7$.
3. $5x + 12y + 169 = 0$ and $12x + 5y = 169$ respectively.
4. The centre of one circle is $A(8, 2)$ and of the other is $B(8, 10)$. Equations are $(x - 8)^2 + (y - 2)^2 = 16$, $(x - 8)^2 + (y - 10)^2 = 16$.
5. 4
6. The centre of the circle is (0, 0) and the radius 8.
   $AT^2 = 11^2 - 8^2 = 57$. $\therefore$ $AT = \sqrt{57}$.
7. Circle centre $C(2, 0)$, radius $\sqrt{20}$. $AC^2 = 4 + 144$, $AT^2 = 148 - 20 = 128$
   Circle centre $D(-3, 0)$, radius 5. $AD^2 = 9 + 144$, $AT^2 = 153 - 25 = 128$
   Note: The squared length of the tangent is also given by substituting the coordinates of A in the equation of the circle.
8. Any parallel line must be of the form $2y + x = c$. Find the intersection of this line with the circle by substituting $x = c - 2y$ to get $5y^2 - 4yc + c^2 - 20 = 0$. Condition for equal roots (i.e. a tangent) $16c^2 = 20\ (c^2 - 20)$. $c = \pm 10$
   Tangents are $2y + x = \pm 10$.
9. (i) $x + y\sqrt{3} = 4$ (ii) $x\sqrt{3} + y = 4$ (iii) $x\sqrt{2} + y\sqrt{2} = 4$
   Notice that these equations all arise from substitution in the equation $xx_1 + yy_1 = 4$, e.g. in (i) $x_1 = 1$, $y_1 = \sqrt{3}$, etc.
10. Substitute $y = mx$. $x^2(1 + m^2) - 12x + 18 = 0$ has equal roots in x if $144 = 72(1 + m^2)$. $m = \pm 1$. The required tangents are $y = \pm x$. Contact points (3, 3), $(3, -3)$.

Exercise 7.1

1. (i) $x + y = 6$ at (1, 5) (2, 4), (3, 3) (ii) $x + 2y = 11$ at (1, 5)
   (iii) $2x + y = 9$ at (4, 1); (3, 3) (iv) $4x + 3y = 21$ at (3, 3)
   (v) $7x + 5y = 36$ at (3, 3) (vi) $5x + 7y = 40$ at (1, 5)

2. The extra points introduced into the solution region are (5, 0), and (0, 6).
 (i) no change (ii) $x + 2y = 12$ at (0, 6)
 (iii) $2x + y = 10$ at (5, 0) (iv) no change
 (v) no change (vi) $5x + 7y = 42$ at (0, 6)
3. (i) $8x + 5y \leqslant 40$ (ii) $2000x + 3000y \leqslant 18000$ or $2x + 3y \leqslant 18$
 (iii) $7x + 5y$ pence. Three cartons of A and four cartons B to give a profit of 36 pence (see Question 2 (v)).
4. $5x + 6y = 48$ at (0, 8)
5. (i) $2x + y = 9$ at (2, 5), (3, 3) (ii) $3x + y = 12$ at (4, 0), (3, 3)
6. Greatest value $x + y = 7$ at (3, 4) Least value $x + y = 4$ at (4, 0)
7. 2·4, −2·4. The lines are on different sides of the origin.
8. Result as for question 7. 9. $1{\cdot}4 - 0{\cdot}9 = 0{\cdot}5$ 10. 3

Exercise 7.2

1. Anywhere on the curve between T and Q.
2. Line PT has the equation $y = x + 6$ (i) $x^2 < x + 6$ for $-2 < x < 3$
 (ii) $x^2 > x + 6$ for $x < -2$ and $3 < x$
3. This will be the reflection of the tangent at (3, 9) in the y axis $y = -6x - 9$.
4. The gradient of the tangent is $\tan\theta = 6$. Hence $\theta = 80° 34'$ approximately. Notice that you cannot measure the angle θ directly from Fig. 25 since the x and y axes are not drawn to the same scale.
5. -8; $y = -8x - 16$. If you obtain gradient m then the equation becomes $y = mx + 16 - 4m$. In finding the gradient remember that 1 cm on the y axis is 5 units while 1 cm on the x axis is 1 unit.
6. (i) $4{\cdot}5 \pm 0{\cdot}2$ (ii) $5{\cdot}5 \pm 0{\cdot}2$ (iii) $7{\cdot}7 \pm 0{\cdot}2$
7. (i) $y = x^2$ raised 3 units on the y axis (see Fig. 1(*c*), page 37)
 (ii) $y = x^2$ lowered 10 units on the y axis
 (iii) $y = x^2$ reflected in the x axis, i.e. all the y coordinates become $-y$.
 (iv) $y = x^2$ rotated clockwise through 90° about the origin (see Fig. 1(*b*), page 37).
 (v) $y = x^2$ rotated anti-clockwise through 90° about the origin.
 (vi) $y = x^2$ rotated clockwise through 90° about the origin and then moved to the right three units on the x axis (see Fig. 1(*d*), page 37).

Exercise 7.3

1. $\sqrt{24}$
2. $m = 1, c = 0, p(2, 3) = \dfrac{3-2}{\sqrt{2}} = \dfrac{1}{\sqrt{2}}; p(3, 2) = \dfrac{2-3}{\sqrt{2}} = -\dfrac{1}{\sqrt{2}}$. The points are equidistant from the line but on opposite sides.
3. $\dfrac{a-b}{\sqrt{2}}$ or $\dfrac{b-a}{\sqrt{2}}$ 4. $\dfrac{a+b}{\sqrt{2}}$
5. 3·5 (correct to one decimal place)
6. (i) $2y + 8x = 32$ or $y + 4x = 16$ (ii) $4y + x = 16$
 The coordinates of any point on $xy = 16$ are either both positive or both negative. The gradient of the tangent at (x_1, y_1) is $-y_1/x_1$ and thus always negative. From $m_1m_2 = -1$ for perpendicularity there is no positive m_1 hence no two tangents are perpendicular. [exclude the asymptotes]
7. (i) $y = x + 5$ (ii) $y = -x - 5$
8. Tangent $9y + x = 30$ Normal $y - 9x = -133\frac{1}{3}$
9. $y + x = 2$ and $y + x = -2$ at $(-1, -1)$, $(1, 1)$. Find the intersection of the line $y = -x + c$ with $xy = 1$, and then apply the condition that both points must coincide.
10. Let $P(x, y)$ be any point on $y^2 = 4ax$. Distance from $(a, 0)$ is given by $(x - a)^2 + y^2 = x^2 - 2xa + a^2 + 4ax = (x + a)^2$.

Exercise 8.1

1. $x = 2, y = 3, z = 5$
2. $x = 2, y = -1, z = 0, w = -3$. Eliminate x and w by (i) + (ii) + (iv) and -10 (i) + (ii) + 8(iii).
3. $x = 1, \ y = -2, \ z = 1, \ w = 1$. Eliminate y and z by (iii) + (iv) and 2(ii) − (i) − (iv).
4. $x = 2, \ y = -1, \ z = 5, \ w = -1$. Eliminate w and z by 4 (ii) − (iv) and 5(i) − 2(ii) + (iii).
5. $x = \frac{4}{15}, \ y = \frac{8}{9}, \ z = -\frac{40}{3}, \ w = \frac{51}{5}$. Eliminate z and w by (i) + (iv) and (ii) − (iii) + (iv).
6. $x = \frac{11}{8}, \ y = -\frac{29}{4}, \ z = \frac{9}{8}, \ w = \frac{13}{4}$. Eliminate x and y by (i) + (iii) and 5(ii) + (iii) − 3(iv).
7. 2(i) − (iv) = (ii). $x = 5 + t, y = \frac{1}{4}(65 - t), z = t$
8. $x = \frac{1}{3}(49 + 23t), \quad y = t, \quad z = 5 + 4t, \quad w = \frac{1}{3}(26 + 7t)$. 2(iv) = (i) + (iii) When $t = 1, x = 24, y = 1, z = 9, w = 11$
9. (i) + (ii) + (iii) gives $0 = 60$
10. $x + y + z = 6, \ x + y - z = 0$. There is more than one solution. In any solution $z = 3, \ x + y = 3$. If the solutions are integers then $x = 1, \ y = 2, z = 3; x = 2, y = 1, z = 3$ are the only solutions.

Exercise 8.2

1. (iv) and (vi). Two parallel planes (Fig. 31). (case 2B)
2. (ii), (iii), (iv). (Fig. 37). (case 7)
3. (i) = (ii) + (v). Fig. 34. Three planes with a common line. (case 5B)
4. (ii), (iii), (v). Fig. 36. (case 6B)
5. $x = -4, y = -2, z = 18$. Fig. 36.
6. Three parallel planes since all pairs of the equations are inconsistent.
7. Three inconsistent and independent, each pair consistent. Fig. 37.

Exercise 9.1

1.
$$x = \frac{\begin{vmatrix} 13 & 4 \\ 8 & 5 \end{vmatrix}}{\begin{vmatrix} 7 & 4 \\ 3 & 5 \end{vmatrix}} = \frac{33}{23} \qquad y = \frac{\begin{vmatrix} 7 & 13 \\ 3 & 8 \end{vmatrix}}{23} = \frac{17}{23}$$

2.
$$x = \frac{\begin{vmatrix} -7 & 19 \\ 3 & 11 \end{vmatrix}}{\begin{vmatrix} 13 & 19 \\ 7 & 11 \end{vmatrix}} = -\frac{134}{10} = -13{\cdot}4 \qquad y = \frac{\begin{vmatrix} 13 & -7 \\ 7 & 3 \end{vmatrix}}{10} = 8{\cdot}8$$

3. $x = -\frac{38}{25}, y = \frac{41}{25}$ 4. $x = \frac{85}{32}, y = \frac{137}{32}$
5. $x = y = -1$ 6. $x = 1, y = -4$
7. The determinant of the detached coefficients is zero

$$x = \frac{12}{0}, y = -\frac{6}{0}$$

both results being meaningless.

Exercise 9.2

1. -5 2. -36 3. 15 4. abc 5. 30 6. $4(x - 1)(x + 2)$

Exercise 9.3

1. 0 2. 0 3. 0

Exercise 9.4

1. 0 2. − 129 3. $r_2 = 2r_1$ 4. $2c_1 + c_3 = c_2$ 5. $c_2 + c_3$
6. $2c_1 + c_3 = c_2$ [c = column, r = row.]

Exercise 9.5

1. $x = 1, y = -2, z = 3, w = -4$
2.
$$\begin{aligned} x + 2y - z + w &= 1 \\ 3x + 3y + 2z + 5w &= 5 \\ 5x + 4y + 3z - 6w &= 7 \\ -7x + 6y - 2z + 7w &= 2 \end{aligned}$$
3. ab, abc
4. $a^3 + b^3 + c^3 - 3abc$
5. $f(1) = 0$ because two rows are identical $\therefore x - 1$ is a factor of $f(x)$.
   $f(2) = 0$ because two rows are identical $\therefore x - 2$ is a factor of $f(x)$.
   $f(x) = (x - 1)(x - 2)$
6. $f(b) = 0, f(a) = 0 \therefore x - b$ and $x - a$ are factors
   Similarly $b - a$ is a factor because putting $b = a$ makes $D = 0$.
   $D = (x - b)(x - a)(b - a)$
7. $2x + y = 7$

Exercise 10.1

1. 6 2. 4 3. 2 4. $\frac{1}{2}$ 5. $\frac{1}{8}$ 6. 8 7. 8
8. 25 9. $\frac{1}{25}$ 10. $\frac{1}{27}$ 11. 2 12. $\frac{1}{16}$ 13. $\frac{1}{16}$ 14. $\frac{1}{16}$
15. 16 16. $x^{\frac{7}{6}}$ 17. $x^{\frac{7}{2}}$ 18. $x^{\frac{1}{4}}$ 19. 64 20. 64
21. $x + 2x^{\frac{1}{2}}y^{\frac{1}{2}} + y$ 22. $x - y$ 23. $2(2)^{\frac{1}{4}}$ 24. $3\sqrt{7}$ 25. $\dfrac{1}{\sqrt{108}} = \dfrac{1}{6\sqrt{3}} = \dfrac{\sqrt{3}}{18}$

Exercise 10.2

1. $\frac{3}{2}$ 2. $\frac{2}{3}$ 3. $\frac{9}{4}$ 4. $\frac{4}{9}$ 5. $\frac{8}{27}$ 6. $\frac{27}{8}$ 7. $\frac{\sqrt{6}}{3}$
8. $\dfrac{4}{3\sqrt{3}} = \dfrac{4\sqrt{3}}{9}$ 9. 1 10. 1 11. $\left(\dfrac{48}{81}\right)^{\frac{1}{4}} = \dfrac{2\sqrt[4]{3}}{3}$
12. $\left(\dfrac{16}{125}\right)^{\frac{3}{4}} = \left(\dfrac{16 \times 5}{5^4}\right)^{\frac{3}{4}} = \dfrac{2\sqrt[4]{5^3}}{5^3}$ 13. $\frac{1}{2}(\sqrt{3} + 1)$ 14. $\frac{1}{2}(\sqrt{3} - 1)$
15. $(\sqrt{3} - \sqrt{2})$ 16. $(\sqrt{3} + \sqrt{2})$ 17. $\dfrac{\sqrt{x} - 2}{x - 4}$ 18. $\dfrac{\sqrt{x} - \sqrt{y}}{x - y}$
19. $\dfrac{\sqrt{y} + \sqrt{x}}{y - x}$ 20. $x + \sqrt{(x^2 - 1)}$
21. $3 - \left(\dfrac{17}{10}\right)^2 = \dfrac{11}{10^2} \cdot 3 - \left(\dfrac{47}{27}\right)^2 = 3 - \dfrac{2209}{729} = -\dfrac{22}{27^2}$.
    The next approximation would be $\dfrac{47 + (3 \times 27)}{47 + 27} = \dfrac{128}{74}$.
22. Use the form $\dfrac{a + 5b}{a + b}$. $a = 11, b = 5$. The next approximation is $\dfrac{36}{16} = \dfrac{9}{4}$.
23. $\dfrac{99}{70}, \dfrac{239}{169}, \dfrac{577}{408}$ (Note: $99 + 70 = 169$, $70 + 169 = 239$, $169 + 239 = 408$, etc.)
24. $\dfrac{17}{10}, \dfrac{47}{27}, \dfrac{128}{74} = \dfrac{64}{37}, \dfrac{175}{101}, \dfrac{478}{276} = \dfrac{239}{138}$ (Note: $17 + 10 = 27$, $10 \times 2 + 27 = 47$, $27 + 47 = 74$ etc.)

Exercise 10.3

1. (i) 1·2345 (ii) 2·3456 (iii) 3·4567 (iv) 4·5678
   (v) 5·6789 (vi) $\bar{1}$·2345 (vii) $\bar{2}$·3456 (viii) $\bar{3}$·4567
   (ix) $\bar{4}$·5678 (x) $\bar{5}$·6789
2. (i) 3 (ii) 2 (iii) 1 (iv) 4 (v) 3
   (vi) 1·0792 (vii) 0 (viii) 0 (ix) 0 (x) 3
3. (i) -3 (ii) -2 (iii) 2 (iv) -2
   (v) -3 (vi) -2 (vii) -2 (viii) 3
4. (i) $100 = 10^2$ (ii) $1000 = 10^3$ (iii) $81 = 3^4$ (iv) $49 = 7^2$
   (v) $128 = 2^7$ (vi) $343 = 7^3$ (vii) $x = 2^y$ (viii) $x = 3^y$
   (ix) $a = b^c$

5. (i) 10 (ii) 0 (iii) 8 (iv) 5·1
 (v) 27 (vi) 2 (vii) 3 (viii) $x = 1$
 (ix) 3 (x) $\frac{1}{2}$
6. (i) $\log 10 = 1$ (ii) $\log 1 = 0$ (iii) $\log\left(\frac{27 \times 4}{36}\right) = \log 3$
 (iv) $\log\left(\frac{3^6}{6^3}\right) = \log\left(\frac{27}{8}\right)$ (v) log 2·5 (vi) log 3

Exercise 10.4

1. (i) 1·149 (ii) 1·682 (iii) 1·585 (iv) 1·073
 (v) 0·6694 (vi) 0·9458 (vii) 0·631
 (viii) $-\frac{1}{2}\log 0{\cdot}4 = -\frac{1}{2}(\bar{1}{\cdot}6021) = -\frac{1}{2}(-0{\cdot}3979) = 0{\cdot}1990$ (corrected).
 $0{\cdot}4^{-\frac{1}{2}} = 1{\cdot}581$,
 (ix) $-\frac{1}{3}\log 0{\cdot}6 = -\frac{1}{3}(\bar{1}{\cdot}7782) = -\frac{1}{3}(-0{\cdot}2218) = (0{\cdot}0739)$. $0{\cdot}6^{-\frac{1}{3}} = 1{\cdot}185$
 (x) $-\frac{2}{5}\log 0{\cdot}8 = -\frac{2}{5}(\bar{1}{\cdot}9031) = -\frac{2}{5}(-0{\cdot}0969) = (0{\cdot}038\ 76)$. $0{\cdot}8^{-\frac{2}{5}} = 1{\cdot}093$
2. (i) 13·93 (ii) 5·656 (iii) 5·644 (iv) 3·322
3. (i) $Y = \log 60 = 1{\cdot}7782$. From the graph $X = 1{\cdot}1$, $x = 12{\cdot}59$.
 (ii) $x = 10$, $X = \log x = 1$. $Y = 1{\cdot}7 = \log 50{\cdot}12$, $y = 50{\cdot}12$
4. $y = 20\,x^{0{\cdot}4}$. $a = 50{\cdot}24$, $b = 32{\cdot}0$
5. (a) $x = -0{\cdot}64$, $y = 0{\cdot}64$, (b) $x = 6{\cdot}23$, $y = 74{\cdot}76$; $x = 0{\cdot}1$, $y = 1{\cdot}2$.
 Draw line $y = -x$ for (a) and $y = 12x$ for (b).
6. (i) 3·563 (ii) 0 (iii) −0·6288
 (iv) (a) −0·55, +0·55 (b) 3, 27 , $x = 0{\cdot}15$, $y = 1{\cdot}35$
7. (i) $1 + y$ (ii) $y + b$ (iii) $b - y$ (iv) $2b$
 (v) $\frac{1}{2}y - b$ (vi) $-1 + b$ (vii) $-b + 1$ (vii) $\frac{1}{2}y + \frac{1}{2}b$

Exercise 11.1

1. $1, \frac{1}{2}, \frac{1}{3}, \frac{1}{4}$ $u_{10} = \frac{1}{10}$ 2. 2, 4, 6, 8 $u_{10} = 20$
3. 3, 5, 7, 9 $u_{10} = 21$ 4. 1, 3, 5, 7 $u_{10} = 19$
5. 1, 8, 27, 64 $u_{10} = 10^3 = 1000$ 6. 1, −1, 1, −1 $u_{10} = -1$
7. $a, a + 1, a + 2, a + 3$ $u_{10} = a + 9$
8. $a, a + d, a + 2d, a + 3d$ $u_{10} = a + 9d$ 9. $1, r, r^2, r^3$ $u_{10} = r^9$
10. a, ar, ar^2, ar^3 $u_{10} = ar^9$ 11. 9, 11 $u_n = 2n - 1$
12. −9, 11 $u_n = (2n - 1)(-1)^n$ 13. $\frac{1}{5}, \frac{1}{6}$ $u_n = \frac{1}{n}$
14. $\frac{1}{9}, \frac{1}{10}$ $u_n = \frac{1}{n + 4}$ 15. 96, 192 $u_n = 3(2)^{n-1}$
16. $\frac{9}{10}, \frac{11}{12}$ $u_n = \frac{2n - 1}{2n}$
17. $\frac{41 + 58}{41 + 29} = \frac{99}{70}, \frac{99 + 140}{99 + 70} = \frac{239}{169}$ (see Exercise 10.2, question 23)
18. $\frac{94 + 162}{94 + 54} = \frac{256}{148}, \frac{256 + 444}{256 + 148} = \frac{700}{404}$ (see Exercise 10.2, question 21)

Exercise 11.2

1. 0 2. 1 3. 0 4. 0
5. 1 6. −1 7. 1 8. $\frac{1}{2}$
9. $\frac{1}{3}$ 10. 0
11. (i) $\frac{1}{9}\left\{1 - \frac{1}{10^n}\right\} \to \frac{1}{9}$ as $n \to \infty$ (ii) $\left\{2 + \frac{2}{3}\left(1 - \frac{1}{10^n}\right)\right\} \to 2\frac{2}{3}$
12. (i) 0·9, 0·99, 0·999, 0·999 9, 0·999 99, $u_n \to 1$ as $n \to \infty$
 (ii) $\frac{1}{2}, \frac{1}{2}, \frac{3}{8}, \frac{4}{16} = \frac{1}{4}, \frac{5}{32}$, $u_n \to 0$ as $n \to \infty$
 (iii) −2, 2, −2, 2, −2. The sequence does not converge to a limit.

Exercise 11.3

1. (i) A.P. $d = 2$ (ii) not A.P. (iii) not A.P.
(iv) A.P. $d = -50$ (v) A.P. $d = -1$ (vi) not A.P.
(vii) A.P. $d = 0{\cdot}1$ (viii) A.P. $d = 2k$ (ix) not A.P.
(x) A.P. $d = 3$
2. (i) 12, 14, $u_n = 6 + (n-1)2 = 2n + 4$
(ii) 4, 2, $u_n = 10 + (n-1)(-2) = 12 - 2n$
(iii) 40, 50, $u_n = 10 + (n-1)10 = 10n$
(iv) 0, -10, $u_n = 30 + (n-1)(-10) = 40 - 10n$
(v) $4d$, $5d$, $u_n = 0 + (n-1)d = d(n-1)$
(vi) $-4k$, $-5k$, $u_n = 0 + (n-1)(-k) = k(1-n)$
(vii) $a + 3d$, $a + 4d$, $u_n = a + (n-1)d$
(viii) $a - 12$, $a - 15$, $u_n = a + (n-1)(-3) = a + 3 - 3n$
(ix) $1\frac{3}{4}$, 2 $u_n = 1 + (n-1)\frac{1}{4} = \frac{1}{4}(n+3)$
(x) $2\frac{5}{8}$, $2\frac{1}{2}$, $u_n = 3 + (n-1)(-\frac{1}{8}) = \frac{1}{8}(25-n)$
3. (i) 3, 10, 66, $d = 7$ (ii) 5, 7, 23, $d = 2$
(iii) 5, 7, 23, $d = 2$ (iv) 4, 2, -14, $d = -2$
(v) 1, -6, -62, $d = -7$ (vi) 2·2, 3·8, 16·6, $d = 1{\cdot}6$
4. (i) $S_{20} = 10\{2 + 19(1)\} = 210$
(ii) $S_n = \frac{n}{2}\left\{2 + (n-1)(1)\right\} = \frac{1}{2}n(n+1)$
(iii) $S_{15} = \frac{15}{2}\left\{60 + (15-1)(-2)\right\} = \frac{15}{2}(60 - 28) = 240$
(iv) $S_{31} = \frac{31}{2}\left\{2 + 30(2)\right\} = 31^2 = 961$
(v) $S_{101} = \frac{101}{2}\left\{8 + 100(4)\right\} = 101 \times 204 = 20\,604$
5. £1500 + £900 = £2400. $S_{10} = \frac{10}{2}\left\{3000 + 9(100)\right\} = £19\,500$
6. $a = 1$, $d = 2$. $S_n = \frac{n}{2}\left\{2 + (n-1)2\right\} = n^2$ always a perfect square
$25 = 1 + 3 + 5 + 7 + 9$; $49 = 1 + 3 + 5 + 7 + 9 + 11 + 13$
7. We are given $S_{1000} = 2 + 4 + 6 \ldots = 1\,001\,000$
(i) $S_{1000} = 1 + 3 + 5 \ldots = 1\,001\,000 - 1000 = 1\,000\,000 = 10^6$
each odd term being one less than the corresponding term of the given result.
(ii) $S_{1000} = 1 + 2 + 3 + 4 + \ldots = \frac{1}{2}(1\,001\,000) = 500\,500$
each term being half the corresponding term of the given result.
8. $a + 10d = 34$; $a + 20d = 64$; $d = 3$, $a = 4$, $S_{15} = 375$

Exercise 11.4

1. (i) G.P. $a = 1$, $r = \frac{1}{4}$, $u_{11} = (0{\cdot}25)^{10}$
(ii) G.P. $a = 2$, $r = 2$, $u_{11} = 2(2)^{10} = 2^{11}$
(iii) A.P. $a = 3$, $d = 2$, $u_{11} = 3 + 10(2) = 23$
(iv) G.P. $a = 1$, $r = 5$, $u_{11} = 5^{10}$
(v) G.P. $a = 1$, $r = -2$, $u_{11} = (-2)^{10}$
(vi) G.P. $a = 243$, $r = -\frac{1}{3}$, $u_{11} = 243(-\frac{1}{3})^{10} = \frac{1}{243}$
2. (i) $\frac{1 - x^{13}}{1 - x}$ (ii) $\frac{1 - (-x)^{13}}{1 - (-x)} = \frac{1 + x^{13}}{1 + x}$ (iii) $\frac{1 - x^{10}}{1 - x}$ (iv) $\frac{1 - (2x)^7}{1 - 2x}$
3. (i) 2 (ii) $2\frac{1}{4}$ (iii) $\frac{1}{1 - x}$ (iv) $\frac{1}{1 + x}$
4. $S_{10} - xS_{10} = 1 + x + x^2 + x^3 + \ldots x^9 - 10x^{10}$

$$= \frac{1 - x^{10}}{1 - x} - 10x^{10} = \frac{1 - 11x^{10} + 10x^{11}}{1 - x}$$

$$S_{10} = \frac{1 - 11x^{10} + 10x^{11}}{(1 - x)^2}$$

5. $\dfrac{x}{x-1}$, because $x > 1$ and $0 < \dfrac{1}{x} < 1$ then $\left(\dfrac{1}{x}\right)^n \to 0$ as $n \to \infty$
6. $u_3 = ar^2 = 50$, $u_5 = ar^4 = 1250 \therefore \dfrac{ar^4}{ar^2} = r^2 = \dfrac{1250}{50} = 25$, $r = 5$
   $\therefore \quad u_3 = a25 = 50$ and $a = 2$. $S_6 = \dfrac{2(1-5^6)}{1-5} = \frac{1}{2}(5^6 - 1) = 7812$

Exercise 11.5

1. (i) $A = 6\frac{1}{2}$, $G = 6$, $A > G$
   (ii) $A = 7\frac{2}{3}$, $G = 6$, $A > G$
   (iii) $A = 19{\cdot}5$, $G = 8$, $A > G$
2. $a = 1$, $r = \dfrac{1}{10}$, $S = \dfrac{1}{1-\frac{1}{10}} = \dfrac{10}{9}$, $S_n = \dfrac{10}{9}\left\{1 - \dfrac{1}{10^n}\right\}$. $S - S_n = \dfrac{1}{9 \times 10^{n-1}}$
   $n = 7$ is required.
3. $1000 \times (1{\cdot}1)^n \geqslant 2000$ is required, i.e. $(1 \cdot 1)^n > 2$
   $n \log 1{\cdot}1 = n(0{\cdot}0414) > 0{\cdot}301$ required. $n = 8$
4. (i) $AC = 4$ (ii) $AG = 3$ (iii) $2 + \dfrac{1}{2} + \dfrac{1}{2^3} + \dfrac{1}{2^5} + \ldots + \dfrac{1}{2^{2n-3}}$
   (iv) $\dfrac{2}{1-\frac{1}{4}} = 2\frac{2}{3}$
5. (i) $3 + 2 + 1 = 6$ (ii) $4 + 3 + 2 + 1 = 10$
6. Top layer $1 = 1$
   next layer $1 + 2 = 3$
   next layer $1 + 2 + 3 = 6$
   next layer $1 + 2 + 3 + 4 = 10$
   bottom layer $1 + 2 + 3 + 4 + 5 = 15$ Total 35
7. $120 + 120(\frac{3}{4}) + 120(\frac{3}{4}) + 120(\frac{3}{4})^2 + 120(\frac{3}{4})^2 + 120(\frac{3}{4})^3 + 120(\frac{3}{4})^3 + \ldots$
   $= 120\,\{1 + 2(\frac{3}{4}) + 2(\frac{3}{4})^2 + 2(\frac{3}{4})^3 + \ldots\}$
   $= 120 + 240\,\{\frac{3}{4} + (\frac{3}{4})^2 + (\frac{3}{4})^3 + \ldots\}$
   $= 120 + 240\left(\dfrac{\frac{3}{4}}{1-\frac{3}{4}}\right) = 120 + 720 = 940$ cm
8. The clock strikes one twice, two twice and so on. Number of strikes on the hour $= 2\{1 + 2 + 3 + 4 + 5 + 6 + \ldots + 12\} = 156$
   Number of strikes on the half-hour $= 24$. Total 180
9. $1 + 2 + 4 + 8 + 16 \ldots S_{10} = \dfrac{1(1-2^{10})}{1-2} = 2^{10} - 1$. £1023

Exercise 12.1

1. $a^6, 6a^5x, 15a^4x^2, 20a^3x^2, 15a^2x^4, 6ax^5, x^6$
   $a^7, 7a^6x, 21a^5x^2, 35a^4x^3, 35a^3x^4, 21a^2x^5, 7ax^6, x^7$
   $(a + x)^6 = a^6 + 6a^5x + 15a^4x^2 + 20a^3x^3 + 15a^2x^4 + 6ax^5 + x^6$
   $(a + x)^7 = a^7 + 7a^6x + 21a^5x^2 + 35a^4x^3 + 35a^3x^4 + 21a^2x^5 + 7ax^6 + x^7$
2. First row $= 1$ (this is not in the pattern of the results for the other rows).
   Second row $= 4 = 2^2$.
   Third row $= 8 = 2^3$.
   Fourth row $= 16 = 2^4$, etc. Tenth row $= 2^{10}$, nth row $= 2^n$.
3. (i) 2^6, (ii) 2^9 Sum of the odds (D) + sum of the evens $(E) = 2^n$ but $D = E$
   $\therefore 2D = 2^n$, $D = 2^{n-1}$.
4. $(a + x)^6 = (x + a)^6 = x^6 + 6x^5a + 15x^4a^2 + 20x^3a^3 + 15x^2a^4 + 6a^5x + a^6$
5. $(3x + 2y)^4 = (2y + 3x)^4$
   $= (2y)^4 + 4(2y)^3(3x) + 6(2y)^2(3x)^2 + 4(2y)(3x)^3 + (3x)^4$
   $= 16y^4 + 96y^3x + 216y^2x^2 + 216yx^3 + 81x^4$
6. $(x - y)^5 = x^5 - 5x^4y + 10x^3y^2 - 10x^2y^3 + 5xy^4 - y^5$

7. $(1 + 0{\cdot}01)^6 = 1^6 + 6(0{\cdot}01) + 15(0{\cdot}01)^2 + 20(0{\cdot}01)^3 + 15(0{\cdot}01)^4 + 6(0{\cdot}01)^5 + (0{\cdot}01)^6$
$= 1 + 0{\cdot}06 + 0{\cdot}0015 + 0{\cdot}000\,02 \ldots$
$= 1{\cdot}0615$ correct to 4 decimal places

8. $(0{\cdot}99)^6 = (1 - 0{\cdot}01)^6$
$= 1 - 0{\cdot}06 + 0{\cdot}0015 - 0{\cdot}000\,02 \ldots$
$= 1{\cdot}0015 - 0{\cdot}060\,02 = 0{\cdot}941\,48$

9. (i) $(1 + 3\sqrt{5} + 3(5) + 5\sqrt{5}) - (1 - 3\sqrt{5} + 3(5) - 5\sqrt{5}) = 16\sqrt{5}$
(ii) $(2\sqrt{2} + 6\sqrt{5} + 15\sqrt{2} + 5\sqrt{5}) - (2\sqrt{2} - 6\sqrt{5} + 15\sqrt{2} - 5\sqrt{5}) = 22\sqrt{5}$
(iii) $(81 + 4(27)\sqrt{2} + 6(9)(2) + 4(3)(2\sqrt{2}) + 4) - (81 - 4(27)\sqrt{2} + 6(9)(2) - 4(3)(2\sqrt{2}) + 4) = 216\sqrt{2} + 48\sqrt{2} = 264\sqrt{2}$

10. (i) $5^{\frac{1}{2}} = (5^3)^{\frac{1}{6}} = (125)^{\frac{1}{6}}$, $8^{\frac{1}{3}} = (8^2)^{\frac{1}{6}} = 64^{\frac{1}{6}}$, $5^{\frac{1}{2}} > 8^{\frac{1}{3}}$ or $5^{\frac{1}{2}} > 2$
(ii) $5^{\frac{1}{2}} = (25)^{\frac{1}{4}}$, $\therefore 5^{\frac{1}{2}} > 24^{\frac{1}{4}}$
(iii) $2^{\frac{1}{2}} = (2^6)^{\frac{1}{12}} = (64)^{\frac{1}{12}}$, $3^{\frac{1}{3}} = (3^4)^{\frac{1}{12}} = (81)^{\frac{1}{12}}$, $4^{\frac{1}{4}} = (4^3)^{\frac{1}{12}} = (64)^{\frac{1}{12}}$
$2^{\frac{1}{2}} \surd 4^{\frac{1}{4}} < 3^{\frac{1}{3}}$
(iv) $5^{\frac{4}{6}} = (5^4)^{\frac{1}{6}} = (625)^{\frac{1}{6}}$, $7^{\frac{1}{2}} = (7^3)^{\frac{1}{6}} = (343)^{\frac{1}{6}}$ $\therefore 5^{\frac{2}{3}} > 7^{\frac{1}{2}}$

Exercise 12.2

1. 15, 2. 35, 3. 15, 4. 35, 5. 1, 6, 56, 7. 56, 8. 91

9. $C_4^{10} + C_3^{10} = \dfrac{10!}{4!6!} + \dfrac{10!}{3!7!} = \dfrac{10!}{3!6!}\left\{\dfrac{1}{4} + \dfrac{1}{7}\right\} = \dfrac{10!}{3!6!}\,\dfrac{11}{4 \times 7} = \dfrac{11!}{4!7!} = C_4^{11}$
$C_r^n + C_{r-1}^n = C_r^{n+1}$

10. (i) $C_4^5 = 5$, (ii) Take (*ab*) out and select combinations of two from the remaining three (Answer 3).

11. $P_4^6 = 360$ (i) Only those numbers which end in 5. Take 5 out and select from the remaining five numbers (Answer $P_3^5 = 60$).
(ii) Those numbers which end in 2, 4 and 6. 60 numbers end in 2 (as in 5), etc. (Answer 180).

12. Only numbers which end in 0 or 5. Each number has four digits. The first place can be filled in one way only (1) the second place ten ways, third place ten ways and the fourth place two ways (0 and 5) Total $1 \times 10 \times 10 \times 2 = 200$ less the number 1000 (Answer 199).

Exercise 12.3

1. (i) $1 + 8x + 28x^2 + 56x^3 + 70x^4 + 56x^5 + 28x^6 + 8x^7 + x^8$
(ii) $1 - 8x + 28x^2 - 56x^3 + 70x^4 - 56x^5 + 28x^6 - 8x^7 + x^8$
(iii) $x^8 - 8x^7 + 28x^6 - 56x^5 + 70x^4 - 56x^3 + 28x^2 - 8x + 1$. This result is the same as (ii)
(iv) $2^7 + C_1^7 2^6 x + C_2^7 2^5 x^2 + C_3^7 2^4 x^3 + C_4^7 2^3 x^4 + C_5^7 2^2 x^5 + C_6^7 2x^6 + x^7$
(v) $2^7 - C_1^7 2^6 x + C_2^7 2^5 x^2 - C_3^7 2^4 x^3 + C_4^7 2^3 x^4 - C_5^7 2^2 x^5 + C_6^7 2x^6 - x^7$
(vi) $x^7 - C_1^7 2x^6 + C_2^7 2^2 x^5 - C_3^7 2^3 x^4 + C_4^7 2^4 x^3 - C_5^7 2^5 x^2 + C_6^7 2^6 x - 2^7$. This result is -1 (v)
(vii) $3^6 + 3^5 C_1^6 2x + 3^4 C_2^6 (2x)^2 + 3^3 C_3^6 (2x)^3 + 3^2 C_4^6 (2x)^4 + 3C_5^6 (2x)^5 + (2x)^6$
(viii) $3^6 - 3^5 C_1^6 2x + 3^4 C_2^6 (2x)^2 - 3^3 C_3^6 (2x)^3 + 3^2 C_4^6 (2x)^4 - 3C_5^6 (2x)^5 + (2x)^6$
(ix) Same as (viii) written in descending powers of x.

2. (i) $C_9^{11} = 55$, (ii) $-C_9^{11} = -55$, (iii) $C_9^{11}\, 2^2(\tfrac{1}{2})^9 = \tfrac{55}{128}$

3. (i) $2^{10} + 10.2^9 x + \dfrac{10.9.2^8.x^2}{2} + \dfrac{10.9.8.2^7 x^3}{3.2} + \ldots$
$= 2^{10} + 10.2^9 x + 45.2^8 x^2 + 120.2^7 x^3 \ldots$
Put $x = 0{\cdot}0001$. $(2{\cdot}0001)^{10} = 1024 + 0{\cdot}512 + 0{\cdot}011\,52 + \ldots$
$= 1024{\cdot}52$ (correct to 2 decimal places)
(ii) $3^{10} + 10.3^9 x + 45.3^8 x^2 + 120.3^7 x^3 + \ldots$
Put $x = 0{\cdot}000\,01$. $(3{\cdot}000\,01)^{10}$
$= 3^{10} + 10.\,3^9(0{\cdot}000\,01) + 45.3^8\,(0{\cdot}000\,01)^2 + \ldots$
$= 590\,49 + 1{\cdot}9683 + 0{\cdot}000\,029\,524\,5 \ldots$
$= 590\,50{\cdot}97$ (correct to 2 decimal places)

4. (i) $[1 + (x + x^2)]^{12} = 1 + 12(x + x^2) + \frac{12.11\,(x + x^2)^2}{2}$
$+ \frac{12.11.10(x + x^2)^3}{2.3} + \ldots$
$= 1 + 12x + 12x^2 + 66x^2 + 132x^3 + 220x^3 +$ (terms in x^4 and above)
$= 1 + 12x + 78x^2 + 352x^3 + \ldots$
(ii) $[(1 + x) - x^2]^{12} = (1 + x)^{12} - 12(1 + x)^{11}x^2 + 66(1 + x)^{10}x^4 \ldots$
$= 1 + 12x + 66x^2 + 220x^3 - 12x^2 - 132x^3 +$ (terms in x^4 and above)
$= 1 + 12x + 54x^2 + 88x^3 + \ldots$
(iii) $[(1 + x) + 2x^2]^{12} = (1 + x)^{12} + 12(1 + x)^{11}(2x^2) + 66(1 + x)^{10}(2x^2)^2 + \ldots$
$= 1 + 12x + 66x^2 + 220x^3 + 24x^2 + 264x^3 +$ (terms in x^4 and above)
$= 1 + 12x + 90x^2 + 484x^3 + \ldots$
5. $C_t^{10}x^{10-t}\,.\,1/x^t$. Put $t = 5$ to get the sixth term which is C_5^{10}. For $(x + 1/x)^{11}$ the $t + 1$ term is $C_t^{11}\,.\,x^{11-t}\,1/x^t$, it is not possible to find a positive integer t so that $11 - t = t$ in order to have a term independent of x.

Exercise 12.4

1. When $n = 1.\ n^2 = 1$ the formula is true for $n = 1$. $S_k = k^2$, $S_{k+1} = S_k + (2k + 1)$.
$\therefore S_{k+1} = k^2 + 2k + 1 = (k + 1)^2$ which is the same as substituting $n = k + 1$ in $S_n = n^2$.
2. $S_1 = 1(1 + 1) = 2$ is correct.
$$S_{k+1} = S_k + 2(k + 1) = k(k + 1) + 2(k + 1) = (k + 1)(k + 2)$$
which is the same as S_n when $n = k + 1$.
3. $S_1 = \frac{1}{4}(1 + 1)^2 = 1$ is true. $S_{k+1} = S_k + (k + 1)^3 = \frac{1}{4}k^2(k + 1)^2 + (k + 1)^3$
$\therefore S_{k+1} = (k + 1)^2\,\{\frac{1}{4}k^2 + k + 1\} = (k + 1)^2\frac{1}{4}(k + 2)^2 = \frac{1}{4}(k + 1)^2(k + 2)^2$
which is S_n with $n = k + 1$.
4. $f(1) = 6 - 5 + 4 = 5$. $f(k + 1) = 6^{k+1} - 5(k + 1) + 4$
$f(k + 1) - f(k) = 6^{k+1} - 6^k - 5(k + 1) + 5k = 6^k\{6 - 1\} - 5 = 5\,\{6^k - 1\}$
$\therefore f(k + 1) = f(k) + 5(6^k - 1)$. If $f(k)$ is divisible by 5 then so is $f(k + 1)$ and the result follows.
5. $S_n = \frac{n(n + 3)}{4(n + 1)(n + 2)}$ $S_1 = \frac{1.4}{4.2.3} = \frac{1}{2.3}$ which is the first term so that the formula is true for $n = 1$.
$$S_{k+1} = \frac{k(k + 3)}{4(k + 1)(k + 2)} + \frac{1}{(k + 1)(k + 2)(k + 3)} = \frac{k(k + 3)^2 + 4}{4(k + 1)(k + 2)(k + 3)}$$
$$= \frac{(k + 1)^2(k + 4)}{4(k + 1)(k + 2)(k + 3)}$$
$\therefore S_{k+1} = \frac{(k + 1)(k + 4)}{4(k + 2)(k + 3)}$ which is S_n with $n = k + 1$
$$S_n \rightarrow \tfrac{1}{4} \text{ as } n \rightarrow \infty$$
6. $f(n + 1) - f(n) = 9^{n+1} - 9^n = 9^n(9 - 1) = 9^n(8)$.
$\therefore f(n + 1) = f(n) + 9^n(8)$ so if $f(n)$ is divisible by 8 then so is $f(n + 1)$. Note that $3^{2n} - 1 = 9^n - 1$.
7. $f(n) = 9^n - 8n - 1$. $f(0) = 1 - 1 = 0$, $f(1) = 0$, $f(2) = 81 - 16 - 1 = 64$
The result really starts at $n = 2$
$f(k + 1) - f(k) = 9^{k+1} - 9^k - 8 = 9^k(9 - 1) - 8 = 8[9^k - 1]$. Now use the result of Question 6.
8. $S_1 = \frac{1}{3}(2 - 1)(2 + 1) = 1$.
$S_{k+1} = S_k + (2k + 1)^2$
$= \frac{k}{3}(2k - 1)(2k + 1) + (2k + 1)^2$

$= (2k+1)\left\{\frac{2k^2 - k + 6k + 3}{3}\right\} = \frac{(2k+1)(2k+3)(k+1)}{3}$

$= \frac{1}{3}(k+1)(2k+1)(2k+3)$

which is S_n with $n = k + 1$.

9. $f(1) = 2, f(k+1) - f(k) = (k+1)(k+2) - k(k+1) = (k+1)2$ which is divisible by 2.

10. $f(n) = n^3 - n = n(n^2 - 1) = n(n+1)(n-1)$. $f(1) = 0$, $f(2) = 6$.
$f(k+1) - f(k) = (k+1)k(k+2) - k(k-1)(k+1)$
$= k(k+1)(k+2-k+1) = 3k(k+1)$ and $k(k+1)$ is divisible by 2 so the result follows.

Exercise 13.1

1. (i) 0·6428, 0·6428
(ii) 0·8391, 1·1918, tan 40° tan 50° = 1
(iii) 0·5, 0·5
(iv) 0·5774, 1·7321, tan 30° tan 60° = 1
2. (i) 30°, (ii) 60°, (iii) 60°, (iv) 20° 13′
3. The sides of the first triangle are 2, $\sqrt{3}$, 1. The sides of the second triangle are $\sqrt{2}$, 1, 1.
4. (i) $(2s-1)(s-2) = 0$. $s = 2$ rejected, $s = \frac{1}{2}$ leads to $x = 30°$
(ii) $(2c-1)(c-2) = 0$. $c = 2$ rejected, $c = \frac{1}{2}$ leads to $x = 60°$
(iii) $\sin^2 x = (1 - \cos x)^2 = 1 - \cos^2 x$ reduces to $(\cos x)(1 - \cos x) = 0$
$\therefore \cos x = 0, x = 90°$ or $\cos x = 1, x = 0$
(iv) $\sin^2 x = (1 + \cos x)^2 = 1 - \cos^2 x$ reduces to $(\cos x)(1 + \cos x) = 0$
$\therefore \cos x = 0, x = 90°$ or $\cos x = -1$ for which no acute x exists
(v) $(3t-1)(t-1) = 0 \therefore \tan x = 1, x = 45°$ or $\tan x = \frac{1}{3}, x = 18° 26'$

Exercise 13.2

1. (i) 46° (ii) 39° (iii) 38° (iv) 14° (v) 15° (vi) 15°
2. ±180°, ±540°, ±900° 3. −90°, −450°, −810°, +270°, +630°, 990°
4. 0°, ±360°, ±720° 5. 45°, 405°, 765°, −315°, −675°
6. 225°, 585°, 945°, −135°, −495°, −855°

Exercise 13.3

1. $\frac{\pi}{6}$ 2. $-\frac{\pi}{3}$ 3. $\frac{47}{36}\pi$ 4. $-\frac{5}{4}\pi$
5. $\frac{5}{2}\pi$ 6. -4π 7. $\frac{\pi}{18}$ 8. $-\frac{\pi}{90}$
9. 45° 10. −60° 11. 30° 12. −540°
13. $\frac{720°}{\pi^2}$ 14. −1620° 15. 5·7° 16. −11·4°

Exercise 13.4

1.

Quadrant	I	II	III	IV
sin	+	+	−	−
cos	+	−	−	+
tan	+	−	+	−
cot	+	−	+	−
sec	+	−	−	+
cosec	+	+	−	−

2. $\tan\theta_1 = -\tan(\pi - \theta_1) = \tan(\pi + \theta_1) = -\tan(2\pi - \theta_1)$
3. (i) $\frac{1}{2}\sqrt{3} = 0·866$ (ii) −1 (iii) $1/\sqrt{2} = \frac{1}{2}\sqrt{2}$
(iv) $\frac{1}{2}$ (v) $\frac{1}{2}\sqrt{3} = 0·866$ (vi) $-\frac{1}{2}$
(vii) $-1/\sqrt{3} = -\frac{1}{3}\sqrt{3}$ (viii) $\sqrt{3}$
4. (i) 30°, 150° (ii) 60°, 300° (iii) 45°, 225°
(iv) 120°, 240° (v) 210°, 330° (vi) 135°, 315°

Exercise 13.5

1. $\cos x = 0{\cdot}5$ or $\cos x = 2$ $x = 60° \pm n(360°)$; $x = 300° \pm n(360°)$; $\cos x = 2$ has no real solution.
2. $\cos x = 0$ or $\cos x = 1$; $x = 90° \pm n(360°)$; $x = 270° \pm n(360°)$ or we may combine both solutions into $x = 90° \pm n(180°)$. From $\cos x = 1$, $x = 0 \pm n(360°) = \pm n(360°)$.
3. $\cos x = 0$ or $\cos x = -1$. $x = 90° \pm n(180°)$ or $x = 180° \pm n(360°)$
4. $\cos x = \sqrt{1{\cdot}5}$. No real solution.
5. $\tan x = \pm 1$. $x = 45° \pm n(360°)$ or $x = 225° \pm n(360°)$
   $x = 135° \pm n(360°)$ or $x = 315° \pm n(360°)$
6. $\sin x = \pm 3$. No real solution.
7. $\sin x = 0$ or $\frac{1}{2}$. $x = \pm n(360°)$ or $180° \pm n(360°)$, $x = 30° \pm n(360°)$ or $150° \pm n(360°)$
8. $\tan x = 1 \pm \sqrt{2}$. $\tan x = 2{\cdot}4142$. $x = 67° \, 30' \pm n(360°)$ or $x = 247° \, 30' \pm n(360°)$
   $\tan x = -0{\cdot}4142$. $x = 167° \, 30' \pm n(360°)$ or $x = 337° \, 30' \pm n(360°)$
9. $\sin x = 0{\cdot}5$. $x = 30° \pm n(360°)$ or $x = 150° \pm n(360°)$. No real solution for $\sin x = -2$.

Exercise 13.6

1. $x = 38° \, 10'$
2. Draw the graph of $y = x$ on the graph of $y = \cos x$. $x = 42° \, 20'$.

Exercise 13.7 See Fig. S.2 on pages 292–3.

1. 0, 60°, 180°, 300°, 360° or 0, $\frac{1}{3}\pi$, π, $\frac{5}{3}\pi$, 2π
2. 45°, 225° or $\frac{1}{4}\pi$, $\frac{5}{4}\pi$ 3. 223° 497° 4. 90°, 270° or $\frac{1}{2}\pi$, $1{\cdot}5\pi$
5. 90°, 270° 6. 90°, 270°

Exercise 14.1

1. $a = \sqrt{124} = 11{\cdot}14$, $2 . 10\sqrt{124} \cos C = 124 + 100 - 144$, $\cos C = 0{\cdot}3592$, $C = 68° \, 57'$, $B = 180° - 60° - 68° \, 57' = 51° \, 03'$
2. $a = \sqrt{364} = 19{\cdot}08$, $2 . 10\sqrt{364} \cos C = 364 + 100 - 144$, $\cos C = 0{\cdot}8386$, $C = 33°$, $B = 180° - 120° - 33° = 27°$
3. $b = \sqrt{1570} = 39{\cdot}62$; $2 . 20\sqrt{1570} \cos C = 400 + 1570 - 900$, $\cos C = 0{\cdot}6751$, $c = 47° \, 33'$, $A = 180° - 103° - 47° \, 33' = 29° \, 27'$
4. $81 = 25 + 36 - 2 . 5 . 6 \cos C$. $\cos C = 0{\cdot}3333$. $C = 70° \, 32'$
   $36 = 25 + 81 - 2.5.9 \cos B$. $\cos B = 0{\cdot}7778$. $B = 38° \, 56'$. $A = 70° \, 32'$
5. $\cos A = \sin (90° + A)$ or $\sin (90° - A)$ or $\cos (360° - A)$ or $-\cos (180° - A)$ for any A
6. $\cos A = \cos (360° - A)$ or $-\sin (270° - A)$ for any A
7. $\cos (A + B) + \cos (A - B) = 2 \cos A \cos B$. Put $A = 60°$, $B = 40°$, $\cos 100° + \cos 20° = 2 \cos 60° \cos 40° = \cos 40°$ since $\cos 60° = 0{\cdot}5$
8. $\cos (A - B) - \cos (A + B) = 2 \sin A \sin B$. Put $A = 60°$, $B = 40°$, $\cos 20° - \cos 100° = 2 \sin 60° \sin 40° = \sqrt{3} \sin 40°$ since $\sin 60° = \frac{1}{2}\sqrt{3}$
9. $\cos (135° + A) = \cos [180° + (A - 45°)] = \cos 180° \cos (A - 45°) - \sin 180° \sin (A - 45°)] = -\cos(A - 45°)$ because $\cos 180° = -1$, $\sin 180° = 0$
10. $\cos 75° = \cos (45° + 30°) = \cos 45° \cos 30° - \sin 45° \sin 30° = \frac{1}{2}\sqrt{2}(\frac{1}{2}\sqrt{3} - \frac{1}{2})$ $= \frac{1}{4}(\sqrt{6} - \sqrt{2})$

Exercise 14.2

1. (i) $\cos 105° = \cos (60° + 45°) = \dfrac{1}{2\sqrt{2}} (1 - \sqrt{3})$

   $\cos 75° = \cos (30° + 45°) = \dfrac{1}{2\sqrt{2}} (\sqrt{3} - 1)$

   (ii) $\sin 15° = \sin (60° - 45°) = \dfrac{1}{2\sqrt{2}}(\sqrt{3} - 1) = \cos 75°$

   $\sin 225° = -\sin 45° = -\frac{1}{2}\sqrt{2}$

(iii) $\cos(45° - 15°) - \cos(45° + 15°) = \cos 30° - \cos 60° = \frac{1}{2}(\sqrt{3} - 1)$

2. $\cos A = \frac{4}{5}$, $\cos B = \frac{12}{13}$, $\cos(A + B) = \frac{48}{65} - \frac{15}{65} = \frac{33}{65}$

$\cos(A - B) = \frac{48}{65} + \frac{15}{65} = \frac{63}{65}$

3. $\cos A = \frac{11}{61}$, $\sin B = \frac{40}{41}$, $\sin(A + B) = \frac{60}{61}\cdot\frac{9}{41} + \frac{11}{61}\cdot\frac{40}{41} = \frac{980}{61.41}$,

$\sin(A - B) = \frac{100}{61.41}$

4. (i) $(1/2 \cos x + \sqrt{3}/2 \sin x)\,2 = 2(\cos 60° \cos x + \sin 60° \sin x)$
$= 2\cos(x - 60°)$. Maximum 2 when $x = 60°$, minimum -2 when $x = 240°$.
(ii) $\sqrt{2}(1/\sqrt{2} \cos x + 1/\sqrt{2} \sin x) = \sqrt{2}(\cos 45° \cos x + \sin 45° \sin x)$
$= \sqrt{2}\cos(x - 45)$. Maximum $\sqrt{2}$ when $x = 45°$, minimum $-\sqrt{2}$ when $x = 225°$.
(iii) $5(\frac{3}{5}\cos x + \frac{4}{5}\sin x) = 5(\cos p \cos x + \sin p \sin x) = 5\cos(x - p)$ where $\tan p = \frac{4}{3}$. Maximum 5 when $x = p = \tan^{-1}\frac{4}{3}$, minimum -5 when $x = 180° + p = 180° + \tan^{-1}\frac{4}{3}$.

5. $\tan(A + B) = \frac{\tan A + \tan B}{1 - \tan A \tan B}$

6. Put $A = 45°$ in the answer to Question 5, $\tan 45° = 1$, etc.

7. Put $B = A$ in the answer to Question 5.

8. $\frac{2}{2\tan A} - \frac{1 - \tan^2 A}{2\tan A} = \frac{1 + \tan^2 A}{2\tan A} = \frac{1}{2\tan A\cos^2 A} = \frac{1}{2\sin A\cos A} = \frac{1}{\sin 2A}$

9. $1 + \cos 30° = 2\cos^2 15 = 1{\cdot}866$, $\cos 15° = \sqrt{0{\cdot}933} = 0{\cdot}9659$

10. $5\{\frac{3}{5}\sin x + \frac{4}{5}\cos x\} = 5\cos(x - p)$ where $\tan p = \frac{3}{4}$ $\therefore p = 36°\,52'$
$\therefore\ \cos(x - p) = 0{\cdot}2 = \cos 78°\,28'$ or $\cos 281°\,32'$
$\therefore\ x - p = 78°\,28'$ or $281°\,32'$, $x = 115°\,20'$ or $318°\,24'$

Exercise 14.3

1. (i) $m_1 = 2$, $m_2 = 3$, $\tan\theta = -\frac{1}{7}$, acute angle $8°\,8'$
(ii) $m_1 = -3$, $m_2 = 4$, $\tan\theta = -\frac{7}{11}$, acute angle $32°\,28'$
(iii) $m_1 = -1$, $m_2 = \frac{1}{3}$, $\tan\theta = -2$, acute angle $63°\,26'$

2. $m_1 = 2$, $\tan 45° = 1 = \frac{m_2 - 2}{1 + 2m_2}$ $\therefore\ m_2 = -3$

3. $\tan 135° = -1 = \frac{m_2 - \frac{1}{2}}{1 + \frac{1}{2}m_2}$ $\therefore\ m_2 = -\frac{1}{3}$, $3y + x = 2$

4. Angle between (i) and (ii) is $26°\,34'$, angle between (i) and (iii) is $90°$, remaining angle is $63°\,26'$.

Exercise 14.4

1. (i) Use $\sin 25° = \cos 65°$, $2\sin 30° = 1$.
(ii) $\sin 40° - \sin 80° = -2\sin 20°\cos 60° = -\sin 20°$.

2. $\sin A + \sin(A + 120°) = 2\sin(A + 60°)\cos 60°$

3. $AC = 15$ cm, $BN = 5$ cm, $\cos A = \frac{33}{65}$, $\sin\frac{1}{2}A = \cos\frac{1}{2}(B + C)$

so $2\sin\frac{1}{2}A\cos\frac{1}{2}(B - C) = \cos B + \cos C = \frac{64}{65}$,

$\sin 2B = \frac{120}{169}$, $\tan\frac{1}{2}B = \frac{\sin\frac{1}{2}B}{\cos\frac{1}{2}B} = \frac{2\sin B}{1 + \cos B} = \frac{2}{3}$

4. — 5. —

6. $\frac{2\sin 2A\cos A + 2\sin 6A\cos A}{2\cos 2A\cos A + 2\cos 6A\cos A} = \frac{\sin 2A + \sin 6A}{\cos 2A + \cos 6A} = \tan 4A$

7. $-\tan A = \tan(B + C) = \frac{\tan B + \tan C}{1 - \tan B\tan C}$ hence result

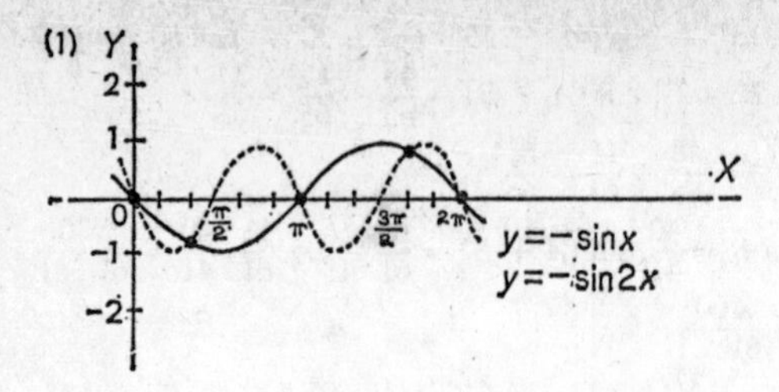
(1) Y
X
y=−sin x
y=−sin 2x

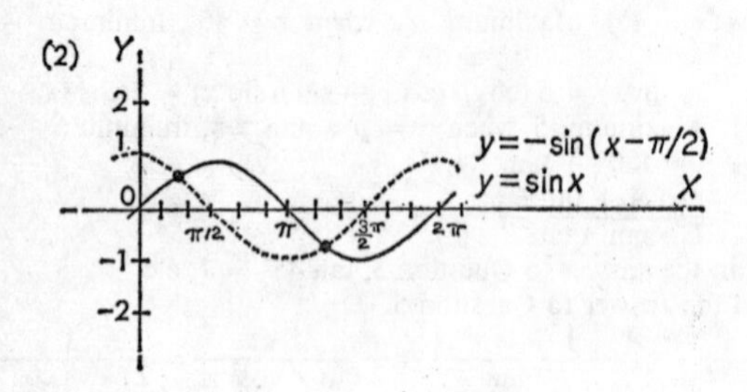
(2) Y
y=−sin (x−π/2)
y=sin x
X

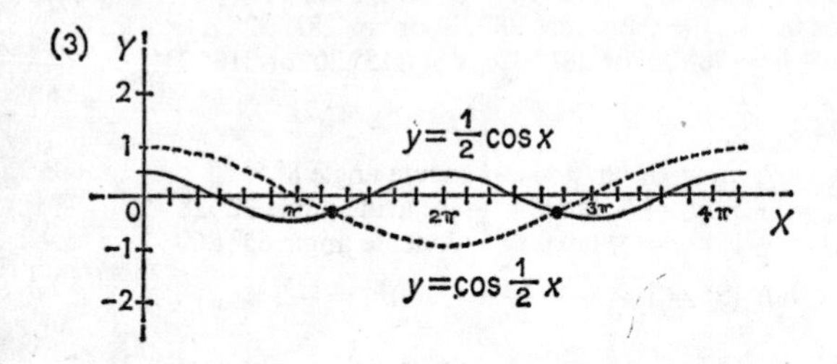
(3) Y
$y=\frac{1}{2}\cos x$
$y=\cos\frac{1}{2}x$
X

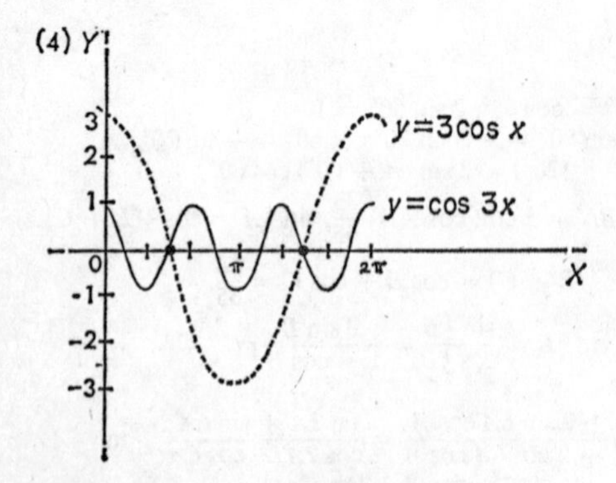
(4) Y
y=3cos x
y=cos 3x
X

Fig. S.2

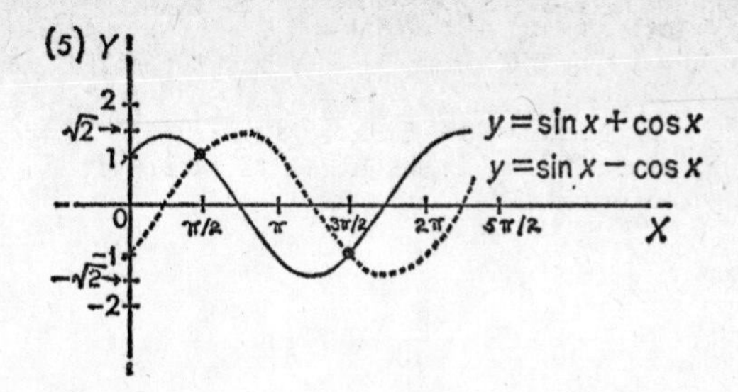
(5) Y
2
√2
1
0
-1
-√2
-2
π/2
π
3π/2
2π
5π/2
X
y=sin x+cos x
y=sin x−cos x

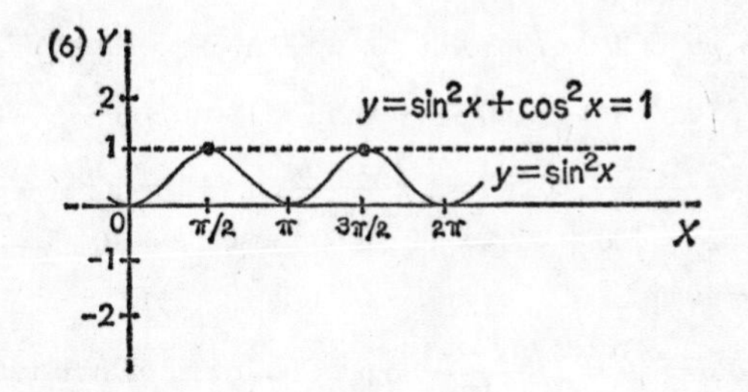
(6) Y
2
1
0
-1
-2
π/2
π
3π/2
2π
X
y=sin²x+cos²x=1
y=sin²x

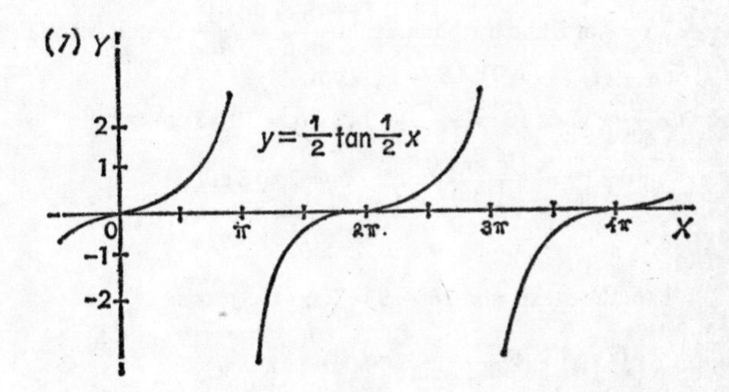
(7) Y
2
1
0
-1
-2
π
2π
3π
4π
X
y=½ tan ½ x

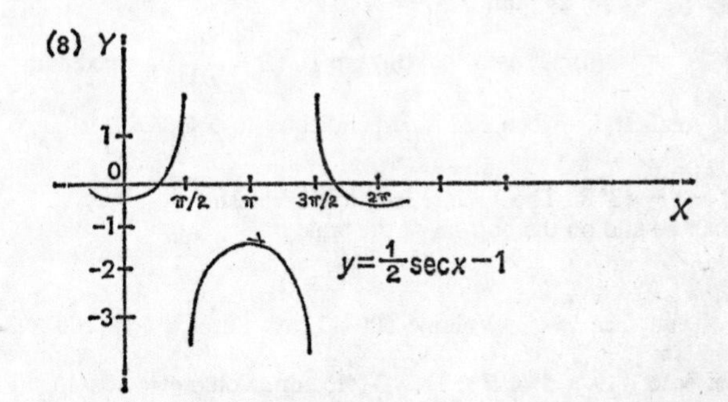
(8) Y
1
0
-1
-2
-3
π/2
π
3π/2
2π
X
y=½ sec x−1

Fig. S.2

8. $2 \sin 3x \cos x = \cos x$, $\cos x (1 - 2 \sin 3x) = 0$
   $\therefore \cos x = 0$, $x = 90°$ or $270°$ or $\sin 3x = \frac{1}{2}$, $3x = 30°$ or $150°$, i.e. $x = 10°$ or $50°$
9. $\sin 2x = \sin (90 - 3x)$ $\therefore 2x = 90 - 3x$, $x = 18°$
10. $10 \{\sin (15° + 2x) - \sin 25°\} = 1$ $\therefore \sin (15° + 2x) = 0{\cdot}1 + \sin 25°$
    $= 0{\cdot}5226$, $x = 8°\ 15'$ or $66°\ 45'$

Exercise 14.5

1. $s = 18$ (i) 48 m^2 (ii) $\frac{1}{\sqrt{10}}$ (ii) $\frac{3}{\sqrt{10}}$ (iv) $2\frac{2}{3}$ m (v) $8\frac{1}{3}$ m
2. $s = 5$ (i) $2\sqrt{5}\text{ m}^2$ (ii) $\sqrt{\frac{1}{6}}$ (iii) $\sqrt{\frac{5}{6}}$ (iv) $\frac{2}{5}\sqrt{5}$ m (v) $\frac{9}{10}\sqrt{5}$ m
3. $s = 15$ (i) $15\sqrt{7}\text{ cm}^2$ (ii) $\sqrt{\frac{7}{32}}$ (iii) $\frac{5\sqrt{2}}{8}$ (iv) $\sqrt{7}$ cm (v) $\frac{16}{7}\sqrt{7}$ cm
4. $s = 20$ (i) $8\sqrt{35}\text{ cm}^2$ (ii) $\frac{1}{6}$ (iii) $\frac{\sqrt{35}}{6}$ (iv) $\frac{2}{5}\sqrt{35}$ cm (v) $\frac{54\sqrt{35}}{35}$ cm
5. $\sin \frac{1}{2}B = \sqrt{\frac{(s-a)(s-c)}{ac}}$, $\cos \frac{1}{2}B = \sqrt{\frac{s(s-b)}{ac}}$; $\tan \frac{1}{2}C = \sqrt{\frac{(s-a)(s-b)}{s(s-c)}}$
6. $\Delta = 50\sqrt{3}\text{ m}^2$. $b^2 = 100 + 400 - 400 \cos 60° = 300$, $b = 10\sqrt{3}$
   $R = \frac{10\sqrt{3}}{2 \sin 60°} = 10$ m
7. $\cos \frac{1}{2}A = \cos 30° = \frac{\sqrt{3}}{2} = \sqrt{\frac{s(s-a)}{bc}}$ Square both sides of the equation to get the required result.
8. $\sin (A + C) = \sin B$ and by the sine rule $\frac{b}{\sin B} = \frac{a}{\sin A}$; result becomes $\frac{1}{2}ab \sin C$.
9. $\frac{1}{2}AB \times 6 = \sqrt{(15.3.5.7)}$ $AB = 5\sqrt{7}$ cm
10. (i) $R = \frac{b}{2 \sin B}$ gives $10 = \frac{b}{2 \sin 60}$ $\therefore$ $b = 10\sqrt{3}$ cm
    (ii) $r = \frac{\Delta}{s}$ gives $10 = \frac{\frac{1}{2}b^2 \sin 60}{\frac{1}{2}(3b)}$ $\therefore$ $b = 20\sqrt{3}$ cm

Exercise 15.1

1. $8 - 5 = 3$ on the x axis and $16 - 9 = 7$ on the y axis.
2. (i) 38
   (ii) $2 \sin^{-1} \left(\frac{12}{38}\right) = 36°\ 48'$, $2 \sin^{-1} \left(\frac{20}{38}\right) = 63°\ 30'$
   (iii) $\sin^{-1} \frac{12}{38} = 18°\ 24'$, $\tan^{-1} \frac{12}{30} = 21°\ 48'$
3. (i) $63°\ 26'$ (ii) $39°\ 48'$ (iii) $\tan DMB = \frac{10}{BM}$, a maximum when BM is at its smallest, i.e. when BM is perpendicular to CA $\therefore BM = \frac{60}{13}$
   $\angle DMB = 65°\ 13'$
4. $\sin^{-1} (0{\cdot}8) = 53°\ 8'$. The 2 metre length is irrelevant.
5. 0°. It can be laid on the bottom of the tank.

Exercise 15.2

1. Area of end face $25\sqrt{3}$. Volume $500\sqrt{3}\text{ cm}^3$. Surface area $200 \times 3 + 50\sqrt{3} = 686{\cdot}6\text{ cm}^2$.
2. Area of base $\sqrt{(9 \times 3 \times 5 \times 1)} = 3\sqrt{15}\text{ cm}^2$. Volume $= 10\sqrt{15}\text{ cm}^3$.
3. (i) $\tan^{-1} \frac{3\sqrt{3}}{2}$ (ii) $\tan^{-1} 3\sqrt{3}$ (iii) $1000\sqrt{3}\text{ cm}^3$

4. In Fig. 64 AE bisects $\angle BAC$, $AE \sin 30° = FE$ $\therefore$ $AE = EC = 2\ FE$ $\therefore$ E divides FC in the ratio 1:2 similarly for the other two medians. Result similarly proved for question 3.
5. (i) $30\sqrt{3}$ cm (ii) $\tan^{-1}\frac{5\sqrt{2}}{2}$ (iii) $\tan^{-1} 5$ (iv) $2\tan^{-1}\frac{\sqrt{2}}{5}$ (v) $6666\frac{2}{3}$ cm³
6. $V = \frac{1}{2} \times 12 \times 12 \times \sin 60° \times \frac{23}{3} = 276\sqrt{3}$ cm³. Area $BCD = 6 \times \sqrt{541}$

   $= 140$ to nearest cm. $\therefore$ Height required is $\frac{207\sqrt{3}}{35}$ cm.
7. $V = 314$ cm³, Area $90\ \pi$ cm²
8. (i) Join A and D to midpoint of CB. Angle DMA is required $AM = 2\sqrt{2}$,

   $\tan DMA = \frac{3}{2\sqrt{2}} = \frac{3\sqrt{2}}{4}$.

   (ii) Perpendicular required is $AM \sin DMA = 2\sqrt{2} \sin 46° 41' = 2{\cdot}06$ m (correct to 2 decimal places).
9. $\frac{216}{360} 2\pi 20 = 2\pi r$ $\therefore$ $r = 12$ cm $h^2 = l^2 - r^2$, $h = 16$ cm $V = 2411{\cdot}52$ cm³
10. $V = 52\ \pi$ cm³

Exercise 15.3

1. 24 cm, $\sin^{-1}\left(\frac{5}{13}\right) = 22° 37'$
2. Small circle distance is 40π cm $= 125{\cdot}60$ cm. Angle $PON = \tan^{-1} 4{\cdot}4444$ $= 77° 19'$. Great circle distance $\frac{154° 38'}{360°} \times 2\pi 41 = 110{\cdot}6$ cm (Take $38'$ as $\frac{38°}{60}$ $= 0{\cdot}6333$) (to the nearest 0·1 cm). Great circle distances are always the shortest distances between two points on the surface of a sphere.
3. (i) $\frac{90}{360} \times 2\pi\ 6368 \cos 60° = 4998{\cdot}88$ km

   (ii) $\frac{120}{360} \times 2\pi\ 6368 = 120 \times 111$ km $= 13320$ km

   (iii) $\frac{60}{360} \times 2\pi\ 6368 \cos 30° = 30\sqrt{3} \times 111$ km $= 576{\cdot}8$ km
4. $\frac{133}{360} \times 2\pi\ 6368 \cos 34° = 122\ 400$ km (logs) along the small circle.
5. (i) 4187 cm³ (ii) $\frac{4}{3}\pi(12^3 - 2^3) = 7201$ cm³
6. 400π cm², 400π cm² 7. 20 cm
8. $\frac{60}{360} \times 2\pi 10 = 10\frac{\pi}{3}$ cm, the shortest surface distance being along a great circle.
9. $6 + \sqrt{132} = 17{\cdot}49$ cm 10. $\sin^{-1}\left\{\frac{25 - 15}{25 + 15} = 0{\cdot}25\right\} = 14° 29'$

Exercise 15.4

1. (i) $\frac{x^2}{5^2} + \frac{y^2}{(5\cos 60)^2} = 1$, $x^2 + 4y^2 = 25$; (ii) $x^2 + 2y^2 = 25$;

   (iii) $\frac{x^2}{5^2} + \frac{y^2}{(5\cos 30°)^2} = 1$ $3x^2 + 4y^2 = 75$
2. $\pi\ 5.5 \cos\alpha$ (i) $12{\cdot}5\pi$ (ii) $12{\cdot}5\pi\sqrt{2}$ (iii) $12{\cdot}5\pi\sqrt{3}$
3. $\cos\alpha = \frac{1}{3}$, 'a' $= 2$, 'b' $= \frac{2}{3}$, Area $= \frac{4\pi}{3}$
4. Area of the circle is 16π, 'a' $= 4$. If $\pi ab = 4\pi$ then $b = 1$. The angle of projection is $\cos^{-1}\left(\frac{1}{4}\right)$. Equation of the ellipse is $x^2 + 16y^2 = 16$.

5. An ellipse inscribed in a parallelogram. Ellipse of area $4{\cdot}5\pi$ parallelogram of area $36 \cos 60° = 18$.

Exercise 16.1

1. (i) 10 (ii) -3 (iii) $\frac{1}{3}$ (iv) $-\frac{1}{3}$
2. $y = 2x + 3$ 3. If $\delta x = 5$ then $\delta y = -15$ 4. 6

Exercise 16.2

1. B is the limit point of the sequence of points $\{M_n\}$. 2. A
3. The circle radius 10 cm. 4. The triangle ABC.
5. The triangle degenerates into the diagonal AC.
6. The two tangents to the circle which are parallel to AB.

Exercise 16.3

1. $f(1{\cdot}1) = \dfrac{1{\cdot}1^2 - 1}{0{\cdot}1} = 2{\cdot}1$; $f(1{\cdot}01) = \dfrac{1{\cdot}0201 - 1}{0{\cdot}01} = 2{\cdot}01$; $f(1{\cdot}001) = 2{\cdot}001$; $f(1{\cdot}0001) = 2{\cdot}0001$. As x gets closer to 1 so $f(x)$ gets closer to 2. $f(1)$ does not exist since it is of the form $\frac{0}{0}$. $\lim\limits_{x\to 1} f(x) = 2$
2. $f(0{\cdot}1) = 2{\cdot}1$; $f(0{\cdot}01) = 2{\cdot}01$; $f(0{\cdot}001) = 2{\cdot}001$. As h gets closer to 0 so $f(h)$ gets closer to 2. $f(0)$ does not exist. $\lim\limits_{h\to 0} f(h) = 2$. Observe that this is the same problem as in Question 1. We have put $x = 1 + h$ so that as $x \to 1$ so $h \to 0$.
3. The gradient is $\dfrac{(1 + h)^2 - 1}{1 + h - 1} = \dfrac{(1 + h)^2 - 1}{h} = \dfrac{1 + 2h + h^2 - 1}{h} \to 2$.
   The line tends to the tangent at P.
4. The x coordinates of P and Q tend to 1. When $x = 1$, $y = 1$ so P and Q both tend to the point (1, 1) but from different sides. The y coordinate of P is $(1 - h)^2$ and the y coordinate of Q is $(1 + h)^2$. Gradient of PQ is $\dfrac{(1 + h)^2 - (1 - h)^2}{(1 + h) - (1 - h)}$ $= \dfrac{4h}{2h} = \dfrac{2h}{h} = 2$. We cannot put $h = 0$ but as $h \to 0$ the limiting position of PQ is the tangent at the point (1, 1).
5. (i) $\dfrac{f(1 + h) - f(1)}{h} = 2 + h \to 2$, i.e. $\lim\limits_{h\to 0} \dfrac{f(1 + h) - f(1)}{h} = 2$
   (ii) $\dfrac{f(1 + h) - f(1)}{h} = \dfrac{k(2 + h)}{h} \to 2k$ as $h \to 0$
   (iii) $\dfrac{f(3 + h) - f(3)}{h} = \dfrac{7h + h^2}{h} \to 7$ as $h \to 0$
   (iv) $\dfrac{f(3 + h) - f(3)}{h} = \dfrac{1}{h}\left\{\dfrac{1}{3 + h} - \dfrac{1}{3}\right\} = \dfrac{-h}{h(3 + h)3} \to -\dfrac{1}{9}$ as $h \to 0$
6. $\dfrac{(t + h)^2 - t^2}{h} = \dfrac{h(2t + h)}{h} \to 2t$ as $h \to 0$. Speed at time t is $2t$ metres per second at $t = 3$, $v = 6$ metres per second.
7. $2\pi r$
8. $V = x^3$; $\dfrac{\delta V}{\delta x} = \dfrac{(x + \delta x)^3 - x^3}{\delta x} = 3x^2 + 3x\,\delta x + (\delta x)^2 \to 3x^2$ as $\delta x \to 0$
   Alternatively $\dfrac{f(x + h) - f(x)}{h} = 3x^2 + 3xh + h^2 \to 3x^2$ as $h \to 0$.
9. $\dfrac{f(x + h) - f(x)}{h} = 2x - 6 + h \to 2x - 6$ as $h \to 0$. When $x = 3$ the gradient is 0, i.e. the tangent is parallel to the x axis. Alternatively
   $\dfrac{f(3 + h) - f(3)}{h} = \dfrac{9 + 6h + h^2 - 18 - 6h + 9}{h} = h \to 0$ as $h \to 0$.

10. $y = mx + c$. 11. 8. 12. 16.

Exercise 17.1

1. (i) $f(x+h) - f(x) = (x+h)^6 - x^6 = 6x^5h +$ (terms in h^2 and higher powers of h). $\dfrac{f(x+h)-f(x)}{h} = 6x^5 +$ (terms in h and higher powers of h)
   $\therefore\ f'(x) = 6x^5,\ f'(1) = 6$
   (ii) $f(x+h) - f(x) = \dfrac{1}{(x+h)^2} - \dfrac{1}{x^2} = \dfrac{x^2-(x+h)^2}{x^2(x+h)^2} = \dfrac{-2xh-h^2}{x^2(x+h)^2}$
   $\dfrac{f(x+h)-f(x)}{h} = \dfrac{-2xh-h^2}{x^2(x+h)^2}$ tends to $-\dfrac{2x}{x^4} = -\dfrac{2}{x^3}$ as h tends to 0.
   $f'(x) = -\dfrac{2}{x^3},\ f'(1) = -2$
2. (i) $21x^{20}$ (ii) $-2x^{-3}$ (iii) $-3x^{-4}$ (iv) $\frac{1}{3}x^{-\frac{2}{3}}$ (v) $\frac{1}{2}x^{-\frac{1}{2}}$
3. (i) $-2kx^{-3}$ (ii) $-3kx^{-4}$ (iii) $\frac{1}{2}kx^{-\frac{1}{2}}$
4. (i) $\dfrac{dy}{dx} = 2x,\ 8$ (ii) $\dfrac{dy}{dx} = -\dfrac{2}{x^3} = -2x^{-3},\ -\dfrac{1}{32}$
   (iii) same as (ii) (iv) $\dfrac{dy}{dx} = f'(x) = \dfrac{1}{2}x^{-\frac{1}{2}},\ \dfrac{1}{4}$
5. — 6. $f'(x) = 6x + 8;\ f'(1) = 14$
7. $f'(x) = 6x + 8,\ f'(0) = 8$
8. $f'(x) = 16 - \dfrac{2}{x^3} = \dfrac{2(8x^3-1)}{x^3}$. $f'(x) = 0$ when $x = \dfrac{1}{2}$
9. $\dfrac{dy}{dx} = 2x$ (i) $x = 1$ (ii) $x = 3$ (iii) $x = -2$
10. $y = x^2 - 7x$. $\dfrac{dy}{dx} = 2x - 7$ (i) $x = 3{\cdot}5$ (ii) $x = 4$ (iii) $x = 3$
11. $f(x) = 2x + k$ where k is any constant
12. On the x axis, $y = 0 = x^2 - x - 2 = (x-2)(x+1)$. Points of intersection are $x = 2,\ x = -1$, i.e. $(2, 0), (-1, 0)$, $\dfrac{dy}{dx} = 2x - 1$. At $(2, 0)$, $\dfrac{dy}{dx} = 3$, at $(-1, 0)$, $\dfrac{dy}{dx} = -3$. $y = 3x - 6$; $y = -3x - 3$.

Exercise 17.2

1. $\dfrac{dy}{dx} = -\dfrac{4}{x^2}$ (i) -4 (ii) -4 (iii) $-\frac{1}{4}$ (iv) $-\frac{1}{4}$, (i) is parallel to (ii), (iii) is parallel to (iv).
2. $\dfrac{dy}{dx} = -\dfrac{k^2}{x^2}$, which is negative for all real values of x.
3. $\dfrac{dy}{dx} = -\dfrac{36}{x^2}$ At $x = \pm 2$. Points are $(2, 18), (-2, -18)$.
4. $y = 2x^{\frac{1}{2}},\ \dfrac{dy}{dx} = x^{-\frac{1}{2}} = \dfrac{2}{y}$ (i) 1 (ii) -1 (iii) $\frac{1}{4}$ (iv) $\frac{1}{2}$
5. $y = 4x^{\frac{1}{2}},\ \dfrac{dy}{dx} = 2x^{-\frac{1}{2}} = \dfrac{8}{y}$ (i) $(1, 4)$ (ii) $(9, 12)$
6. $\dfrac{dy}{dx} = 8x;\ \dfrac{dy}{dx} = -\dfrac{32}{x^2}$. Gradient of tangent to parabola is 16. Gradient of tangent to rectangular hyperbola is -8. $\tan^{-1}\dfrac{24}{127}$.
7. $\dfrac{dy}{dx} = -\dfrac{10}{x^2} = -2{\cdot}5$ at $(2, 5)$, i.e. y is decreasing at the rate of 5 units for every 2 unit increase in x.

8. $\frac{dy}{dx} = \frac{x}{2}$. Gradient of the normal is $-\frac{2}{x}$. The normal at $(-2, 1)$ is $\frac{y-1}{x+2} = 1$; $y = x + 3$. The normal at $(-4, 4)$ is $\frac{y-4}{x+4} = \frac{1}{2}$; $2y = x + 12$.
Point of intersection is $(6, 9)$ which lies on the same parabola $4y = x^2$.
9. $f'(x) = 12 - 2x = 2(6 - x)$
10. On $y = x^2 + k$, $\frac{dy}{dx} = 2x$. On $y = x^2$, $\frac{dy}{dx} = 2x$. For the same x the gradients are equal and thereby the tangents are parallel.

Exercise 17.3

1. (i) $v = 6t^2 - 2t = 2t(3t - 1)$, $v = 0$ when $t = 0$, $t = \frac{1}{3}$
(ii) $v = 5 - 2t$. $v = 0$ when $t = 2{\cdot}5$
(iii) $v = 3$ velocity is constant for all t and never zero.
2. (i) $v = 3$; (iv) $v = 6$
3. (i) The speed increases as the time increases, i.e. the body moves faster and faster.
(ii) The speed is always positive, the distance from O is getting greater and greater. Note that if $s = -20 + t$ then $\frac{ds}{dt} = 1 > 0$ but up to $t = 20$ the body is approaching O from the left. After $t = 20$ then the distance from O increases. Of course s is negative before $t = 20$ so that $\frac{ds}{dt} > 0$; implying that s is always increasing as t increases; refers to the value of s increasing and not the gap between the body and the point O.
(iii) If $v =$ constant at $1\ \text{m s}^{-1}$ then $\frac{dv}{dt} = 0$.
4. $\frac{dv}{dt} = -8t$. $v = 0$ at $t = \pm 1$ but since we only consider a positive time then $t = 1$ is the only solution. When $t = 3$, $v = -32$, and the body is moving in the opposite direction to when t was in between 0 and 1.
5. (i) $v = \frac{ds}{dt} = at$; $\frac{dv}{dt} = a$ (iii) $3/b$
(ii) $v^2 = a^2t^2 = 2as$ $\therefore$ $v = s^{\frac{1}{2}}\sqrt{2a}$ (We write it this way to prepare for the differentiation with respect to s) $\therefore$ $\frac{dv}{ds} = \frac{1}{2}s^{-\frac{1}{2}}\sqrt{2a}$ $\therefore$ $\frac{v\,dv}{ds} = a$.
6. When $t = 1$, $s = 16$; when $t + \delta t = 3$, $s + \delta s = 144$

$$\frac{\delta s}{\delta t} = \frac{144 - 16}{3 - 1} = \frac{128}{2} = 64\ \text{m s}^{-1}$$

$$\frac{ds}{dt} = 32t \text{ at } t = 1,\ v = 32;\ \text{at } t = 3,\ v = 96$$

7. (i) $s = 0$ when $t = 0$ or 8 (ii) $s = 60$ m
(iii) $v = 40 - 10t = 0$ when $t = 4$ (iv) when $t = 4$, $s = 80$ m
(v) 65 m
8. (i) $v = 10\ \text{m s}^{-1}$
(ii) 200 m
(iii) $-\frac{1}{3}\ \text{ms}^{-2}$

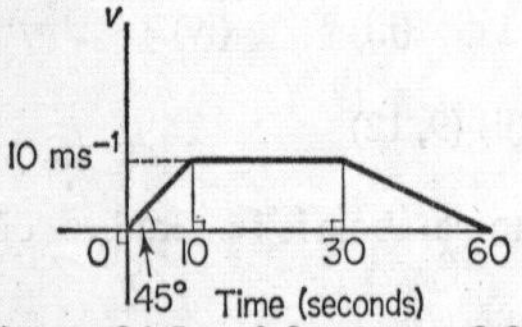

v
10 ms^{-1}
0 10 30 60
45°
Time (seconds)

9. (i) $v = 24{\cdot}5 - 9{\cdot}8t$, $a = -9{\cdot}8\ \text{m s}^{-2}$. The acceleration is constant.
(ii) $v = 0$ when $24{\cdot}5 = 9{\cdot}8t$, $t = 2{\cdot}5$ s. The greatest height $30{\cdot}625$ m.
It takes $2\frac{1}{2}$ s to reach the greatest height and $2\frac{1}{2}$ s to return to the point of projection.

10. (i) $v = 12t^2 - 48t + 45 = 3(4t^2 - 16t + 15) = 3(2t - 3)(2t - 5)$. $t = 1$
$v = 9$; $t = 2$, $v = -3$; $t = 3$; $v = 9$
(ii) $t = 1{\cdot}5$ or $2{\cdot}5$
(iii) The body starts by moving left to right with velocity 45. It stops at $t = 1\frac{1}{2}$ and returns to the starting point, overshoots and stops again at $t = 2{\cdot}5$; it then returns to the starting point, passes through and continues moving to the right with increasing speed.
11. (i) $t = 3$, $s = 18 + 14{\cdot}4 = 32{\cdot}4$ m, height $67{\cdot}6$ m
$v = 6 + 3{\cdot}2t$; $t = 3$, $v = 15{\cdot}6\ \text{m s}^{-1}$
(ii) $1{\cdot}6t^2 + 6t - 100 = 0$; $t = 6{\cdot}25$ s
(iii) $\frac{dv}{dt} = 3{\cdot}2$ a constant

Exercise 18.1

1. $f'(x) = 2x + 1 = 0$ at $x = -\frac{1}{2}$. Least value $0{\cdot}75$.
2. Least value of $0{\cdot}75$ at $x = \frac{1}{2}$.
3. $f'(x) = -4x + 8 = 0$ at $x = 2$. Greatest value 7.
4. $f'(x) = 32x + 64 = 0$ at $x = -2$. Least value -54.
5. $f(x) = x^2 + x - 2$. $f'(x) = 2x + 1 = 0$ at $x = -\frac{1}{2}$. Least value $-2\frac{1}{4}$.
6. $f(x) = -x^2 + 2x + 8$. $f'(x) = -2x + 2 = 0$ at $x = 1$. Greatest value 9.
7. $\frac{ds}{dt} = 30 - 10t = 0$ when $t = 3$. $s = 45$ m is greatest height.
8. A (i) $= x(600 - x) = 600x - x^2$. $\frac{dA}{dx} = 600 - 2x = 0$ when $x = 300$. Greatest area, 90 000 m^2.
9. $\frac{dy}{dx} = 6x + 4$ (independent of c) $= 0$ when $x = -\frac{2}{3}$. Least value when $x = -\frac{2}{3}$ for any c. (ii) $A = x(1200 - 2x)$; $x = 300$, $A = 18000\ \text{m}^2$.
10. $\frac{dA}{du} = 42 - 14u = 0$ when $u = 3$. Greatest value 59.

Exercise 18.2

1. $x = 0$ gives a point of inflexion.
2. $x = 0$ gives a point of inflexion.
3. $f'(x) = 6x(x - 1)$. $x = 0$, maximum 0; $x = 1$, minimum -1. Stationary values 1, and -1.
4. $f'(x) = 3x^2 - 3 = 0$ when $x = 1$ or -1. $x = 1$, minimum 5; $x = -1$ maximum 9. Stationary values 5, 9.
5. $f'(x) = 12 - 3x^2 = 3(4 - x^2) = 0$ when $x = \pm 2$. $x = 2$, maximum -8; $x = -2$, minimum -40. Stationary values -8 and 40.
6. $f'(x) = 1 - \frac{1}{x^2} = 0$ when $x = 1$ or -1. $x = 1$, minimum 2; $x = -1$, maximum -2. Stationary values 2 and -2.
7. $f'(x) = 2x^3 - 2x = 2x(x^2 - 1) = 2x(x - 1)(x + 1)$. $x = 1$, minimum $-\frac{1}{2}$; $x = 0$, maximum 0; $x = -1$, minimum $-\frac{1}{2}$. Stationary values $-\frac{1}{2}, 0, -\frac{1}{2}$.
8. $\frac{dy}{dx} = 9x^2 - 9 = 9(x^2 - 1) = 0$ at $x = \pm 1$. Turning points are minimum point $(1, 1)$, maximum point $(-1, 13)$.
9. $\frac{dy}{dx} = 3x^2 - 6x + 3 = 3(x - 1)^2$. Turning point $(1, 5)$, a point of inflexion $\frac{dy}{dx}$ is positive for all real x, positive just to the left and right of $x = 1$.
10. $\frac{dy}{dx} = 2x - 1 - x^2 = -(x - 1)^2$. Turning point $(1, -2)$, a point of inflexion $\frac{dy}{dx}$ is negative to the left and right of $x = 1$.

11. $\frac{dy}{dx} = x^3 + 3x^2 + 3x + 1 = (x + 1)^3$. Turning point $(-1, -\frac{1}{4})$.

12. $V = 160x - 52x^2 + 4x^3$. $\frac{dV}{dx} = 12x^2 - 104x + 160 = 4(x - 2)(3x - 20)$.
Maximum for $x = 2$. $V = 12 \times 6 \times 2 = 144$ cm³.

13. $V = (24 - 2x)(9 - 2x)x = 216x - 66x^2 + 4x^3$
$\frac{dV}{dx} = 216 - 132x + 12x^2 = 12(x^2 - 11x + 18) = 12(x - 9)(x - 2)$.
Maximum for $x = 2$, $V = 20 \times 5 \times 2 = 200$ cm³.

14. $V = 1 \times x \times b = 2\text{ m}^3 \quad \therefore \quad b = \frac{2}{x}$
$A = x \times \frac{2}{x} + 2\left(x + \frac{2}{x}\right) = 2 + 2x + \frac{4}{x}$
$\frac{dA}{dx} = 2 - \frac{4}{x^2} = 0$ when $x = \pm\sqrt{2}$. Minimum area for square base of side $\sqrt{2}$ m.

15. $\frac{dA}{dr} = 4\pi r - \frac{2V}{r^2} = 0$ when $r^3 = \frac{V}{2\pi} = \frac{\pi r^2 h}{2\pi}$. Hence $r = \frac{h}{2}$
(iv) $r^3 = \frac{269{\cdot}5 \times 7}{2 \times 22} = \frac{24{\cdot}5 \times 7}{2 \times 2} = \frac{49 \times 7}{2 \times 2 \times 2}$. $r = \frac{7}{2}$ cm

Exercise 18.3

1. $f'(x) = 12x^3 - 12x^2 = 12x^2(x - 1) = 0$ at $x = 0$ or 1
$f''(x) = 36x^2 - 24x = 12x(3x - 2)$
$f''(1)$ positive hence $f(1)$ a minimum. $f(1) = 0$.
$f''(0) = 0$, no conclusions.
On the left of $x = 0$, $f'(x)$ is negative; on the right of $x = 0$, $f'(x)$ is negative
$\therefore x = 0$ gives a point of inflexion.

2. $x = 0$ is a point of inflexion. 3. $x = 1$ is a point of inflexion.

4. $x = 0$ gives a minimum $f(0) = 0$.

5. $f'(x) = x^2 + 6x - 40 = (x + 10)(x - 4) = 0$ when $x = 4$ or -10
$f''(x) = 2x + 6$
$f''(-10)$ is negative. $f(-10) = 526\frac{2}{3}$ is a maximum.
$f''(4)$ is positive. $f(4) = 69\frac{1}{3}$ is a minimum.

6. $A = \frac{x}{2}(20 - x) = 10x - \frac{x^2}{2}$
$\frac{dA}{dx} = 10 - x = 0$ when $x = 10$. $\frac{d^2A}{dx^2} = -1$, negative, hence, $A = 50$ cm² is a maximum.

7. $y = a^2 + (10 - a)^2 = 2a^2 - 20a + 100$
$\frac{dy}{da} = 4a - 20$. $\frac{d^2y}{da^2} = 4$, positive, and so $a = 5$ gives a minimum of $y = 50$.

8. $V = \pi r^2 h = \pi r^2(30 - r) = 30\pi r^2 - \pi r^3$
$\frac{dV}{dr} = 60\pi r - 3\pi r^2 = 0$ when $r = 0$ or $r = 20$ $(h = 10)$
$\frac{d^2V}{dr^2} = 60\pi - 6\pi r$ which is negative when $r = 20$ which therefore gives a maximum.
Maximum $V = 4000\pi$ cubic centimetres.

9. Let the numbers be x and $2n - x$
$S = x^3 + (2n - x)^3 = 8n^3 - 12n^2x + 6nx^2$
$\frac{dS}{dx} = -12n^2 + 12nx = 0$ when $x = n$
$\frac{d^2S}{dx^2} = 12n$, positive, and so $x = n$ gives a minimum, i.e. both numbers are equal to n.

10. $\frac{ds}{dt} = 3t^2 - 24t + 45 = 3(t^2 - 8t + 15) = 3(t-3)(t-5) = 0$ when $t = 3$ or 5

$\frac{d^2s}{dt^2} = 6t - 24$, positive when $t = 5$, a minimum for s, negative when $t = 3$, a maximum for s. As t increases after 5 so s increases indefinitely and the body gets further away, i.e. the maximum s at $t = 3$ is not the greatest possible value of s.

Exercise 19.1

4. $f'(x) = \cos x - \sin x = 0$ when $\tan x = 1$, $x = \frac{\pi}{4}$ radians

$f''(x) = -\sin x - \cos x$ is negative when $x = \frac{\pi}{4}$

$\sin\frac{\pi}{4} = \cos\frac{\pi}{4} = \frac{1}{\sqrt{2}}$. Maximum $\sqrt{2}$.

5. $f'(x) = 3\cos x - 4\sin x = 0$ when $\tan x = \frac{3}{4}$

$\left(\text{in which case } \sin x = \frac{3}{5}, \cos x = \frac{4}{5}\right)\left(0 < x < \frac{\pi}{2}\right)$

$f''(x) = -3\sin x - 4\cos x$ is negative, for $0 < x < \frac{\pi}{2}$. Maximum 5.

6. Maximum 5 for $\tan x = \frac{4}{3}$.

7. $f'(x) = \cos x + \sin x = 0$ when $\tan x = -1$, $x = \frac{3\pi}{4}$ radians or $\sin x = \frac{1}{\sqrt{2}}$, $\cos x = -\frac{1}{\sqrt{2}}$.

$f''(x) = -\sin x + \cos x$ is negative when $x = \frac{3\pi}{4}$ radians. Maximum $\sqrt{2}$.

8. $f'(x) = 1 - 2\cos x = 0$ when $\cos x = \frac{1}{2}$; $x = \frac{\pi}{3}$

$f''(x) = +2\sin x$ is positive when $x = \frac{\pi}{3}$

Minimum of $\frac{\pi}{3} - \sqrt{3}$ when $x = \frac{\pi}{3}$

9. $f'(x) = 1 - 2\sin x = 0$ when $\sin x = \frac{1}{2}$; $x = \frac{\pi}{6}$ or $\frac{5\pi}{6}$

$f''(x) = -2\cos x$ is negative when $x = \frac{\pi}{6}$ maximum $\frac{\pi}{6} + \sqrt{3}$

$f''(x) = -2\cos x$ is positive when $x = \frac{5\pi}{6}$ minimum $\frac{5\pi}{6} - \sqrt{3}$

10. —

11. $f'(x) = 1 - 2\cos 2x = 0$ when $\cos 2x = \frac{1}{2}$; $x = \frac{\pi}{6}, \frac{5\pi}{6}$.

Just to the left of $x = \frac{\pi}{6}$, $f'(x)$ is negative.

Just to the right of $x = \frac{\pi}{6}$, $f'(x)$ is positive.

$f(x)$ has a minimum of $\frac{\pi}{6} - \frac{\sqrt{3}}{2}$ at $x = \frac{\pi}{6}$.

Exercise 19.2

1. $y = x^6 + 2x^3 + 1$; $\frac{dy}{dx} = 6x^2(x^3 + 1)$

2. $y = 8x^3 + 12x^2 + 6x + 1$; $\frac{dy}{dx} = 24x^2 + 24x + 6 = 6(2x+1)^2$

3. $y = 6 + x^6 + 2x^3 + 1; \frac{dy}{dx} = 6x^2(x^3 + 1)$

4. $y = 6 + 5x^3 - 15x^2 + 15x - 5; \frac{dy}{dx} = 15x^2 - 30x + 15 = 15(x - 1)^2$

5. $y = \frac{1}{x^2}, \frac{dy}{dx} = -\frac{2}{x^3}; \frac{dy}{du} = -\frac{1}{u^2}; \frac{du}{dx} = 2x$

6. $y = 1 - 10x^3 + 25x^6; \frac{dy}{dx} = -30x^2 + 150x^5 = -30x^2(1 - 5x^3)$

7. $y = x^{\frac{5}{2}}; \frac{dy}{dx} = \frac{5}{2}x^{\frac{3}{2}}$

Exercise 19.3

1. $20x(x^2 + 1)^9$
2. $30x(3x^2 + 4)^4$
3. $-6x(4 - x^2)^2$
4. $-\frac{4}{(x + 1)^5}; f(-1)$ not defined.
5. $2x \cos(x^2)$
6. $5(2x + 1)(x^2 + x + 1)^4$
7. $-\frac{2x}{(x^2 + 1)^2}$
8. $-\frac{3x^2}{(x^3 + 1)^2}; f(-1)$ not defined.
9. $\frac{1}{2\sqrt{(x + 1)}}$
10. $\frac{x}{\sqrt{(x^2 + 1)}}$
11. $-2 \sin x \cos x$
12. $3 \sin^2 x \cos x$
13. 0
14. (i) $f'(x) = 1 - \frac{1}{(x - 1)^2} = 0$ when $(x - 1)^2 = 1, x = 0$ or 2

    $f''(x) = +\frac{2}{(x - 1)^3}$, negative for $x = 0$, positive for $x = 2$

    Maximum $f(0) = -1$, minimum $f(2) = 3$

    We exclude $x = 1$ because $f(1)$ is not defined.

    (ii) $f'(x) = 3(x + 1)^2 - \frac{3}{(x + 1)^2} = \frac{3}{(x + 1)^2}\{(x + 1)^4 - 1\} = 0$

    when $(x + 1)^4 = 1, x = 0$ or -2

    $f''(x) = 6(x + 1) + \frac{6}{(x + 1)^3}$, positive for $x = 0$

    $f(0) = 4$ a minimum

    $f''(x)$ negative for $x = -2, f(-2) = -4$ a maximum.

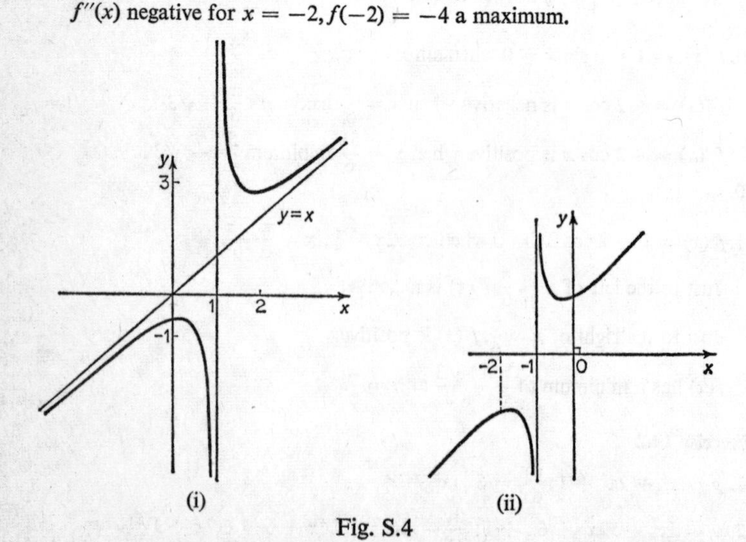

Fig. S.4

Exercise 19.4

1. $(x+2)+(x+1)=2x+3$
2. $(2-x)-x=2-2x$
3. $3(x+4)+(3x+1)=6x+13$
4. $2x\sin x+x^2\cos x$
5. $\sin(x^2)+2x^2\cos(x^2)$
6. $\cos^2 x-\sin^2 x$
7. $4(5x+2)(6x+3)+(4x+1)\,5(6x+3)+(4x+1)(5x+2)6$
8. $4x(x^2+1)(x^3+1)+(x^2+1)^2\,3x^2$

Exercise 19.5

1. $\dfrac{1-x^2}{(1+x^2)^2}$ (ii) $\dfrac{x(2-x^3)}{(1+x^3)^2}$ (iii) $-\dfrac{41}{(5x-7)^2}$ (iv) $-\dfrac{4x}{(x^2-1)^2}$
(v) $-\operatorname{cosec}^2\theta$

2. (i) $f'(x)=\dfrac{(1-3x)(3+x)}{(1+x^2)^2}=0$ when $x=-3$ or $+\frac{1}{3}$
$f'(x)$ is negative on the left of $x=-3$ and positive on the right of $x=-3$
$\therefore\ f(-3)=-0{\cdot}5$ is a minimum
$f'(x)$ is positive on the left of $x=\frac{1}{3}$ and negative on the right of $x=\frac{1}{3}$
$\therefore\ f(\frac{1}{3})=4{\cdot}5$ is a maximum

(ii) $f'(x)=-\dfrac{2(2x+1)(x+2)}{(x^2-1)^2}=0$ when $x=-\frac{1}{2}$ or -2
On the left of $x=-2$, $f'(x)$ is negative, on the right of $x=-2$, $f'(x)$ is positive, $f(-2)=-1$ is a minimum. On the left of $x=-\frac{1}{2}$, $f'(x)$ is positive, on the right of $x=-\frac{1}{2}$ is negative $f(-\frac{1}{2})=-4$ is a maximum.

Exercise 19.6

1. $-\dfrac{1}{2y}$
2. $\dfrac{2x-1}{2y}$
3. $\dfrac{2x-1}{\cos y}$
4. $-\dfrac{(y+\cos x)}{x}$
5. $\dfrac{\sin x-y^2}{2xy}$
6. $\dfrac{1-y^2\cos x}{2y\sin x}$
7. $\dfrac{dy}{dx}=1,\ y=x+2$
8. $\dfrac{dy}{dx}=\dfrac{1}{2},\ 2y=x+4$
9. —

10. $x^m\dfrac{dy}{dx}+mx^{m-1}y=0\ \therefore\ \dfrac{dy}{dx}=-\dfrac{mx^{m-1}y}{x^m}=-\dfrac{mx^m y}{x^{m+1}}=-\dfrac{m}{x^{m+1}}=-mx^{-m-1}$
But $y=x^{-m}\ \therefore\ \dfrac{d(x^{-m})}{dx}=-mx^{-m-1}$ and $\dfrac{d(x^n)}{dx}=nx^{n-1}$ holds when n is a negative integer $-m$.

11. $\dfrac{dv}{dt}=\dfrac{dv}{ds}\cdot\dfrac{ds}{dt}=\dfrac{dv}{ds}v=v\dfrac{dv}{ds}$

12. (i) $\dfrac{dy}{dx}=\dfrac{dy}{du}\cdot\dfrac{du}{dx}=\dfrac{4}{3}(x^2+1)^{\frac{1}{3}}\,2x=\dfrac{8x}{3}(x^2+1)^{\frac{1}{3}}$
(ii) $-5(2x+1)^{-6}\,2=-10(2x+1)^{-6}$ (iii) $-\dfrac{4x}{(x^2+1)^3}$

13. $y=x-x^2,\ \dfrac{dy}{dx}=1-2x$, y is a maximum for the number $x=\frac{1}{2}$

14. $h+2\pi r=3,\ V=\pi r^2h=\pi r^2(3-2\pi r),\ \dfrac{dV}{dr}=6\pi r-6\pi^2r^2=0$ for $r=0$ or $\dfrac{1}{\pi}$
$\dfrac{d^2v}{dr^2}=6\pi-12\pi^2r$ which is negative for $r=\dfrac{1}{\pi}$ so giving V a maximum of $\dfrac{1}{\pi}$ m^3

15. $\dfrac{dV}{dt}=\dfrac{dV}{ds}\cdot\dfrac{ds}{dt}\ \therefore\ 50=(6s^2+4s+9)\dfrac{ds}{dt}$, when $s=3$, $\dfrac{ds}{dt}=\dfrac{50}{75}=\dfrac{2}{3}$ cm per sec

16. $\dfrac{dV}{dt}=\dfrac{dV}{dr}\cdot\dfrac{dr}{dt}\ \therefore\ 10=4\pi r^2\dfrac{dr}{dt}\ \therefore\ \dfrac{dr}{dt}=\dfrac{1}{10\pi}$ cm s^{-1} when $r=5$ cm

Exercise 20.1 (Arbitrary constants omitted in answers 1 and 2.)

1. (i) x^2 (ii) x^3 (iii) x^4 (iv) x^5 (v) x^6 (vi) x^n
2. (i) $t^2 + t^3$ (ii) $8t^3 + \sin t$ (iii) $6t^4 + \cos t$ (iv) $-t^{-1}$ (v) $-t^{-2}$
(vi) $-t^{-1} = -\frac{1}{t}$
3. $y = 8x^2 + 3$
4. $2x^3 + \frac{x^2}{2} + 1 = y$
5. $y = \frac{ax^3}{3} + bx + c$, c an arbitrary constant

Exercise 20.2

1. (i) $4x^2 + c$ (ii) $4x^2 + 5x + c$ (iii) $\frac{8x^3}{3} + \frac{5x^2}{2} + c$
(iv) $\frac{x^3}{3} + \frac{x^2}{2} + c$ (v) $s = 40t + \frac{3t^2}{2} + c$ (vi) $8t - 2{\cdot}45t^2$
2. $v = 4t + 10$, $s = 2t^2 + 10t$
3. $y = 2x^3 + 9x^2 + 12x + 4$
4. (i) $x^3 - x^2 + 3$ (ii) $y = \sin x$
(iii) $\frac{dy}{dx} = 3x^2 + \cos x + c$, $c = 4$ $y = x^3 + \sin x + 4x - 2$
(iv) $A = \sin x - \cos x + 4$
5. $v = 49 - 4{\cdot}9t$ $s = 49t - 2{\cdot}45t^2$
greatest height when $v = 0$, $t = 10$, $s = 245$ m
$t = 1$, $s = 46{\cdot}5$ m, $t = 20$, $s = 0$, i.e. back at 0
6. $\frac{dv}{dt} = 0{\cdot}5 - \frac{1}{10}t$; $v = 0{\cdot}5t - \frac{1}{20}t^2$ ($v = 0$ when $t = 0$) when $t = 2$, $v = 0{\cdot}8$ m s^{-1}
$s = \frac{t^2}{4} - \frac{t^3}{60}$ when $t = 5$, $s = 4\frac{1}{6}$ m
7. 70 m
8. $s = t^3 - 8t^2 + 9$, 272 m

Exercise 20.3

1. $\frac{x^2}{2} + c$
2. $4x + c$
3. $\frac{x^2}{2} + 4x + c$
4. $\frac{x^2}{2} - 4x + c$
5. $x^2 + c$
6. $\sin x + c$
7. $x^2 + \sin x + c$
8. $x^2 - \sin x + c$
9. $-\frac{1}{x} + c$
10. $x \sin x + c$

Exercise 20.4

1. (i) 3·75 (ii) 0 (iii) 0·8 (iv) 18
2. (i) 0·4
(ii) The definite integral is 0 because part of the area is below the x axis, area is $\frac{1}{2}$.
(iii) $23\frac{1}{4}$ (iv) $\left[\sin x\right]_0^{\pi/2} = 1$
3. (i) $\left[\frac{x^4}{4} - x^3\right]_{-1}^{0} = -1\frac{1}{4}$
The whole area is below the x axis. Area measures $1\frac{1}{4}$ square units.
(ii) $\left[\frac{x^4}{4} - x^3\right]_0^3 = -6\frac{3}{4}$
The whole area is below the x axis. Area measures $6\frac{3}{4}$ square units.
(iii) $\left[\frac{x^4}{4} - x^3\right]_2^3 = -\frac{11}{4}$; $\left[\frac{x^4}{4} - x^3\right]_3^4 = \frac{27}{4}$
The area measures $9\frac{1}{2}$ square units.

4. 479·52 m.

5. $\dfrac{dA}{dx} = -\dfrac{4x}{(x^2-1)^2} \quad \displaystyle\int_2^3 -\dfrac{4x\,dx}{(x^2-1)^2} = \left[\dfrac{x^2+1}{x^2-1}\right]_2^3 = -\dfrac{5}{12}$

6. $\dfrac{dA}{dx} = \dfrac{4x}{(x^2+1)^2} \quad \displaystyle\int \dfrac{4x}{(x^2+1)^2}\,dx = \left[\dfrac{x^2-1}{x^2+1}\right]_0^1 = 1$

7. (i) 32 (ii) $117\frac{1}{3}$ (iii) $85\frac{1}{3}$

8. $x = 0, x = 4, 10\frac{2}{3}$

9. $10\frac{2}{3}$ m. The same problem as in question 8.

10. Let t_1 be the time. We need $\left[\dfrac{t^3}{3}\right]_0^{t_1} = \left[2t^2\right]_0^{t_1}$

$\therefore \quad 2t_1^2 = \dfrac{t_1^3}{3} \qquad t_1 = 6$

Distance is 72 metres.

TABLES

N	0	1	2	3	4	5	6	7	8	9	1	2	3	4	5	6	7	8	9
10	·0000	0043	0086	0128	0170	0212	0253	0294	0334	0374	4	8	12	17	21	25	29	33	37
11	·0414	0453	0492	0531	0569	0607	0645	0682	0719	0755	4	8	11	15	19	23	26	30	34
12	·0792	0828	0864	0899	0934	0969	1004	1038	1072	1106	3	7	10	14	17	21	24	28	31
13	·1139	1173	1206	1239	1271	1303	1335	1367	1399	1430	3	6	10	13	16	19	23	26	29
14	·1461	1492	1523	1553	1584	1614	1644	1673	1703	1732	3	6	9	12	15	18	21	24	27
15	·1761	1790	1818	1847	1875	1903	1931	1959	1987	2014	3	6	8	11	14	17	20	22	25
16	·2041	2068	2095	2122	2148	2175	2201	2227	2253	2279	3	5	8	11	13	16	18	21	24
17	·2304	2330	2355	2380	2405	2430	2455	2480	2504	2529	2	5	7	10	12	15	17	20	22
18	·2553	2577	2601	2625	2648	2672	2695	2718	2742	2765	2	5	7	9	12	14	16	19	21
19	·2788	2810	2833	2856	2878	2900	2923	2945	2967	2989	2	4	7	9	11	13	16	18	20
20	·3010	3032	3054	3075	3096	3118	3139	3160	3181	3201	2	4	6	8	11	13	15	17	19
21	·3222	3243	3263	3284	3304	3324	3345	3365	3385	3404	2	4	6	8	10	12	14	16	18
22	·3424	3444	3464	3483	3502	3522	3541	3560	3579	3598	2	4	6	8	10	12	14	15	17
23	·3617	3636	3655	3674	3692	3711	3729	3747	3766	3784	2	4	6	7	9	11	13	15	17
24	·3802	3820	3838	3856	3874	3892	3909	3927	3945	3962	2	4	5	7	9	11	12	14	16
25	·3979	3997	4014	4031	4048	4065	4082	4099	4116	4133	2	3	5	7	9	10	12	14	15
26	·4150	4166	4183	4200	4216	4232	4249	4265	4281	4298	2	3	5	7	8	10	11	13	15
27	·4314	4330	4346	4362	4378	4393	4409	4425	4440	4456	2	3	5	6	8	9	11	13	14
28	·4472	4487	4502	4518	4533	4548	4564	4579	4594	4609	2	3	5	6	8	9	11	12	14
29	·4624	4639	4654	4669	4683	4698	4713	4728	4742	4757	1	3	4	6	7	9	10	12	13
30	·4771	4786	4800	4814	4829	4843	4857	4871	4886	4900	1	3	4	6	7	9	10	11	13
31	·4914	4928	4942	4955	4969	4983	4997	5011	5024	5038	1	3	4	6	7	8	10	11	12
32	·5051	5065	5079	5092	5105	5119	5132	5145	5159	5172	1	3	4	5	7	8	9	11	12
33	·5185	5198	5211	5224	5237	5250	5263	5276	5289	5302	1	3	4	4	6	8	9	10	12
34	·5315	5328	5340	5353	5366	5378	5391	5403	5416	5428	1	3	4	5	6	8	9	10	11
35	·5441	5453	5465	5478	5490	5502	5514	5527	5539	5551	1	2	4	5	6	7	9	10	11
36	·5563	5575	5587	5599	5611	5623	5635	5647	5658	5670	1	2	4	5	6	7	8	10	11
37	·5682	5694	5705	5717	5729	5740	5752	5763	5775	5786	1	2	3	5	6	7	8	9	10
38	·5798	5809	5821	5832	5843	5855	5866	5877	5888	5899	1	2	3	5	6	7	8	9	10
39	·5911	5922	5933	5944	5955	5966	5977	5988	5999	6010	1	2	3	4	5	7	8	9	10
40	·6021	6031	6042	6053	6064	6075	6085	6096	6107	6117	1	2	3	4	5	6	8	9	10
41	·6128	6138	6149	6160	6170	6180	6191	6201	6212	6222	1	2	3	4	5	6	7	8	9
42	·6232	6243	6253	6263	6274	6284	6294	6304	6314	6325	1	2	3	4	5	6	7	8	9
43	·6335	6345	6355	6365	6375	6385	6395	6405	6415	6425	1	2	3	4	5	6	7	8	9
44	·6435	6444	6454	6464	6474	6484	6493	6503	6513	6522	1	2	3	4	5	6	7	8	9
45	·6532	6542	6551	6561	6571	6580	6590	6599	6609	6618	1	2	3	4	5	6	7	8	9
46	·6628	6637	6646	6656	6665	6675	6684	6693	6702	6712	1	2	3	4	5	6	7	7	8
47	·6721	6730	6739	6749	6758	6767	6776	6785	6794	6803	1	2	3	4	5	5	6	7	8
48	·6812	6821	6830	6839	6848	6857	6866	6875	6884	6893	1	2	3	4	4	5	6	7	8
49	·6902	6911	6920	6928	6937	6946	6955	6964	6972	6981	1	2	3	4	4	5	6	7	8
50	·6990	6998	7007	7016	7024	7033	7042	7050	7059	7067	1	2	3	3	4	5	6	7	8
51	·7076	7084	7093	7101	7110	7118	7126	7135	7143	7152	1	2	3	3	4	5	6	7	8
52	·7160	7168	7177	7185	7193	7202	7210	7218	7226	7235	1	2	3	3	4	5	6	7	7
53	·7243	7251	7259	7267	7275	7284	7292	7300	7308	7316	1	2	2	3	4	5	6	6	7
54	·7324	7332	7340	7348	7356	7364	7372	7380	7388	7396	1	2	2	3	4	5	6	6	7
	0	1	2	3	4	5	6	7	8	9	1	2	3	4	5	6	7	8	9

N	0	1	2	3	4	5	6	7	8	9	1	2	3	4	5	6	7	8	9
55	·7404	7412	7419	7427	7435	7443	7451	7459	7466	7474	1	2	2	3	4	5	5	6	7
56	·7482	7490	7497	7505	7513	7520	7528	7536	7543	7551	1	2	2	3	4	5	5	6	7
57	·7559	7566	7574	7582	7589	7597	7604	7612	7619	7627	1	2	2	3	4	5	5	6	7
58	·7634	7642	7649	7657	7664	7672	7679	7686	7694	7701	1	1	2	3	4	4	5	6	7
59	·7709	7716	7723	7731	7738	7745	7752	7760	7767	7774	1	1	2	3	4	4	5	6	7
60	·7782	7789	7796	7803	7810	7818	7825	7832	7839	7846	1	1	2	3	4	4	5	6	6
61	·7853	7860	7868	7875	7882	7889	7896	7903	7910	7917	1	1	2	3	4	4	5	6	6
62	·7924	7931	7938	7945	7952	7959	7966	7973	7980	7987	1	1	2	3	3	4	5	6	6
63	·7993	8000	8007	8014	8021	8028	8035	8041	8048	8055	1	1	2	3	3	4	5	5	6
64	·8062	8069	8075	8082	8089	8096	8102	8109	8116	8122	1	1	2	3	3	4	5	5	6
65	·8129	8136	8142	8149	8156	8162	8169	8176	8182	8189	1	1	2	3	3	4	5	5	6
66	·8195	8202	8209	8215	8222	8228	8235	8241	8248	8254	1	1	2	3	3	4	5	5	6
67	·8261	8267	8274	8280	8287	8293	8299	8306	8312	8319	1	1	2	3	3	4	5	5	6
68	·8325	8331	8338	8344	8351	8357	8363	8370	8376	8382	1	1	2	3	3	4	4	5	6
69	·8388	8395	8401	8407	8414	8420	8426	8432	8439	8445	1	1	2	2	3	4	4	5	6
70	·8451	8457	8463	8470	8476	8482	8488	8494	8500	8506	1	1	2	2	3	4	4	5	6
71	·8513	8519	8525	8531	8537	8543	8549	8555	8561	8567	1	1	2	2	3	4	4	5	5
72	·8573	8579	8585	8591	8597	8603	8609	8615	8621	8627	1	1	2	2	3	4	4	5	5
73	·8633	8639	8645	8651	8657	8663	8669	8675	8681	8686	1	1	2	2	3	4	4	5	5
74	·8692	8698	8704	8710	8716	8722	8727	8733	8739	8745	1	1	2	2	3	4	4	5	5
75	·8751	8756	8762	8768	8774	8779	8785	8791	8797	8802	1	1	2	2	3	3	4	5	5
76	·8808	8814	8820	8825	8831	8837	8842	8848	8854	8859	1	1	2	2	3	3	4	5	5
77	·8865	8871	8876	8882	8887	8893	8899	8904	8910	8915	1	1	2	2	3	3	4	4	5
78	·8921	8927	8932	8938	8943	8949	8954	8960	8965	8971	1	1	2	2	3	3	4	4	5
79	·8976	8982	8987	8993	8998	9004	9009	9015	9020	9025	1	1	2	2	3	3	4	4	5
80	·9031	9036	9042	9047	9053	9058	9063	9069	9074	9079	1	1	2	2	3	3	4	4	5
81	·9085	9090	9096	9101	9106	9112	9117	9122	9128	9133	1	1	2	2	3	3	4	4	5
82	·9138	9143	9149	9154	9159	9165	9170	9175	9180	9186	1	1	2	2	3	3	4	4	5
83	·9191	9196	9201	9206	9212	9217	9222	9227	9232	9238	1	1	2	2	3	3	4	4	5
84	·9243	9248	9253	9258	9263	9269	9274	9279	9284	9289	1	1	2	2	3	3	4	4	5
85	·9294	9299	9304	9309	9315	9320	9325	9330	9335	9340	1	1	2	2	3	3	4	4	5
86	·9345	9350	9355	9360	9365	9370	9375	9380	9385	9390	1	1	1	2	3	3	4	4	5
87	·9395	9400	9405	9410	9415	9420	9425	9430	9435	9440	0	1	1	2	2	3	3	4	4
88	·9445	9450	9455	9460	9465	9469	9474	9479	9484	9489	0	1	1	2	2	3	3	4	4
89	·9494	9499	9504	9509	9513	9518	9523	9528	9533	9538	0	1	1	2	2	3	3	4	4
90	·9542	9547	9552	9557	9562	9566	9571	9576	9581	9586	0	1	1	2	2	3	3	4	4
91	·9590	9595	9600	9605	9609	9614	9619	9624	9628	9633	0	1	1	2	2	3	3	4	4
92	·9638	9643	9647	9652	9657	9661	9666	9671	9675	9680	0	1	1	2	2	3	3	4	4
93	·9685	9689	9694	9699	9703	9708	9713	9717	9722	9727	0	1	1	2	2	3	3	4	4
94	·9731	9736	9741	9745	9750	9754	9759	9763	9768	9773	0	1	1	2	2	3	3	4	4
95	·9777	9782	9786	9791	9795	9800	9805	9809	9814	9818	0	1	1	2	2	3	3	4	4
96	·9823	9827	9832	9836	9841	9845	9850	9854	9859	9863	0	1	1	2	2	3	3	4	4
97	·9868	9872	9877	9881	9886	9890	9894	9899	9903	9908	0	1	1	2	2	3	3	4	4
98	·9912	9917	9921	9926	9930	9934	9939	9943	9948	9952	0	1	1	2	2	3	3	4	4
99	·9956	9961	9965	9969	9974	9978	9983	9987	9991	9996	0	1	1	2	2	3	3	3	4
N	0	1	2	3	4	5	6	7	8	9	1	2	3	4	5	6	7	8	9

N	0	1	2	3	4	5	6	7	8	9	1	2	3	4	5	6	7	8	9
·00	1000	1002	1005	1007	1009	1012	1014	1016	1019	1021	0	0	1	1	1	1	2	2	2
·01	1023	1026	1028	1030	1033	1035	1038	1040	1042	1045	0	0	1	1	1	1	2	2	2
·02	1047	1050	1052	1054	1057	1059	1062	1063	1067	1069	0	0	1	1	1	1	2	2	2
·03	1072	1074	1076	1079	1081	1084	1086	1089	1091	1094	0	0	1	1	1	1	2	2	2
·04	1096	1099	1102	1104	1107	1109	1112	1114	1117	1119	0	1	1	1	1	2	2	2	2
·05	1122	1125	1127	1130	1132	1135	1138	1140	1143	1146	0	1	1	1	1	2	2	2	2
·06	1148	1151	1153	1156	1159	1161	1164	1167	1169	1172	0	1	1	1	1	2	2	2	2
·07	1175	1178	1180	1183	1186	1189	1191	1194	1197	1199	0	1	1	1	1	2	2	2	2
·08	1202	1205	1208	1211	1213	1216	1219	1222	1225	1227	0	1	1	1	1	2	2	2	3
·09	1230	1233	1236	1239	1242	1245	1247	1250	1253	1256	0	1	1	1	1	2	2	2	3
·10	1259	1262	1265	1268	1271	1274	1276	1279	1282	1285	0	1	1	1	1	2	2	2	3
·11	1288	1291	1294	1297	1300	1303	1306	1309	1312	1315	0	1	1	1	2	2	2	2	3
·12	1318	1321	1324	1327	1330	1334	1337	1340	1343	1346	0	1	1	1	2	2	2	2	3
·13	1349	1352	1355	1358	1361	1365	1368	1371	1374	1377	0	1	1	1	2	2	2	3	3
·14	1380	1384	1387	1390	1393	1396	1400	1403	1406	1409	0	1	1	1	2	2	2	3	3
·15	1413	1416	1419	1422	1426	1429	1432	1435	1439	1442	0	1	1	1	2	2	2	3	3
·16	1445	1449	1452	1455	1459	1462	1466	1469	1472	1476	0	1	1	1	2	2	2	3	3
·17	1479	1483	1486	1489	1493	1496	1500	1503	1507	1510	0	1	1	1	2	2	2	3	3
·18	1514	1517	1521	1524	1528	1531	1535	1538	1542	1545	0	1	1	1	2	2	2	3	3
·19	1549	1552	1556	1560	1563	1567	1570	1574	1578	1581	0	1	1	1	2	2	3	3	3
·20	1585	1589	1592	1596	1600	1603	1607	1611	1614	1618	0	1	1	1	2	2	3	3	3
·21	1622	1626	1629	1633	1637	1641	1644	1648	1652	1656	0	1	1	2	2	2	3	3	3
·22	1660	1663	1667	1671	1675	1679	1683	1687	1690	1694	0	1	1	2	2	2	3	3	3
·23	1698	1702	1706	1710	1714	1718	1722	1726	1730	1734	0	1	1	2	2	2	3	3	4
·24	1738	1742	1746	1750	1754	1758	1762	1766	1770	1774	0	1	1	2	2	2	3	3	4
·25	1778	1782	1786	1791	1795	1799	1802	1807	1811	1816	0	1	1	2	2	2	3	3	4
·26	1820	1824	1828	1832	1837	1841	1845	1849	1854	1858	0	1	1	2	2	3	3	3	4
·27	1862	1866	1871	1875	1879	1884	1888	1892	1897	1901	0	1	1	2	2	3	3	3	4
·28	1905	1910	1914	1919	1923	1928	1932	1936	1941	1945	0	1	1	2	2	3	3	4	4
·29	1950	1954	1959	1963	1968	1972	1977	1982	1986	1991	0	1	1	2	2	3	3	4	4
·30	1995	2000	2004	2009	2014	2018	2023	2028	2032	2037	0	1	1	2	2	3	3	4	4
·31	2042	2046	2051	2056	2061	2065	2070	2075	2080	2084	0	1	1	2	2	3	3	4	4
·32	2089	2094	2099	2104	2109	2113	2118	2123	2128	2133	0	1	1	2	2	3	3	4	4
·33	2138	2143	2148	2153	2158	2163	2168	2173	2178	2183	0	1	1	2	2	3	3	4	4
·34	2188	2193	2198	2203	2208	2213	2218	2223	2228	2234	1	1	2	2	3	3	4	4	5
·35	2239	2244	2249	2254	2259	2265	2270	2275	2280	2286	1	1	2	2	3	3	4	4	5
·36	2291	2296	2301	2307	2312	2317	2323	2328	2333	2339	1	1	2	2	3	3	4	4	5
·37	2344	2350	2355	2360	2366	2371	2377	2382	2388	2393	1	1	2	2	3	3	4	4	5
·38	2399	2404	2410	2415	2421	2427	2432	2438	2443	2449	1	1	2	2	3	3	4	4	5
·39	2455	2460	2466	2472	2477	2483	2489	2495	2500	2506	1	1	2	2	3	3	4	5	5
·40	2512	2518	2523	2529	2535	2541	2547	2553	2559	2564	1	1	2	2	3	4	4	5	5
·41	2570	2576	2582	2588	2594	2600	2606	2612	2618	2624	1	1	2	2	3	4	4	5	5
·42	2630	2636	2642	2649	2655	2661	2667	2673	2679	2685	1	1	2	2	3	4	4	5	6
·43	2692	2698	2704	2710	2716	2723	2729	2735	2742	2748	1	1	2	3	3	4	4	5	6
·44	2754	2761	2767	2773	2780	2786	2793	2799	2805	2812	1	1	2	3	3	4	4	5	6
·45	2818	2825	2831	2838	2844	2851	2858	2864	2871	2877	1	1	2	3	3	4	5	5	6
·46	2884	2891	2897	2904	2911	2917	2924	2931	2938	2944	1	1	2	3	3	4	5	5	6
·47	2951	2958	2965	2972	2979	2985	2992	2999	3006	3013	1	1	2	3	3	4	5	5	6
·48	3020	3027	3034	3041	3048	3055	3062	3069	3076	3083	1	1	2	3	4	4	5	6	6
·49	3090	3097	3105	3112	3119	3126	3133	3141	3148	3155	1	1	2	3	4	4	5	6	6
N	0	1	2	3	4	5	6	7	8	9	1	2	3	4	5	6	7	8	9

N	0	1	2	3	4	5	6	7	8	9	1	2	3	4	5	6	7	8	9
·50	3162	3170	3177	3184	3192	3199	3206	3214	3221	3228	1	1	2	3	4	4	5	6	7
·51	3236	3243	3251	3258	3266	3273	3281	3289	3296	3304	1	2	2	3	4	5	5	6	7
·52	3311	3319	3327	3334	3342	3350	3357	3365	3373	3381	1	2	2	3	4	5	5	6	7
·53	3388	3396	3404	3412	3420	3428	3436	3443	3451	3459	1	2	2	3	4	5	6	6	7
·54	3467	3475	3483	3491	3499	3508	3516	3524	3532	3540	1	2	2	3	4	5	6	6	7
·55	3548	3556	3565	3573	3581	3589	3597	3606	3614	3622	1	2	2	3	4	5	6	7	7
·56	3631	3639	3648	3656	3664	3673	3681	3690	3698	3707	1	2	3	3	4	5	6	7	8
·57	3715	3724	3733	3741	3750	3758	3767	3776	3784	3793	1	2	3	3	4	5	6	7	8
·58	3802	3811	3819	3828	3837	3846	3855	3864	3873	3882	1	2	3	4	4	5	6	7	8
·59	3890	3899	3908	3917	3926	3936	3945	3954	3963	3972	1	2	3	4	5	5	6	7	8
·60	3981	3990	3999	4009	4018	4027	4036	4046	4055	4064	1	2	3	4	5	6	6	7	8
·61	4074	4083	4093	4102	4111	4121	4130	4140	4150	4159	1	2	3	4	5	6	7	8	9
·62	4169	4178	4188	4198	4207	4217	4227	4236	4246	4256	1	2	3	4	5	6	7	8	9
·63	4266	4276	4285	4295	4305	4315	4325	4335	4345	4355	1	2	3	4	5	6	7	8	9
·64	4365	4375	4385	4395	4406	4416	4426	4436	4446	4457	1	2	3	4	5	6	7	8	9
·65	4467	4477	4487	4498	4508	4519	4529	4539	4550	4560	1	2	3	4	5	6	7	8	9
·66	4571	4581	4592	4603	4613	4624	4634	4645	4656	4667	1	2	3	4	5	6	7	9	10
·67	4677	4688	4699	4710	4721	4732	4742	4753	4764	4775	1	2	3	4	5	7	8	9	10
·68	4786	4797	4808	4819	4831	4842	4853	4864	4875	4887	1	2	3	4	6	7	8	9	10
·69	4898	4909	4920	4932	4943	4955	4966	4977	4989	5000	1	2	3	5	6	7	8	9	10
·70	5012	5023	5035	5047	5058	5070	5082	5093	5105	5117	1	2	4	5	6	7	8	9	11
·71	5129	5140	5152	5164	5176	5188	5200	5212	5224	5236	1	2	4	5	6	7	8	10	11
·72	5248	5260	5272	5284	5297	5309	5321	5333	5346	5358	1	2	4	5	6	7	9	10	11
·73	5370	5383	5395	5408	5420	5433	5445	5458	5470	5483	1	3	4	5	6	8	9	10	11
·74	5495	5508	5521	5534	5546	5559	5572	5585	5598	5610	1	3	4	5	6	8	9	10	12
·75	5623	5636	5649	5662	5675	5689	5702	5715	5728	5741	1	3	4	5	7	8	9	10	12
·76	5754	5768	5781	5794	5808	5821	5834	5848	5861	5875	1	3	4	5	7	8	9	11	12
·77	5888	5902	5916	5929	5943	5957	5970	5984	5998	6012	1	3	4	5	7	8	10	11	12
·78	6026	6039	6053	6067	6081	6095	6109	6124	6138	6152	1	3	4	6	7	8	10	11	13
·79	6166	6180	6194	6209	6223	6237	6252	6266	6281	6295	1	3	4	6	7	9	10	11	13
·80	6310	6324	6339	6353	6368	6383	6397	6412	6427	6442	1	3	4	6	7	9	10	12	13
·81	6457	6471	6486	6501	6516	6531	6546	6561	6577	6592	2	3	5	6	8	9	11	12	14
·82	6607	6622	6637	6653	6668	6683	6699	6714	6730	6745	2	3	5	6	8	9	11	12	14
·83	6761	6776	6792	6808	6823	6839	6855	6871	6887	6902	2	3	5	6	8	9	11	13	14
·84	6918	6934	6950	6966	6982	6998	7015	7031	7047	7063	2	3	5	6	8	10	11	13	15
·85	7079	7096	7112	7129	7145	7161	7178	7194	7211	7228	2	3	5	7	8	10	12	13	15
·86	7244	7261	7278	7295	7311	7328	7345	7362	7379	7396	2	3	5	7	8	10	12	13	15
·87	7413	7430	7447	7464	7482	7499	7516	7534	7551	7568	2	3	5	7	9	10	12	14	16
·88	7586	7603	7621	7638	7656	7674	7691	7709	7727	7745	2	4	5	7	9	11	12	14	16
·89	7762	7780	7798	7816	7834	7852	7870	7889	7907	7925	2	4	5	7	9	11	13	14	16
·90	7943	7962	7980	7998	8017	8035	8054	8072	8091	8110	2	4	6	7	9	11	13	15	17
·91	8128	8147	8166	8185	8204	8222	8241	8260	8279	8299	2	4	6	8	9	11	13	15	17
·92	8318	8337	8356	8375	8395	8414	8433	8453	8472	8492	2	4	6	8	10	12	14	15	17
·93	8511	8531	8551	8570	8590	8610	8630	8650	8670	8690	2	4	6	8	10	12	14	16	18
·94	8710	8730	8750	8770	8790	8810	8831	8851	8872	8892	2	4	6	8	10	12	14	16	18
·95	8913	8933	8954	8974	8995	9016	9036	9057	9078	9099	2	4	6	8	10	12	15	17	19
·96	9120	9141	9162	9183	9204	9226	9247	9268	9290	9311	2	4	6	8	11	13	15	17	19
·97	9333	9354	9376	9397	9419	9441	9462	9484	9506	9528	2	4	7	9	11	13	15	17	20
·98	9550	9572	9594	9616	9638	9661	9683	9705	9727	9750	2	4	7	9	11	13	16	18	20
·99	9772	9795	9817	9840	9863	9886	9908	9931	9954	9977	2	5	7	9	11	14	16	18	20
N	0	1	2	3	4	5	6	7	8	9	1	2	3	4	5	6	7	8	9

	0′	6′	12′	18′	24′	30′	36′	42′	48′	54′	1′	2′	3′	4′	5′
0°	$-\infty$	$\bar{3}$·242	$\bar{3}$·543	$\bar{3}$·719	$\bar{3}$·844	$\bar{3}$·941	$\bar{2}$·020	$\bar{2}$·087	$\bar{2}$·145	$\bar{2}$·196					
1	$\bar{2}$·2419	2832	3210	3558	3880	4179	4459	4723	4971	5206					
2	$\bar{2}$·5428	5640	5842	6035	6220	6397	6567	6731	6889	7041					
3	$\bar{2}$·7188	7330	7468	7602	7731	7857	7979	8098	8213	8326	21	41	62	83	103
4	$\bar{2}$·8436	8543	8647	8749	8849	8946	9042	9135	9226	9315	16	32	48	64	81
5	$\bar{2}$·9403	9489	9573	9655	9736	9816	9894	9970	**0046**	**0120**	13	26	39	53	66
6	$\bar{1}$·0192	0264	0334	0403	0472	0539	0605	0670	0734	0797	11	22	33	44	55
7	$\bar{1}$·0859	0920	0981	1040	1099	1157	1214	1271	1326	1381	10	19	29	38	48
8	$\bar{1}$·1436	1489	1542	1594	1646	1697	1747	1797	1847	1895	8	17	25	34	42
9	$\bar{1}$·1943	1991	2038	2085	2131	2176	2221	2266	2310	2353	8	15	23	30	38
10	$\bar{1}$·2397	2439	2482	2524	2565	2606	2647	2687	2727	2767	7	14	20	27	34
11	$\bar{1}$·2806	2845	2883	2921	2959	2997	3034	3070	3107	3143	6	12	19	25	31
12	$\bar{1}$·3179	3214	3250	3284	3319	3353	3387	3421	3455	3488	6	11	17	23	28
13	$\bar{1}$·3521	3554	3586	3618	3650	3682	3713	3745	3775	3806	5	11	16	21	26
14	$\bar{1}$·3837	3867	3897	3927	3957	3986	4015	4044	4073	4102	5	10	15	20	24
15	$\bar{1}$·4130	4158	4186	4214	4242	4269	4296	4323	4350	4377	5	9	14	18	23
16	$\bar{1}$·4403	4430	4456	4482	4508	4533	4559	4584	4609	4634	4	9	13	17	21
17	$\bar{1}$·4659	4684	4709	4733	4757	4781	4805	4829	4853	4876	4	8	12	16	20
18	$\bar{1}$·4900	4923	4946	4969	4992	5015	5037	5060	5082	5104	4	8	11	15	19
19	$\bar{1}$·5126	5148	5170	5192	5213	5235	5256	5278	5299	5320	4	7	11	14	18
20	$\bar{1}$·5341	5361	5382	5402	5423	5443	5463	5484	5504	5523	3	7	10	14	17
21	$\bar{1}$·5543	5563	5583	5602	5621	5641	5660	5679	5698	5717	3	6	10	13	16
22	$\bar{1}$·5736	5754	5773	5792	5810	5828	5847	5865	5883	5901	3	6	9	12	15
23	$\bar{1}$·5919	5937	5954	5972	5990	6007	6024	6042	6059	6076	3	6	9	12	15
24	$\bar{1}$·6093	6110	6127	6144	6161	6177	6194	6210	6227	6243	3	6	8	11	14
25	$\bar{1}$·6259	6276	6292	6308	6324	6340	6356	6371	6387	6403	3	5	8	11	13
26	$\bar{1}$·6418	6434	6449	6465	6480	6495	6510	6526	6541	6556	3	5	8	10	13
27	$\bar{1}$·6570	6585	6600	6615	6629	6644	6659	6673	6687	6702	2	5	7	10	12
28	$\bar{1}$·6716	6730	6744	6759	6773	6787	6801	6814	6828	6842	2	5	7	9	12
29	$\bar{1}$·6856	6869	6883	6896	6910	6923	6937	6950	6963	6977	2	4	7	9	11
30	$\bar{1}$·6990	7003	7016	7029	7042	7055	7068	7080	7093	7106	2	4	6	9	11
31	$\bar{1}$·7118	7131	7144	7156	7168	7181	7193	7205	7218	7230	2	4	6	8	10
32	$\bar{1}$·7242	7254	7266	7278	7290	7302	7314	7326	7338	7349	2	4	6	8	10
33	$\bar{1}$·7361	7373	7384	7396	7407	7419	7430	7442	7453	7464	2	4	6	8	10
34	$\bar{1}$·7476	7487	7498	7509	7520	7531	7542	7553	7564	7575	2	4	6	7	9
35	$\bar{1}$·7586	7597	7607	7618	7629	7640	7650	7661	7671	7682	2	4	5	7	9
36	$\bar{1}$·7692	7703	7713	7723	7734	7744	7754	7764	7774	7785	2	3	5	7	9
37	$\bar{1}$·7795	7805	7815	7825	7835	7844	7854	7864	7874	7884	2	3	5	7	8
38	$\bar{1}$·7893	7903	7913	7922	7932	7941	7951	7960	7970	7979	2	3	5	6	8
39	$\bar{1}$·7989	7998	8007	8017	8026	8035	8044	8053	8063	8072	2	3	5	6	8
40	$\bar{1}$·8081	8090	8099	8108	8117	8125	8134	8143	8152	8161	1	3	4	6	7
41	$\bar{1}$·8169	8178	8187	8195	8204	8213	8221	8230	8238	8247	1	3	4	6	7
42	$\bar{1}$·8255	8264	8272	8280	8289	8297	8305	8313	8322	8330	1	3	4	6	7
43	$\bar{1}$·8338	8346	8354	8362	8370	8378	8386	8394	8402	8410	1	3	4	5	7
44	$\bar{1}$·8418	8426	8433	8441	8449	8457	8464	8472	8480	8487	1	3	4	5	6
	0′	6′	12′	18′	24′	30′	36′	42′	48′	54′	1′	2′	3′	4′	5′

The black type indicates that the integer changes.

	0′	6′	12′	18′	24′	30′	36′	42′	48′	54′	1′	2′	3′	4′	5′
45°	$\bar{1}$·8495	8502	8510	8517	8525	8532	8540	8547	8555	8562	1	2	4	5	6
46	$\bar{1}$·8569	8577	8584	8591	8598	8606	8613	8620	8627	8634	1	2	4	5	6
47	$\bar{1}$·8641	8648	8655	8662	8669	8676	8683	8690	8697	8704	1	2	3	5	6
48	$\bar{1}$·8711	8718	8724	8731	8738	8745	8751	8758	8765	8771	1	2	3	4	6
49	$\bar{1}$·8778	8784	8791	8797	8804	8810	8817	8823	8830	8836	1	2	3	4	5
50	$\bar{1}$·8843	8849	8855	8862	8868	8874	8880	8887	8893	8899	1	2	3	4	5
51	$\bar{1}$·8905	8911	8917	8923	8929	8935	8941	8947	8953	8959	1	2	3	4	5
52	$\bar{1}$·8965	8971	8977	8983	8989	8995	9000	9006	9012	9018	1	2	3	4	5
53	$\bar{1}$·9023	9029	9035	9041	9046	9052	9057	9063	9069	9074	1	2	3	4	5
54	$\bar{1}$·9080	9085	9091	9096	9101	9107	9112	9118	9123	9128	1	2	3	4	5
55	$\bar{1}$·9134	9139	9144	9149	9155	9160	9165	9170	9175	9181	1	2	3	3	4
56	$\bar{1}$·9186	9191	9196	9201	9206	9211	9216	9221	9226	9231	1	2	3	3	4
57	$\bar{1}$·9236	9241	9246	9251	9255	9260	9265	9270	9275	9279	1	2	2	3	4
58	$\bar{1}$·9284	9289	9294	9298	9303	9308	9312	9317	9322	9326	1	2	2	3	4
59	$\bar{1}$·9331	9335	9340	9344	9349	9353	9358	9362	9367	9371	1	1	2	3	4
60	$\bar{1}$·9375	9380	9384	9388	9393	9397	9401	9406	9410	9414	1	1	2	3	4
61	$\bar{1}$·9418	9422	9427	9431	9435	9439	9443	9447	9451	9455	1	1	2	3	3
62	$\bar{1}$·9459	9463	9467	9471	9475	9479	9483	9487	9491	9495	1	1	2	3	3
63	$\bar{1}$·9499	9503	9506	9510	9514	9518	9522	9525	9529	9533	1	1	2	3	3
64	$\bar{1}$·9537	9540	9544	9548	9551	9555	9558	9562	9566	9569	1	1	2	2	3
65	$\bar{1}$·9573	9576	9580	9583	9587	9590	9594	9597	9601	9604	1	1	2	2	3
66	$\bar{1}$·9607	9611	9614	9617	9621	9624	9627	9631	9634	9637	1	1	2	2	3
67	$\bar{1}$·9640	9643	9647	9650	9653	9656	9659	9662	9666	9669	1	1	2	2	3
68	$\bar{1}$·9672	9675	9678	9681	9684	9687	9690	9693	9696	9699	0	1	1	2	2
69	$\bar{1}$·9702	9704	9707	9710	9713	9716	9719	9722	9724	9727	0	1	1	2	2
70	$\bar{1}$·9730	9733	9735	9738	9741	9743	9746	9749	9751	9754	0	1	1	2	2
71	$\bar{1}$·9757	9759	9762	9764	9767	9770	9772	9775	9777	9780	0	1	1	2	2
72	$\bar{1}$·9782	9785	9787	9789	9792	9794	9797	9799	9801	9804	0	1	1	2	2
73	$\bar{1}$·9806	9808	9811	9813	9815	9817	9820	9822	9824	9826	0	1	1	1	2
74	$\bar{1}$·9828	9831	9833	9835	9837	9839	9841	9843	9845	9847	0	1	1	1	2
75	$\bar{1}$·9849	9851	9853	9855	9857	9859	9861	9863	9865	9867	0	1	1	1	2
76	$\bar{1}$·9869	9871	9873	9875	9876	9878	9880	9882	9884	9885	0	1	1	1	2
77	$\bar{1}$·9887	9889	9891	9892	9894	9896	9897	9899	9901	9902	0	1	1	1	1
78	$\bar{1}$·9904	9906	9907	9909	9910	9912	9913	9915	9916	9918	0	1	1	1	1
79	$\bar{1}$·9919	9921	9922	9924	9925	9927	9928	9929	9931	9932	0	0	1	1	1
80	$\bar{1}$·9934	9935	9936	9937	9939	9940	9941	9943	9944	9945	0	0	1	1	1
81	$\bar{1}$·9946	9947	9949	9950	9951	9952	9953	9954	9955	9956	0	0	1	1	1
82	$\bar{1}$·9958	9959	9960	9961	9962	9963	9964	9965	9966	9967	0	0	0	1	1
83	$\bar{1}$·9968	9968	9969	9970	9971	9972	9973	9974	9975	9975					
84	$\bar{1}$·9976	9977	9978	9978	9979	9980	9981	9981	9982	9983					
85	$\bar{1}$·9983	9984	9985	9985	9986	9987	9987	9988	9988	9989					
86	$\bar{1}$·9989	9990	9990	9991	9991	9992	9992	9993	9993	9994					
87	$\bar{1}$·9994	9994	9995	9995	9996	9996	9996	9996	9997	9997					
88	$\bar{1}$·9997	9998	9998	9998	9998	9999	9999	9999	9999	9999					
89	$\bar{1}$·9999	9999	**0000**	**0000**	**0000**	**0000**	**0000**	**0000**	**0000**	**0000**					
	0′	6′	12′	18′	24′	30′	36′	42′	48′	54	1′	2′	3′	4′	5′

The black type indicates that the integer changes.

	0′	6′	12′	18′	24′	30′	36′	42′	48′	54′	1′	2′	3′	4′	5′
0°	0·0000	0000	0000	0000	0000	0000	0000	0000	0000	**9999**					
1	$\bar{1}$·9999	9999	9999	9999	9999	9999	9998	9998	9998	9998					
2	$\bar{1}$·9997	9997	9997	9996	9996	9996	9996	9995	9995	9994					
3	$\bar{1}$·9994	9994	9993	9993	9992	9992	9991	9991	9990	9990					
4	$\bar{1}$·9989	9989	9988	9988	9987	9987	9986	9985	9985	9984					
5	$\bar{1}$·9983	9983	9982	9981	9981	9980	9979	9978	9978	9977					
6	$\bar{1}$·9976	9975	9975	9974	9973	9972	9971	9970	9969	9968					
7	$\bar{1}$·9968	9967	9966	9965	9964	9963	9962	9961	9960	9959	0	0	0	1	1
8	$\bar{1}$·9958	9956	9955	9954	9953	9952	9951	9950	9949	9947	0	0	1	1	1
9	$\bar{1}$·9946	9945	9944	9943	9941	9940	9939	9937	9936	9935	0	0	1	1	1
10	$\bar{1}$·9934	9932	9931	9929	9928	9927	9925	9924	9922	9921	0	0	1	1	1
11	$\bar{1}$·9919	9918	9916	9915	9913	9912	9910	9909	9907	9906	0	1	1	1	1
12	$\bar{1}$·9904	9902	9901	9899	9897	9896	9894	9892	9891	9889	0	1	1	1	1
13	$\bar{1}$·9887	9885	9884	9882	9880	9878	9876	9875	9873	9871	0	1	1	1	2
14	$\bar{1}$·9869	9867	9865	9863	9861	9859	9857	9855	9853	9851	0	1	1	1	2
15	$\bar{1}$·9849	9847	9845	9843	9841	9839	9837	9835	9833	9831	0	1	1	1	2
16	$\bar{1}$·9828	9826	9824	9822	9820	9817	9815	9813	9811	9808	0	1	1	1	2
17	$\bar{1}$·9806	9804	9801	9799	9797	9794	9792	9789	9787	9785	0	1	1	2	2
18	$\bar{1}$·9782	9780	9777	9775	9772	9770	9767	9764	9762	9759	0	1	1	2	2
19	$\bar{1}$·9757	9754	9751	9749	9746	9743	9741	9738	9735	9733	0	1	1	2	2
20	$\bar{1}$·9730	9727	9724	9722	9719	9716	9713	9710	9707	9704	0	1	1	2	2
21	$\bar{1}$·9702	9699	9696	9693	9690	9687	9684	9681	9678	9675	0	1	1	2	2
22	$\bar{1}$·9672	9669	9666	9662	9659	9656	9653	9650	9647	9643	1	1	2	2	3
23	$\bar{1}$·9640	9637	9634	9631	9627	9624	9621	9617	9614	9611	1	1	2	2	3
24	$\bar{1}$·9607	9604	9601	9597	9594	9590	9587	9583	9580	9576	1	1	2	2	3
25	$\bar{1}$·9573	9569	9566	9562	9558	9555	9551	9548	9544	9540	1	1	2	2	3
26	$\bar{1}$·9537	9533	9529	9525	9522	9518	9514	9510	9506	9503	1	1	2	3	3
27	$\bar{1}$·9499	9495	9491	9487	9483	9479	9475	9471	9467	9463	1	1	2	3	3
28	$\bar{1}$·9459	9455	9451	9447	9443	9439	9435	9431	9427	9422	1	1	2	3	3
29	$\bar{1}$·9418	9414	9410	9406	9401	9397	9393	9388	9384	9380	1	1	2	3	4
30	$\bar{1}$·9375	9371	9367	9362	9358	9353	9349	9344	9340	9335	1	1	2	3	4
31	$\bar{1}$·9331	9326	9322	9317	9312	9308	9303	9298	9294	9289	1	2	2	3	4
32	$\bar{1}$·9284	9279	9275	9270	9265	9260	9255	9251	9246	9241	1	2	2	3	4
33	$\bar{1}$·9236	9231	9226	9221	9216	9211	9206	9201	9196	9191	1	2	3	3	4
34	$\bar{1}$·9186	9181	9175	9170	9165	9160	9155	9149	9144	9139	1	2	3	3	4
35	$\bar{1}$·9134	9128	9123	9118	9112	9107	9101	9096	9091	9085	1	2	3	4	5
36	$\bar{1}$·9080	9074	9069	9063	9057	9052	9046	9041	9035	9029	1	2	3	4	5
37	$\bar{1}$·9023	9018	9012	9006	9000	8995	8989	8983	8977	8971	1	2	3	4	5
38	$\bar{1}$·8965	8959	8953	8947	8941	8935	8929	8923	8917	8911	1	2	3	4	5
39	$\bar{1}$·8905	8899	8893	8887	8880	8874	8868	8862	8855	8849	1	2	3	4	5
40	$\bar{1}$·8843	8836	8830	8823	8817	8810	8804	8797	8791	8784	1	2	3	4	5
41	$\bar{1}$·8778	8771	8765	8758	8751	8745	8738	8731	8724	8718	1	2	3	4	6
42	$\bar{1}$·8711	8704	8697	8690	8683	8676	8669	8662	8655	8648	1	2	3	5	6
43	$\bar{1}$·8641	8634	8627	8620	8613	8606	8598	8591	8584	8577	1	2	4	5	6
44	$\bar{1}$·8569	8562	8555	8547	8540	8532	8525	8517	8510	8502	1	2	4	5	6
	0′	**6′**	**12′**	**18′**	**24′**	**30′**	**36′**	**42′**	**48′**	**54′**	**1′**	**2′**	**3′**	**4′**	**5′**

The black type indicates that the integer changes.

	0′	6′	12′	18′	24′	30′	36′	42′	48′	54′	1′	2′	3′	4′	5′
45°	$\bar{1}$·8495	8487	8480	8472	8464	8457	8449	8441	8433	8426	1	3	4	5	6
46	$\bar{1}$·8418	8410	8402	8394	8386	8378	8370	8362	8354	8346	1	3	4	5	7
47	$\bar{1}$·8338	8330	8322	8313	8305	8297	8289	8280	8272	8264	1	3	4	6	7
48	$\bar{1}$·8255	8247	8238	8230	8221	8213	8204	8195	8187	8178	1	3	4	6	7
49	$\bar{1}$·8169	8161	8152	8143	8134	8125	8117	8108	8099	8090	1	3	4	6	7
50	$\bar{1}$·8081	8072	8063	8053	8044	8035	8026	8017	8007	7998	2	3	5	6	8
51	$\bar{1}$·7989	7979	7970	7960	7951	7941	7932	7922	7913	7903	2	3	5	6	8
52	$\bar{1}$·7893	7884	7874	7864	7854	7844	7835	7825	7815	7805	2	3	5	7	8
53	$\bar{1}$·7795	7785	7774	7764	7754	7744	7734	7723	7713	7703	2	3	5	7	9
54	$\bar{1}$·7692	7682	7671	7661	7650	7640	7629	7618	7607	7597	2	4	5	7	9
55	$\bar{1}$·7586	7575	7564	7553	7542	7531	7520	7509	7498	7487	2	4	5	7	9
56	$\bar{1}$·7476	7464	7453	7442	7430	7419	7407	7396	7384	7373	2	4	6	8	10
57	$\bar{1}$·7361	7349	7338	7326	7314	7302	7290	7278	7266	7254	2	4	6	8	10
58	$\bar{1}$·7242	7230	7218	7205	7193	7181	7168	7156	7144	7131	2	4	6	8	10
59	$\bar{1}$·7118	7106	7093	7080	7068	7055	7042	7029	7016	7003	2	4	6	9	11
60	$\bar{1}$·6990	6977	6963	6950	6937	6923	6910	6896	6883	6869	2	4	7	9	11
61	$\bar{1}$·6856	6842	6828	6814	6801	6787	6773	6759	6744	6730	2	5	7	9	12
62	$\bar{1}$·6716	6702	6687	6673	6659	6644	6629	6615	6600	6585	2	5	7	10	12
63	$\bar{1}$·6570	6556	6541	6526	6510	6495	6480	6465	6449	6434	3	5	8	10	13
64	$\bar{1}$·6418	6403	6387	6371	6356	6340	6324	6308	6292	6276	3	5	8	11	13
65	$\bar{1}$·6259	6243	6227	6210	6194	6177	6161	6144	6127	6110	3	6	8	11	14
66	$\bar{1}$·6093	6076	6059	6042	6024	6007	5990	5972	5954	5937	3	6	9	12	15
67	$\bar{1}$·5919	5901	5883	5865	5847	5828	5810	5792	5773	5754	3	6	9	12	15
68	$\bar{1}$·5736	5717	5698	5679	5660	5641	5621	5602	5583	5563	3	6	10	13	16
69	$\bar{1}$·5543	5523	5504	5484	5463	5443	5423	5402	5382	5361	3	7	10	14	17
70	$\bar{1}$·5341	5320	5299	5278	5256	5235	5213	5192	5170	5148	4	7	11	14	18
71	$\bar{1}$·5126	5104	5082	5060	5037	5015	4992	4969	4946	4923	4	8	11	15	19
72	$\bar{1}$·4900	4876	4853	4829	4805	4781	4757	4733	4709	4684	4	8	12	16	20
73	$\bar{1}$·4659	4634	4609	4584	4559	4533	4508	4482	4456	4430	4	9	13	17	21
74	$\bar{1}$·4403	4377	4350	4323	4296	4269	4242	4214	4186	4158	5	9	14	18	23
75	$\bar{1}$·4130	4102	4073	4044	4015	3986	3957	3927	3897	3867	5	10	15	20	24
76	$\bar{1}$·3837	3806	3775	3745	3713	3682	3650	3618	3586	3554	5	11	16	21	26
77	$\bar{1}$·3521	3488	3455	3421	3387	3353	3319	3284	3250	3214	6	11	17	23	28
78	$\bar{1}$·3179	3143	3107	3070	3034	2997	2959	2921	2883	2845	6	12	19	25	31
79	$\bar{1}$·2806	2767	2727	2687	2647	2606	2565	2524	2482	2439	7	14	20	27	34
80	$\bar{1}$·2397	2353	2310	2266	2221	2176	2131	2085	2038	1991	8	15	23	30	38
81	$\bar{1}$·1943	1895	1847	1797	1747	1697	1646	1594	1542	1489	8	17	25	34	42
82	$\bar{1}$·1436	1381	1326	1271	1214	1157	1099	1040	0981	0920	10	19	29	38	48
83	$\bar{1}$·0859	0797	0734	0670	0605	0539	0472	0403	0334	0264	11	22	33	44	55
84	$\bar{1}$·0192	0120	0046	**9970**	**9894**	**9816**	**9736**	**9655**	**9573**	**9489**	13	26	39	53	66
85	$\bar{2}$·9403	9315	9226	9135	9042	8946	8849	8749	8647	8543	16	32	48	64	81
86	$\bar{2}$·8436	8326	8213	8098	7979	7857	7731	7602	7468	7330	21	41	62	83	103
87	$\bar{2}$·7188	7041	6889	6731	6567	6397	6220	6035	5842	5640	Differences untrustworthy here				
88	$\bar{2}$·5428	5206	4971	4723	4459	4179	3880	3558	3210	2832					
89	$\bar{2}$·242	$\bar{2}$·196	$\bar{2}$·145	$\bar{2}$·087	$\bar{2}$·020	$\bar{3}$·941	$\bar{3}$·844	$\bar{3}$·719	$\bar{3}$·543	$\bar{3}$·242					
	0′	6′	12′	18′	24′	30′	36′	42′	48′	54′	1′	2′	3′	4′	5′

The black type indicates that the integer changes.

	0′	6′	12′	18′	24′	30′	36′	42′	48′	54′	1′	2′	3′	4′	5′
0°	$-\infty$	$\bar{3}$·242	$\bar{3}$·543	$\bar{3}$·719	$\bar{3}$·844	$\bar{3}$·941	$\bar{2}$·020	$\bar{2}$·087	$\bar{2}$·145	$\bar{2}$·196					
1	$\bar{2}$·2419	2833	3211	3559	3881	4181	4461	4725	4973	5208					
2	$\bar{2}$·5431	5643	5845	6038	6223	6401	6571	6736	6894	7046					
3	$\bar{2}$·7194	7337	7475	7609	7739	7865	7988	8107	8223	8336	21	42	63	83	104
4	$\bar{2}$·8446	8554	8659	8762	8862	8960	9056	9150	9241	9331	16	32	48	65	81
5	$\bar{2}$·9420	9506	9591	9674	9756	9836	9915	9992	**0068**	**0143**	13	26	40	53	66
6	$\bar{1}$·0216	0289	0360	0430	0499	0567	0633	0699	0764	0828	11	22	34	45	56
7	$\bar{1}$·0891	0954	1015	1076	1135	1194	1252	1310	1367	1423	10	20	29	39	49
8	$\bar{1}$·1478	1533	1587	1640	1693	1745	1797	1848	1898	1948	9	17	26	35	43
9	$\bar{1}$·1997	2046	2094	2142	2189	2236	2282	2328	2374	2419	8	16	23	31	39
10	$\bar{1}$·2463	2507	2551	2594	2637	2680	2722	2764	2805	2846	7	14	21	28	35
11	$\bar{1}$·2887	2927	2967	3006	3046	3085	3123	3162	3200	3237	6	13	19	26	32
12	$\bar{1}$·3275	3312	3349	3385	3422	3458	3493	3529	3564	3599	6	12	18	24	30
13	$\bar{1}$·3634	3668	3702	3736	3770	3804	3837	3870	3903	3935	6	11	17	22	28
14	$\bar{1}$·3968	4000	4032	4064	4095	4127	4158	4189	4220	4250	5	10	16	21	26
15	$\bar{1}$·4281	4311	4341	4371	4400	4430	4459	4488	4517	4546	5	10	15	20	25
16	$\bar{1}$·4575	4603	4632	4660	4688	4716	4744	4771	4799	4826	5	9	14	19	23
17	$\bar{1}$·4853	4880	4907	4934	4961	4987	5014	5040	5066	5092	4	9	13	18	22
18	$\bar{1}$·5118	5143	5169	5195	5220	5245	5270	5295	5320	5345	4	8	13	17	21
19	$\bar{1}$·5370	5394	5419	5443	5467	5491	5516	5539	5563	5587	4	8	12	16	20
20	$\bar{1}$·5611	5634	5658	5681	5704	5727	5750	5773	5796	5819	4	8	12	15	19
21	$\bar{1}$·5842	5864	5887	5909	5932	5954	5976	5998	6020	6042	4	7	11	15	19
22	$\bar{1}$·6064	6086	6108	6129	6151	6172	6194	6215	6236	6257	4	7	11	14	18
23	$\bar{1}$·6279	6300	6321	6341	6362	6383	6404	6424	6445	6465	3	7	10	14	17
24	$\bar{1}$·6486	6506	6527	6547	6567	6587	6607	6627	6647	6667	3	7	10	13	17
25	$\bar{1}$·6687	6706	6726	6746	6765	6785	6804	6824	6843	6863	3	7	10	13	16
26	$\bar{1}$·6882	6901	6920	6939	6958	6977	6996	7015	7034	7053	3	6	9	13	16
27	$\bar{1}$·7072	7090	7109	7128	7146	7165	7183	7202	7220	7238	3	6	9	12	15
28	$\bar{1}$·7257	7275	7293	7311	7330	7348	7366	7384	7402	7420	3	6	9	12	15
29	$\bar{1}$·7438	7455	7473	7491	7509	7526	7544	7562	7579	7597	3	6	9	12	15
30	$\bar{1}$·7614	7632	7649	7667	7684	7701	7719	7736	7753	7771	3	6	9	12	14
31	$\bar{1}$·7788	7805	7822	7839	7856	7873	7890	7907	7924	7941	3	6	9	11	14
32	$\bar{1}$·7958	7975	7992	8008	8025	8042	8059	8075	8092	8109	3	6	8	11	14
33	$\bar{1}$·8125	8142	8158	8175	8191	8208	8224	8241	8257	8274	3	5	8	11	14
34	$\bar{1}$·8290	8306	8323	8339	8355	8371	8388	8404	8420	8436	3	5	8	11	14
35	$\bar{1}$·8452	8468	8484	8501	8517	8533	8549	8565	8581	8597	3	5	8	11	13
36	$\bar{1}$·8613	8629	8644	8660	8676	8692	8708	8724	8740	8755	3	5	8	11	13
37	$\bar{1}$·8771	8787	8803	8818	8834	8850	8865	8881	8897	8912	3	5	8	10	13
38	$\bar{1}$·8928	8944	8959	8975	8990	9006	9022	9037	9053	9068	3	5	8	10	13
39	$\bar{1}$·9084	9099	9115	9130	9146	9161	9176	9192	9207	9223	3	5	8	10	13
40	$\bar{1}$·9238	9254	9269	9284	9300	9315	9330	9346	9361	9376	3	5	8	10	13
41	$\bar{1}$·9392	9407	9422	9438	9453	9468	9483	9499	9514	9529	3	5	8	10	13
42	$\bar{1}$·9544	9560	9575	9590	9605	9621	9636	9651	9666	9681	3	5	8	10	13
43	$\bar{1}$·9697	9712	9727	9742	9757	9772	9788	9803	9818	9833	3	5	8	10	13
44	$\bar{1}$·9848	9864	9879	9894	9909	9924	9939	9955	9970	9985	3	5	8	10	13
	0′	6′	12′	18′	24′	30′	36′	42′	48′	54′	1′	2′	3′	4′	5′

The black type indicates that the integer changes.

	0′	6′	12′	18′	24′	30′	36′	42′	48′	54′	1′	2′	3′	4′	5′
45°	0·0000	0015	0030	0045	0061	0076	0091	0106	0121	0136	3	5	8	10	13
46	0·0152	0167	0182	0197	0212	0228	0243	0258	0273	0288	3	5	8	10	13
47	0·0303	0319	0334	0349	0364	0379	0395	0410	0425	0440	3	5	8	10	13
48	0·0456	0471	0486	0501	0517	0532	0547	0562	0578	0593	3	5	8	10	13
49	0·0608	0624	0639	0654	0670	0685	0700	0716	0731	0746	3	5	8	10	13
50	0·0762	0777	0793	0808	0824	0839	0854	0870	0885	0901	3	5	8	10	13
51	0·0916	0932	0947	0963	0978	0994	1010	1025	1041	1056	3	5	8	10	13
52	0·1072	1088	1103	1119	1135	1150	1166	1182	1197	1213	3	5	8	10	13
53	0·1229	1245	1260	1276	1292	1308	1324	1340	1356	1371	3	5	8	11	13
54	0·1387	1403	1419	1435	1451	1467	1483	1499	1516	1532	3	5	8	11	13
55	0·1548	1564	1580	1596	1612	1629	1645	1661	1677	1694	3	5	8	11	14
56	0·1710	1726	1743	1759	1776	1792	1809	1825	1842	1858	3	5	8	11	14
57	0·1875	1891	1908	1925	1941	1958	1975	1992	2008	2025	3	6	8	11	14
58	0·2042	2059	2076	2093	2110	2127	2144	2161	2178	2195	3	6	9	11	14
59	0·2212	2229	2247	2264	2281	2299	2316	2333	2351	2368	3	6	9	12	14
60	0·2386	2403	2421	2438	2456	2474	2491	2509	2527	2545	3	6	9	12	15
61	0·2562	2580	2598	2616	2634	2652	2670	2689	2707	2725	3	6	9	12	15
62	0·2743	2762	2780	2798	2817	2835	2854	2872	2891	2910	3	6	9	12	15
63	0·2928	2947	2966	2985	3004	3023	3042	3061	3080	3099	3	6	9	13	16
64	0·3118	3137	3157	3176	3196	3215	3235	3254	3274	3294	3	7	10	13	16
65	0·3313	3333	3353	3373	3393	3413	3433	3453	3473	3494	3	7	10	13	17
66	0·3514	3535	3555	3576	3596	3617	3638	3659	3679	3700	3	7	10	14	17
67	0·3721	3743	3764	3785	3806	3828	3849	3871	3892	3914	4	7	11	14	18
68	0·3936	3958	3980	4002	4024	4046	4068	4091	4113	4136	4	7	11	15	19
69	0·4158	4181	4204	4227	4250	4273	4296	4319	4342	4366	4	8	12	15	19
70	0·4389	4413	4437	4461	4484	4509	4533	4557	4581	4606	4	8	12	16	20
71	0·4630	4655	4680	4705	4730	4755	4780	4805	4831	4857	4	8	13	17	21
72	0·4882	4908	4934	4960	4986	5013	5039	5066	5093	5120	4	9	13	18	22
73	0·5147	5174	5201	5229	5256	5284	5312	5340	5368	5397	5	9	14	19	23
74	0·5425	5454	5483	5512	5541	5570	5600	5629	5659	5689	5	10	15	20	25
75	0·5719	5750	5780	5811	5842	5873	5905	5936	5968	6000	5	10	16	21	26
76	0·6032	6065	6097	6130	6163	6196	6230	6264	6298	6332	6	11	17	22	28
77	0·6366	6401	6436	6471	6507	6542	6578	6615	6651	6688	6	12	18	24	30
78	0·6725	6763	6800	6838	6877	6915	6954	6994	7033	7073	6	13	19	26	32
79	0·7113	7154	7195	7236	7278	7320	7363	7406	7449	7493	7	14	21	28	35
80	0·7537	7581	7626	7672	7718	7764	7811	7858	7906	7954	8	16	23	31	39
81	0·8003	8052	8102	8152	8203	8255	8307	8360	8413	8467	9	17	26	35	43
82	0·8522	8577	8633	8690	8748	8806	8865	8924	8985	9046	10	20	29	39	49
83	0·9109	9172	9236	9301	9367	9433	9501	9570	9640	9711	11	22	34	45	56
84	0·9784	9857	9932	**0008**	**0085**	**0164**	**0244**	**0326**	**0409**	**0494**	13	26	40	53	66
85	1·0580	0669	0759	0850	0944	1040	1138	1238	1341	1446	16	32	48	65	81
86	1·1554	1664	1777	1893	2012	2135	2261	2391	2525	2663	21	42	63	83	104
87	1·2806	2954	3106	3264	3429	3599	3777	3962	4155	4357					
88	1·4569	4792	5027	5275	5539	5819	6119	6441	6789	7167	Differences untrustworthy here				
89	1·758	1·804	1·855	1·913	1·980	2·059	2·156	2·281	2·457	2·758					
	0′	6′	12′	18′	24′	30′	36′	42′	48′	54′	1′	2′	3′	4′	5′

The black type indicates that the integer changes.

	0′	6′	12′	18′	24′	30′	36′	42′	48′	54′	1′	2′	3′	4′	5′
0°	·0000	0017	0035	0052	0070	0087	0105	0122	0140	0157	3	6	9	12	15
1	·0175	0192	0209	0227	0244	0262	0279	0297	0314	0332	3	6	9	12	15
2	·0349	0366	0384	0401	0419	0436	0454	0471	0488	0506	3	6	9	12	15
3	·0523	0541	0558	0576	0593	0610	0628	0645	0663	0680	3	6	9	12	15
4	·0698	0715	0732	0750	0767	0785	0802	0819	0837	0854	3	6	9	12	14
5	·0872	0889	0906	0924	0941	0958	0976	0993	1011	1028	3	6	9	12	14
6	·1045	1063	1080	1097	1115	1132	1149	1167	1184	1201	3	6	9	12	14
7	·1219	1236	1253	1271	1288	1305	1323	1340	1357	1374	3	6	9	12	14
8	·1392	1409	1426	1444	1461	1478	1495	1513	1530	1547	3	6	9	12	14
9	·1564	1582	1599	1616	1633	1650	1668	1685	1702	1719	3	6	9	11	14
10	·1736	1754	1771	1788	1805	1822	1840	1857	1874	1891	3	6	9	11	14
11	·1908	1925	1942	1959	1977	1994	2011	2028	2045	2062	3	6	9	11	14
12	·2079	2096	2113	2130	2147	2164	2181	2198	2215	2233	3	6	9	11	14
13	·2250	2267	2284	2300	2317	2334	2351	2368	2385	2402	3	6	8	11	14
14	·2419	2436	2453	2470	2487	2504	2521	2538	2554	2571	3	6	8	11	14
15	·2588	2605	2622	2639	2656	2672	2689	2706	2723	2740	3	6	8	11	14
16	·2756	2773	2790	2807	2823	2840	2857	2874	2890	2907	3	6	8	11	14
17	·2924	2940	2957	2974	2990	3007	3024	3040	3057	3074	3	6	8	11	14
18	·3090	3107	3123	3140	3156	3173	3190	3206	3223	3239	3	6	8	11	14
19	·3256	3272	3289	3305	3322	3338	3355	3371	3387	3404	3	5	8	11	14
20	·3420	3437	3453	3469	3486	3502	3518	3535	3551	3567	3	5	8	11	14
21	·3584	3600	3616	3633	3649	3665	3681	3697	3714	3730	3	5	8	11	14
22	·3746	3762	3778	3795	3811	3827	3843	3859	3875	3891	3	5	8	11	13
23	·3907	3923	3939	3955	3971	3987	4003	4019	4035	4051	3	5	8	11	13
24	·4067	4083	4099	4115	4131	4147	4163	4179	4195	4210	3	5	8	11	13
25	·4226	4242	4258	4274	4289	4305	4321	4337	4352	4368	3	5	8	11	13
26	·4384	4399	4415	4431	4446	4462	4478	4493	4509	4524	3	5	8	10	13
27	·4540	4555	4571	4586	4602	4617	4633	4648	4664	4679	3	5	8	10	13
28	·4695	4710	4726	4741	4756	4772	4787	4802	4818	4833	3	5	8	10	13
29	·4848	4863	4879	4894	4909	4924	4939	4955	4970	4985	3	5	8	10	13
30	·5000	5015	5030	5045	5060	5075	5090	5105	5120	5135	3	5	8	10	13
31	·5150	5165	5180	5195	5210	5225	5240	5255	5270	5284	2	5	7	10	12
32	·5299	5314	5329	5344	5358	5373	5388	5402	5417	5432	2	5	7	10	12
33	·5446	5461	5476	5490	5505	5519	5534	5548	5563	5577	2	5	7	10	12
34	·5592	5606	5621	5635	5650	5664	5678	5693	5707	5721	2	5	7	10	12
35	·5736	5750	5764	5779	5793	5807	5821	5835	5850	5864	2	5	7	9	12
36	·5878	5892	5906	5920	5934	5948	5962	5976	5990	6004	2	5	7	9	12
37	·6018	6032	6046	6060	6074	6088	6101	6115	6129	6143	2	5	7	9	12
38	·6157	6170	6184	6198	6211	6225	6239	6252	6266	6280	2	5	7	9	11
39	·6293	6307	6320	6334	6347	6361	6374	6388	6401	6414	2	4	7	9	11
40	·6428	6441	6455	6468	6481	6494	6508	6521	6534	6547	2	4	7	9	11
41	·6561	6574	6587	6600	6613	6626	6639	6652	6665	6678	2	4	7	9	11
42	·6691	6704	6717	6730	6743	6756	6769	6782	6794	6807	2	4	6	9	11
43	·6820	6833	6845	6858	6871	6884	6896	6909	6921	6934	2	4	6	8	11
44	·6947	6959	6972	6984	6997	7009	7022	7034	7046	7059	2	4	6	8	10
	0′	6′	12′	18′	24′	30′	36′	42′	48′	54′	1′	2′	3′	4′	5′

	0′	6′	12′	18′	24′	30′	36′	42′	48′	54′	1′	2′	3′	4′	5′
45°	·7071	7083	7096	7108	7120	7133	7145	7157	7169	7181	2	4	6	8	10
46	·7193	7206	7218	7230	7242	7254	7266	7278	7290	7302	2	4	6	8	10
47	·7314	7325	7337	7349	7361	7373	7385	7396	7408	7420	2	4	6	8	10
48	·7431	7443	7455	7466	7478	7490	7501	7513	7524	7536	2	4	6	8	10
49	·7547	7559	7570	7581	7593	7604	7615	7627	7638	7649	2	4	6	8	9
50	·7660	7672	7683	7694	7705	7716	7727	7738	7749	7760	2	4	6	7	9
51	·7771	7782	7793	7804	7815	7826	7837	7848	7859	7869	2	4	5	7	9
52	·7880	7891	7902	7912	7923	7934	7944	7955	7965	7976	2	4	5	7	9
53	·7986	7997	8007	8018	8028	8039	8049	8059	8070	8080	2	3	5	7	9
54	·8090	8100	8111	8121	8131	8141	8151	8161	8171	8181	2	3	5	7	8
55	·8192	8202	8211	8221	8231	8241	8251	8261	8271	8281	2	3	5	7	8
56	·8290	8300	8310	8320	8329	8339	8348	8358	8368	8377	2	3	5	6	8
57	·8387	8396	8406	8415	8425	8434	8443	8453	8462	8471	2	3	5	6	8
58	·8480	8490	8499	8508	8517	8526	8536	8545	8554	8563	2	3	5	6	8
59	·8572	8581	8590	8599	8607	8616	8625	8634	8643	8652	1	3	4	6	7
60	·8660	8669	8678	8686	8695	8704	8712	8721	8729	8738	1	3	4	6	7
61	·8746	8755	8763	8771	8780	8788	8796	8805	8813	8821	1	3	4	6	7
62	·8829	8838	8846	8854	8862	8870	8878	8886	8894	8902	1	3	4	5	7
63	·8910	8918	8926	8934	8942	8949	8957	8965	8973	8980	1	3	4	5	6
64	·8988	8996	9003	9011	9018	9026	9033	9041	9048	9056	1	3	4	5	6
65	·9063	9070	9078	9085	9092	9100	9107	9114	9121	9128	1	2	4	5	6
66	·9135	9143	9150	9157	9164	9171	9178	9184	9191	9198	1	2	3	5	6
67	·9205	9212	9219	9225	9232	9239	9245	9252	9259	9265	1	2	3	4	6
68	·9272	9278	9285	9291	9298	9304	9311	9317	9323	9330	1	2	3	4	5
69	·9336	9342	9348	9354	9361	9367	9373	9379	9385	9391	1	2	3	4	5
70	·9397	9403	9409	9415	9421	9426	9432	9438	9444	9449	1	2	3	4	5
71	·9455	9461	9466	9472	9478	9483	9489	9494	9500	9505	1	2	3	4	5
72	·9511	9516	9521	9527	9532	9537	9542	9548	9553	9558	1	2	3	4	4
73	·9563	9568	9573	9578	9583	9588	9593	9598	9603	9608	1	2	2	3	4
74	·9613	9617	9622	9627	9632	9636	9641	9646	9650	9655	1	2	2	3	4
75	·9659	9664	9668	9673	9677	9681	9686	9690	9694	9699	1	1	2	3	4
76	·9703	9707	9711	9715	9720	9724	9728	9732	9736	9740	1	1	2	3	3
77	·9744	9748	9751	9755	9759	9763	9767	9770	9774	9778	1	1	2	3	3
78	·9781	9785	9789	9792	9796	9799	9803	9806	9810	9813	1	1	2	2	3
79	·9816	9820	9823	9826	9829	9833	9836	9839	9842	9845	1	1	2	2	3
80	·9848	9851	9854	9857	9860	9863	9866	9869	9871	9874	0	1	1	2	2
81	·9877	9880	9882	9885	9888	9890	9893	9895	9898	9900	0	1	1	2	2
82	·9903	9905	9907	9910	9912	9914	9917	9919	9921	9923	0	1	1	2	2
83	·9925	9928	9930	9932	9934	9936	9938	9940	9942	9943	0	1	1	1	2
84	·9945	9947	9949	9951	9952	9954	9956	9957	9959	9960	0	1	1	1	1
85	·9962	9963	9965	9966	9968	9969	9971	9972	9973	9974	0	0	1	1	1
86	·9976	9977	9978	9979	9980	9981	9982	9983	9984	9985	0	0	0	0	1
87	·9986	9987	9988	9989	9990	9990	9991	9992	9993	9993					
88	·9994	9995	9995	9996	9996	9997	9997	9997	9998	9998					
89	·9998	9999	9999	9999	9999	1·000	1·000	1·000	1·000	1·000					
	0′	6′	12′	18′	24′	30′	36′	42′	48′	54′	1′	2′	3′	4′	5′

SUBTRACT

	0′	6′	12′	18′	24′	30′	36′	42′	48′	54′	1′	2′	3′	4′	5′
0°	1·0000	1·000	1·000	1·000	1·000	1·000	9999	9999	9999	9999					
1	·9998	9998	9998	9997	9997	9997	9996	9996	9995	9995					
2	·9994	9993	9993	9992	9991	9990	9990	9989	9988	9987					
3	·9986	9985	9984	9983	9982	9981	9980	9979	9978	9977					
4	·9976	9974	9973	9972	9971	9969	9968	9966	9965	9963					
5	·9962	9960	9959	9957	9956	9954	9952	9951	9949	9947					
6	·9945	9943	9942	9940	9938	9936	9934	9932	9930	9928	0	1	1	1	2
7	·9925	9923	9921	9919	9917	9914	9912	9910	9907	9905	0	1	1	2	2
8	·9903	9900	9898	9895	9893	9890	9888	9885	9882	9880	0	1	1	2	2
9	·9877	9874	9871	9869	9866	9863	9860	9857	9854	9851	0	1	1	2	2
10	·9848	9845	9842	9839	9836	9833	9829	9826	9823	9820	1	1	2	2	3
11	·9816	9813	9810	9806	9803	9799	9796	9792	9789	9785	1	1	2	2	3
12	·9781	9778	9774	9770	9767	9763	9759	9755	9751	9748	1	1	2	3	3
13	·9744	9740	9736	9732	9728	9724	9720	9715	9711	9707	1	1	2	3	3
14	·9703	9699	9694	9690	9686	9681	9677	9673	9668	9664	1	1	2	3	4
15	·9659	9655	9650	9646	9641	9636	9632	9627	9622	9617	1	2	2	3	4
16	·9613	9608	9603	9598	9593	9588	9583	9578	9573	9568	1	2	2	3	4
17	·9563	9558	9553	9548	9542	9537	9532	9527	9521	9516	1	2	3	3	4
18	·9511	9505	9500	9494	9489	9483	9478	9472	9466	9461	1	2	3	4	5
19	·9455	9449	9444	9438	9432	9426	9421	9415	9409	9403	1	2	3	4	5
20	·9397	9391	9385	9379	9373	9367	9361	9354	9348	9342	1	2	3	4	5
21	·9336	9330	9323	9317	9311	9304	9298	9291	9285	9278	1	2	3	4	5
22	·9272	9265	9259	9252	9245	9239	9232	9225	9219	9212	1	2	3	4	6
23	·9205	9198	9191	9184	9178	9171	9164	9157	9150	9143	1	2	3	5	6
24	·9135	9128	9121	9114	9107	9100	9092	9085	9078	9070	1	2	4	5	6
25	·9063	9056	9048	9041	9033	9026	9018	9011	9003	8996	1	3	4	5	6
26	·8988	8980	8973	8965	8957	8949	8942	8934	8926	8918	1	3	4	5	6
27	·8910	8902	8894	8886	8878	8870	8862	8854	8846	8838	1	3	4	5	7
28	·8829	8821	8813	8805	8796	8788	8780	8771	8763	8755	1	3	4	6	7
29	·8746	8738	8729	8721	8712	8704	8695	8686	8678	8669	1	3	4	6	7
30	·8660	8652	8643	8634	8625	8616	8607	8599	8590	8581	1	3	4	6	7
31	·8572	8563	8554	8545	8536	8526	8517	8508	8499	8490	2	3	5	6	8
32	·8480	8471	8462	8453	8443	8434	8425	8415	8406	8396	2	3	5	6	8
33	·8387	8377	8368	8358	8348	8339	8329	8320	8310	8300	2	3	5	6	8
34	·8290	8281	8271	8261	8251	8241	8231	8221	8211	8202	2	3	5	7	8
35	·8192	8181	8171	8161	8151	8141	8131	8121	8111	8100	2	3	5	7	8
36	·8090	8080	8070	8059	8049	8039	8028	8018	8007	7997	2	3	5	7	9
37	·7986	7976	7965	7955	7944	7934	7923	7912	7902	7891	2	4	5	7	9
38	·7880	7869	7859	7848	7837	7826	7815	7804	7793	7782	2	4	5	7	9
39	·7771	7760	7749	7738	7727	7716	7705	7694	7683	7672	2	4	6	7	9
40	·7660	7649	7638	7627	7615	7604	7593	7581	7570	7559	2	4	6	8	9
41	·7547	7536	7524	7513	7501	7490	7478	7466	7455	7443	2	4	6	8	10
42	·7431	7420	7408	7396	7385	7373	7361	7349	7337	7325	2	4	6	8	10
43	·7314	7302	7290	7278	7266	7254	7242	7230	7218	7206	2	4	6	8	10
44	·7193	7181	7169	7157	7145	7133	7120	7108	7096	7083	2	4	6	8	10

SUBTRACT

SUBTRACT

	0′	6′	12′	18′	24′	30′	36′	42′	48′	54′	1′	2′	3′	4′	5′
45°	·7071	7059	7046	7034	7022	7009	6997	6984	6972	6959	2	4	6	8	10
46	·6947	6934	6921	6909	6896	6884	6871	6858	6845	6833	2	4	6	8	11
47	·6820	6807	6794	6782	6769	6756	6743	6730	6717	6704	2	4	6	9	11
48	·6691	6678	6665	6652	6639	6626	6613	6600	6587	6574	2	4	7	9	11
49	·6561	6547	6534	6521	6508	6494	6481	6468	6455	6441	2	4	7	9	11
50	·6428	6414	6401	6388	6374	6361	6347	6334	6320	6307	2	4	7	9	11
51	·6293	6280	6266	6252	6239	6225	6211	6198	6184	6170	2	5	7	9	11
52	·6157	6143	6129	6115	6101	6088	6074	6060	6046	6032	2	5	7	9	12
53	·6018	6004	5990	5976	5962	5948	5934	5920	5906	5892	2	5	7	9	12
54	·5878	5864	5850	5835	5821	5807	5793	5779	5764	5750	2	5	7	9	12
55	·5736	5721	5707	5693	5678	5664	5650	5635	5621	5606	2	5	7	10	12
56	·5592	5577	5563	5548	5534	5519	5505	5490	5476	5461	2	5	7	10	12
57	·5446	5432	5417	5402	5388	5373	5358	5344	5329	5314	2	5	7	10	12
58	·5299	5284	5270	5255	5240	5225	5210	5195	5180	5165	2	5	7	10	12
59	·5150	5135	5120	5105	5090	5075	5060	5045	5030	5015	3	5	8	10	13
60	·5000	4985	4970	4955	4939	4924	4909	4894	4879	4863	3	5	8	10	13
61	·4848	4833	4818	4802	4787	4772	4756	4741	4726	4710	3	5	8	10	13
62	·4695	4679	4664	4648	4633	4617	4602	4586	4571	4555	3	5	8	10	13
63	·4540	4524	4509	4493	4478	4462	4446	4431	4415	4399	3	5	8	10	13
64	·4384	4368	4352	4337	4321	4305	4289	4274	4258	4242	3	5	8	11	13
65	·4226	4210	4195	4179	4163	4147	4131	4115	4099	4083	3	5	8	11	13
66	·4067	4051	4035	4019	4003	3987	3971	3955	3939	3923	3	5	8	11	13
67	·3907	3891	3875	3859	3843	3827	3811	3795	3778	3762	3	5	8	11	13
68	·3746	3730	3714	3697	3681	3665	3649	3633	3616	3600	3	5	8	11	14
69	·3584	3567	3551	3535	3518	3502	3486	3469	3453	3437	3	5	8	11	14
70	·3420	3404	3387	3371	3355	3338	3322	3305	3289	3272	3	5	8	11	14
71	·3256	3239	3223	3206	3190	3173	3156	3140	3123	3107	3	6	8	11	14
72	·3090	3074	3057	3040	3024	3007	2990	2974	2957	2940	3	6	8	11	14
73	·2924	2907	2890	2874	2857	2840	2823	2807	2790	2773	3	6	8	11	14
74	·2756	2740	2723	2706	2689	2672	2656	2639	2622	2605	3	6	8	11	14
75	·2588	2571	2554	2538	2521	2504	2487	2470	2453	2436	3	6	8	11	14
76	·2419	2402	2385	2368	2351	2334	2317	2300	2284	2267	3	6	8	11	14
77	·2250	2233	2215	2198	2181	2164	2147	2130	2113	2096	3	6	9	11	14
78	·2079	2062	2045	2028	2011	1994	1977	1959	1942	1925	3	6	9	11	14
79	·1908	1891	1874	1857	1840	1822	1805	1788	1771	1754	3	6	9	11	14
80	·1736	1719	1702	1685	1668	1650	1633	1616	1599	1582	3	6	9	11	14
81	·1564	1547	1530	1513	1495	1478	1461	1444	1426	1409	3	6	9	12	14
82	·1392	1374	1357	1340	1323	1305	1288	1271	1253	1236	3	6	9	12	14
83	·1219	1201	1184	1167	1149	1132	1115	1097	1080	1063	3	6	9	12	14
84	·1045	1028	1011	0993	0976	0958	0941	0924	0906	0889	3	6	9	12	14
85	·0872	0854	0837	0819	0802	0785	0767	0750	0732	0715	3	6	9	12	14
86	·0698	0680	0663	0645	0628	0610	0593	0576	0558	0541	3	6	9	12	15
87	·0523	0506	0488	0471	0454	0436	0419	0401	0384	0366	3	6	9	12	15
88	·0349	0332	0314	0297	0279	0262	0244	0227	0209	0192	3	6	9	12	15
89	·0175	0157	0140	0122	0105	0087	0070	0052	0035	0017	3	6	9	12	15

SUBTRACT

	0′	6′	12′	18′	24′	30′	36′	42′	48′	54′	1′	2′	3′	4′	5′
0°	0·0000	0017	0035	0052	0070	0087	0105	0122	0140	0157	3	6	9	12	15
1	0·0175	0192	0209	0227	0244	0262	0279	0297	0314	0332	3	6	9	12	15
2	0·0349	0367	0384	0402	0419	0437	0454	0472	0489	0507	3	6	9	12	15
3	0·0524	0542	0559	0577	0594	0612	0629	0647	0664	0682	3	6	9	12	15
4	0·0699	0717	0734	0752	0769	0787	0805	0822	0840	0857	3	6	9	12	15
5	0·0875	0892	0910	0928	0945	0963	0981	0998	1016	1033	3	6	9	12	15
6	0·1051	1069	1086	1104	1122	1139	1157	1175	1192	1210	3	6	9	12	15
7	0·1228	1246	1263	1281	1299	1317	1334	1352	1370	1388	3	6	9	12	15
8	0·1405	1423	1441	1459	1477	1495	1512	1530	1548	1566	3	6	9	12	15
9	0·1584	1602	1620	1638	1655	1673	1691	1709	1727	1745	3	6	9	12	15
10	0·1763	1781	1799	1817	1835	1853	1871	1890	1908	1926	3	6	9	12	15
11	0·1944	1962	1980	1998	2016	2035	2053	2071	2089	2107	3	6	9	12	15
12	0·2126	2144	2162	2180	2199	2217	2235	2254	2272	2290	3	6	9	12	15
13	0·2309	2327	2345	2364	2382	2401	2419	2438	2456	2475	3	6	9	12	15
14	0·2493	2512	2530	2549	2568	2586	2605	2623	2642	2661	3	6	9	12	16
15	0·2679	2698	2717	2736	2754	2773	2792	2811	2830	2849	3	6	9	13	16
16	0·2867	2886	2905	2924	2943	2962	2981	3000	3019	3038	3	6	9	13	16
17	0·3057	3076	3096	3115	3134	3153	3172	3191	3211	3230	3	6	10	13	16
18	0·3249	3269	3288	3307	3327	3346	3365	3385	3404	3424	3	6	10	13	16
19	0·3443	3463	3482	3502	3522	3541	3561	3581	3600	3620	3	7	10	13	16
20	0·3640	3659	3679	3699	3719	3739	3759	3779	3799	3819	3	7	10	13	17
21	0·3839	3859	3879	3899	3919	3939	3959	3979	4000	4020	3	7	10	13	17
22	0·4040	4061	4081	4101	4122	4142	4163	4183	4204	4224	3	7	10	14	17
23	0·4245	4265	4286	4307	4327	4348	4369	4390	4411	4431	3	7	10	14	17
24	0·4452	4473	4494	4515	4536	4557	4578	4599	4621	4642	4	7	11	14	18
25	0·4663	4684	4706	4727	4748	4770	4791	4813	4834	4856	4	7	11	14	18
26	0·4877	4899	4921	4942	4964	4986	5008	5029	5051	5073	4	7	11	15	18
27	0·5095	5117	5139	5161	5184	5206	5228	5250	5272	5295	4	7	11	15	18
28	0·5317	5340	5362	5384	5407	5430	5452	5475	5498	5520	4	8	11	15	19
29	0·5543	5566	5589	5612	5635	5658	5681	5704	5727	5750	4	8	12	15	19
30	0·5774	5797	5820	5844	5867	5890	5914	5938	5961	5985	4	8	12	16	20
31	0·6009	6032	6056	6080	6104	6128	6152	6176	6200	6224	4	8	12	16	20
32	0·6249	6273	6297	6322	6346	6371	6395	6420	6445	6469	4	8	12	16	20
33	0·6494	6519	6544	6569	6594	6619	6644	6669	6694	6720	4	8	13	17	21
34	0·6745	6771	6796	6822	6847	6873	6899	6924	6950	6976	4	9	13	17	21
35	0·7002	7028	7054	7080	7107	7133	7159	7186	7212	7239	4	9	13	18	22
36	0·7265	7292	7319	7346	7373	7400	7427	7454	7481	7508	5	9	14	18	23
37	0·7536	7563	7590	7618	7646	7673	7701	7729	7757	7785	5	9	14	18	23
38	0·7813	7841	7869	7898	7926	7954	7983	8012	8040	8069	5	9	14	19	24
39	0·8098	8127	8156	8185	8214	8243	8273	8302	8332	8361	5	10	15	20	24
40	0·8391	8421	8451	8481	8511	8541	8571	8601	8632	8662	5	10	15	20	25
41	0·8693	8724	8754	8785	8816	8847	8878	8910	8941	8972	5	10	16	21	26
42	0·9004	9036	9067	9099	9131	9163	9195	9228	9260	9293	5	11	16	21	27
43	0·9325	9358	9391	9424	9457	9490	9523	9556	9590	9623	6	11	17	22	28
44	0·9657	9691	9725	9759	9793	9827	9861	9896	9930	9965	6	11	17	23	29
	0′	6′	12′	18′	24′	30′	36′	42′	48′	54′	1′	2′	3′	4′	5′

°	0′	6′	12′	18′	24′	30′	36′	42′	48′	54′	1′	2′	3′	4′	5′
45	1·0000	0035	0070	0105	0141	0176	0212	0247	0283	0319	6	12	18	24	30
46	1·0355	0392	0428	0464	0501	0538	0575	0612	0649	0686	6	12	18	25	31
47	1·0724	0761	0799	0837	0875	0913	0951	0990	1028	1067	6	13	19	25	32
48	1·1106	1145	1184	1224	1263	1303	1343	1383	1423	1463	7	13	20	26	33
49	1·1504	1544	1585	1626	1667	1708	1750	1792	1833	1875	7	14	21	28	34
50	1·1918	1960	2002	2045	2088	2131	2174	2218	2261	2305	7	14	22	29	36
51	1·2349	2393	2437	2482	2527	2572	2617	2662	2708	2753	8	15	23	30	38
52	1·2799	2846	2892	2938	2985	3032	3079	3127	3175	3222	8	16	24	31	39
53	1·3270	3319	3367	3416	3465	3514	3564	3613	3663	3713	8	16	25	33	41
54	1·3764	3814	3865	3916	3968	4019	4071	4124	4176	4229	9	17	26	34	43
55	1·4281	4335	4388	4442	4496	4550	4605	4659	4715	4770	9	18	27	36	45
56	1·4826	4882	4938	4994	5051	5108	5166	5224	5282	5340	10	19	29	38	48
57	1·5399	5458	5517	5577	5637	5697	5757	5818	5880	5941	10	20	30	40	50
58	1·6003	6066	6128	6191	6255	6319	6383	6447	6512	6577	11	21	32	43	53
59	1·6643	6709	6775	6842	6909	6977	7045	7113	7182	7251	11	23	34	45	56
60	1·7321	7391	7461	7532	7603	7675	7747	7820	7893	7966	12	24	36	48	60
61	1·8040	8115	8190	8265	8341	8418	8495	8572	8650	8728	13	26	38	51	64
62	1·8807	8887	8967	9047	9128	9210	9292	9375	9458	9542	14	27	41	55	68
63	1·9626	9711	9797	9883	9970	**0057**	**0145**	**0233**	**0323**	**0413**	15	29	44	58	73
64	2·0503	0594	0686	0778	0872	0965	1060	1155	1251	1348	16	31	47	63	78
65	2·1445	1543	1642	1742	1842	1943	2045	2148	2251	2355	17	34	51	68	85
66	2·2460	2566	2673	2781	2889	2998	3109	3220	3332	3445	18	37	55	73	91
67	2·3559	3673	3789	3906	4023	4142	4262	4383	4504	4627	20	40	60	79	99
68	2·4751	4876	5002	5129	5257	5386	5517	5649	5782	5916	22	43	65	87	108
69	2·6051	6187	6325	6464	6605	6746	6889	7034	7179	7326	24	47	71	95	119
70	2·7475	7625	7776	7929	8083	8239	8397	8556	8716	8878	26	52	78	104	130
71	2·9042	9208	9375	9544	9714	9887	**0061**	**0237**	**0415**	**0595**	29	58	87	116	144
72	3·0777	0961	1146	1334	1524	1716	1910	2106	2305	2506	32	64	97	129	161
73	3·2709	2914	3122	3332	3544	3759	3977	4197	4420	4646	36	72	108	144	180
74	3·4874	5105	5339	5576	5816	6059	6305	6554	6806	7062	41	81	122	163	203
75	3·7321	7583	7848	8118	8391	8667	8947	9232	9520	9812	46	93	139	186	232
76	4·0108	0408	0713	1022	1335	1653	1976	2303	2635	2972	53	107	160	214	267
77	4·3315	3662	4015	4373	4737	5107	5483	5864	6252	6646	62	124	186	248	310
78	4·7046	7453	7867	8288	8716	9152	9594	**0045**	**0504**	**0970**	73	146	220	293	366
79	5·1446	1929	2422	2924	3435	3955	4486	5026	5578	6140	87	175	263	350	438
80	5·671	5·730	5·789	5·850	5·912	5·976	6·041	6·107	6·174	6·243					
81	6·314	6·386	6·460	6·535	6·612	6·691	6·772	6·855	6·940	7·026					
82	7·115	7·207	7·300	7·396	7·495	7·596	7·700	7·806	7·916	8·028					
83	8·144	8·264	8·386	8·513	8·643	8·777	8·915	9·058	9·205	9·357					
84	9·51	9·68	9·84	10·02	10·20	10·39	10·58	10·78	10·99	11·20	Differences untrustworthy here				
85	11·43	11·66	11·91	12·16	12·43	12·71	13·00	13·30	13·62	13·95					
86	14·30	14·67	15·06	15·46	15·89	16·35	16·83	17·34	17·89	18·46					
87	19·08	19·74	20·45	21·20	22·02	22·90	23·86	24·90	26·03	27·27					
88	28·64	30·14	31·82	33·69	35·80	38·19	40·92	44·07	47·74	52·08					
89	57·29	63·66	71·62	81·85	95·49	114·6	143·2	191·0	286·5	573·0					
	0′	6′	12′	18′	24′	30′	36′	42′	48′	54′	1′	2′	3′	4′	5′

The black type indicates that the integer changes.

	0	1	2	3	4	5	6	7	8	9	1	2	3	4	5	6	7	8	9
10	1000	1020	1040	1061	1082	1103	1124	1145	1166	1188	2	4	6	8	10	13	15	17	19
11	1210	1232	1254	1277	1300	1323	1346	1369	1392	1416	2	5	7	9	11	14	16	18	21
12	1440	1464	1488	1513	1538	1563	1588	1613	1638	1664	2	5	7	10	12	15	17	20	22
13	1690	1716	1742	1769	1796	1823	1850	1877	1904	1932	3	5	8	11	13	16	19	22	24
14	1960	1988	2016	2045	2074	2103	2132	2161	2190	2220	3	6	9	12	14	17	20	23	26
15	2250	2280	2310	2341	2372	2403	2434	2465	2496	2528	3	6	9	12	15	19	22	25	28
16	2560	2592	2624	2657	2690	2723	2756	2789	2822	2856	3	7	10	13	16	20	23	26	30
17	2890	2924	2958	2993	3028	3063	3098	3133	3168	3204	3	7	10	14	17	21	24	28	31
18	3240	3276	3312	3349	3386	3423	3460	3497	3534	3572	4	7	11	15	18	22	26	30	33
19	3610	3648	3686	3725	3764	3803	3842	3881	3920	3960	4	8	12	16	19	23	27	31	35
20	4000	4040	4080	4121	4162	4203	4244	4285	4326	4368	4	8	12	16	20	25	29	33	37
21	4410	4452	4494	4537	4580	4623	4666	4709	4752	4796	4	9	13	17	21	26	30	34	39
22	4840	4884	4928	4973	5018	5063	5108	5153	5198	5244	4	9	13	18	22	27	31	36	40
23	5290	5336	5382	5429	5476	5523	5570	5617	5664	5712	5	9	14	19	23	28	33	38	42
24	5760	5808	5856	5905	5954	6003	6052	6101	6150	6200	5	10	15	20	24	29	34	39	44
25	6250	6300	6350	6401	6452	6503	6554	6605	6656	6708	5	10	15	20	25	31	36	41	46
26	6760	6812	6864	6917	6970	7023	7076	7129	7182	7236	5	11	16	21	26	32	37	42	48
27	7290	7344	7398	7453	7508	7563	7618	7673	7728	7784	5	11	16	22	28	33	38	44	49
28	7840	7896	7952	8009	8066	8123	8180	8237	8294	8352	6	11	17	23	28	34	40	46	51
29	8410	8468	8526	8585	8644	8703	8762	8821	8880	8940	6	12	18	24	30	35	41	47	53
30	9000	9060	9120	9181	9242	9303	9364	9425	9486	9548	6	12	18	24	31	37	43	49	55
31	9610	9672	9734	9797	9860	9923	9986				6	13	19	25	31	38	44	50	56
31								1005	1011	1018	1	1	2	3	3	4	5	5	6
32	1024	1030	1037	1043	1050	1056	1063	1069	1076	1082	1	1	2	3	3	4	5	5	6
33	1089	1096	1102	1109	1116	1122	1129	1136	1142	1149	1	1	2	3	3	4	5	5	6
34	1156	1163	1170	1176	1183	1190	1197	1204	1211	1218	1	1	2	3	3	4	5	6	6
35	1225	1232	1239	1246	1253	1260	1267	1274	1282	1289	1	1	2	3	4	4	5	6	6
36	1296	1303	1310	1318	1325	1332	1340	1347	1354	1362	1	1	2	3	4	4	5	6	7
37	1369	1376	1384	1391	1399	1406	1414	1421	1429	1436	1	2	2	3	4	5	5	6	7
38	1444	1452	1459	1467	1475	1482	1490	1498	1505	1513	1	2	2	3	4	5	5	6	7
39	1521	1529	1537	1544	1552	1560	1568	1576	1584	1592	1	2	2	3	4	5	6	6	7
40	1600	1608	1616	1624	1632	1640	1648	1656	1665	1673	1	2	2	3	4	5	6	6	7
41	1681	1689	1697	1706	1714	1722	1731	1739	1747	1756	1	2	2	3	4	5	6	7	7
42	1764	1772	1781	1789	1798	1806	1815	1823	1832	1840	1	2	3	3	4	5	6	7	8
43	1849	1858	1866	1875	1884	1892	1901	1910	1918	1927	1	2	3	3	4	5	6	7	8
44	1936	1945	1954	1962	1971	1980	1989	1998	2007	2016	1	2	3	4	5	5	6	7	8
45	2025	2034	2043	2052	2061	2070	2079	2088	2098	2107	1	2	3	4	5	5	6	7	8
46	2116	2125	2134	2144	2153	2162	2172	2181	2190	2200	1	2	3	4	5	6	7	7	8
47	2209	2218	2228	2237	2247	2256	2266	2275	2285	2294	1	2	3	4	5	6	7	8	9
48	2304	2314	2323	2333	2343	2352	2362	2372	2381	2391	1	2	3	4	5	6	7	8	9
49	2401	2411	2421	2430	2440	2450	2460	2470	2480	2490	1	2	3	4	5	6	7	8	9
50	2500	2510	2520	2530	2540	2550	2560	2570	2581	2591	1	2	3	4	5	6	7	8	9
51	2601	2611	2621	2632	2642	2652	2663	2673	2683	2694	1	2	3	4	5	6	7	8	9
52	2704	2714	2725	2735	2746	2756	2767	2777	2788	2798	1	2	3	4	5	6	7	8	9
53	2809	2820	2830	2841	2852	2862	2873	2884	2894	2905	1	2	3	4	5	6	7	9	10
54	2916	2927	2938	2948	2959	2970	2981	2992	3003	3014	1	2	3	4	6	7	8	9	10
	0	**1**	**2**	**3**	**4**	**5**	**6**	**7**	**8**	**9**	**1**	**2**	**3**	**4**	**5**	**6**	**7**	**8**	**9**

The position of the decimal point must be determined by inspection.

	0	1	2	3	4	5	6	7	8	9	1	2	3	4	5	6	7	8	9
55	3025	3036	3047	3058	3069	3080	3091	3102	3114	3125	1	2	3	4	6	7	8	9	10
56	3136	3147	3158	3170	3181	3192	3204	3215	3226	3238	1	2	3	5	6	7	8	9	10
57	3249	3260	3272	3283	3295	3306	3318	3329	3341	3352	1	2	3	5	6	7	8	9	10
58	3364	3376	3387	3399	3411	3422	3434	3446	3457	3469	1	2	4	5	6	7	8	9	11
59	3481	3493	3505	3516	3528	3540	3552	3564	3576	3588	1	2	4	5	6	7	8	10	11
60	3600	3612	3624	3636	3648	3660	3672	3684	3697	3709	1	2	4	5	6	7	8	10	11
61	3721	3733	3745	3758	3770	3782	3795	3807	3819	3832	1	2	4	5	6	7	9	10	11
62	3844	3856	3869	3881	3894	3906	3919	3931	3944	3956	1	3	4	5	6	8	9	10	11
63	3969	3982	3994	4007	4020	4032	4045	4058	4070	4083	1	3	4	5	6	8	9	10	11
64	4096	4109	4122	4134	4147	4160	4173	4186	4199	4212	1	3	4	5	6	8	9	10	12
65	4225	4238	4251	4264	4277	4290	4303	4316	4330	4343	1	3	4	5	7	8	9	10	12
66	4356	4369	4382	4396	4409	4422	4436	4449	4462	4476	1	3	4	5	7	8	9	11	12
67	4489	4502	4516	4529	4543	4556	4570	4583	4597	4610	1	3	4	5	7	8	9	11	12
68	4624	4638	4651	4665	4679	4692	4706	4720	4733	4747	1	3	4	5	7	8	10	11	12
69	4761	4775	4789	4802	4816	4830	4844	4858	4872	4886	1	3	4	6	7	8	10	11	13
70	4900	4914	4928	4942	4956	4970	4984	4998	5013	5027	1	3	4	6	7	8	10	11	13
71	5041	5055	5069	5084	5098	5112	5127	5141	5155	5170	1	3	4	6	7	9	10	11	13
72	5184	5198	5213	5227	5242	5256	5271	5285	5300	5314	1	3	4	6	7	9	10	11	13
73	5329	5344	5358	5373	5388	5402	5417	5432	5446	5461	1	3	4	6	7	9	10	12	13
74	5476	5491	5506	5520	5535	5550	5565	5580	5595	5610	1	3	4	6	7	9	10	12	13
75	5625	5640	5655	5670	5685	5700	5715	5730	5746	5761	2	3	5	6	8	9	11	12	14
76	5776	5791	5806	5822	5837	5852	5868	5883	5898	5914	2	3	5	6	8	9	11	12	14
77	5929	5944	5960	5975	5991	6006	6022	6037	6053	6068	2	3	5	6	8	9	11	12	14
78	6084	6100	6115	6131	6147	6162	6178	6194	6209	6225	2	3	5	6	8	9	11	13	14
79	6241	6257	6273	6288	6304	6320	6336	6352	6368	6384	2	3	5	6	8	10	11	13	14
80	6400	6416	6432	6448	6464	6480	6496	6512	6529	6545	2	3	5	6	8	10	11	13	14
81	6561	6577	6593	6610	6626	6642	6659	6675	6691	6708	2	3	5	7	8	10	11	13	15
82	6724	6740	6757	6773	6790	6806	6823	6839	6856	6872	2	3	5	7	8	10	12	13	15
83	6889	6906	6922	6939	6956	6972	6989	7006	7022	7039	2	3	5	7	8	10	12	13	15
84	7056	7073	7090	7106	7123	7140	7157	7174	7191	7208	2	3	5	7	8	10	12	14	15
85	7225	7242	7259	7276	7293	7310	7327	7344	7362	7379	2	3	5	7	9	10	12	14	15
86	7396	7413	7430	7448	7465	7482	7500	7517	7534	7552	2	3	5	7	9	10	12	14	16
87	7569	7586	7604	7621	7639	7656	7674	7691	7709	7726	2	4	5	7	9	11	12	14	16
88	7744	7762	7779	7797	7815	7832	7850	7868	7885	7903	2	4	5	7	9	11	12	14	16
89	7921	7939	7957	7974	7992	8010	8028	8046	8064	8082	2	4	5	7	9	11	13	14	16
90	8100	8118	8136	8154	8172	8190	8208	8226	8245	8263	2	4	5	7	9	11	13	14	16
91	8281	8299	8317	8336	8354	8372	8391	8409	8427	8446	2	4	5	7	9	11	13	15	16
92	8464	8482	8501	8519	8538	8556	8575	8593	8612	8630	2	4	6	7	9	11	13	15	17
93	8649	8668	8686	8705	8724	8742	8761	8780	8798	8817	2	4	6	7	9	11	13	15	17
94	8836	8855	8874	8892	8911	8930	8949	8968	8987	9006	2	4	6	8	9	11	13	15	17
95	9025	9044	9063	9082	9101	9120	9139	9158	9178	9197	2	4	6	8	10	11	13	15	17
96	9216	9235	9254	9274	9293	9312	9332	9351	9370	9390	2	4	6	8	10	12	14	15	17
97	9409	9428	9448	9467	9487	9506	9526	9545	9565	9584	2	4	6	8	10	12	14	16	18
98	9604	9624	9643	9663	9683	9702	9722	9742	9761	9781	2	4	6	8	10	12	14	16	18
99	9801	9821	9841	9860	9880	9900	9920	9940	9960	9980	2	4	6	8	10	12	14	16	18
	0	**1**	**2**	**3**	**4**	**5**	**6**	**7**	**8**	**9**	**1**	**2**	**3**	**4**	**5**	**6**	**7**	**8**	**9**

The position of the decimal point must be determined by inspection.

	0	1	2	3	4	5	6	7	8	9	1	2	3	4	5	6	7	8	9
10	1000	1005	1010	1015	1020	1025	1030	1034	1039	1044	0	1	1	2	2	3	3	4	4
	3162	3178	3194	3209	3225	3240	3256	3271	3286	3302	2	3	5	6	8	9	11	12	14
11	1049	1054	1058	1063	1068	1072	1077	1082	1086	1091	0	1	1	2	2	3	3	4	4
	3317	3332	3347	3362	3376	3391	3406	3421	3435	3450	1	3	4	6	7	9	10	12	13
12	1095	1100	1105	1109	1114	1118	1122	1127	1131	1136	0	1	1	2	2	3	3	4	4
	3464	3479	3493	3507	3521	3536	3550	3564	3578	3592	1	3	4	6	7	8	10	11	13
13	1140	1145	1149	1153	1158	1162	1166	1170	1175	1179	0	1	1	2	2	3	3	3	4
	3606	3619	3633	3647	3661	3674	3688	3701	3715	3728	1	3	4	5	7	8	10	11	12
14	1183	1187	1192	1196	1200	1204	1208	1212	1217	1221	0	1	1	2	2	3	3	3	4
	3742	3755	3768	3782	3795	3808	3821	3834	3847	3860	1	3	4	5	7	8	9	11	12
15	1225	1229	1233	1237	1241	1245	1249	1253	1257	1261	0	1	1	2	2	3	3	3	4
	3873	3886	3899	3912	3924	3937	3950	3962	3975	3987	1	3	4	5	6	8	9	10	11
16	1265	1269	1273	1277	1281	1285	1288	1292	1296	1300	0	1	1	2	2	3	3	3	4
	4000	4012	4025	4037	4050	4062	4074	4087	4099	4111	1	2	4	5	6	7	9	10	11
17	1304	1308	1311	1315	1319	1323	1327	1330	1334	1338	0	1	1	2	2	2	3	3	3
	4123	4135	4147	4159	4171	4183	4195	4207	4219	4231	1	2	4	5	6	7	8	10	11
18	1342	1345	1349	1353	1356	1360	1364	1367	1371	1375	0	1	1	1	2	2	3	3	3
	4243	4254	4266	4278	4290	4301	4313	4324	4336	4347	1	2	3	5	6	7	8	9	10
19	1378	1382	1386	1389	1393	1396	1400	1404	1407	1411	0	1	1	1	2	2	3	3	3
	4359	4370	4382	4393	4405	4416	4427	4438	4450	4461	1	2	3	5	6	7	8	9	10
20	1414	1418	1421	1425	1428	1432	1435	1439	1442	1446	0	1	1	1	2	2	2	3	3
	4472	4483	4494	4506	4517	4528	4539	4550	4561	4572	1	2	3	4	5	7	8	9	10
21	1449	1453	1456	1459	1463	1466	1470	1473	1476	1480	0	1	1	1	2	2	2	3	3
	4583	4593	4604	4615	4626	4637	4648	4658	4669	4680	1	2	3	4	5	6	8	9	10
22	1483	1487	1490	1493	1497	1500	1503	1507	1510	1513	0	1	1	1	2	2	2	3	3
	4690	4701	4712	4722	4733	4743	4754	4764	4775	4785	1	2	3	4	5	6	7	8	9
23	1517	1520	1523	1526	1530	1533	1536	1539	1543	1546	0	1	1	1	2	2	2	3	3
	4796	4806	4817	4827	4837	4848	4858	4868	4879	4889	1	2	3	4	5	6	7	8	9
24	1549	1552	1556	1559	1562	1565	1568	1572	1575	1578	0	1	1	1	2	2	2	3	3
	4899	4909	4919	4930	4940	4950	4960	4970	4980	4990	1	2	3	4	5	6	7	8	9
25	1581	1584	1587	1591	1594	1597	1600	1603	1606	1609	0	1	1	1	2	2	2	3	3
	5000	5010	5020	5030	5040	5050	5060	5070	5079	5089	1	2	3	4	5	6	7	8	9
26	1612	1616	1619	1622	1625	1628	1631	1634	1637	1640	0	1	1	1	2	2	2	2	3
	5099	5109	5119	5128	5138	5148	5158	5167	5177	5187	1	2	3	4	5	6	7	8	9
27	1643	1646	1649	1652	1655	1658	1661	1664	1667	1670	0	1	1	1	2	2	2	2	3
	5196	5206	5215	5225	5235	5244	5254	5263	5273	5282	1	2	3	4	5	6	7	8	9
28	1673	1676	1679	1682	1685	1688	1691	1694	1697	1700	0	1	1	1	1	2	2	2	3
	5292	5301	5310	5320	5329	5339	5348	5357	5367	5376	1	2	3	4	5	6	7	7	8
29	1703	1706	1709	1712	1715	1718	1720	1723	1726	1729	0	1	1	1	1	2	2	2	3
	5385	5394	5404	5413	5422	5431	5441	5450	5459	5468	1	2	3	4	5	5	6	7	8
30	1732	1735	1738	1741	1744	1746	1749	1752	1755	1758	0	1	1	1	1	2	2	2	3
	5477	5486	5495	5505	5514	5523	5532	5541	5550	5559	1	2	3	4	4	5	6	7	8
31	1761	1764	1766	1769	1772	1775	1778	1780	1783	1786	0	1	1	1	1	2	2	2	3
	5568	5577	5586	5595	5604	5612	5621	5630	5639	5648	1	2	3	3	4	5	6	7	8
32	1789	1792	1794	1797	1800	1803	1806	1808	1811	1814	0	1	1	1	1	2	2	2	2
	5657	5666	5675	5683	5692	5701	5710	5718	5727	5736	1	2	3	3	4	5	6	7	8
	0	1	2	3	4	5	6	7	8	9	1	2	3	4	5	6	7	8	9

The first significant figure and the position of the decimal point must be determined by inspection.

	0	1	2	3	4	5	6	7	8	9	1	2	3	4	5	6	7	8	9
33	1817	1819	1822	1825	1828	1830	1833	1836	1838	1841	0	1	1	1	1	2	2	2	2
	5745	5753	5762	5771	5779	5788	5797	5805	5814	5822	1	2	3	3	4	5	6	7	8
34	1844	1847	1849	1852	1855	1857	1860	1863	1865	1868	0	1	1	1	1	2	2	2	2
	5831	5840	5848	5857	5865	5874	5882	5891	5899	5908	1	2	3	3	4	5	6	7	8
35	1871	1873	1876	1879	1881	1884	1887	1889	1892	1895	0	1	1	1	1	2	2	2	2
	5916	5925	5933	5941	5950	5958	5967	5975	5983	5992	1	2	2	3	4	5	6	7	8
36	1897	1900	1903	1905	1908	1910	1913	1916	1918	1921	0	1	1	1	1	2	2	2	2
	6000	6008	6017	6025	6033	6042	6050	6058	6066	6075	1	2	2	3	4	5	6	7	7
37	1924	1926	1929	1931	1934	1936	1939	1942	1944	1947	0	1	1	1	1	2	2	2	2
	6083	6091	6099	6107	6116	6124	6132	6140	6148	6156	1	2	2	3	4	5	6	7	7
38	1949	1952	1954	1957	1960	1962	1965	1967	1970	1972	0	1	1	1	1	2	2	2	2
	6164	6173	6181	6189	6197	6205	6213	6221	6229	6237	1	2	2	3	4	5	6	6	7
39	1975	1977	1980	1982	1985	1987	1990	1992	1995	1997	0	1	1	1	1	2	2	2	2
	6245	6253	6261	6269	6277	6285	6293	6301	6309	6317	1	2	2	3	4	5	6	6	7
40	2000	2002	2005	2007	2010	2012	2015	2017	2020	2022	0	0	1	1	1	1	2	2	2
	6325	6332	6340	6348	6356	6364	6372	6380	6387	6395	1	2	2	3	4	5	6	6	7
41	2025	2027	2030	2032	2035	2037	2040	2042	2045	2047	0	0	1	1	1	1	2	2	2
	6403	6411	6419	6427	6434	6442	6450	6458	6465	6473	1	2	2	3	4	5	5	6	7
42	2049	2052	2054	2057	2059	2062	2064	2066	2069	2071	0	0	1	1	1	1	2	2	2
	6481	6488	6496	6504	6512	6519	6527	6535	6542	6550	1	2	2	3	4	5	5	6	7
43	2074	2076	2078	2081	2083	2086	2088	2090	2093	2095	0	0	1	1	1	1	2	2	2
	6557	6565	6573	6580	6588	6595	6603	6611	6618	6626	1	2	2	3	4	5	5	6	7
44	2098	2100	2102	2105	2107	2110	2112	2114	2117	2119	0	0	1	1	1	1	2	2	2
	6633	6641	6648	6656	6663	6671	6678	6686	6693	6701	1	2	2	3	4	4	5	6	7
45	2121	2124	2126	2128	2131	2133	2135	2138	2140	2142	0	0	1	1	1	1	2	2	2
	6708	6716	6723	6731	6738	6745	6753	6760	6768	6775	1	1	2	3	4	4	5	6	7
46	2145	2147	2149	2152	2154	2156	2159	2161	2163	2166	0	0	1	1	1	1	2	2	2
	6782	6790	6797	6804	6812	6819	6826	6834	6841	6848	1	1	2	3	4	4	5	6	7
47	2168	2170	2173	2175	2177	2179	2182	2184	2186	2189	0	0	1	1	1	1	2	2	2
	6856	6863	6870	6877	6885	6892	6899	6907	6914	6921	1	1	2	3	4	4	5	6	7
48	2191	2193	2195	2198	2200	2202	2205	2207	2209	2211	0	0	1	1	1	1	2	2	2
	6928	6935	6943	6950	6957	6964	6971	6979	6986	6993	1	1	2	3	4	4	5	6	6
49	2214	2216	2218	2220	2223	2225	2227	2229	2232	2234	0	0	1	1	1	1	2	2	2
	7000	7007	7014	7021	7029	7036	7043	7050	7057	7064	1	1	2	3	4	4	5	6	6
50	2236	2238	2241	2243	2245	2247	2249	2252	2254	2256	0	0	1	1	1	1	2	2	2
	7071	7078	7085	7092	7099	7106	7113	7120	7127	7134	1	1	2	3	4	4	5	6	6
51	2258	2261	2263	2265	2267	2269	2272	2274	2276	2278	0	0	1	1	1	1	2	2	2
	7141	7148	7155	7162	7169	7176	7183	7190	7197	7204	1	1	2	3	4	4	5	6	6
52	2280	2283	2285	2287	2289	2291	2293	2296	2298	2300	0	0	1	1	1	1	2	2	2
	7211	7218	7225	7232	7239	7246	7253	7259	7266	7273	1	1	2	3	3	4	5	6	6
53	2302	2304	2307	2309	2311	2313	2315	2317	2319	2322	0	0	1	1	1	1	2	2	2
	7280	7287	7294	7301	7308	7314	7321	7328	7335	7342	1	1	2	3	3	4	5	5	6
54	2324	2326	2328	2330	2332	2335	2337	2339	2341	2343	0	0	1	1	1	1	1	2	2
	7348	7355	7362	7369	7376	7382	7389	7396	7403	7409	1	1	2	3	3	4	5	5	6
	0	1	2	3	4	5	6	7	8	9	1	2	3	4	5	6	7	8	9

The first significant figure and the position of the decimal point must be determined by inspection.

	0	1	2	3	4	5	6	7	8	9	1	2	3	4	5	6	7	8	9
55	2345	2347	2349	2352	2354	2356	2358	2360	2362	2364	0	0	1	1	1	1	1	2	2
	7416	7423	7430	7436	7443	7450	7457	7463	7470	7477	1	1	2	3	3	4	5	5	6
56	2366	2369	2371	2373	2375	2377	2379	2381	2383	2385	0	0	1	1	1	1	1	2	2
	7483	7490	7497	7503	7510	7517	7523	7530	7537	7543	1	1	2	3	3	4	5	5	6
57	2387	2390	2392	2394	2396	2398	2400	2402	2404	2406	0	0	1	1	1	1	1	2	2
	7550	7556	7563	7570	7576	7583	7589	7596	7603	7609	1	1	2	3	3	4	5	5	6
58	2408	2410	2412	2415	2417	2419	2421	2423	2425	2427	0	0	1	1	1	1	1	2	2
	7616	7622	7629	7635	7642	7649	7655	7662	7668	7675	1	1	2	3	3	4	5	5	6
59	2429	2431	2433	2435	2437	2439	2441	2443	2445	2447	0	0	1	1	1	1	1	2	2
	7681	7688	7694	7701	7707	7714	7720	7727	7733	7740	1	1	2	3	3	4	5	5	6
60	2449	2452	2454	2456	2458	2460	2462	2464	2466	2468	0	0	1	1	1	1	1	2	2
	7746	7752	7759	7765	7772	7778	7785	7791	7797	7804	1	1	2	3	3	4	4	5	6
61	2470	2472	2474	2476	2478	2480	2482	2484	2486	2488	0	0	1	1	1	1	1	2	2
	7810	7817	7823	7829	7836	7842	7849	7855	7861	7868	1	1	2	3	3	4	4	5	6
62	2490	2492	2494	2496	2498	2500	2502	2504	2506	2508	0	0	1	1	1	1	1	2	2
	7874	7880	7887	7893	7899	7906	7912	7918	7925	7931	1	1	2	3	3	4	4	5	6
63	2510	2512	2514	2516	2518	2520	2522	2524	2526	2528	0	0	1	1	1	1	1	2	2
	7937	7944	7950	7956	7962	7969	7975	7981	7987	7994	1	1	2	3	3	4	4	5	6
64	2530	2532	2534	2536	2538	2540	2542	2544	2546	2548	0	0	1	1	1	1	1	2	2
	8000	8006	8012	8019	8025	8031	8037	8044	8050	8056	1	1	2	2	3	4	4	5	6
65	2550	2551	2553	2555	2557	2559	2561	2563	2565	2567	0	0	1	1	1	1	1	2	2
	8062	8068	8075	8081	8087	8093	8099	8106	8112	8118	1	1	2	2	3	4	4	5	5
66	2569	2571	2573	2575	2577	2579	2581	2583	2585	2587	0	0	1	1	1	1	1	2	2
	8124	8130	8136	8142	8149	8155	8161	8167	8173	8179	1	1	2	2	3	4	4	5	5
67	2588	2590	2592	2594	2596	2598	2600	2602	2604	2606	0	0	1	1	1	1	1	2	2
	8185	8191	8198	8204	8210	8216	8222	8228	8234	8240	1	1	2	2	3	4	4	5	5
68	2608	2610	2612	2613	2615	2617	2619	2621	2623	2625	0	0	1	1	1	1	1	2	2
	8246	8252	8258	8264	8270	8276	8283	8289	8295	8301	1	1	2	2	3	4	4	5	5
69	2627	2629	2631	2632	2634	2636	2638	2640	2642	2644	0	0	1	1	1	1	1	2	2
	8307	8313	8319	8325	8331	8337	8343	8349	8355	8361	1	1	2	2	3	4	4	5	5
70	2646	2648	2650	2651	2653	2655	2657	2659	2661	2663	0	0	1	1	1	1	1	2	2
	8367	8373	8379	8385	8390	8396	8402	8408	8414	8420	1	1	2	2	3	4	4	5	5
71	2665	2666	2668	2670	2672	2674	2676	2678	2680	2681	0	0	1	1	1	1	1	1	2
	8426	8432	8438	8444	8450	8456	8462	8468	8473	8479	1	1	2	2	3	3	4	5	5
72	2683	2685	2687	2689	2691	2693	2694	2696	2698	2700	0	0	1	1	1	1	1	1	2
	8485	8491	8497	8503	8509	8515	8521	8526	8532	8538	1	1	2	2	3	3	4	5	5
73	2702	2704	2706	2707	2709	2711	2713	2715	2717	2718	0	0	1	1	1	1	1	1	2
	8544	8550	8556	8562	8567	8573	8579	8585	8591	8597	1	1	2	2	3	3	4	5	5
74	2720	2722	2724	2726	2728	2729	2731	2733	2735	2737	0	0	1	1	1	1	1	1	2
	8602	8608	8614	8620	8626	8631	8637	8643	8649	8654	1	1	2	2	3	3	4	5	5
75	2739	2740	2742	2744	2746	2748	2750	2751	2753	2755	0	0	1	1	1	1	1	1	2
	8660	8666	8672	8678	8683	8689	8695	8701	8706	8712	1	1	2	2	3	3	4	5	5
76	2757	2759	2760	2762	2764	2766	2768	2769	2771	2773	0	0	1	1	1	1	1	1	2
	8718	8724	8729	8735	8741	8746	8752	8758	8764	8769	1	1	2	2	3	3	4	5	5
77	2775	2777	2778	2780	2782	2784	2786	2787	2789	2791	0	0	1	1	1	1	1	1	2
	8775	8781	8786	8792	8798	8803	8809	8815	8820	8826	1	1	2	2	3	3	4	4	5
	0	1	2	3	4	5	6	7	8	9	1	2	3	4	5	6	7	8	9

The first significant figure and the position of the decimal point must be determined by inspection.

	0	1	2	3	4	5	6	7	8	9	1 2 3	4 5 6	7 8 9
78	2793	2795	2796	2798	2800	2802	2804	2805	2807	2809	0 0 1	1 1 1	1 1 2
	8832	8837	8843	8849	8854	8860	8866	8871	8877	8883	1 1 2	2 3 3	4 4 5
79	2811	2812	2814	2816	2818	2820	2821	2823	2825	2827	0 0 1	1 1 1	1 1 2
	8888	8894	8899	8905	8911	8916	8922	8927	8933	8939	1 1 2	2 3 3	4 4 5
80	2828	2830	2832	2834	2835	2837	2839	2841	2843	2844	0 0 1	1 1 1	1 1 2
	8944	8950	8955	8961	8967	8972	8978	8983	8989	8994	1 1 2	2 3 3	4 4 5
81	2846	2848	2850	2851	2853	2855	2857	2858	2860	2862	0 0 1	1 1 1	1 1 2
	9000	9006	9011	9017	9022	9028	9033	9039	9044	9050	1 1 2	2 3 3	4 4 5
82	2864	2865	2867	2869	2871	2872	2874	2876	2877	2879	0 0 1	1 1 1	1 1 2
	9055	9061	9066	9072	9077	9083	9088	9094	9099	9105	1 1 2	2 3 3	4 4 5
83	2881	2883	2884	2886	2888	2890	2891	2893	2895	2897	0 0 1	1 1 1	1 1 2
	9110	9116	9121	9127	9132	9138	9143	9149	9154	9160	1 1 2	2 3 3	4 4 5
84	2898	2900	2902	2903	2905	2907	2909	2910	2912	2914	0 0 1	1 1 1	1 1 2
	9165	9171	9176	9182	9187	9192	9198	9203	9209	9214	1 1 2	2 3 3	4 4 5
85	2915	2917	2919	2921	2922	2924	2926	2927	2929	2931	0 0 1	1 1 1	1 1 2
	9220	9225	9230	9236	9241	9247	9252	9257	9263	9268	1 1 2	2 3 3	4 4 5
86	2933	2934	2936	2938	2939	2941	2943	2944	2946	2948	0 0 1	1 1 1	1 1 2
	9274	9279	9284	9290	9295	9301	9306	9311	9317	9322	1 1 2	2 3 3	4 4 5
87	2950	2951	2953	2955	2956	2958	2960	2961	2963	2965	0 0 1	1 1 1	1 1 2
	9327	9333	9338	9343	9349	9354	9359	9365	9370	9375	1 1 2	2 3 3	4 4 5
88	2966	2968	2970	2972	2973	2975	2977	2978	2980	2982	0 0 1	1 1 1	1 1 2
	9381	9386	9391	9397	9402	9407	9413	9418	9423	9429	1 1 2	2 3 3	4 4 5
89	2983	2985	2987	2988	2990	2992	2993	2995	2997	2998	0 0 1	1 1 1	1 1 2
	9434	9439	9445	9450	9455	9460	9466	9471	9476	9482	1 1 2	2 3 3	4 4 5
90	3000	3002	3003	3005	3007	3008	3010	3012	3013	3015	0 0 0	1 1 1	1 1 1
	9487	9492	9497	9503	9508	9513	9518	9524	9529	9534	1 1 2	2 3 3	4 4 5
91	3017	3018	3020	3022	3023	3025	3027	3028	3030	3032	0 0 0	1 1 1	1 1 1
	9539	9545	9550	9555	9560	9566	9571	9576	9581	9586	1 1 2	2 3 3	4 4 5
92	3033	3035	3036	3038	3040	3041	3043	3045	3046	3048	0 0 0	1 1 1	1 1 1
	9592	9597	9602	9607	9612	9618	9623	9628	9633	9638	1 1 2	2 3 3	4 4 5
93	3050	3051	3053	3055	3056	3058	3059	3061	3063	3064	0 0 0	1 1 1	1 1 1
	9644	9649	9654	9659	9664	9670	9675	9680	9685	9690	1 1 2	2 3 3	4 4 5
94	3066	3068	3069	3071	3072	3074	3076	3077	3079	3081	0 0 0	1 1 1	1 1 1
	9695	9701	9706	9711	9716	9721	9726	9731	9737	9742	1 1 2	2 3 3	4 4 5
95	3082	3084	3085	3087	3089	3090	3092	3094	3095	3097	0 0 0	1 1 1	1 1 1
	9747	9752	9757	9762	9767	9772	9778	9783	9788	9793	1 1 2	2 3 3	4 4 5
96	3098	3100	3102	3103	3105	3106	3108	3110	3111	3113	0 0 0	1 1 1	1 1 1
	9798	9803	9808	9813	9818	9823	9829	9834	9839	9844	1 1 2	2 3 3	4 4 5
97	3114	3116	3118	3119	3121	3122	3124	3126	3127	3129	0 0 0	1 1 1	1 1 1
	9849	9854	9859	9864	9869	9874	9879	9884	9889	9894	1 1 2	2 3 3	4 4 5
98	3130	3132	3134	3135	3137	3138	3140	3142	3143	3145	0 0 0	1 1 1	1 1 1
	9899	9905	9910	9915	9920	9925	9930	9935	9940	9945	0 1 1	2 2 3	3 4 4
99	3146	3148	3150	3151	3153	3154	3156	3158	3159	3161	0 0 0	1 1 1	1 1 1
	9950	9955	9960	9965	9970	9975	9980	9985	9990	9995	0 1 1	2 2 3	3 4 4
	0	**1**	**2**	**3**	**4**	**5**	**6**	**7**	**8**	**9**	**1 2 3**	**4 5 6**	**7 8 9**

The first significant figure and the position of the decimal point must be determined by inspection.

	0	1	2	3	4	5	6	7	8	9	SUBTRACT 1	2	3	4	5	6	7	8	9
1·0	1·0000	·9901	·9804	·9709	·9615	·9524	·9434	·9346	·9259	·9174	9	18	28	37	46	55	64	73	83
1·1	·9091	·9009	·8929	·8850	·8772	·8696	·8621	·8547	·8475	·8403	8	15	23	31	38	46	54	61	69
1·2	·8333	·8264	·8197	·8130	·8065	·8000	·7937	·7874	·7813	·7752	6	13	19	26	32	38	45	51	58
1·3	·7692	·7634	·7576	·7519	·7463	·7407	·7353	·7299	·7246	·7194	5	11	16	22	27	33	38	44	49
1·4	·7143	·7092	·7042	·6993	·6944	·6897	·6849	·6803	·6757	·6711	5	10	14	19	24	29	33	38	43
1·5	·6667	·6623	·6579	·6536	·6494	·6452	·6410	·6369	·6329	·6289	4	8	13	17	21	25	29	33	38
1·6	·6250	·6211	·6173	·6135	·6098	·6061	·6024	·5988	·5952	·5917	4	7	11	15	18	22	26	29	33
1·7	·5882	·5848	·5814	·5780	·5747	·5714	·5682	·5650	·5618	·5587	3	7	10	13	16	20	23	26	30
1·8	·5556	·5525	·5495	·5464	·5435	·5405	·5376	·5348	·5319	·5291	3	6	9	12	15	18	20	23	26
1·9	·5263	·5236	·5208	·5181	·5155	·5128	·5102	·5076	·5051	·5025	3	5	8	11	13	16	18	21	24
2·0	·5000	·4975	·4950	·4926	·4902	·4878	·4854	·4831	·4808	·4785	2	5	7	10	12	14	17	19	21
2·1	·4762	·4739	·4717	·4695	·4673	·4651	·4630	·4608	·4587	·4566	2	4	7	9	11	13	15	17	20
2·2	·4545	·4525	·4505	·4484	·4464	·4444	·4425	·4405	·4386	·4367	2	4	6	8	10	12	14	16	18
2·3	·4348	·4329	·4310	·4292	·4274	·4255	·4237	·4219	·4202	·4184	2	4	5	7	9	11	13	14	16
2·4	·4167	·4149	·4132	·4115	·4098	·4082	·4065	·4049	·4032	·4016	2	3	5	7	8	10	12	13	15
2·5	·4000	·3984	·3968	·3953	·3937	·3922	·3906	·3891	·3876	·3861	2	3	5	6	8	9	11	12	14
2·6	·3846	·3831	·3817	·3802	·3788	·3774	·3759	·3745	·3731	·3717	1	3	4	6	7	8	10	11	13
2·7	·3704	·3690	·3676	·3663	·3650	·3636	·3623	·3610	·3597	·3584	1	3	4	5	7	8	9	11	12
2·8	·3571	·3559	·3546	·3534	·3521	·3509	·3497	·3484	·3472	·3460	1	2	4	5	6	7	9	10	11
2·9	·3448	·3436	·3425	·3413	·3401	·3390	·3378	·3367	·3356	·3344	1	2	3	5	6	7	8	9	10
3·0	·3333	·3322	·3311	·3300	·3289	·3279	·3268	·3257	·3247	·3236	1	2	3	4	5	6	7	9	10
3·1	·3226	·3215	·3205	·3195	·3185	·3175	·3165	·3155	·3145	·3135	1	2	3	4	5	6	7	8	9
3·2	·3125	·3115	·3106	·3096	·3086	·3077	·3067	·3058	·3049	·3040	1	2	3	4	5	6	7	8	9
3·3	·3030	·3021	·3012	·3003	·2994	·2985	·2976	·2967	·2959	·2950	1	2	3	4	4	5	6	7	8
3·4	·2941	·2933	·2924	·2915	·2907	·2899	·2890	·2882	·2874	·2865	1	2	3	3	4	5	6	7	8
3·5	·2857	·2849	·2841	·2833	·2825	·2817	·2809	·2801	·2793	·2786	1	2	2	3	4	5	6	6	7
3·6	·2778	·2770	·2762	·2755	·2747	·2740	·2732	·2725	·2717	·2710	1	2	2	3	4	5	5	6	7
3·7	·2703	·2695	·2688	·2681	·2674	·2667	·2660	·2653	·2646	·2639	1	1	2	3	4	4	5	6	6
3·8	·2632	·2625	·2618	·2611	·2604	·2597	·2591	·2584	·2577	·2571	1	1	2	3	3	4	5	5	6
3·9	·2564	·2558	·2551	·2545	·2538	·2532	·2525	·2519	·2513	·2506	1	1	2	3	3	4	4	5	6
4·0	·2500	·2494	·2488	·2481	·2475	·2469	·2463	·2457	·2451	·2445	1	1	2	2	3	4	4	5	5
4·1	·2439	·2433	·2427	·2421	·2415	·2410	·2404	·2398	·2392	·2387	1	1	2	2	3	3	4	5	5
4·2	·2381	·2375	·2370	·2364	·2358	·2353	·2347	·2342	·2336	·2331	1	1	2	2	3	3	4	4	5
4·3	·2326	·2320	·2315	·2309	·2304	·2299	·2294	·2288	·2283	·2278	1	1	2	2	3	3	4	4	5
4·4	·2273	·2268	·2262	·2257	·2252	·2247	·2242	·2237	·2232	·2227	1	1	2	2	3	3	4	4	5
4·5	·2222	·2217	·2212	·2208	·2203	·2198	·2193	·2188	·2183	·2179	0	1	1	2	2	3	3	4	4
4·6	·2174	·2169	·2165	·2160	·2155	·2151	·2146	·2141	·2137	·2132	0	1	1	2	2	3	3	4	4
4·7	·2128	·2123	·2119	·2114	·2110	·2105	·2101	·2096	·2092	·2088	0	1	1	2	2	3	3	4	4
4·8	·2083	·2079	·2075	·2070	·2066	·2062	·2058	·2053	·2049	·2045	0	1	1	2	2	3	3	3	4
4·9	·2041	·2037	·2033	·2028	·2024	·2020	·2016	·2012	·2008	·2004	0	1	1	2	2	2	3	3	4
5·0	·2000	·1996	·1992	·1988	·1984	·1980	·1976	·1972	·1969	·1965	0	1	1	2	2	2	3	3	4
5·1	·1961	·1957	·1953	·1949	·1946	·1942	·1938	·1934	·1931	·1927	0	1	1	2	2	2	3	3	3
5·2	·1923	·1919	·1916	·1912	·1908	·1905	·1901	·1898	·1894	·1890	0	1	1	1	2	2	3	3	3
5·3	·1887	·1883	·1880	·1876	·1873	·1869	·1866	·1862	·1859	·1855	0	1	1	1	2	2	3	3	3
5·4	·1852	·1848	·1845	·1842	·1838	·1835	·1832	·1828	·1825	·1821	0	1	1	1	2	2	2	3	3
	0	**1**	**2**	**3**	**4**	**5**	**6**	**7**	**8**	**9**	**1**	**2**	**3**	**4**	**5**	**6**	**7**	**8**	**9**

											SUBTRACT								
	0	1	2	3	4	5	6	7	8	9	1	2	3	4	5	6	7	8	9
5·5	·1818	·1815	·1812	·1808	·1805	·1802	·1799	·1795	·1792	·1789	0	1	1	1	2	2	2	3	3
5·6	·1786	·1783	·1779	·1776	·1773	·1770	·1767	·1764	·1761	·1757	0	1	1	1	2	2	2	3	3
5·7	·1754	·1751	·1748	·1745	·1742	·1739	·1736	·1733	·1730	·1727	0	1	1	1	2	2	2	2	3
5·8	·1724	·1721	·1718	·1715	·1712	·1709	·1706	·1704	·1701	·1698	0	1	1	1	1	2	2	2	3
5·9	·1695	·1692	·1689	·1686	·1684	·1681	·1678	·1675	·1672	·1669	0	1	1	1	1	2	2	2	3
6·0	·1667	·1664	·1661	·1658	·1656	·1653	·1650	·1647	·1645	·1642	0	1	1	1	1	2	2	2	3
6·1	·1639	·1637	·1634	·1631	·1629	·1626	·1623	·1621	·1618	·1616	0	1	1	1	1	2	2	2	2
6·2	·1613	·1610	·1608	·1605	·1603	·1600	·1597	·1595	·1592	·1590	0	1	1	1	1	2	2	2	2
6·3	·1587	·1585	·1582	·1580	·1577	·1575	·1572	·1570	·1567	·1565	0	0	1	1	1	1	2	2	2
6·4	·1563	·1560	·1558	·1555	·1553	·1550	·1548	·1546	·1543	·1541	0	0	1	1	1	1	2	2	2
6·5	·1538	·1536	·1534	·1531	·1529	·1527	·1524	·1522	·1520	·1517	0	0	1	1	1	1	2	2	2
6·6	·1515	·1513	·1511	·1508	·1506	·1504	·1502	·1499	·1497	·1495	0	0	1	1	1	1	2	2	2
6·7	·1493	·1490	·1488	·1486	·1484	·1481	·1479	·1477	·1475	·1473	0	0	1	1	1	1	2	2	2
6·8	·1471	·1468	·1466	·1464	·1462	·1460	·1458	·1456	·1453	·1451	0	0	1	1	1	1	2	2	2
6·9	·1449	·1447	·1445	·1443	·1441	·1439	·1437	·1435	·1433	·1431	0	0	1	1	1	1	1	2	2
7·0	·1429	·1427	·1425	·1422	·1420	·1418	·1416	·1414	·1412	·1410	0	0	1	1	1	1	1	2	2
7·1	·1408	·1406	·1404	·1403	·1401	·1399	·1397	·1395	·1393	·1391	0	0	1	1	1	1	1	2	2
7·2	·1389	·1387	·1385	·1383	·1381	·1379	·1377	·1376	·1374	·1372	0	0	1	1	1	1	1	2	2
7·3	·1370	·1368	·1366	·1364	·1362	·1361	·1359	·1357	·1355	·1353	0	0	1	1	1	1	1	2	2
7·4	·1351	·1350	·1348	·1346	·1344	·1342	·1340	·1339	·1337	·1335	0	0	1	1	1	1	1	1	2
7·5	·1333	·1332	·1330	·1328	·1326	·1325	·1323	·1321	·1319	·1318	0	0	1	1	1	1	1	1	2
7·6	·1316	·1314	·1312	·1311	·1309	·1307	·1305	·1304	·1302	·1300	0	0	1	1	1	1	1	1	2
7·7	·1299	·1297	·1295	·1294	·1292	·1290	·1289	·1287	·1285	·1284	0	0	0	1	1	1	1	1	1
7·8	·1282	·1280	·1279	·1277	·1276	·1274	·1272	·1271	·1269	·1267	0	0	0	1	1	1	1	1	1
7·9	·1266	·1264	·1263	·1261	·1259	·1258	·1256	·1255	·1253	·1252	0	0	0	1	1	1	1	1	1
8·0	·1250	·1248	·1247	·1245	·1244	·1242	·1241	·1239	·1238	·1236	0	0	0	1	1	1	1	1	1
8·1	·1235	·1233	·1232	·1230	·1229	·1227	·1225	·1224	·1222	·1221	0	0	0	1	1	1	1	1	1
8·2	·1220	·1218	·1217	·1215	·1214	·1212	·1211	·1209	·1208	·1206	0	0	0	1	1	1	1	1	1
8·3	·1205	·1203	·1202	·1200	·1199	·1198	·1196	·1195	·1193	·1192	0	0	0	1	1	1	1	1	1
8·4	·1190	·1189	·1188	·1186	·1185	·1183	·1182	·1181	·1179	·1178	0	0	0	1	1	1	1	1	1
8·5	·1176	·1175	·1174	·1172	·1171	·1170	·1168	·1167	·1166	·1164	0	0	0	1	1	1	1	1	1
8·6	·1163	·1161	·1160	·1159	·1157	·1156	·1155	·1153	·1152	·1151	0	0	0	1	1	1	1	1	1
8·7	·1149	·1148	·1147	·1145	·1144	·1143	·1142	·1140	·1139	·1138	0	0	0	1	1	1	1	1	1
8·8	·1136	·1135	·1134	·1133	·1131	·1130	·1129	·1127	·1126	·1125	0	0	0	1	1	1	1	1	1
8·9	·1124	·1122	·1121	·1120	·1119	·1117	·1116	·1115	·1114	·1112	0	0	0	1	1	1	1	1	1
9·0	·1111	·1110	·1109	·1107	·1106	·1105	·1104	·1103	·1101	·1100	0	0	0	1	1	1	1	1	1
9·1	·1099	·1098	·1096	·1095	·1094	·1093	·1092	·1091	·1089	·1088	0	0	0	0	1	1	1	1	1
9·2	·1087	·1086	·1085	·1083	·1082	·1081	·1080	·1079	·1078	·1076	0	0	0	0	1	1	1	1	1
9·3	·1075	·1074	·1073	·1072	·1071	·1070	·1068	·1067	·1066	·1065	0	0	0	0	1	1	1	1	1
9·4	·1064	·1063	·1062	·1060	·1059	·1058	·1057	·1056	·1055	·1054	0	0	0	0	1	1	1	1	1
9·5	·1053	·1052	·1050	·1049	·1048	·1047	·1046	·1045	·1044	·1043	0	0	0	0	1	1	1	1	1
9·6	·1042	·1041	·1040	·1038	·1037	·1036	·1035	·1034	·1033	·1032	0	0	0	0	1	1	1	1	1
9·7	·1031	·1030	·1029	·1028	·1027	·1026	·1025	·1024	·1022	·1021	0	0	0	0	1	1	1	1	1
9·8	·1020	·1029	·1018	·1017	·1016	·1015	·1014	·1013	·1012	·1011	0	0	0	0	1	1	1	1	1
9·9	·1010	·1009	·1008	·1007	·1006	·1005	·1004	·1003	·1002	·1001	0	0	0	0	0	1	1	1	1
	0	1	2	3	4	5	6	7	8	9	1	2	3	4	5	6	7	8	9

INDEX